Wolfgang Trier

Glass Furnaces

Design Construction and Operation

Translated by

K. L. Loewenstein

SOCIETY OF GLASS TECHNOLOGY
9 Churchill Way, Chapeltown, Sheffield S35 2PY, UK
Reprinted 2026

Glass furnaces: design, construction and operation by Wolfgang Trier

Translated from
Glasschmelzöfen: Konstruktion und Betreibsverhalten
originally published by Springer-Verlag 1984

Translated by
K. L. Loewenstein

First published 1987
Reprinted 2000
Reprinted 2026 (print on demand)

Cover image is taken from a Furnace Solutions Training Day presentation by Geoff Evans FSGT. Geoff was the driving force in the establishment of the Furnace Solutions conference series held by the Society of Glass Technology, this reprint is dedicated to his memory.

The objects of the Society of Glass Technology are to encourage and advance the study of the history, art, science, design, manufacture, after treatment, distribution and end use of glass of any and every kind. These aims are furthered by meetings, publications, the maintenance of a library and the promotion of association with other interested persons and organisations.

Society of Glass Technology
9 Churchill Way
Chapeltown
Sheffield
S35 2PY
UK
Tel +44(0)114 263 4455
Email info@sgt.org
Web www.sgt.org

Society of Glass Technology is a registered charity no. 237438.

ISBN 978-0-900682-94-0

Contents

1	**Introduction**	**3**
2	**Technical criteria, descriptions and designs.**	**5**
2.1	Technical criteria	5
2.2	Other criteria	7
2.3	Nomenclature.	7
2.4	Types of furnace	7
2.4.1	Tank furnaces for the production of glass containers	7
2.4.2	Tank furnaces for the production of flat glass	10
2.4.3	Tank furnaces for the production of special glasses	12
3	**Construction elements and dimensions**	**14**
3.1	The tank	14
3.1.1	Geometrical data	14
3.1.1.1	Melting area	14
3.1.1.2	Length-to-width ratio	14
3.1.1.3	Depth.	15
3.1.2	Construction	16
3.1.2.1	Sidewalls	16
3.1.2.2	The tank bottom	17
3.1.2.3	The throat	19
3.1.2.4	The doghouse	21
3.2	Superstructure	22
3.2.1	The combustion chamber	22
3.2.2	Sidewalls	22
3.2.3	The crown.	23
3.2.3.1	Forces in the crown	24
3.2.3.2	Lines of thrust	24
3.2.3.3	The effect of thermal expansion	27
3.2.3.4	The construction of crowns and their behaviour during operation	28
3.2.4	End walls and doghouse	29
3.2.5	Separation of melting and working (or conditioning) ends	30
3.2.6	Ports	31
3.3	Regenerators	34
3.3.1	Layout	34
3.3.2	Dimensions	35
3.3.3	Gas flow through the regenerator	35
3.3.3.1	Regenerators with smooth plain packing	35
3.3.3.2	Regenerators with straight packing	37
13.3.3	Comparison between different regenerator packings	38

3.3.3.4 The influence of temperature distribution ... 39
3.3.4 Operating data for regenerators ... 39
3.14.1 Types of checker packing ... 40
3.3.4.1.1 Smooth plain packing ... 40
3.3.4.1.2 Straight and staggered packing ... 40
3.3.4.1.3 Basket weave packing ... 40
3.3.4.1.4 Brick sizes ... 41
3.3.4.2 Data for checkers ... 41
3.3.5 Regenerator crowns and walls ... 44
3.3.6 Bearer arches and checkerwork support bricks ... 44
3.4 Recuperators ... 45
3.4.1 Ceramic recuperators ... 45
3.4.2 Metal recuperators ... 47
3.5 Flues ... 48
3.6 The reversal mechanism ... 50
3.7 Chimneys ... 52
3.7.1 Sizing of a chimney ... 52
3.7.2 Parts of a chimney ... 52
3.7.3 Chimney equipment ... 53
3.8 The supporting steelwork ... 53
3.8.1 The bracing of the crown ... 53
3.8.1.1 Buckstays and their hinged supports ... 53
3.8.1.2 Rigid crown bracing ... 55
3.8.2 Bracing steelwork for walls ... 56
3.8.2.1 Superstructure sidewalls ... 56
3.8.2.2 Tank blocks ... 57
3.8.2.3 Steelwork for end and gable walls ... 57
3.8.2.4 Steelwork securing the doghouse ... 57
3.8.3 Bracing of ports ... 57
3.8.4 Regenerator bracing ... 58
3.8.5 Calculations ... 59

4 Selection and use of refractories in glass tank furnaces ... 60
4.1 Selection criteria ... 60
4.2 Compositions and properties of the more important refractories ... 60
4.2.1 Clay blocks, firebrick ... 60
4.2.2 Sillimanite and mullite ... 60
4.2.3 Zircon ... 60
4.2.4 Fusion cast refractories ... 62
4.2.4.1 Ramming materials used with fusion cast blocks ... 64
4.2.5 Silica ... 66
4.2.6 Basic refractories ... 66
4.3 Refractories, their selection and use in particular cases ... 66
4.3.1 Tank furnace crowns ... 66
4.3.2 The crown of the working and conditioning ends ... 68
4.3.3 Melting end sidewalls ... 69
4.3.4 Sidewalls of working and conditioning ends ... 71
4.3.5 The melting end ... 71
4.3.6 Working and conditioning ends ... 73

4.3.7 Port necks . . . 73
4.3.8 Regenerator chambers . . . 73
4.3.9 Flues . . . 75

5 Operation of glass tank furnaces . . . 76
5.1 The heating process . . . 76
5.1.1 Temperature distribution . . . 76
5.1.2 Heat flux density and fuel distribution . . . 77
5.1.3 Efficiencies . . . 78
5.1.3.1 Fuel efficiency . . . 78
5.1.3.2 The theoretical heat requirement of molten glasses . . . 80
5.1.4 Heat consumption . . . 80
5.1.4.1 The influence of furnace age and ambient temperature on fuel consumption . . 81
5.1.4.2 The influence of excess air and induced air on fuel consumption . . . 82
5.1.5 Heat balances . . . 83
5.1.5.1 Subdivision for balances . . . 83
5.1.5.2 Terms used for the construction of heat balances . . . 83
5.1.5.3 Reference units . . . 83
5.1.5.3.1 Heat introduced stated as a percentage . . . 83
5.1.5.3.2 Useful heat in the glass stated as a percentage . . . 84
5.1.5.3.3 Heat balance using unit weight of glass as reference parameter, . . . 84
5.1.5.3.4 Unit of time . . . 84
5.1.5.3.5 Examples . . . 85
5.1.5.4 Multi boundary balances . . . 85
5.1.5.5 Energy balance,s . . . 86
5.1.6 Thermal insulation and cooling . . . 86
5.1.6.1 Thermal insulation . . . 86
5.1.6.1.1 Calculation of thermal insulation . . . 87
5.1.6.1.2 Installation of thermal insulation . . . 89
5.1.6.1.2.1 Glass contact refractories . . . 89
5.1.6.1.2.2 Crown and sidewalls . . . 90
5.1.6.2 Cooling of tank blocks . . . 90
5.1.6.2.1 The effect of cooling . . . 92
5.1.6.2.2 Design of cooling system . . . 92
5.2 Measurement and control . . . 93
5.2.1 Temperature . . . 93
5.2.2 Furnace pressure . . . 97
5.2.3 Volume and fuel/air mixture . . . 98
5.2.4 Glass level . . . 99
5.2.5 Composition of waste gases . . . 100
5.2.6 Instrumentation and control installations . . . 101
5.3 Reversal . . . 102
5.3.1 Reversal as a periodic process . . . 102
5.3.2 Control of the reversal process . . . 104
5.4 The heating and cooling of glass furnaces . . . 105
5.4.1 Expansion of refractories . . . 105
5.4.2 Temperature gradients and thermal stresses in refractories . . . 106
5.4.3 The heating-up curve . . . 107
5.4.4 The heating-up process . . . 107

5.4.4.1	Location of burners	107
5.4.4.2	Burners	108
5.4.4.3	Heating up	108
5.4.4.4	Control of the thermal expansion of refractories	109
5.4.4.5	Charging the furnace after heating up	111
5.4.4.6	Instrumentation and control during heat-up	112
5.4.5	The cooling down of glass furnaces	112
5.4.6	The draining of tank furnaces	112
5.5	Hot repairs	113
5.5.1	Increasing the cooling	113
5.5.2	Refractories for hot repairs	114
5.5.3	Refractory compounds for hot repairs	114
5.5.4	Gunning	116
5.5.5	Hot repairs using fusion cast refractories	117
5.5.6	Hot repairs to regenerators	118
6	**Tank furnaces - their behaviour in operation**	**119**
6.1	Flow of combustion gases in the combustion space	119
6.1.1	Cross-fired tank furnaces	119
6.1.2	U shaped flame tank furnaces	120
6.2	Flame formation in the combustion space	122
6.2.1	Types and arrangement of flames	122
6.2.2	Turbulent jet diffusion flames	123
6.2.2.1	Free jet	123
6.2.2.2	Diffusion flame	124
6.2.2.2.1	Free jet diffusion flame	124
6.2.2.2.2	Diffusion flames in a limited space	125
6.3	Heat emission in the combustion chamber	127
6.3.1	The combustion of jet flames	127
6.3.2	Heat emission and flame length	127
6.3.3	Heat emission of flames by radiation	128
6.3.3.1	Radiation of gases	128
6.3.3.2	Radiation of luminous flames	130
6.3.3.3	Interplay between flame and melting chamber	131
6.3.4	Heat transfer by convection between flame and bath of molten glass	133
6.4	The charging and melting of batch	134
6.4.1	The melting process	134
6.4.2	Dust and volatilisation	138
6.4.3	Methods of batch charging	139
6.4.3.1	Cross-fired tank furnaces	139
6.4.3.2	U shaped flame tanks	141
6.4.3.3	Small, continuously fired tank furnaces	142
6.4.3.4	Electrically heated tank furnaces	142
6.5	Glass currents	144
6.5.1	The origin of glass currents	144
6.5.2	Calculation of glass currents	147
6.5.3	The effect of glass currents	148
6.5.3.1	The boundary between melting and refining zones	148
6.5.3.1.1	Increasing the withdrawal current	149

6.5.3.1.2 Electric boosting in the refining zone ... 149
6.5.3.1.3 Electric boosting in the melting zone ... 150
6.5.3.1.4 Bubbling ... 150
6.5.3.1.5 Weirs ... 153
6.5.4.6 The provision of drain holes in a tank bottom ... 153
6.5.3.1.7 Depth and colour of glass, and insulation of the tank ... 154
6.5.3.1.8 Currents along the sidewalls of a tank ... 155
6.5.3.2 Delimitation of working and conditioning zones ... 155
6.5.3.3 Currents in the throat of a container glass tank ... 156
6.5.3.4 Currents in the working end and conditioning zone ... 157
6.5.3.5 Glass currents in electric furnaces ... 158
6.6 The homogenisation and refining of molten glass ... 158
6.7 Mixing processes in glass tank furnaces ... 161
6.7.1 Experimental results ... 161
6.7.2 The performance of a perfect mixer ... 162
6.7.3 The mixing process during accidental variations in the composition of batch or cullet ... 163
6.8 Changing glass colour in tank furnaces ... 164
6.8.1 Draining and refilling of the tank ... 164
6.8.2 Colour changes during operation ... 165
6.8.2.1 The mixing characteristics of glass tank furnaces ... 165
6.8.2.2 Influencing a change in colour ... 165
6.8.2.3 Procedures for changing colour ... 166
6.8.2.3.1 Examples of colour change procedures ... 167
6.8.2.3.1.1 Container glass ... 167
6.8.2.3.1.2 Flat glass ... 168
6.9 The behaviour of refractories in contact with molten glass ... 169
6.9.1 The corrosion of tank blocks by diffusion without convection ... 169
6.9.2 The corrosion of tank blocks by diffusion and forced convection ... 170
6.9.3 Corrosion of tank blocks by diffusion and density convection ... 171
6.9.4 The influence of interfacial and surface tensions ... 171
6.9.4.1 Flux line corrosion ... 172
6.9.4.2 Upward and downward drilling, pitting ... 173
6.9.5 Possibilities of reducing corrosion ... 173
6.10 The performance of refractories in the superstructure and waste gas regions of a furnace ... 174
6.11 The stirring of glass ... 175
6.11.1 The stirring mechanism ... 175
6.11.2 Stirrers ... 176

7 Mathematical models of glass tank furnaces ... 178
7.1 Models of the firing system ... 178
7.2 Models of glass currents ... 179
7.3 Models of glass tanks ... 179
7.3.1 General survey ... 179
7.3.2 The Trier & Voss glass tank model; its construction ... 179
7.13 The Trier & Voss glass tank model, results ... 183
7.3.3.1 Fuel input, specific energy consumption, specific melting rate (related to melting end area), depth of glass ... 183

7.3.3.2 Radiation of the superstructure, regenerator efficiency and crown temperature . . . 186
7.33.2.1 The influence of the emissivity of the combustion chamber, ε_G . . . 187
7.3.3.2.2 Influence of heat losses through the crown . . . 187
7.3.3.2.3 Influence of crown emissivity, ε_G . . . 188
7.3.3.2.4 The influence of the surface temperature of the molten glass, ϑ_O . . . 189
7.3.3.2.5 The influence of the emissivity of the molten glass, ε_O . . . 189
7.3.3.2.6 The influence of the regenerator efficiency, η_k . . . 190
7.3.3.2.7 The influence of the difference in melting down rate of the batch blanket between upper and lower surfaces (factor φ) . . . 190
7.3.3.3 Results for the tank furnace as a whole . . . 191
7.3.3.3.1 The application of the mathematical model to process economics . . . 193
7.3.3.3.2 The effect of electric boosting . . . 194

8 Electric tank furnaces . . . 195
8.1 Electrical properties of molten glass . . . 195
8.1.1 Electrical conductivity . . . 195
8.1.2 Thermal instability . . . 197
8.2 Transfer of energy from electrodes to molten glass . . . 199
8.3 Electrode circuits . . . 201
8.3.1 Heating circuits connected in series . . . 201
8.3.2 Heating circuits connected in parallel . . . 202
8.3.3 Delta connections used with alternating current . . . 202
8.3.4 Star connections used with alternating current . . . 202
8.3.5 Special connections . . . 203
8.4 The electrical resistance of different arrangements of electrodes . . . 203
8.4.1 Rod electrodes installed in parallel ignoring tank boundaries and using single phase power supply . . . 203
8.4.1.1 Two rod electrodes in parallel . . . 203
8.4.1.2 Four rod electrodes in parallel . . . 203
8.4.1.3 Six rod electrodes in parallel . . . 204
8.4.2 Rod electrodes in parallel ignoring tank boundaries and using a three-phase network . . . 204
8.4.2.1 Three rods in parallel connected as a delta network . . . 204
8.4.2.2 Three parallel electrodes in series . . . 204
8.4.2.3 Four parallel electrodes connected in series . . . 205
8.4.3 Rod electrodes in a finite furnace volume . . . 205
8.4.3.1 Two rod electrodes in parallel . . . 205
8.4.3.2 Two rod electrodes installed identically but opposing one another . . . 205
8.4.3.3 Rod electrodes forming a rectangle and inserted through the floor using a two-phase supply . . . 206
8.4.3.4 Rod electrodes inserted through the floor arranged on a triangle and using a three-phase supply . . . 206
8.5 Electrode materials . . . 207
8.5.1 Molybdenum . . . 207
8.5.2 Graphite . . . 209
8.5.3 Tin oxide . . . 209
8.5.4 Platinum . . . 210
8.6 Electrode positioning . . . 210

8.6.1 Horizontal installation of rod electrodes 210
8.6.2 Vertical installation of rod electrodes 211
8.6.3 Plate electrodes 211
8.7 Electrode holders 211
8.7.1 Rod electrodes 211
8.7.2 Plate electrodes 213
8.7.3 Tin oxide electrodes 213
8.8 Design and operation of electric tank furnaces 214
8.8.1 Design 214
8.8.2 Operation 218
8.8.2.1 The melting process 218
8.8.2.2 Instrumentation for measurement and control 219
8.8.2.2.1 Measurement of temperature 220
3.8.2.2.2 Glass level 221
8.9 Glass tank furnaces with electric boosting 222
8.10 Safety features 224
8.10.1 Existing regulations 224
8.10.2 Regulations covering prevention of accidents 225

9 Day tanks 226
9.1 Temperature schedule and melting process 226
9.2 Firing system 227
9.3 Fuel consumption 227
9.4 Melting rate 228
9.5 Glass quality 228
9.6 Refractories 228

10 Pot furnaces 229
10.1 Design 229
10.2 Furnace operation 231
10.3 The heating process 234
10.4 Electrically heated pot furnaces 235
10.5 Refractories 236

11 Fuels and combustion 239
11.1 Fundamentals 239
11.1.1 Excess air 240
11.1.2 Development of heat 241
11.1.3 The combustion process 242
11.1.4 Thd eew point 244
11.2 Industrial fuels 245
11.2.1 Gaseous fuels 245
11.2.2 Liquid fuels 246
1L2.2.1 Fuel oils 246
11.2.2.2 Liquefied gases 247

12 Waste heat recovery 249
12.1 Recuperators and regenerators: design calculations 249
12.1.1 Recuperators 249

12.1.1.1 Local changes in temperature and heat transmission coefficient 250
12.1.1.2 Efficiency 250
12.1.1.3 Temperature relationships.................................... 251
12.1.2 Regenerators 251
12.1.2.1 Determinations according to Hausen 252
12.1.2.1.1 Heat transmission coefficient 252
12.1.2.1.2 Temperature changes and checkerwork temperature.............. 252
12.1.2.1.3 Heat transfer coefficients 252
12.1.2.1.4 Pressure losses.. 254
12.1.2.1.5 Influence of flow through the checkerwork on the quantity of heat exchanged 254
12.1.2.2 Regenerator calculations using the computer program by HVG 255
12.1.2.2.1 Changes in temperature of regenerator refractories according to location and time.. 255
12.2 Waste heat boilers... 256
12.2.1 Waste heat boilers located downstream from regenerative glass furnaces 256
12.2.1.1 Waste heat boilers for general heating.......................... 256
12.2.1.2 Waste heat electric power plant 257
12.2.1.3 Design of waste heat boilers 259
12.2.2 Waste heat boilers linked to a recuperatively fired glass tank furnaces....... 259
12.2.2.1 Design.. 260
12.2.2.2 Operation 262
12.2.3 Energy management and energy distribution 263

13 Emissions from glass furnaces 265
13.1 Breakdown of materials emitted...................................... 265
13.2 Volatilisation from molten glass..................................... 266
13.3 Evaporation from melting batch...................................... 267
13.4 Dust evolved from melting batch..................................... 268
13.5 Deposits and reactions taking place in regenerator chambers............... 268
13.6 Chimney emissions.. 269
13.6.1 Dust.. 269
13.6.2 Harmful gases 271
13.7 Visibility of waste gas plumes....................................... 272
13.8 Equipment for the reduction of emissions 273
13.8.1 Fabric filters 273
13.8.2 Electrostatic filters... 274
13.8.3 Filtration equipment using dry adsorption 275

References... 277

Index 281

Translator's Foreword

Fate had a hand in my being asked to review W. Trier's book on glass furnaces as well as R. Günther's book of over thirty years ago entitled *Glass-melting tank furnaces*. As important as Günther's book was in presenting an overall review of the technology as it then stood, a fact borne out by the decision of the Society of Glass Technology to publish an English version (translated by John Currie), so it became equally clear to me that Professor Trier's book, dealing as it does not only with tank furnaces but all types of glass furnace and covering not only the technology but also much of the science underlying it, was an equally important milestone in the literature of glass technology.

Accordingly, I strongly recommended that it be translated into English and, to emphasise my conviction, I offered to undertake this task.

The work of translation started auspiciously enough, namely during a tour of China with the Group of British Glass Machinery Manufacturers in May 1985. During that trip and during many subsequent journeys, long hours of tedious waiting in airports and boredom during flights were relieved by working on the translation. Colleagues in the industry, looking over my shoulder, so to speak, were able to help in advising the correct terms to use in their specialist fields of interest, and several of them allowed themselves 'to be volunteered' to check whole sections in draft. With others I established contact by letter or telephone so that specific points and terms could be checked. Finally, the Frank Wood Joint Library of Glass Technology at Sheffield appeared to be on 'instant action', ready to send me copies of relevant papers so that I could check terminology and other details. For all this support and friendly help I am most sincerely grateful.

At the same time, and throughout the preparation of the translation, I enjoyed the full assistance and cooperation of the author who also permitted certain sections to be amplified where more recent advances made this desirable and possible. The author made some additions to the text (e.g. The stirring of glass) and at the same time some errors and omissions of the German edition were corrected.

Whenever possible I have used the terms already used by John Currie in his translation of Günther's book, as well as those given in the *Dictionary of glassmaking* prepared by Sub-Committee A1 of the International Commission on Glass and published by Elsevier Scientific Publishing Company, Amsterdam, in 1983. For the remainder I relied on more recent technical literature and the advice of colleagues in the industry; in the end, a few terms were left which had to be described or English terms invented.

Professor Trier remarks in his foreword that the preparation of his book while in full-time employment took many years; he has now retired and it appears that his book is a presentation to the glass industry at large of his many years of experience as Professor of Glass Technology at the Technical University of Aachen and as Director of both the Hüttentechnischen Vereinigung der Glasindustrie and the Deutsche Glastechnische Gesellschaft in Frankfurt. It is a worthy summary of a distinguished career.

I am very pleased to have been able to assist in making this book available to English-speaking readers in the field of glass technology and related technologies, and hope that it proves both useful and stimulating.

Camberley, Surrey, February 1986

K. L. Loewenstein

Foreword

Over recent years technical development and growth of scientific understanding has greatly accelerated in virtually all branches of technology. The result of this is that technology is no longer covered mainly in books but is found in specialised publications dealing with specific problems. As a result it has become increasingly difficult for the student as well as the working technologist to retain an overall view of his subject. This is especially the case in the field of glass technology since so many technologies impinge on it. The glass furnace, the cornerstone of glass technology, is only barely covered in existing books.

The objective of this book is not only to deal with the glass furnace in its basic construction and mode of operation, but to include the relevant technical background knowledge. It is based on Günther's well-known book, *Glass-melting tank furnaces* but differs in covering the technical background in far greater depth. The increase in the field of knowledge is reflected in the much larger volume of literature covering this subject.

The knowledge here assembled originates largely in the work of the Technical Committees (Fachausschüsse) of the Deutsche Glastechnische Gesellschaft, publications in *Glastechnische Berichte*, and information available internally from the Hüttentechnischen Vereinigung der Glasindustrie. In addition, relevant foreign literature has also been evaluated.

It was not always easy to decide logical borderlines in certain technically related areas, such as refractories, heat technology, combustion, firing, and flow theory. The guiding principle used was to include the most important parameters, which implies that the relevant reference books and standard works should be used in conjunction with this book.

The assembly of a comprehensive manuscript, in addition to working in normal employment, unfortunately occupied several years, and therefore, because of the rate of technical development, it has been possible to incorporate the specialised technical literature only to a limited extent.

I must express my thanks to the Hüttentechnische Vereinigung and the Deutsche Glastechnische Gesellschaft for permission to use the available material and for financial support during the book's publication. Only this made it possible to set the price of the book at an acceptable level. Thanks are also due to Herr U. Dohle at Aachen for systematically working through the manuscript and to Frau Kressner, Librarian of the Deutsche Glastechnische Gesellschaft, for checking and completing the literature index. I also thank the publishers for the good appearance of the book and its publication on schedule. Many thanks also for much valuable advice from many colleagues in this field of technology.

Frankfurt am Main, September 1983 **W. Trier**

1 Introduction

Glass, as a material, and the many articles manufactured from it, play an important part in daily life, in industry, in technology, and in science. Its transparency, its chemical resistance, and its relatively high hardness are the reasons for its unique position among materials. Its peculiarities, such as, for example, its brittleness and its relatively high weight, must be borne in mind when considering its applications.

Glass consumption in modern industrial countries is in the region of 60–80 kg per head of population per year. In terms of quantity, container glass dominates, followed by flat glass, and both are soda–lime glasses whose compositions are varied by the addition of different components. There are in addition tableware glasses based on potash, lead glasses, a large group of special glasses for chemical and technical purposes and glasses for electrotechnical applications, optical glasses, and glasses for fibres. The chemical composition of such glasses varies widely (Table 1.1). Special mention must be made of borosilicate glasses for chemical and technical glassware, alkali-free glasses for textile glass fibres, and alkali borosilicate glasses for glass fibre insulation.

All the above glasses are manufactured from carefully selected raw materials, referred to as batch, being melted in furnaces and then manufactured into the desired articles. The manufacturing processes vary widely as do the quantities of glass required. In consequence, the melting capacities of individual furnaces also vary widely, covering a range from 100 kg/day for special glasses up to 800 t/day for flat glass. Most glasses are melted by continuous processes; in special cases, however, especially for expensive industrial glassware, the noncontinuous processes maintain their viability.

Glass furnaces are constructed as hearth furnaces using fossil fuels such as oil or natural gas. The direct use of electricity is possible, either as an additional source of energy or, as in all-electric furnaces, as the sole source of heating energy. As an additional source of energy, i.e. boosting, it is gaining ground in the container glass industry; fully electric furnaces are currently limited to the manufacture of special glasses.

Glass furnaces are considered either in relation to the type of glass melted or in relation to their type of construction. In line with the types of glass, one differentiates between furnaces for container glass, flat glass, and special glasses. This concept of the type of glass covers not only the chemical composition of the glass itself but also the quality requirement for the manufactured products (Table 1.2). References to glass quality relate to its degree of homogeneity (or rather the level of cord or inhomogeneity), bubble and seed content, colour, and chemical resistance. The criterion by which these properties are judged is that they should not vary either with position or with time. In the case of special glasses many other properties must be considered such as coefficient of thermal expansion, refractive index, etc.

The properties and quality of the glass to be melted are determined in part by the products to be made therefrom and in part by the manufacturing technology. For example, an important property from the point of view of manufacturing technology is the temperature–viscosity relationship of the molten glass, which gives rise to the concept of long and short working ranges.

The technology of glass melting comprises the feeding of raw materials (e.g. sand, limestone, and soda ash), the melting itself, the fining and homogenisation, and the supply of glass for processing at a suitable temperature. In continuously operating furnaces these various stages are present simultaneously in different zones of the furnace; in intermittently operating furnaces these stages occur in the same place at different times.

The most important type of continuously operating glass furnace is the tank furnace, the construction of which can be traced back to Siemens;[2,3] the most important intermittently operating glass furnace is the pot furnace. The so-called day tank can be considered as an intermediate type; in construction it is similar to the tank furnace but in operation similar to the pot furnace.

Table 1.1. *Chemical composition of important glasses (wt%)*

Component	*Flat glass (Float)*	*Container glass*	*Tableware (Bohemian crystal)*	*Lead crystal*	*Opal glass (lighting)*	*Laboratory glass*	*Optical glass*	*Glass fibres Centrifugal spinning process*	*Flame attenuation process*	*Textile fibres*
SiO_2	72·8	73·3	74	60	67·5	80·4	61–70	62·0	45·6	53·2
Al_2O_3	0·7	1·5	0·18	0·08	5·0	2·27	0·3–3	3·4	13·5	14·2
$Fe_2O_3 + TiO_2$	0·09	0·06	0·02	0·02	0·15	0·03	–	0·5	15·5	0·34
CaO	8·6	9·8	5·3	–	9·4	–	<3	8·0	10·4	22·6
PbO	–	–	2·8	24·0	–	–	–	–	–	–
MgO	3·61	0·34	–	–	–	–	–	3·7	10·3	0·42
Na_2O	13·7	14·2	5·0	1·0	13·6	3·8	<5	14·0	3·5	0·26
K_2O	0·2	0·6	12·7	14·9	1·8	0·6	12–18	1·8	1·1	0·21
SO_3	0·3	0·2	–	–	0·2	–	–	0·2	0·1	–
F	–	–	–	–	4·0	–	–	–	–	–
B_2O_3	–	–	–	–	–	12·9	6–20	5·0	–	8·55
BaO	–	–	–	–	–	–	<10	1·4	–	$SrO + ZrO + P_2O_5 = 0{\cdot}22$

Table 1.2. *Deviation of SiO_2 content (wt%) taken as limiting values for inhomogeneity of glasses below which rejections do not occur*[1]

Glass type	*Product*	*Deviation*
Flat glass	Windscreens	±0·05
	Construction glass	±0·10
	Greenhouse glass	±0·35
Container glass	Bottles	±1·0
Pressed glass	Street lighting etc.	±0·5
Glass for the electrical industry	Tubing, bulbs	±0·3

2 Technical criteria, descriptions, and designs

2.1 Technical criteria

There are many types of glass tank furnace, differing in design and technical detail, but they can be considered as falling into a few major groups, and these are given in Table 2.1. They differ in the main in the method of working the glass (whether continuously or intermittently), the barrier between the melting and working ends, the direction of the flames, and the method of heat recovery.

With the exception of day tanks and pot furnaces, all tank furnaces are worked continuously. The use of regenerators involves a periodic change of flame direction.

Continuous working demands exact control of the melting process with respect to both time and place, of the fining process, and of the transfer of the molten glass to the working or conditioning section of the tank and its cooling in that section. Later chapters of this book will show that furnace construction and methods of operation are largely governed by these problems. Even at high melting rates it is necessary to ensure that the correct technical sequence of each of the phases in the glass melting process is maintained. The boundaries between the melting and refining sections and the working end of a tank furnace are therefore of major importance.

With open tanks the working end is only separated by locally reducing the width of the tank and lowering the crown; this separates the hot melting and refining sections from the cooler conditioning section. In this instance the cooling process is relatively slow and therefore demands a large conditioning zone; the advantage of this construction is the layered configuration of the molten glass, a factor of great value in the manufacture of flat glass.

A channel near the floor, referred to as the throat, gives, in terms of temperature, a more definite separation and the working end, as the conditioning section is here called, may then then be significantly smaller. The separation between the melting and fining sections of the furnace structure can be made very effective by means of a two-chamber construction. If a sunken throat is used the return flow of cool glass from the working to the melting end is reduced, improving the thermal efficiency of the furnace. Working ends have, in the course of evolution, become progressively smaller; in container glass tanks and tanks for some special glasses requiring forehearths and feeders, the working end functions only as a distributor of the molten glass.

With respect to the position and direction of the flames within the furnace, there are two types, namely cross-fired and U shaped flame. Cross firing is particularly suitable for large wide furnaces with melting areas of 80 m^2 or more but for furnaces whose widths are insufficient to achieve good flames and complete combustion it is usual to employ the U flame arrangement. Furnaces fired in this way often fail to achieve the standard of temperature distribution required for tanks for special glasses and this has led to specially designed and constructed furnaces.

A special position is occupied by furnaces with recuperative heating, which operate with a constant flame along the length of the furnace. These tanks operate with premixing burners which give short flames.

The high operating temperatures of glass tank furnaces which can lie between 1450 and 1650°C, depending on the type of glass, demand efficient heat recovery from the waste gases in conjunction with good heat transfer to provide combustion air at a high preheat temperature; this ensures that the process operates with good thermal efficiency. The regenerator system, first developed 120 years ago by Siemens, has played a valuable part in achieving such efficiency for tank furnaces. The continuously operating recuperator does not give such high air preheat temperatures as can be obtained with regenerators but for small units, where steady flames and simple constructions are particularly important, the recuperator often has the advantage.

Table 2.1. *Technical characteristics of glass tank furnaces*

Elevation of furnace (→ direction of glass flow)	*Type*	*Glass working*	*Separation of melting and working ends*	*Flame direction*	*Heat recovery*
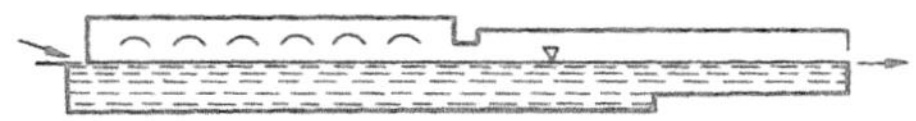 Melting end Working end	Open cross-fired tank	Continuous	Lowering of crown and curtain	Across glass flow	Regenerator
 Melting end Working end	Cross-fired tank with throat	Continuous	Sunken throat	Across glass flow	Regenerator
 Melting end Working end	End-fired tank with U shaped flame	Continuous	Sunken throat	U shaped flame and reversal	Regenerator
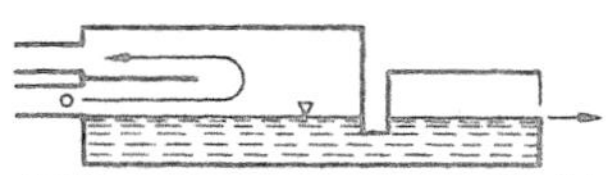 Melting end Working end	End-fired tank with double crown	Continuous	Straight throat	Constant U shaped flame	Recuperator
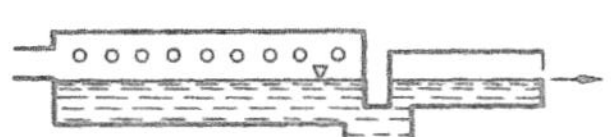 Melting end Working end	Side-fired tank	Continuous	Sunken throat	Across and parallel to glass flow	Recuperator
 Glass removed through gathering holes	End-fired day tank	Intermittent	None, melting and working ends one unit	Parallel to glass flow	Recuperator

Additional heat recovery methods, such as waste heat boilers, have so far only been used in exceptional cases but high energy costs are giving rise to an increase in the use of such equipment.

2.2 Other Criteria

Certain criteria have been developed for describing and comparing glass furnaces, and these should be used in a consistent manner if misunderstandings are to be avoided.

Stating the kind of glass to be melted, for example clear container glass, will indicate to an expert the glass composition together with its quality with respect to homogeneity, seed count, iron content, etc.

The most important factor for the furnace itself is its pull, i.e. its melting rate, expressed as tons of glass per day, and other important factors, especially from the point of view of the economics of operation, are the glass melted per furnace campaign and the length of the campaign itself. A factor of lesser importance, but none the less useful on occasions, is the volume of the furnace expressed as tons of glass.

The most important geometrical factor is the melting area, which is the area from the doghouse to the throat or, for open tanks, to the waist which separates the working end from the melting and refining end.

Although the above criteria consider the furnace *in toto*, other related or specific parameters are needed which can be used to compare one installation with another. The most important are the melting rate, which is expressed as kg of glass/m^2 of glass surface in the melting end, and the specific energy consumption, expressed as kJ/t of glass. For furnaces with mixed heating, for example those with electric boosting, these data may be defined in a number of different ways. Some give only the energy from fossil fuels to state the specific energy consumption and give the electrical energy as a separate figure, some convert the electricity consumption into energy terms and add it to the energy derived from the fossil fuel, while some, before adding the figure for electrically derived energy adjust it by the conversion efficiency from fossil fuels. Each method has its own merits.

In the absence of a uniform approach, clear definitions of data are essential. In parts of the English speaking world it is common to use not the melting rate related to surface area, but its reciprocal value, i.e. the area required per ton of glass melted in units of ft^2/t d. Furthermore, it should be noted that the US ton differs from the English ton; Table 2.2 gives conversion factors for the relevant units.

Table 2.2. *Conversion table between metric and English or American units for the melting rate of tank furnaces*

	English or American units	*To convert: UK or US units to metric system*	*Metric system to UK or US units*
		x ft^2/(UK/US)t.d =	x t/m^2.d =
English	ft^2/UKt.d	$\frac{10{\cdot}95}{x}$ t/m^2.d	$\frac{10{\cdot}95}{x}$ ft^2/UKt.d
American	ft^2/USt.d	$\frac{9{\cdot}77}{x}$ t/m^2.d	$\frac{9{\cdot}77}{x}$ ft^2/USt.d

2.3 Nomenclature

The use of a uniform nomenclature for parts of furnaces is important and useful, and Report No. 49[(4)] of the DGG suggested a uniform nomenclature which is now generally used. The terms are summarised in Figure 2.1 allowing for the current state of furnace technology. It was laid down that the wall which, for cross-fired tank furnaces, contained the doghouse and, for tanks with U shaped flames the ports, should be referred to as the front and that the working end or conditioning tank and its associated parts be referred to as the back. The terms left and right are relative to an observer facing the front wall of the furnace or to one facing a side such that left to right is the direction in which the batch moves as it is converted into glass. However, these terms are not always uniformly used even in German glass plants since, in many cases, the direction of observation is from the machines to the furnace and thus the doghouse end is referred to as the back.

2.4 Types of furnace

2.4.1 Tank furnaces for the production of glass containers

In relation to the total production of the glass industry the production of glass containers forms the largest proportion. By and large the most economic solution has been to use a regeneratively

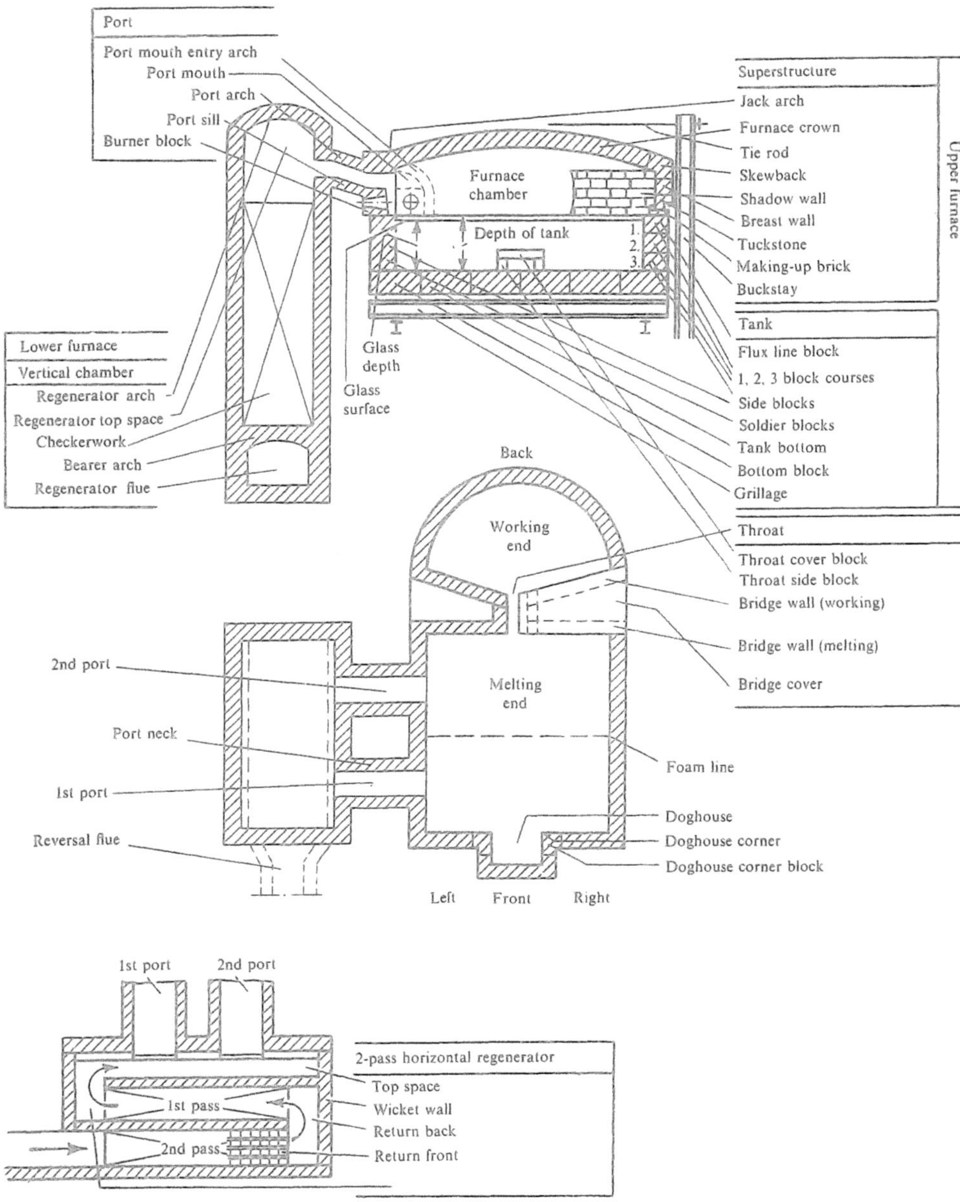

Figure 2.1. Individual parts of a tank furnace[4]

heated tank with a throat and with burners which fire across the melting end or which give a U shaped flame (Figure 2.2). A soda–lime glass is used and its composition (see Table 1.1) varies slightly depending on available raw materials and colour (white, amber, green).[5]

The melting rate of container glass tanks has increased over recent years in parallel with the increase in the production rate of forming machines. Pulls of 100–300 t/d are required and are being met at a melting rate of between 2 and 3 t/m^{2}d. Changes in the production programmes are often necessary so that the tank must be flexible as far as melting rate is concerned; variations of 50 to 100% in glass demand within one week must be accommodated without difficulty.

Glass quality must meet the requirements of the forming machines and the strength demanded for bottles and other containers. The most important property is homogeneity: slight cords, uniformly distributed, are acceptable but thick cords are dangerous as they cause local stresses. The quanti-

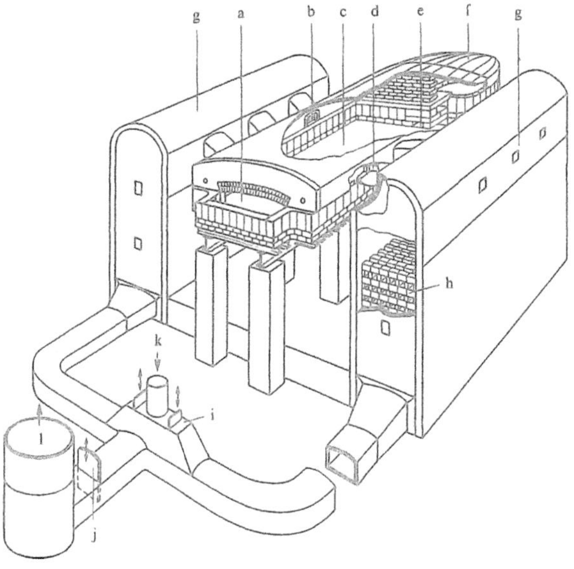

Figure 2.2. Side-fired tank furnace heated by oil for container glass

a doghouse
b burner with atomiser
c melting end
d burner block
e throat
f working end
g regenerators
h checkerwork
i reversal mechanism
j chimney damper
k air intake
l waste gas exit

fication of homogeneity is difficult although it is possible to use local variations in glass composition or measure the stress caused thereby. Table 1.2 shows that the degree of homogeneity required of container glass can be lower than that required of glasses for the electrical industry or, particularly, flat glass. The homogeneity of a container glass is monitored via stress optical measurements on ring samples,[6] a method which determines phase differences between normal and retarded light and is normalised to a glass thickness of 1 cm. The upper limit for critical applications is about 17·5 N/mm^2, which is equivalent to 2·5% of the breaking strength of soda–lime glasses.

In amber container glasses an important additional property is absorption in the biologically active region of 450 nm. The glass should show high absorption in this region in order to protect its contents from the chemically active ultraviolet radiation.

Visible stones in the glass are not acceptable and seeds above 1·5 mm diameter are noted while those above 4·0 mm lead to rejection. A total of not more than 12 faults per ton of glass are permitted.

The temperature–viscosity relationship is important for the forming process, which is monitored via certain specific temperature–viscosity reference points. Overall, the quality requirements for container glass, compared to other types of glass, are not too stringent; the most important aspect is the cost of production.

Since energy forms the major part of the total cost, emphasis is placed on the operation of container tanks at a low energy consumption, and a consumption of 4000–5400 kJ/kg glass gives container tanks the lowest specific energy consumption of all glass furnaces. This is achieved by a combination of high specific melting rate, the separation of melting and working ends via one or more sunken throats, extensive thermal insulation of the furnace, and a high air preheat temperature. Container tanks now achieve compaign lengths of 6 to 8 years and the total melting rate over the whole of the campaign is about 5000–8000 t glass/m^2 of melter area.

The maximum temperature achieved in the melting and refining end is 1600°C; this temperature is only rarely exceeded and temperatures generally lie between 1550 and 1580°C. The glass temperature in the working end is in the region of 1300–1350°C and it is necessary to heat the working end only at low glass throughput.

The requirements for container tank refractories depend largely on operating conditions but in terms of the total tonnage of glass melted in the course of a campaign the following quantities of refractories would have been used: fusion cast corundum–zircon (AZS) 0·5–1·0 kg of refractory/t of glass, silica 0·5–1·5, magnesite and similar types for regenerator packing 1–3, and clay and insulation 0·5–1·0. Steelwork for supporting the furnace works out at about 0·6 kg steel/t glass.

The quality requirements for glass for tableware (e.g. wine glasses, tumblers, etc.) are significantly higher than for container glass with respect to seed content and absence of cord. The glass is of a soda–lime composition of low iron content with added potash. The tanks used for tableware are significantly smaller and have to match the production rate of the machinery; one is satisfied with melting rates of about 1 t/m^2/day. Figure 2.3 shows a cross section of a tableware tank heated by a U shaped flame and using regenerators. Figure 2.4 shows another tank but with recuperator and using U shaped flames which return upwards; the doghouse located in the sidewall is not shown.

The risk of alumina–zircon cords in tableware is avoided by the provision of drainage points in the throat region (see Section 6.5.3.1.6).

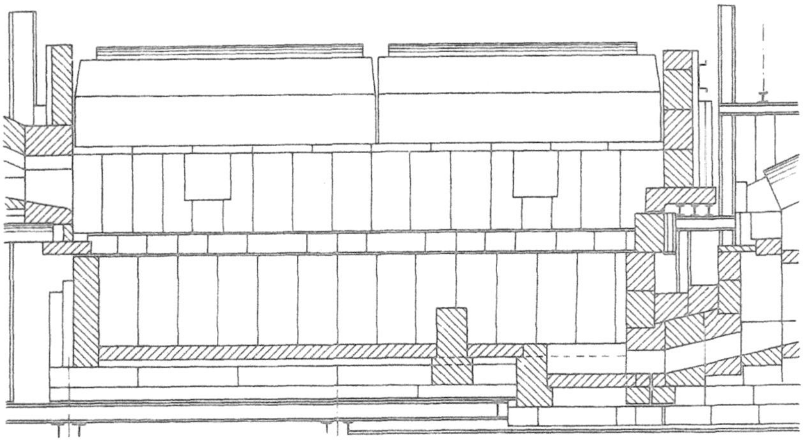

Figure 2.3. U shaped flame fired tank furnace for tableware (melting area 19 m², pull 19 t/d)

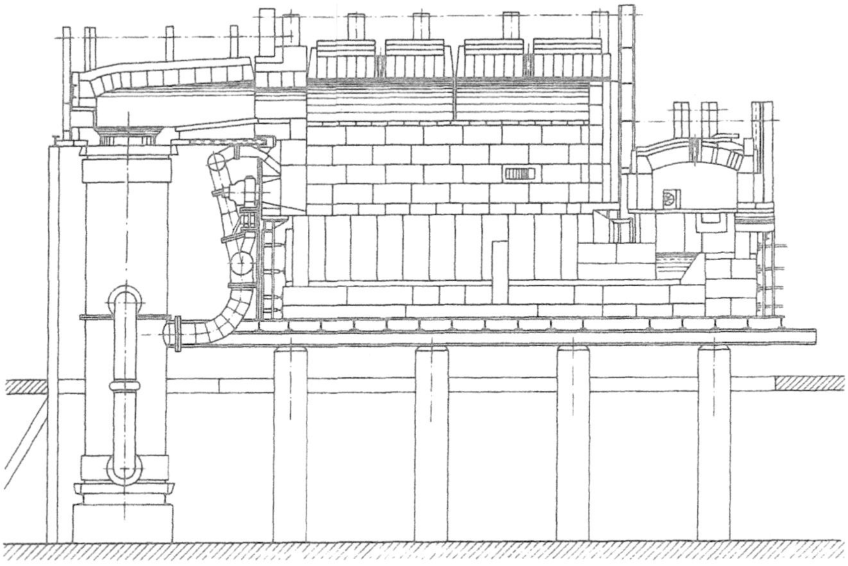

Figure 2.4. Tank furnace for tableware with U shaped flame returning above burner into recuperator (Sorg)

2.4.2 Tank furnaces for the production of flat glass

With the almost complete replacement by Float glass of mechanically ground and polished mirror glass and drawn window glass, virtually all tanks for flat glass are Float glass tanks. In terms of dimensions and melting rate they represent the largest current installations in the glass industry. Melting rates of 700–800 t glass/day are normal, and tanks melting 1000 t/d are technically feasible and economically viable.

Flat glass, like container glass, is a soda–lime glass which has been adapted to the needs of the Float process. The quality requirements with respect of homogeneity and permitted seed content are higher than those for container glass. Of particular importance is the requirement for cords to lie parallel to the glass surface and for this reason Float glass tanks have so far not been constructed with a throat. It is generally known that the constricted passage through a throat leads to a randomisation of the layer structure of the glass which, in its turn, leads to optical disturbance of the flat glass. Consequently, flat glass tanks are open and without throats. It follows that the separation of the hot melting and refining zones from the working end is not as precise as for container glass tanks. The fuel consumption is higher than for container glass and it is necessary to provide a large conditioning zone in order to reduce the glass temperature to one suitable for forming into sheet.

By providing a waist at the end of the melting and refining sections, by lowering the crown, and by the provision of a curtain wall it has been possible in recent years to reduce the energy consumption of Float glass tanks to values in the range 5·8–6·3 kJ/kg glass and specific melting rates have reached 2 t glass/m^2 d.

Figure 2.5 shows the basic design of a flat glass tank.[8] The superstructure is similar to that of a container tank, allowance being made for the larger dimensions. Since there is no throat, the conditioning section is relatively large. To influence the cooling and to prevent any of the cordy glass originating along the glass/refractory interface from being worked, the conditioning section is provided with skimming pockets on each side. Due to the large dimensions of the tank only cross-firing is possible. There is one regenerator to each port, different from tanks for containers which use a simpler construction with common regenerators. An impression of the size of flat glass tanks can be gained from Figures 2.6 and 2.7.

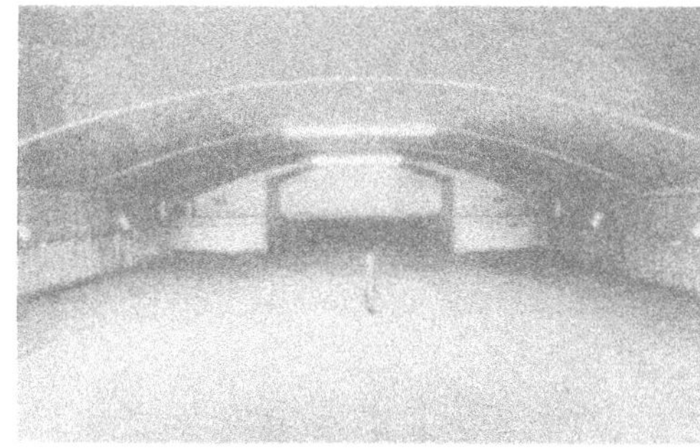

Figure 2.7. View from the working end into the conditioning zone of a Float glass tank

Due to higher quality requirements, the operating temperatures of flat glass tanks are higher than those of container glass tanks, with 1600°C being normal while 1620°C is sometimes reached.

Batch silos, such as those normally used in conjunction with container glass tanks for the supply of batch to batch chargers, are not used for

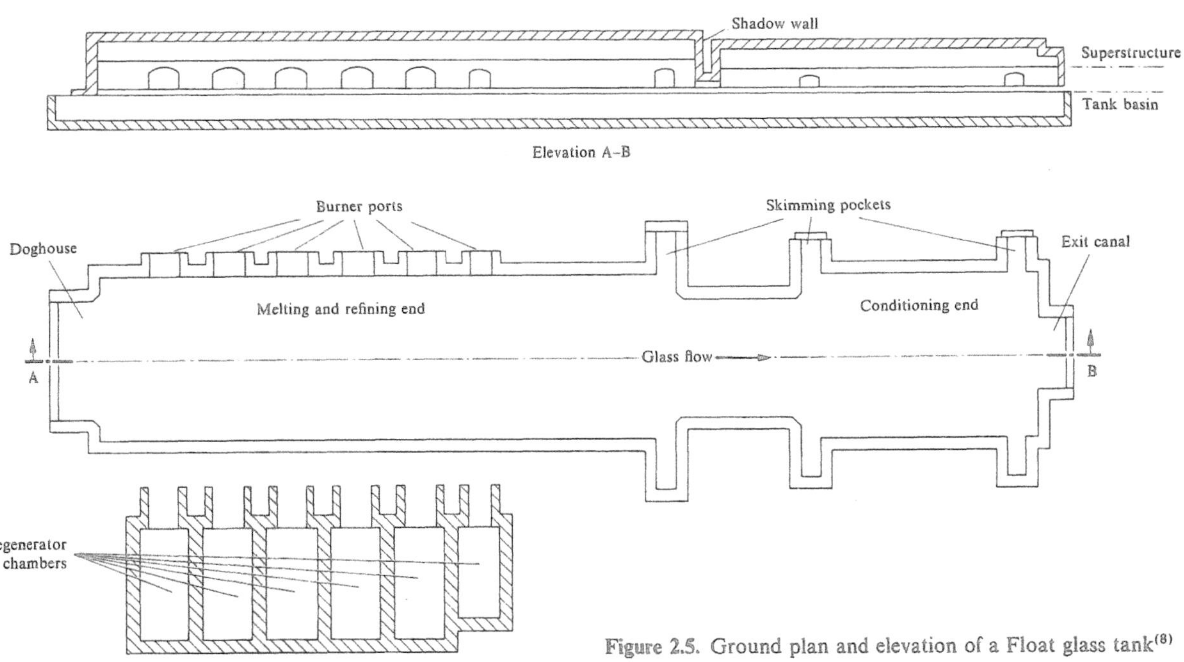

Figure 2.5. Ground plan and elevation of a Float glass tank[8]

Figure 2.6. View from the doghouse into the melting and refining zones of a Float glass tank

flat glass tanks. Instead batch cans are used so that 'fresh batch' is always fed into the furnace.

Campaign lengths of flat glass tanks are similar to those of container glass tanks.

Furnaces used for the manufacture of figured glass should be regarded as special since the quality level for the glass is lower than for Float glass.

2.4.3 Tank furnaces for the production of special glasses

When compared to container and flat glass tanks, tank furnaces for special glasses vary significantly in size and construction; in addition, the quality requirements are usually much higher although they differ between types of glass made.

As an example, Figure 2.8 shows a tank for the manufacture of glass for tubing, which is similar in construction to a container glass tank. To achieve higher glass quality furnaces for special glasses are constructed with cross-fired oil burners and compartmented regenerators, the higher investment being considered to be justified. Since control of the temperature profile along the longitudinal axis is regarded as important, cross-fired tanks are used down to a width of 3·5 m.

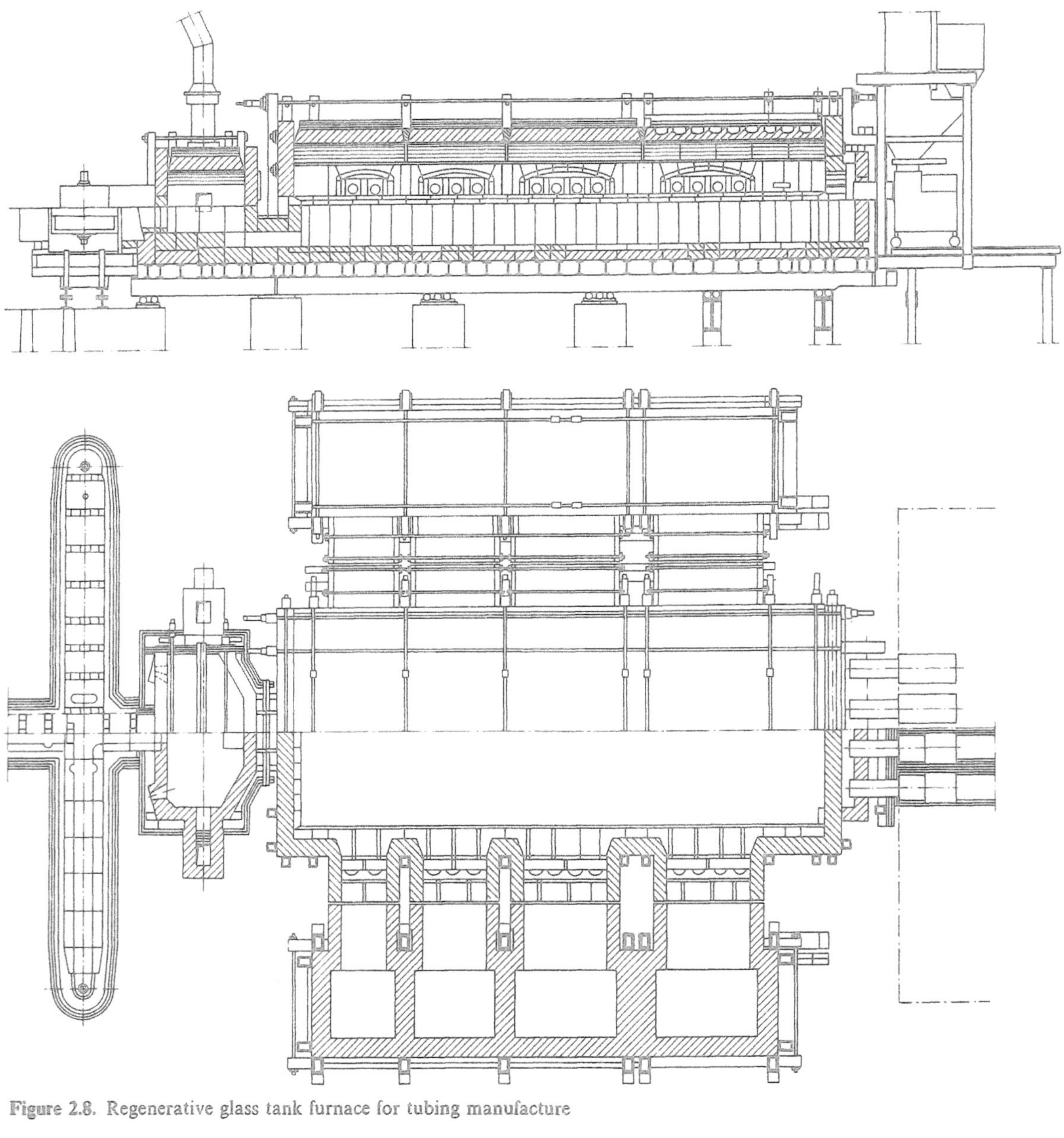

Figure 2.8. Regenerative glass tank furnace for tubing manufacture

The operating temperatures of these tanks vary according to the compositions of the glasses melted, tanks for instrument glasses, for example, operating at crown temperatures of 1650°C and above. The construction of a tank for borosilicate glass is shown in Figure 2.9 and of special interest is the provision of a weir between the melting and refining sections. The working end must be heated separately as a result of the high melting end temperatures. Combustion air is preheated by a small steel recuperator. Operating temperatures are lower for lead glasses, while alkali borosilicate glasses are very corrosive to the waste gas ducts, making the use of regenerators more difficult so that steel recuperators are preferred.

Similar considerations apply to alkali-free glasses such as E glass, and a cross-fired recuperative tank for the manufacture of glass fibres is shown in Figure 2.10. The waste gases are collected at the doghouse end and pass upwards through a steel recuperator. For glass contact it is necessary to use special dense chrome refractories which can be manufactured only within certain size limits.

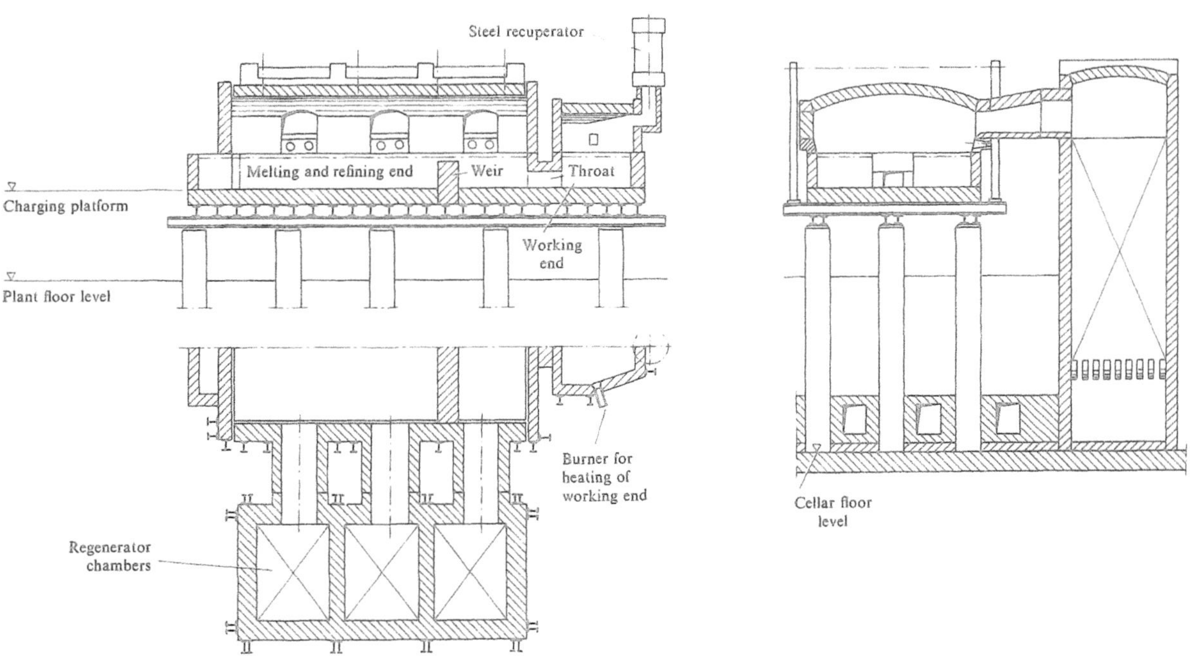

Figure 2.9. Tank for borosilicate glass with regenerative heating of the melting and refining end

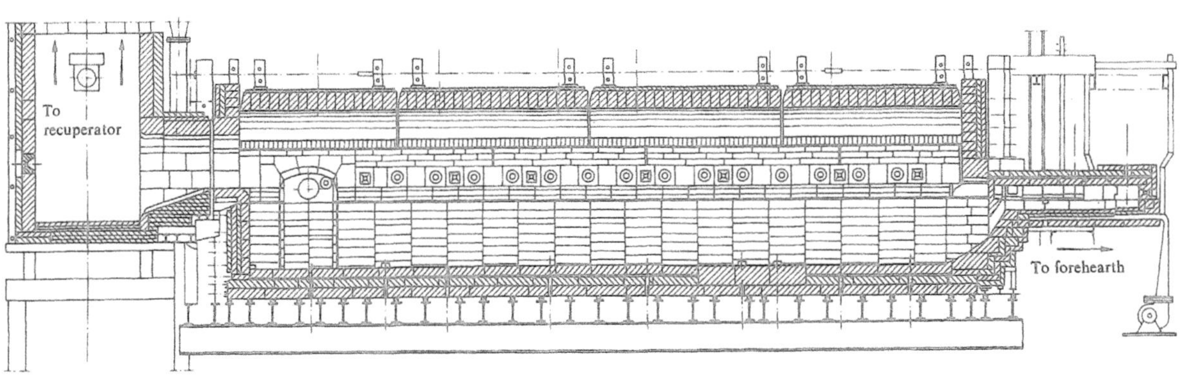

Figure 2.10. Glass tank furnace with recuperator for the manufacture of E glass fibre with burners located in the sidewall

3 Construction elements and dimensions

3.1 The tank

The tank, i.e. that part which contains the liquid glass and melting raw materials, is the main part of a glass furnace. Its geometrical dimensions such as bottom area, length-to-width ratio, and depth, must be carefully decided on the basis of construction and operational requirements. Depending on which stages in the process are taking place in the tank, one must differentiate between melting and refining zones and the working or conditioning end; these parts may need to be separated from one another to a greater or lesser extent. In terms of construction, the melting and refining zones are usually one unit while the working or conditioning end is usually totally separate and its design is governed by the method by which the glass is removed and worked.

3.1.1 Geometrical data

3.1.1.1 Melting area

The combined bottom areas of the melting and refining zones are usually referred to as the melting area and in any particular tank this will depend on the melting rate required and the expected or target specific melting rate: current specific melting rates lie between 1 and 3 t/m^2 d. Actual values depend on the intensity of heating and thereby on the operating temperature of the furnace which is, in turn, related to the type of glass being melted and its quality requirements as well as to the control of batch feeding and of the thermal currents in the glass. Container glass tanks have melting areas in the range 50–100 m^2, flat glass tanks nowadays reach 150–300 m^2 but, because of the large number of special glasses, the range of these tanks is particularly large, being from 5 to 80 m^2.

3.1.1.2 Length-to-width ratio

The choice of this ratio depends largely on the method of heating, the size of the tank, the method of batch feeding, and the construction of the crown.

Using current standard port and burner designs, the method of heating must be such as to ensure complete combustion of the flame and good heat transfer from the flame to the molten glass. The lowest value for the width of cross-fired furnaces is 4·50 m while, except for some special applications, the highest value is about 12 m, a figure governed by the structural loading of the crown and the problems of controlling long flames. The usual range of length-to-width ratios for melting areas of U shaped flame and cross-fired tanks can be determined from Figure 3.1, which shows that small tanks, usually end-fired with horseshoe-type flames, have a ratio of about 1·3 while larger tanks, cross-fired in most cases, reach ratios of above 2. Float glass tanks reach values above 3, because this is the only way by which sufficient melting area can be obtained within permitted width limits.

Cross-fired constant firing from both sides is usually based on burners giving short flames and the length-to-width ratio is correspondingly large, as is the case for unit melters.

Tanks with throats may be shorter than tanks without throats since the refining zone does not have to extend into the conditioning zone in order

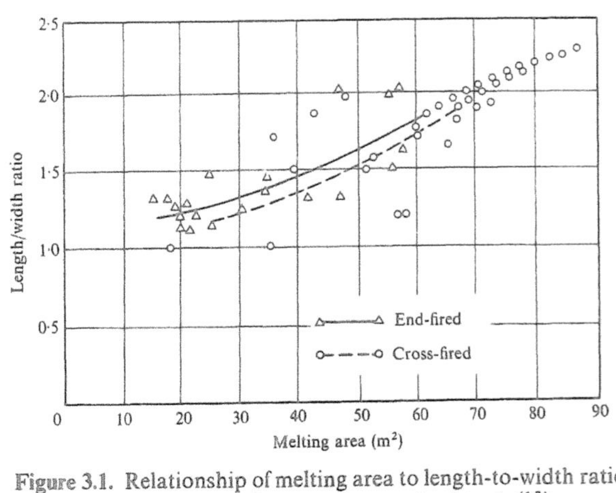

Figure 3.1. Relationship of melting area to length-to-width ratio for end-fired (U shaped flame) and cross-fired tanks[13]

to achieve the temperature reduction needed for working. Flat glass tanks are still constructed without throats and in these instances the melting and refining zones are separated from the conditioning zone by a waist, as each serves a different technical requirement. For tanks without throats it is therefore desirable to exclude the conditioning zone in any consideration of the length-to-width ratio. A comparison with other types of furnace would then not be possible. Considerations of energy losses favour square melting areas but improved techniques of thermal insulation have made this less important.

3.1.1.3 Depth

The depth of the melting and refining sections of the tank bath has a strong influence on the temperature of the tank bottom and the currents in the glass. A second factor is the ratio of the volume of glass in the tank to the melting rate. Small variations in the operation of the furnace or in batch composition have greater adverse effects on small furnaces with high melting rates than on large furnaces. A small glass volume is, however, an advantage for furnaces frequently subjected to colour changes. The details described show that there is no optimum glass depth. The controlling factors are many and each case needs to be examined on its own merits but general guidelines can be given.

Since colour influences infrared transmission, tanks for coloured glasses are shallower than those for clear glass. The differences are not large since shallow tanks melting dark glasses run the risk of operating with insufficient convection currents, which leads to a reduction in the melting rate of the underside of the batch blanket.

An increase in the specific melting rate and the associated increase in superstructure temperature have led to an increase in the thickness of the tank bottom in order to control the rate of corrosion of the floor; this tendency was further advanced by insulating the bottom to reduce energy costs.

Convection currents affect fuel efficiency in different ways. An increase in convection currents leads to an increase in the specific melting rate; this may lead to a reduction or even to an increase in the specific fuel consumption, depending on the circumstances of each case, and it is not possible to predict, even tentatively, which will happen.

Low tank bottom temperatures and low melting rates may cause stagnant layers of glass to be formed, which can lead to devitrification. Fluctuating working conditions or pull will cause the zones of stagnant glass to move around the tank, and this may give rise to faulty glassware.

At present the depths of container glass tanks lie between 1100 and 1300 mm for clear glass and between 900 and 1100 mm for coloured glass, those for green glass being shallower than those for amber glass; depths of 1600 mm for clear glass have remained exceptions. Working ends are generally shallower and have a depth of some 700–900 mm. Bottoms are always insulated. Unpleasant surprises can be avoided if, before deciding on the depth of a new tank, consideration is given to the effect that the anticipated melting chamber temperature would have on the thermal resistance of the bottom and the actual temperatures of each layer of bottom construction likely to arise during operation.

The depth for the melting and refining zones of flat glass tanks in Europe has been standardised at 1200 mm, a figure which originated with Fourcault tanks. In other countries there are some Float glass

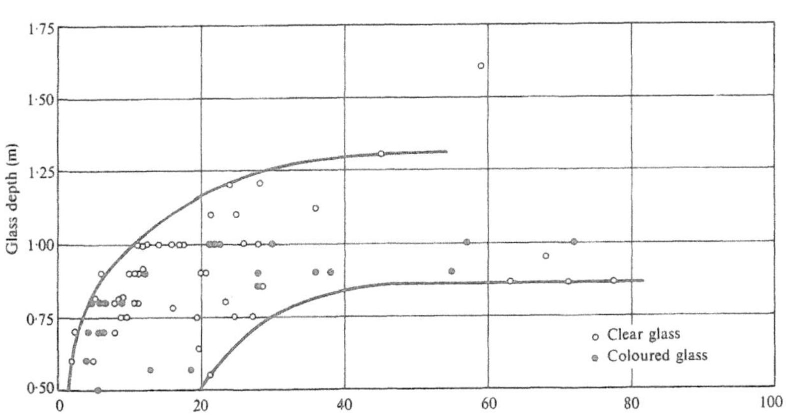

Figure 3.2. Glass depth in relation to melting area for tanks manufacturing container glass and pressed glassware[14]

tanks which reach 1500 mm in parts and the depth of the old Libbey-Owens tanks was 1600 mm. The depth of the conditioning end of a Float glass tank is similar to that of the melting and refining zones.

There are many variations with tanks for special glasses due to the many types of glass melted and the large range of furnace sizes. Results of an early investigation are shown in Figure 3.2 where the large range and the tendency for smaller tanks to be shallower can be seen. High operating temperatures sometimes preclude thermal insulation of glass contact refractories.

3.1.2 Construction

Ignoring the bottom and insulation, the tank is constructed from high quality refractories without the use of cements; a difference is generally drawn between sidewalls and floor. Layers of insulation are added to reduce heat losses. In order to reduce corrosion and thereby extend the furnace campaign, the area near the flux line is artificially cooled.

3.1.2.1 Sidewalls

As will be discussed in detail in Section 6.9, the attack by glass proceeds preferentially at horizontal joints. These are therefore avoided in the hotter sections of the furnace and blocks are used which extend in one piece from above the flux line to the bottom. These must be machined accurately to give joints of minimum width, since vertical joints are also areas of weakness. High performance furnaces use blocks with ground surfaces. A tight fit of joints is required not so much to prevent glass leakage as to reduce glass attack on the refractory blocks.

Except for furnaces for special glasses, fusion cast AZS blocks are used for glass contact refractories in the sidewalls; void-oriented blocks, in which the voids are located at the lower outside ends, are used. The wall thickness used to be 300 mm and is now usually 250 mm, sometimes only 200 mm. To reduce heat losses, all but the upper third of the block is thermally insulated.

A typical sidewall construction for container glass tanks is given in Figure 3.3b.[1] The three-layer insulation is constructed in such a way that the upper joints remain exposed while the middle and lower joints can be covered provided that the sides of the blocks are machined and the joints are very tight. As the blocks wear near the flux line, the upper sections of insulation are removed and replaced by void-free fusion cast refractory tiles. In order to reduce attack by upward drilling (see Section 6.9), the use of tank blocks with an outward sloping inside wall or blocks with parallel walls but installed at an angle to achieve the same effect was suggested but expectations were not fulfilled in practice and the procedures were discontinued. Good contact between blocks and insulation in the middle and lower sections is achieved by a thin layer of cement or by a tissue or felt of ceramic fibre.

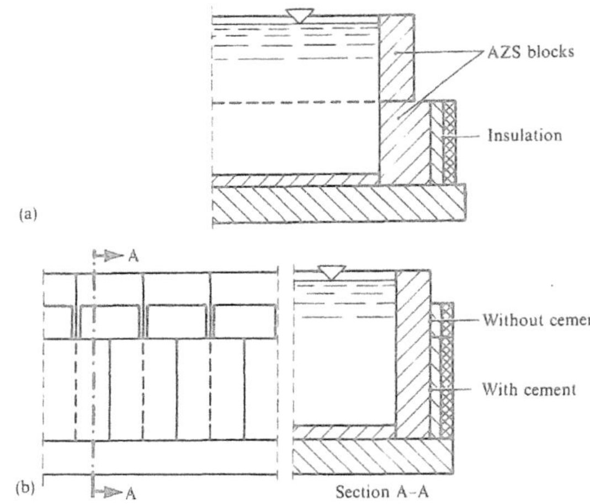

Figure 3.3. Sidewall insulation of a melting furnace[1]
(a) insulation of the lower course of blocks
(b) use of soldier blocks

With deep tanks and tanks for melting glasses of high iron content a multilayer wall construction is used instead of single soldier blocks. The objective is to reduce investment by using cheaper blocks in the lower parts of the sidewalls: the lower layers could then be used again for a second campaign. Only the lower areas of multilayer sidewalls should be insulated, otherwise corrosion would proceed rapidly in the horizontal joints. The upper section of glass contact refractories is often made a little narrower so that a base is created for placing the backing refractory tiles (see Figure 3.3a), otherwise a steel retaining structure must be provided. Sidewalls of working ends are now invariably fully insulated.

The flux line is usually cooled by air although in some special cases water is also used. In relation to the flux line length, the air flow is 0·8–1·0 m^3/s m at 9·8 mbar. The exit speeds are about 30 m/s for nozzles with a slot width of 10–25 mm. The cooling nozzles are arranged side by side along the flux line, the distance between them being 30–50 mm horizontally and 80–900 mm vertically; their inclination to the vertical is 15–30°. Cooling can be augmented by introducing atomised water into the air stream. The cooling of new tank blocks has few advantages.[11] Water cooling is used only in exceptional circumstances, especially when glass leakage has to be prevented. The old cooling boxes have been replaced by a series of steel pipes of square cross section welded together. To prevent scale deposits soft water must be used, which implies a recirculating water cooling system. To improve heat transfer it is recommended that cement to which silicon carbide has been added to improve thermal conductivity should be used between the cooling box and the refractory.

Uninsulated sidewalls made from fusion cast blocks have heat losses of 18 kW/m^2 but good insulation can reduce this to 2·9 kW/m^2.

3.1.2.2 The tank bottom

In earlier days the tank bottom was constructed of clay blocks laid on a steel grid and presented no particular problems. The increase in operating temperatures, melting rates, and length of campaigns and the need to save energy have led to the construction of thick multilayer bottoms as shown in Figure 3.4.[1] The uppermost layer is of void-free paving blocks 75–120 mm thick laid on top of one or two layers of clay blocks 200–300 mm thick, followed by several layers of insulation; the paving is laid on an AZS cement. A second layer of paving or a continuous layer of ramming mix can also be used in high performance furnaces. The sidewall blocks are positioned on the clay blocks; extending the paving under the sidewall blocks is possible but is carried out only rarely.[2] If the sidewall blocks are positioned on top of the paving, the insulation should not be extended beyond the layer of paving blocks (Figure 3.5). Tanks for special glasses sometimes use special ramming mixes of zirconium silicate, either in addition to or instead of paving blocks.

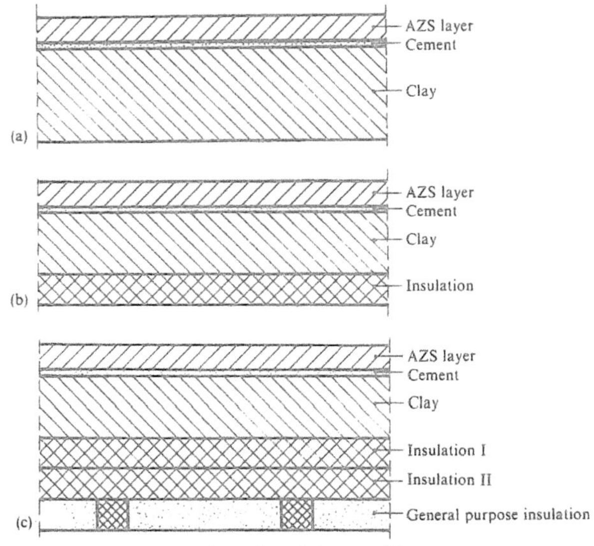

Figure 3.4. Construction of tank bottom[1]
(a) fusion cast blocks on clay
(b) clay with insulation added
(c) usual insulation on tanks for container glass

Well insulated bottoms should have all joints sealed with cement, including the vertical joints of the paving and the blocks used for bubblers, electrodes, and weirs. Cements based on aluminium zirconium silicic acid have been developed for this purpose;[3] at ambient temperatures these cements are relatively fluid, which assists in the filling of the joints, and their thermal expansion matches that of AZS paving blocks. Above 1200°C the cement becomes plastic again, and follows the expansion of the bottom thus providing good adhesion; starting at 1200°C mullite begins to crystallise from the liquid phase, improving the ceramic binding and giving the cement a high strength.

Bottoms with a double layer of paving should have offset joints to obstruct the ingress of glass. Insulation should not be taken too far for clear, low-iron glasses or when a high proportion of cullet is being used. The operating temperature of the upper paving must not exceed 1400°C. It is also important that the temperature at the interface between the clay blocks and the insulation should not exceed the devitrification temperature of the glass so as to prevent a breakthrough of glass. Float tank bottoms have been insulated but progress has been gradual.

Energy losses from tank bottoms without insulation are 4600–6000 W/m^2; insulation reduces this to about 1300 W/m^2.

Technical requirements often call for transverse rows of bubblers in the refining section and bubbler

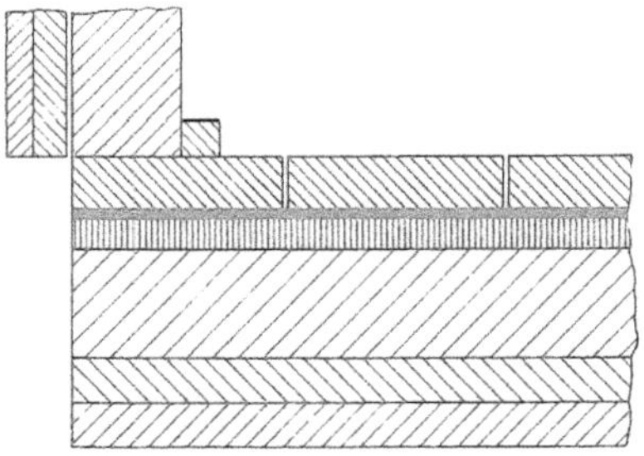

Figure 3.5. Floor tiles under sidewalls

blocks must therefore be installed. For reasons of access and life, these are not insulated and their lower ends are recessed into the tank bottom as shown in Figure 3.6.[2] Even in multilayer bottoms these blocks are of fusion cast material to give longer life by more effective cooling. The additional glass currents caused by bubblers lead to increased attack on sidewalls and bottoms. The use of thicker paving reduces corrosion in this area (Figure 3.7).

More and more tanks for container glass and special glasses are being provided with weirs, whose function is to influence glass currents.[4] Such weirs are built from a single thickness of blocks in contact with glass on both sides, inserted into the bottom, and secured independently of the sidewalls in order to form tight joints. The useful life of such weirs leaves much to be desired despite all precautions and the avoidance of bottom insulation in their proximity. For relatively short campaigns (Figure 3.8) or for glasses with low rates of corrosion (Figure 3.9) reasonably good life has been obtained.

Improvements in their corrosion resistance and their suitability for use in small joints has led to bottom cements being used not only in the vertical joints of the paving, but also with the blocks used for weirs, bubblers, and electrodes. Although insulation must be sacrificed in areas where blocks for bubblers and electrodes are installed, weirs are, compared to previous practice, now often insulated. Bottoms of this type must be heated up very carefully and full account must be taken of the different rates of thermal expansion.

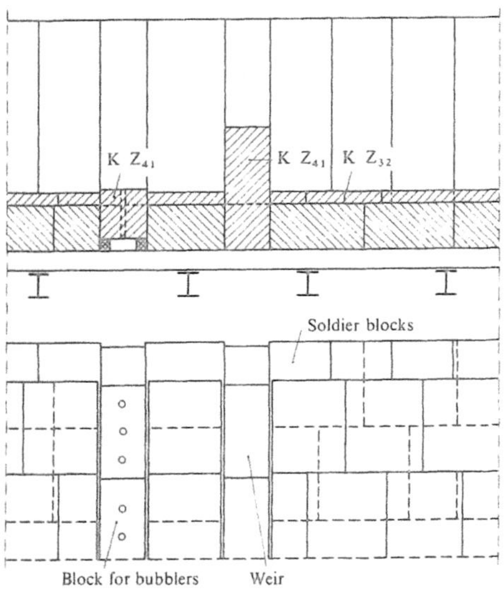

Figure 3.6. Example of a design for a weir and a row of bubblers for a tank bottom with little insulation[7]

Figure 3.8. Weir in a cross-fired tank furnace after 23 months in operation[4]

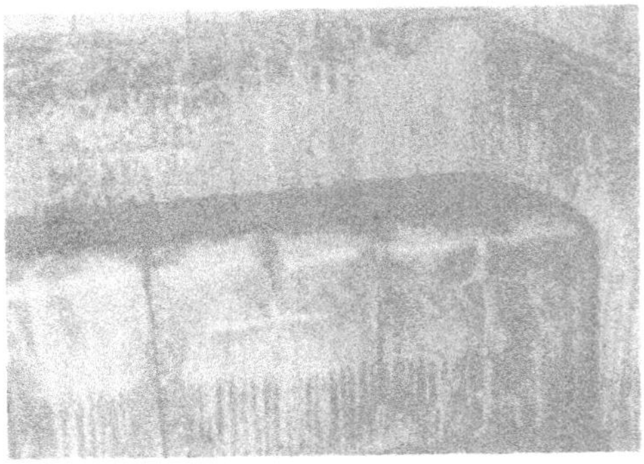

Figure 3.9. Weir in a tank making hotel glassware after four years in operation

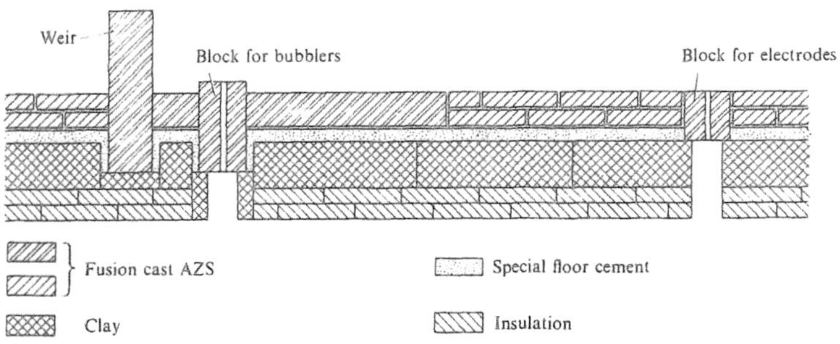

Figure 3.7. Design of an insulated bottom using special cements

3.1.2.3 The throat

In tanks for container glass or special glasses the melting end is connected to the working end by a throat whose primary function is to accommodate the relatively large temperature difference between these sections in a short distance and, at the same time, to form a physical separation between them. As it is heated from above, the body of the glass exhibits a temperature gradient which depends primarily on the colour of the glass. The temperature of the glass flowing into the working end is influenced by the depth and position of the throat. For energy reasons and in order to keep the flow in the throat to a minimum for a given melting rate, the throat design aims at minimising the return current from the working end to the melting end. For dark glasses, and their associated large temperature decrease with glass depth, these requirements are met by a straight throat at the same depth as the melting end. After passing through the throat there is usually a step because the working end is normally shallower than the melting end. The depth obtained with a straight throat is not sufficient for clear glasses and it is normal to use a sunken throat. Figure 3.10 shows both types of throat construction. Throat cross sections in the flow direction usually have a height-to-width ratio of 1:2 and lie between 0·1 and 0·3 m^2, the smaller values being for clear and the larger for coloured glasses. Ignoring possible return currents this leads to average flow rates of 5–20 m/h. In almost all cases there is only one throat; the literature describes tanks with two or more throats but they have remained exceptions since the construction is very costly and the advantages claimed, which are in any case disputed, do not justify the investment.

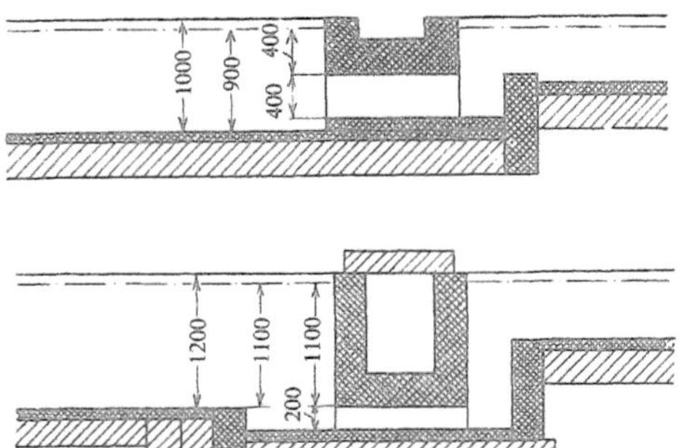

Figure 3.10. Different designs of throat[5] (dimensions in mm)
(a) straight (b) sunken

The throat is one of the most critical parts of the tank; its correct functioning decisively governs the efficient operation of the plant as well as the glass quality and the length of the furnace campaign. The rule is to select the best possible materials and to exercise great care during assembly. Since horizontal joints always constitute sources of problems, it is usual to use facing blocks which reach down to the upper edge of the throat. For very deep throats the cover blocks are under the facing blocks and are often set in pairs (Figure 3.11). Adjoining block surfaces must be ground carefully to achieve maximum joint closure. Normally void-free fusion cast blocks are used. If available block dimensions permit it, the facing blocks should reach right down to the upper edge of the throat in order to avoid the problems inherent in horizontal joints.

Since molten glass strongly attacks horizontally laid blocks by upward drilling (Figure 3.12), it has been suggested that the throat roof might be given an upward slope in the direction of glass flow so that gas bubbles might be carried away; however, this is only achieved if the slope is over 15° and the surfaces of the blocks are smooth.

Setting the throat back towards the working end removes it from zones of high radiation in the melting end and extends the life of this critical part of the furnace.[7] The design of a recessed throat, which has proved itself for smaller tank furnaces, can be seen in Figure 3.13; the superstructure end wall must be secured in the throat area by means of a suitable small arch.

Throats must be cooled intensively to keep corrosion to within acceptable limits; this is usually

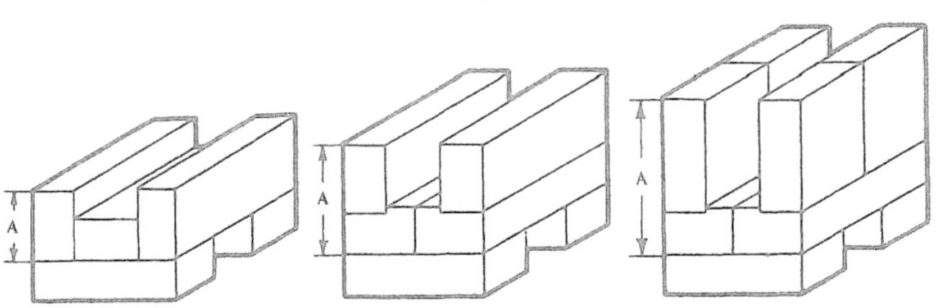

(a) shallow (b) normal (c) deep

Figure 3.11. Examples of throat construction

Figure 3.12. Corrosion of a recessed throat. The two side blocks and the cover block show severe corrosion originating from the horizontal joint[6]

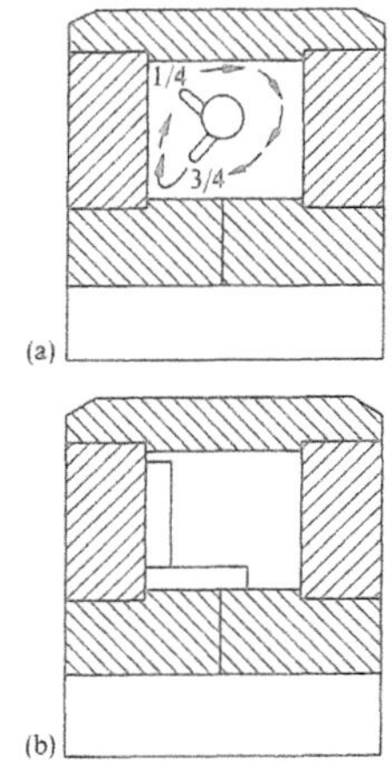

Figure 3.14. Throat cooling[8]
(a) air (b) water

done by air and special attention has to be paid to the flow of air in and out.[8-10] The arrangement shown in Figure 3.15 is very effective but has the disadvantage that the block surfaces can no longer be observed. From this point of view, the arrangement of cooling nozzles shown in Figure 3.14(a) is better. Water cooling is rarely used, since good thermal contact between cooling boxes and throat blocks is difficult to achieve. Throat cooling usually operates at airflows of 3–5 m^3/s at a pressure of 6–10 mbar.

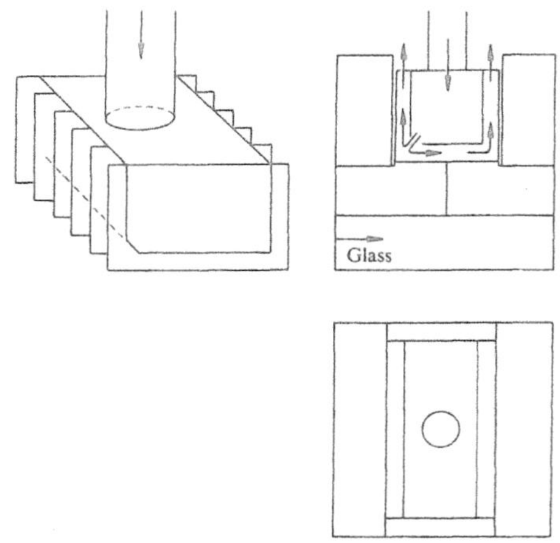

Figure 3.15. Improved cooling by air[8]

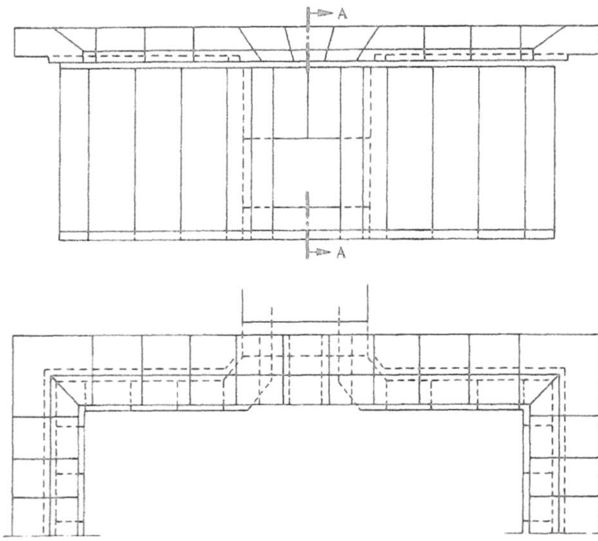

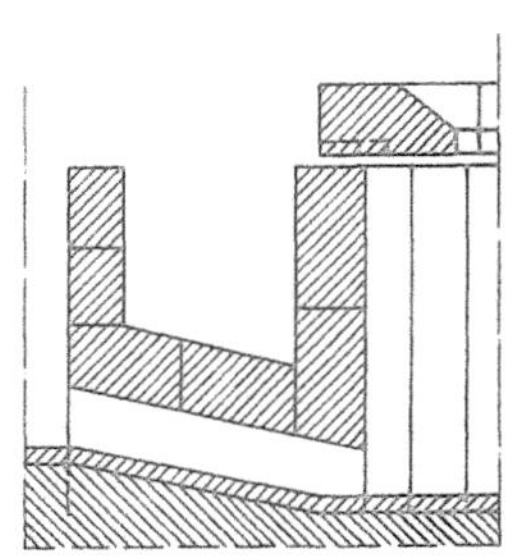

Section A–A

Figure 3.13. Design of a recessed throat[7]

3.1.2.4 The doghouse

An extension of the tank, called a doghouse, is used for receiving batch. Depending on the method of batch feeding one small doghouse, several small doghouses, or one wide doghouse can be used (Figure 3.16).

In cross-fired furnaces the doghouse is located in the end wall. A large wide doghouse permits the batch to be fed as a thin layer which is, from a thermal point of view, the most favourable. For very wide furnaces, e.g. tanks for Float glass, two wide doghouses are used to limit the width of the arches. Three-cornered doghouses are now very rare. In end-fired furnaces operating with U shaped flames the doghouse is located at the side as there is seldom room between the burners. One doghouse may be insufficient for high throughput end-fired furnaces, and two doghouses are provided, one in each side. In comparison to cross-fired furnaces, the doghouse of end-fired tanks is relatively narrow because a wide doghouse may give insufficient separation from the refining section of the tank.

The corner blocks between doghouse and tank proper are subject to higher thermal demands than the other blocks, since heat loss to the outside is greatly restricted, and the use of hexagonal blocks (Figure 3.17) increases the life of blocks in this position. The use of highly corrosion resistant refractories has reduced the necessity for hexagonal blocks and normal square corner blocks are now used.

Despite insulation, the outside blocks of a doghouse operate at too low a temperature. The strong

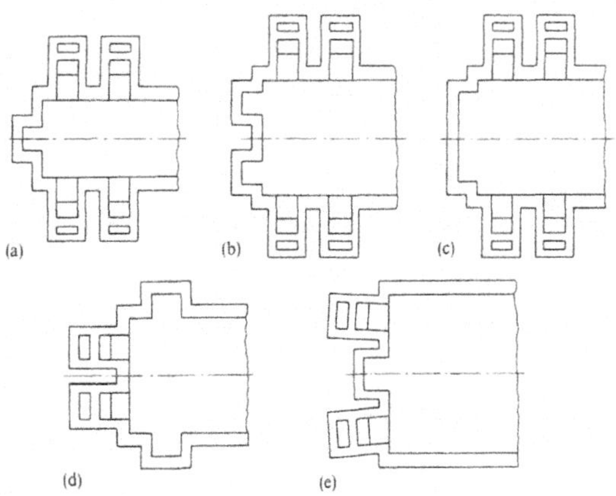

Figure 3.16. Doghouse layouts
(a) side-fired tank with one doghouse
(b) side-fired tank with two doghouses
(c) side-fired tank with one wide doghouse
(d) end-fired tank with one doghouse on each side
(e) end-fired tank with one doghouse between the burners

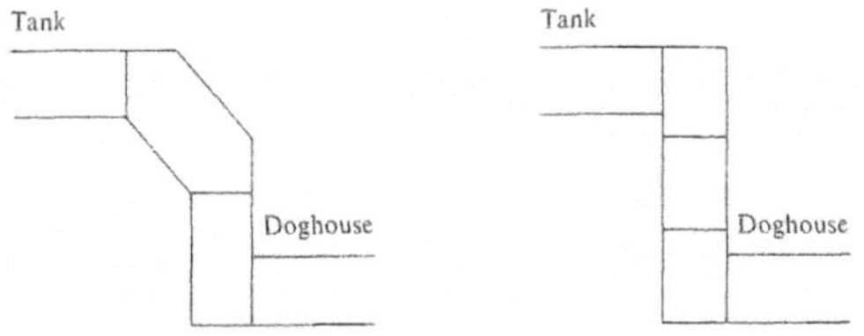

Figure 3.17. Doghouse corner construction

cooling action of the cold air being drawn into the furnace and the deposition of cold batch by water-cooled chutes causes the top surface of these blocks to crack, so that whole pieces of block fall into the furnace. The situation can be improved by insulation or by the use of sloping blocks, see Figure 3.18,[12] which creates a layer of hot glass between the feeding chute and the top corner of the block. This design has been found very useful for small tanks melting special glasses.

Attention must be paid during construction to the need for tank blocks, during both heat-up and operation, to be free to expand. The blocks should therefore be arranged as shown in Figure 3.19.[7]

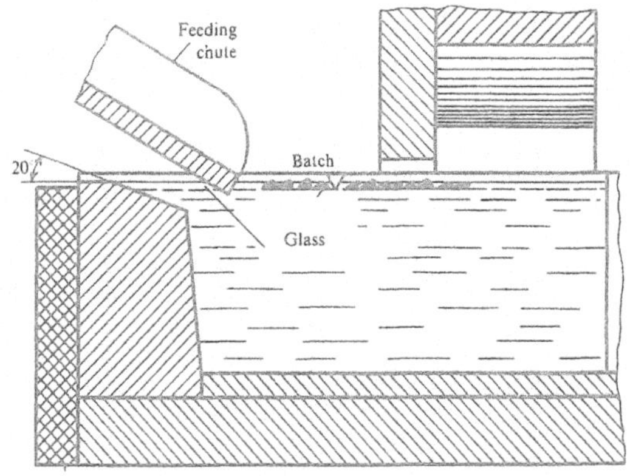

Figure 3.18. Doghouse construction with double tapered block at the outside wall [7]

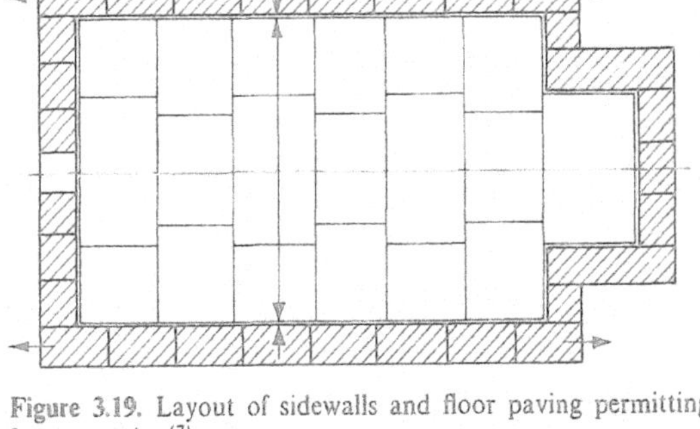

Figure 3.19. Layout of sidewalls and floor paving permitting free expansion[7]

3.2 Superstructure

The functions of the superstructure are to seal the combustion chamber upwards and sideways, to enable fuel and air to be introduced for combustion, to facilitate the removal of waste gases, and to achieve good thermal contact between the flames and the glass.

3.2.1 The combustion chamber

The combustion chamber loading for fuel-fired furnaces, expressed as the ratio of energy supplied per unit volume of combustion chamber and unit time, is 60–210 kW/m^3. The energy introduced is considered in terms of the enthalpies of the fuel and the combustion air. The loading of cross-fired tanks is usually higher than that of end-fired (U shaped flame) tanks.

The average height of combustion chambers increases with melting area and lies, despite wide variation, between 1 and 3 m; examples for various furnaces are given in Figure 3.20,[13] in which the trend curves for end-fired tanks have a steeper gradient than for cross-fired tanks. The dividing of the fuel along the length of the tank through multiple ports and the related improvement in controlling the positions of the flames clearly leads to smaller combustion chambers but there is considerable variation in practice. Introducing all the fuel at a single location (U shaped flame) creates the problem that, as the melting area increases, there must be a corresponding increase in the height of the combustion chamber.

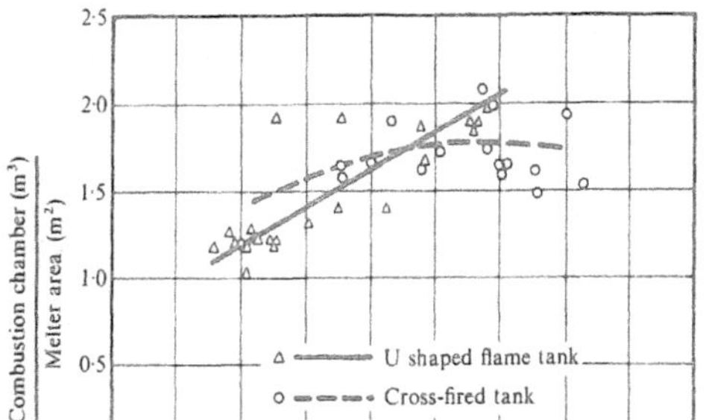

Figure 3.20. Relationship between combustion chamber/melter area and melter area[13]

3.2.2 Sidewalls

Sidewalls are built on top of the tank blocks only in very small tanks or for special reasons. The normal method is to erect sidewalls on separate angle supports attached to buckstays, which prevents the sidewalls from sagging if tank blocks wear rapidly and also allows further refractories to be added on the outside of the tank blocks without restriction. The sidewalls, nowadays usually 375 mm thick, are insulated on the outside to reduce heat loss and thermal stress: the heat loss of an insulated wall with a hot face temperature of 1550°C lies between 2·2 and 3·5 kW/m^2. The construction of a sidewall with its thermal insulation and showing how it is connected to the crown is illustrated in Figure 3.21. Depending on the expected operating conditions either silica, or mullite, or fusion cast materials are used.

The sidewalls are connected to the tank blocks by the tuckstone, and two designs are given in Figure 3.22.[7] Because of the risk of cracking, the nose of the tuckstone should be as small as possible. The tuckstone should project behind the tank blocks far enough for any facing blocks which may be added later to be covered and for the steelwork carrying the sidewalls to be fully protected. The slope on the hot face prevents accumulated glassy materials from running down the refractories. The joint between tuckstone and tank block must be

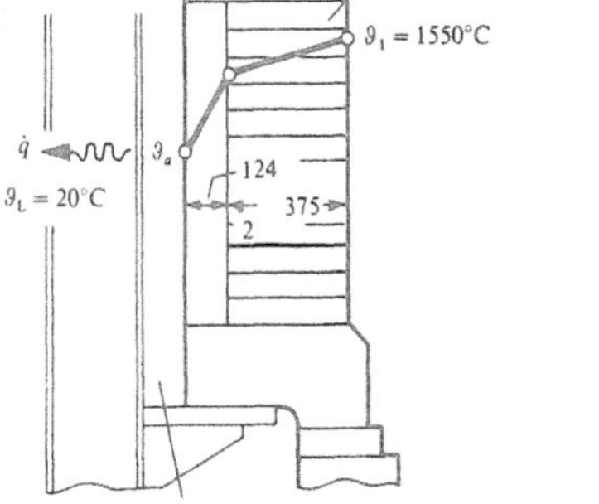

Figure 3.21. Sealing of superstructure to sidewalls of glass tank furnaces (Didier)

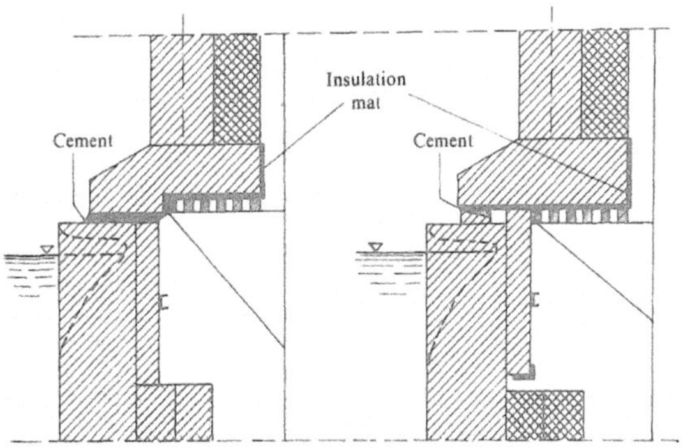

Figure 3.22. Installation of tuckstones[7]

sealed tight to prevent an inflow of cold air or a sting-out of hot gases depending on furnace pressure. Temperature fluctuations arising from these causes lead to cracking of the refractories but joints can be sealed with cements or ceramic fibre tiles.

Connection to the crown is by the skewback protecting brick, and sealing, which is also important here, is best achieved by using ceramic fibre mats. A need for the total avoidance of leaks applies not only to the sidewalls and their connections both upwards and downwards, but also to the superstructure as a whole. Leaks cause accelerated corrosion due to the condensation of alkalis in the colder sections of the structure as well as glass faults which originate in the run-off of slag.

Special care is needed in connecting port mouths to the sidewalls of cross-fired furnaces. Depending on the type of construction, there are port mouths (i) whose arch forms part of the furnace crown, (ii) whose arch butts against the crown, or (iii) where the port is constructed below the skewback angle irons. The last is the simplest and is being increasingly used for container glass tanks, but requires a larger combustion chamber. An example of construction (ii) is given in Figure 3.23. The stress release arch in the crown transfers the crown load from the port to the supporting steelwork: other details will be considered later when dealing with firing. An established construction for port and burner blocks below the skewback blocks is given in Figure 3.24.

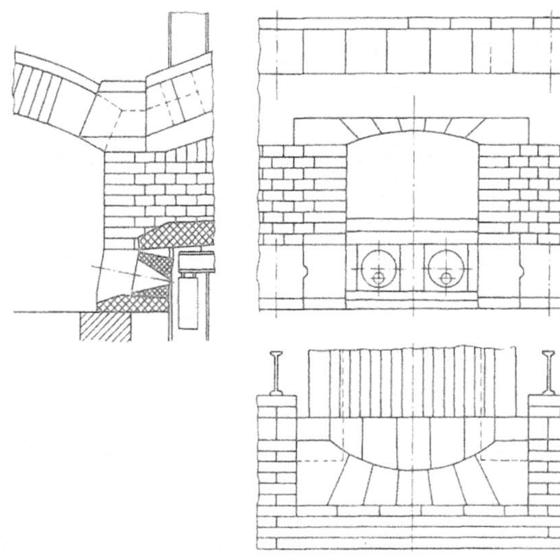

Figure 3.23. Port mouth with stress relieving arch for glass tanks (Didier)

3.2.3 The crown

The function of the crown is to cover the melting and working ends and the regenerator chambers. They are a supported structure and suspended crowns are used only in exceptional cases near the doghouse. The height of the crown determines the size of the combustion chamber. The radius and span of a crown are often the same, which gives an enclosed angle at the centre of 60° and a crown rise of 13·4%; regenerator chamber crowns often have a rise of 12–16% and in some cases barrel-shaped arches are used.

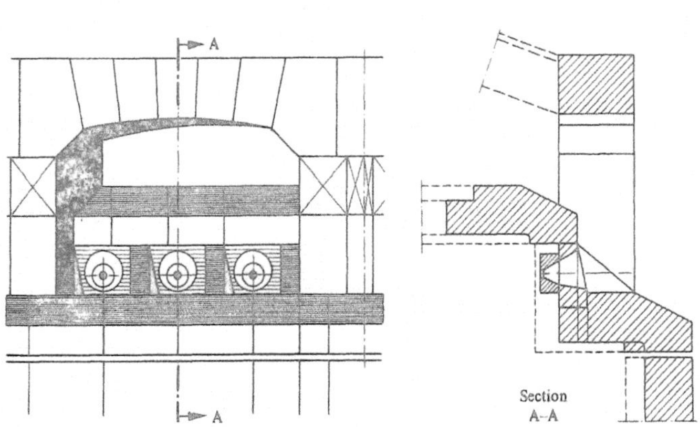

Figure 3.24. Proven burner construction for under-siege firing. Width of port mouth 1300 mm

Self-supporting crowns must be designed and constructed so that only compressive forces are created and excessive pressure at the edges is avoided. The maximum permitted specific compressive strength of the refractory blocks used must not be exceeded, not only when the crown is cold, but also during heat-up and when the crown is hot. The calculations are governed by the theory of structures.[15]

3.2.3.1 Forces in the crown

The forces arising in a crown and the geometrical relationships are given in Figure 3.25, where d is the thickness of the arch and γ is the included angle at the centre of the radius. The weight of the crown is exerted in the direction of gravity and is given by G; it is the force resulting from the weight of individual bricks and can be split into halves, designated $G/2$, which operate at the centres of gravity of the half-arches.

The force at each point of the arch can be considered as resulting force which gives rise to the line of thrust; for reasons of symmetry, the resulting force at the centre line of the crown is horizontal but at the skewback blocks it is inclined at an angle and can be split into two forces, F_h and F_v.

If the line of thrust in any section is located at the centre of the brick, then the pressure is constant over the whole cross section of the brick but if the line of thrust, and thereby the force, F, deviates

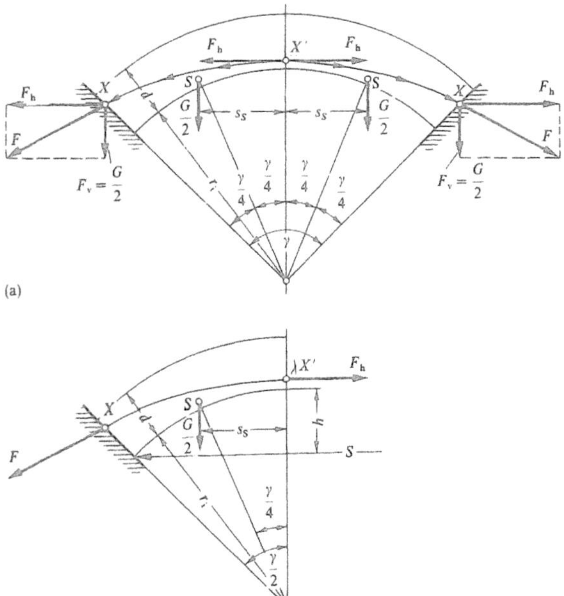

Figure 3.25. Forces in a crown arch held in compression[15]

from the centre, then the pressure distribution varies. The equilibrium of forces and moments in the area of the apex of the crown gives a pressure at the edges for a line of thrust within the inner third of the brick of

$$\sigma_r = \frac{F}{db}\left(1 \pm \frac{6e}{d}\right) \qquad (3.1)$$

where b is the width of the arch and e is the deviation from the centre of the line of force (see Figure 3.26(a)): for $e = d/6$ the tension at the edge is zero. If the force, F, moves out of the inner third, then tensile forces rise. If tensile forces are ignored, compressive forces are distributed as shown in Figure 3.26(b), the maximum compression being given by

$$\sigma_{max} = \frac{2F}{3cb} \qquad (3.2)$$

where c is the distance of the force from the surface. These relationships disclose the importance of the position of the line of thrust for the distribution of compressive forces in a crown.

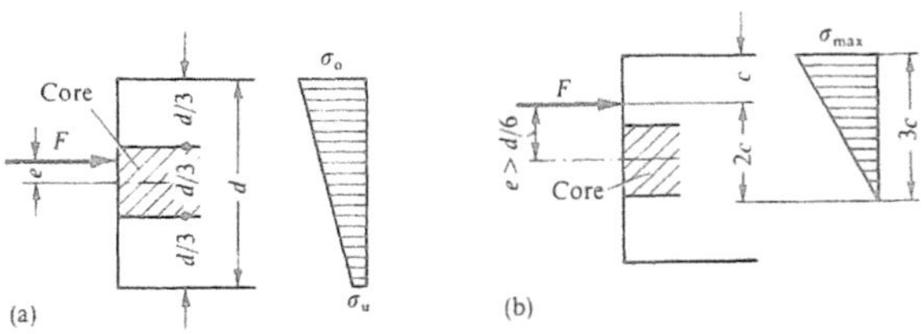

Figure 3.26. Compression stresses for forces applied off-centre

3.2.3.2 Lines of thrust

Excessive compressive forces near the edges and the resulting danger of bricks spalling can be avoided if the line of thrust runs through the central third of the block. Within this limit three conditions can be observed which can be called the central, maximum, and minimum lines of thrust (Figure 3.27).

The central line of thrust runs through the centre of the blocks and the crown should be designed and constructed so that this is the situation when the crown is cold.

The minimum line of thrust lies at the top of the upper third of the brick at the apex of the crown and in the lower third of the brick at the skewback blocks. Here the horizontal forces are the smallest since F is the steepest at the skewback blocks. This condition can arise if the tie rods are released too much during heat-up, opening the top of the joints

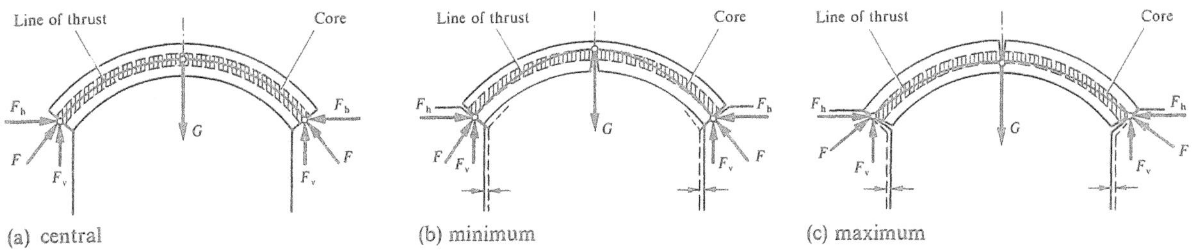

Figure 3.27. Lines of thrust

at the skewbacks so that the line of thrust is forced to move downwards.

The maximum line of thrust is the flattest, giving rise to the highest horizontal shear. During heat-up such a situation can arise from excessive expansion of the bricks. To compensate, the crown rises at its apex and if each half of the crown behaves as a rigid disc, to accommodate this rise they must pivot about the outer skewback point, opening up the joint on the inside of the skewback (Figure 3.28). The maximum line of thrust must be considered in relation to the heat-up in order to avoid this undesirable change in the distribution of forces, and the crown and skewbacks must be designed accordingly.

When designing the crown it is necessary to predetermine the location of the line of thrust and

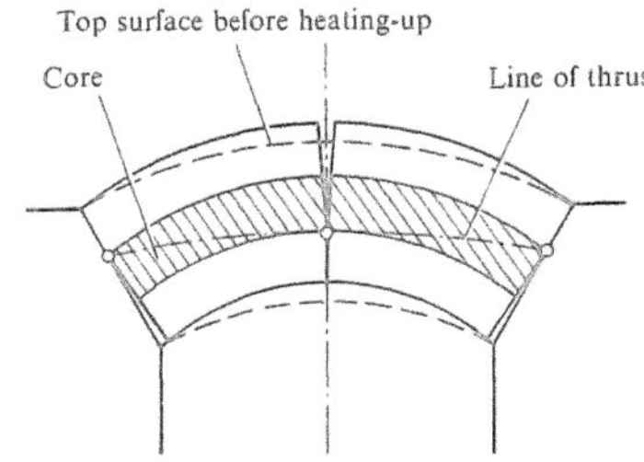

Figure 3.28. Rise of a crown heated from one side

this can be done relatively simply by structural analysis making use of suitable assumptions. Figure 3.29 shows the determination for a circular arch. The first step is to divide the arch into sections of equal load, calculate the gravitational forces, and enter them on the drawing at their centres of gravity. The forces G_1 to G_4 and the arbitrarily

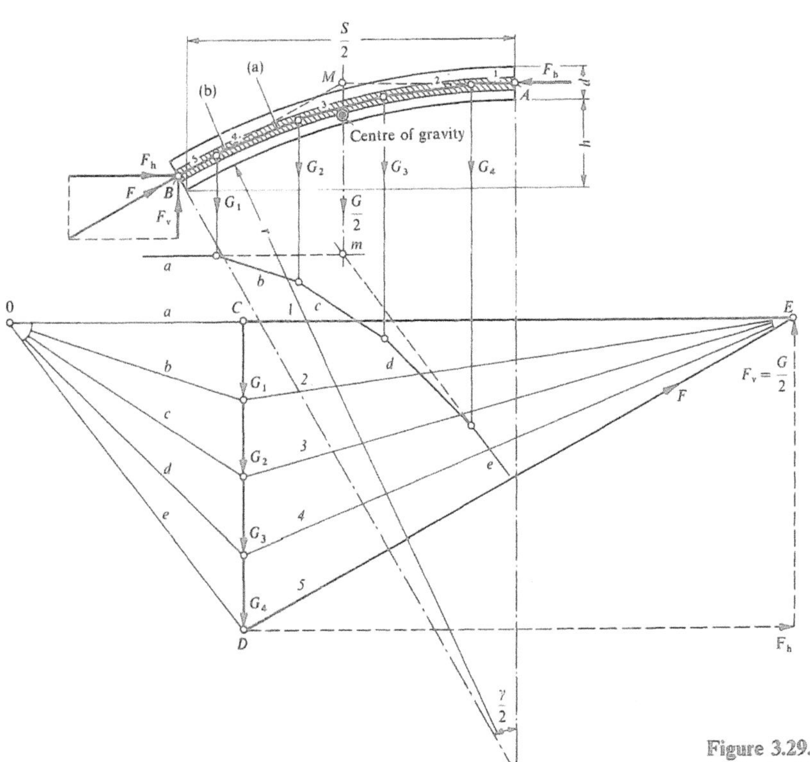

Figure 3.29. Graphic representation of a circular arch
(a) is the line of thrust (b) is the core

chosen focus O are used to form the triangle COD. Using the sketch of the polygon section a to e, the location of the centre of gravity is determined by backward projection of a and e to their intersection at m; the line of the centre of gravity of half the arch goes vertically through this point. The shear force F_h at the apex of the arch and the force F at the skewback must intersect on this line to satisfy conditions of equilibrium.

If one assumes the condition of the central line of thrust and permits the forces to pass through the points A and B of the centre line, then they intersect at M. This fixes the direction of F. By transferring the directions of F and F_h to the lines of force G_1 to G_4, the triangle of forces CED is formed. This determines the magnitudes of F and F_h. To determine the line of thrust the projections 1 to 5 of the triangle CED are drawn. The polygon section through points A and B gives the position of the central line of thrust. The magnitude of the forces which arise in each section of the arch can then be deduced from the triangle CED and are given by the lengths of the projections 1 to 5 measured on the same scale as the forces. The course of the line of thrust discloses the point at which the force acts in relation to the centre line of the section of the arch.

The distance, s_S, of the centre of gravity can also be calculated from the formula given in Figure 3.30 or the corresponding diagram.[15]

The line of thrust does not follow the arch of a crown. If line of thrust and centre line of the arch are combined to form a stable crown, then the optimum arch centre line is an inverted catenary (Figure 3.31), which is generally valid for a given ratio of span width, s_M, to rise h_M. This ratio gives the required centre line of the arch via the upper part of the curve in Figure 3.31; the corresponding

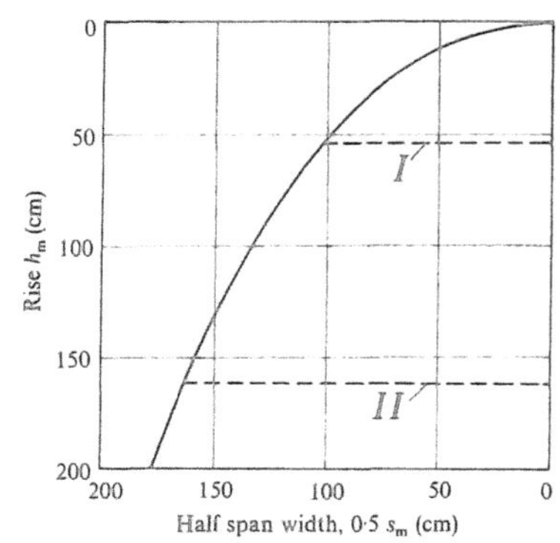

Figure 3.31. Rise of an arch related to half the span width, both measured from the centre line of the arch (inverted catenary)[15]

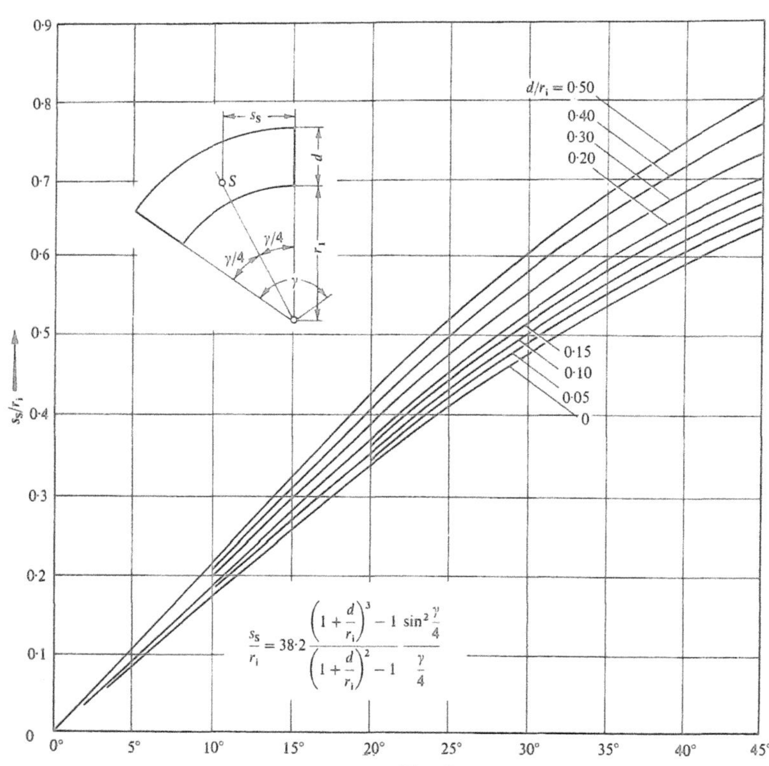

Figure 3.30. Distance of centre of gravity normalised with respect to the interior radius s_S/r_i as a function of $\gamma/4$ and d/r_i[15]

graphical development of such a crown is given in Figure 3.32 and this type is especially recommended for crowns with a high rise. Generally speaking, a thick crown is more stable than a thin crown. Circular (barrel) 180° crowns are structurally unstable and should have a thickness of at least $0{\cdot}18r_i$.[15]

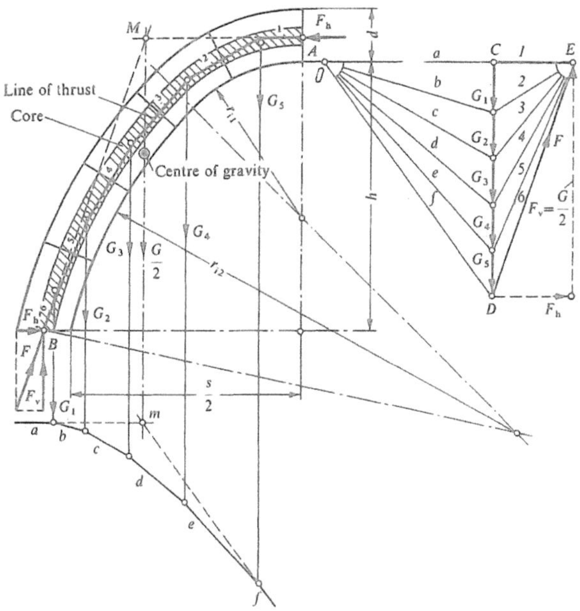

Figure 3.32. Designing a crown based on a catenary

3.2.3.3 The effect of thermal expansion

Crowns of glass melting furnaces, as distinct from those of other furnaces, require special attention during heating-up and in operation due to the effects of thermal expansion. From a theoretical point of view one can regard the ideally insulated crown and the uninsulated crown heated from one side as the limiting conditions. The expansion characteristics are shown schematically in Figure 3.33. In the case of the ideally insulated crown the angle subtended at the centre and, thereby, the angle of the skewback remain unaltered, but for the crown heated from one side the subtended angle grows larger and the skewbacks tip.

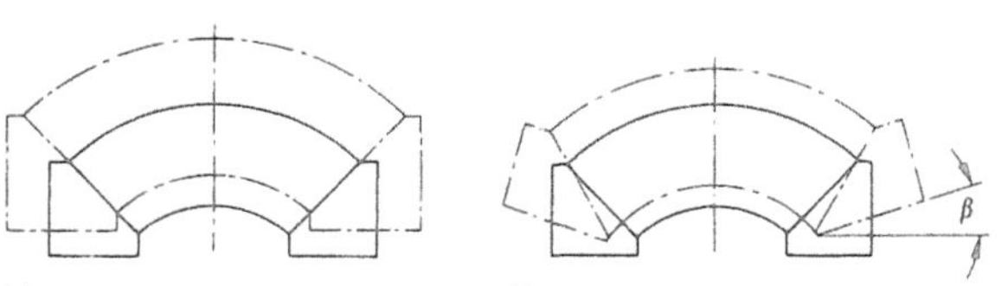

Figure 3.33. Expansion of a crown
(a) heated from both sides (insulated)
(b) heated from one side (uninsulated)

In practice, however, there is a temperature difference between inside and outside, even within an insulated crown, and this is sufficient to move the line of thrust from the centre towards its maximum. It follows that, at the apex of the crown, the shear forces move downwards and, as F'_h, locate at point A_u (Figure 3.34) while, on the other hand, the pressure at the skewback moves upwards and passes, as F', through the point B_0. Ignoring small inaccuracies in the calculations for forces in the hot crown and the fact that the skewbacks are moved outwards in parallel by

$$\frac{\Delta s}{2} = \frac{s}{2}\lambda, \tag{3.3}$$

where λ is the correction factor for the thermal expansion, then the distances of the centre of gravity y' and x' are given with sufficient accuracy by

$$y' = y + \Delta y$$

$$x' = \frac{s}{2} + \frac{\Delta s}{2} + \frac{2}{3}w - y'. \tag{3.4}$$

If the rise of the crown is ignored then, according to Figure 3.34, the horizontal pressure F'_h is

$$F'_h = \frac{G}{2}\frac{x'}{h_1 - \frac{h_w}{6} - \frac{d}{6}} = \frac{G}{2}\frac{x'}{h_2}. \tag{3.5}$$

Since the maximum line of thrust for silica crowns occurs primarily during the transition of β-cristobalite into the α modification, that is at about 300°C, it is accurate enough to carry out the calculations using an expansion of 1%.

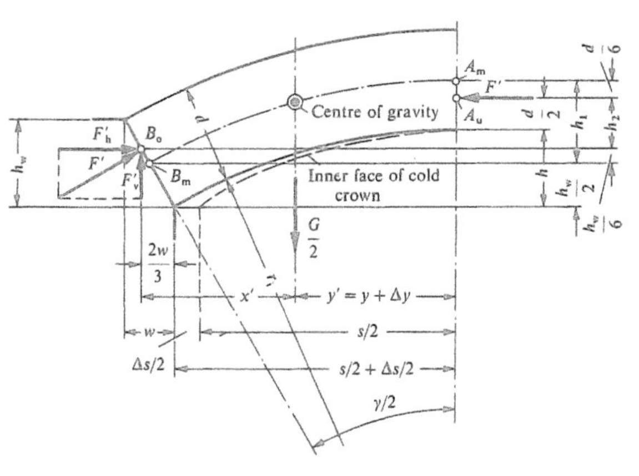

Figure 3.34. Diagram of forces in a hot crown

3.2.3.4 The construction of crowns and their behaviour during operation

Crowns are constructed in the glass industry almost exclusively of end arch bricks; side arch bricks are only used in less critical parts. Joints should be mortared and as small as possible in order to obtain gas-tight brickwork; the quantity of cement used is about 6% of that of the bricks. The thickness of joints depends on the dimensions of the bricks, i.e. their deviation from standard and, possibly, differences between average dimensions and design dimensions;[16] one should therefore use bricks with the best dimensional tolerances. The dimensions of standardised arch bricks are given in DIN 1082 and those of arch bricks used for crowns of glass furnaces are given in Figure 3.35; the height, *h*, is related to the span of the crown and Table 3.1 gives recommended heights.

In most cases crowns are made from silica bricks. Only well converted bricks should be used as residual quartz content above 10% leads to strong after-growth above 1450°C: checking dimensions is a reliable method of eliminating badly converted bricks.[18]

During heat-up, crowns should be covered with insulating blankets which should not be removed until the operating temperature is reached; insulation reduces the temperature gradient in the bricks and thereby evens out thermal expansion between inside and outside. The behaviour of insulated and uninsulated silica crowns is shown in Figure 3.36.[17] Thermal insulation of silica bricks is

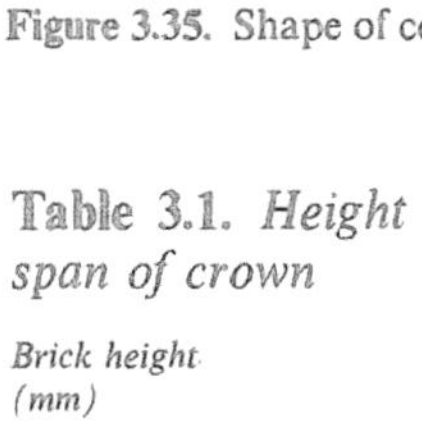

a	b	h	l
66	62	250	250
100	95	350	250
78	75	375	150
124	120	450	150

Figure 3.35. Shape of commonly used brick (dimensions in mm)

Table 3.1. *Height of crown bricks as a function of span of crown*

Brick height (mm)	*Span of crown (mm)*
250	< 3000
300	3000–4900
375	4900–6700
450	6700–9750
500	> 9750

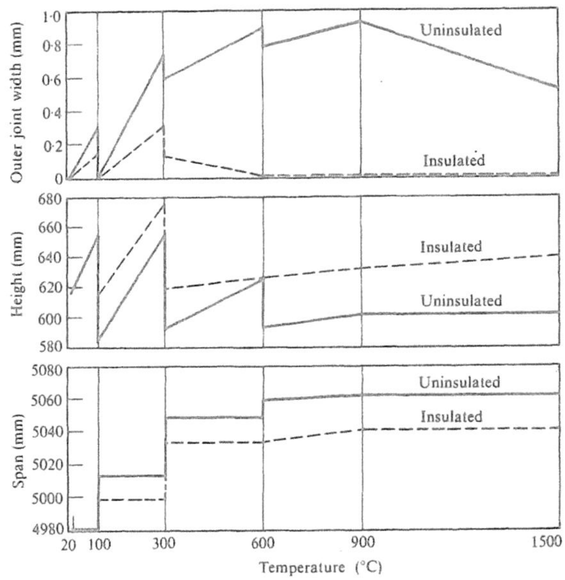

Figure 3.36. Effect of temperature on a silica crown[17]

achieved with light weight silica bricks. As a final covering insulating mats with aluminium foil on the outside or insulating cements are used.

Like the sidewalls the crown is supported and braced on its own and it does not rest on the sidewalls; as well as offering advantages during construction, this also simplifies heat-up. Furthermore, when tank repairs are necessary the crown, if still in good condition, can be used for a second campaign even if sidewalls and burners have to be renewed. The first requirement of a crown is that it must be free from leaks, especially if thermal insulation is applied which is now normal practice for all but exceptionally large spans. The construction of an insulated crown in the area of the skewback is shown in Figure 3.37; insulation reduces heat losses which are about 5·7 kW/m^2 for an uninsulated crown to about 2·4 kW/m^2.[19]

The crowns of furnaces for special glasses sometimes rest on the sidewalls, which avoids the problem of obtaining a leakproof joint to the sidewall in the area of the skewback (see Figure 3.37). This is particularly important if silica bricks have to operate at the limit of their thermal performance when, because they cannot be adequately cooled, slag is likely to form and run from the brick sealing the sidewall to the crown. To allow for the thermal expansion of the crown, expansion joints must be provided every 3 to 5 m across the crown, and when well converted silica bricks are used these joints should account for about 1·4% of the length of the

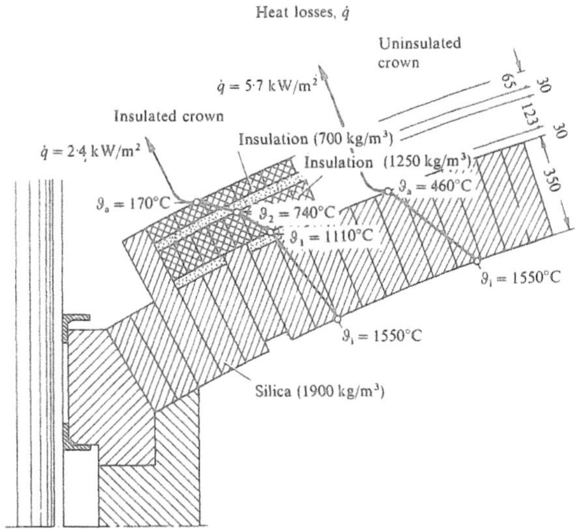

Figure 3.37. Example of crown insulation (mm)[19]

crown. Gas-tight closure of all joints is assured when the size of the expansion joints is such as to ensure that the crown bricks are pressed together from the ends. Since the bricks arch during heat-up, special bricks for expansion joints have been developed (Figure 3.38). Such bricks are used not only between crown joints but also at the ends of the crown as closures against the end walls. The row of closure bricks against the end wall must always be vertical, not horizontal, as gas-tight joints can only be obtained in this way. The V shaped joint formed in the region of the special expansion joint bricks is sealed with V shaped bricks after heat-up.

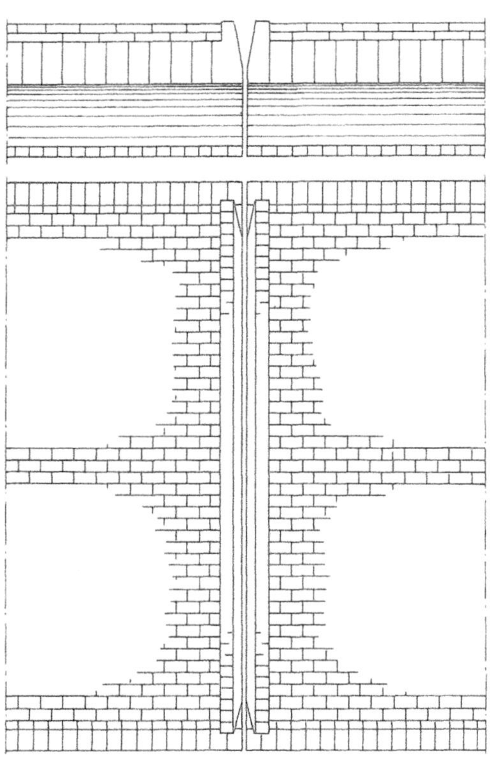

Figure 3.38. Example of the closing of a crown of a glass tank (Didier)

During heat-up the skewbacks must be able to move along the supporting angle iron and the surfaces on which this sliding takes place should therefore be ground smooth.

Slag is especially likely to form on the inside of the crown in the melting zone and should not be allowed to flow over sidewalls or tank blocks which contain alumina as this will cause glass faults. Because of its high SiO_2 content, slag does little damage to the crown and it is normal practice to construct drip edges to prevent slag from running down the sidewalls. Alternative constructions of drip edges are given in Figure 3.39. Design (a) is preferred since protruding bricks in a crown tend to break off and can also be melted down when overheated. The drip edges must be sufficiently far from the sidewalls for dripping onto the tuckstones to be avoided.

Crowns operating above 1630°C cannot be constructed of silica; zircon or mullite bricks must then be used.

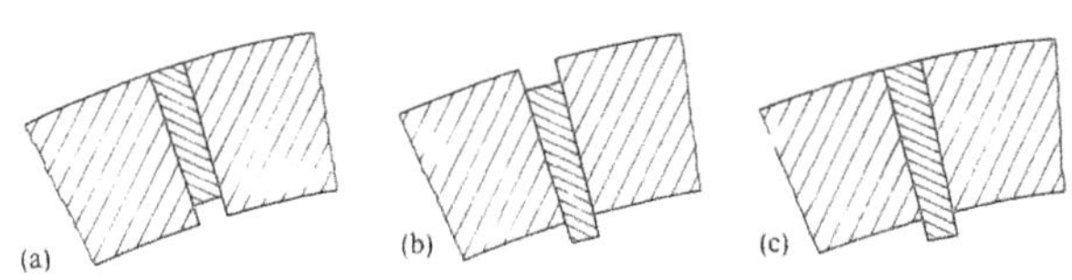

Figure 3.39. Three types of drip edge in crowns

3.2.4 End walls and doghouse

At each end of the combustion chamber which extends from doghouse to glass withdrawal at the working end there is an end, or gable, wall which nowadays is generally insulated and carefully sealed to the crown and sidewalls. Special care must be taken when designing a crown over the doghouse, especially with wide doghouses such as are used in conjunction with the thermally efficient method of feeding batch as a blanket. The rise of the doghouse crown should be small so that heat losses and dusting to the outside are minimised. Flat arches, which were often used in the past, are now generally avoided because the lower parts of the bricks are not under compression and pieces are likely to break off when cracks form in the arch due to changing temperature. The arches covering the doghouse must be anchored independently of the end wall so that they can expand without hindrance. This requires the top of the arch to be flat between anchoring points so that the section of end wall above it is free to move. Figure 3.40 shows a wide doghouse structure using fusion cast blocks, whose resistance to thermal shock can be improved if they are backed by suspended clay blocks with a water-cooled beam beneath. In many cases up to three arches are used, the rise of each arch increasing towards the interior of the furnace (Figure 3.41).

3.2.5 Separation of melting and working (or conditioning) ends

Container glass tanks with throats can be constructed with one or two compartments. For high throughputs the two-compartment design is preferred because it is easier to control the melting and working ends separately and there is little energy transfer between the two.

One-compartment tanks, which are important in special circumstances, have their crowns extended over both parts of the tank, which are separated by a shadow wall erected on the throat blocks (Figure 3.42). Depending on the heating effect desired in the working end, the shadow wall may be of open or closed construction. The open construction, which is to be preferred since the closed construction can cause flame erosion at the crown, may be uniformly open or mainly closed but with 'windows'.

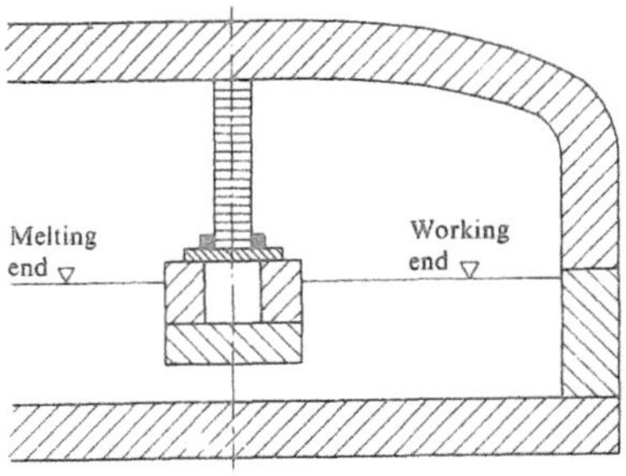

Figure 3.42. Shadow wall

Flat glass tanks have no throat and construction in two compartments is therefore not possible, nor can a shadow wall be erected. Low set crowns (Figure 3.43) may be used if there are only small differences in temperature between the zones or if there is a large conditioning section. Suspended crowns, such as those used for many years with tableware tanks, allow adjustments to be made in line with technical requirements.

If these methods give insufficient separation, as is the case for most plants nowadays, then so-called curtain walls must be used. A two sided air-cooled suspended crown such as that shown in Figure 3.44

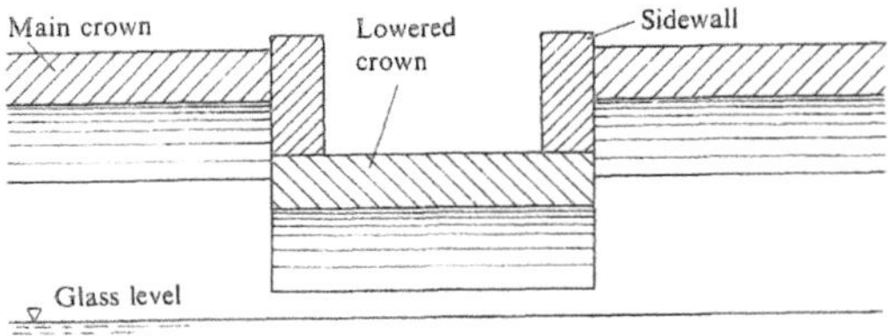

Figure 3.43. Separation by lowering of crown

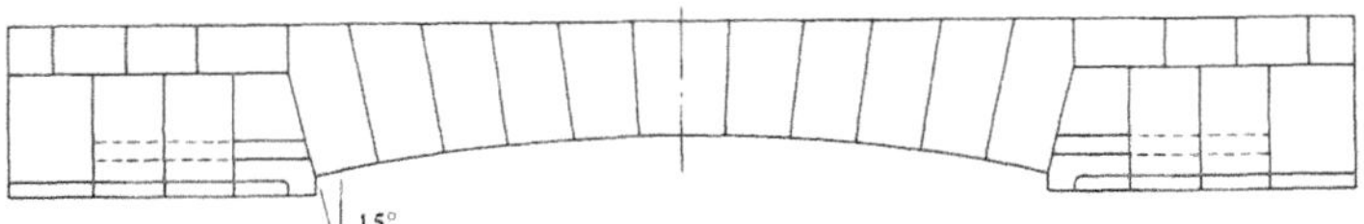

Figure 3.40. Wide doghouse construction with one arch (SEPR)

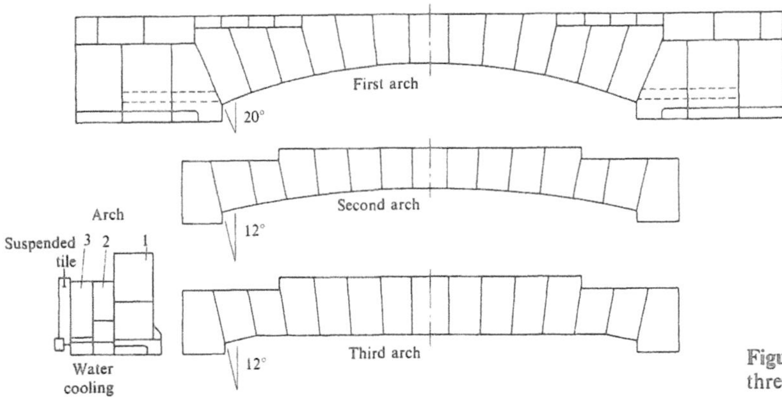

Figure 3.41. Wide doghouse construction with three arches (SEPR)

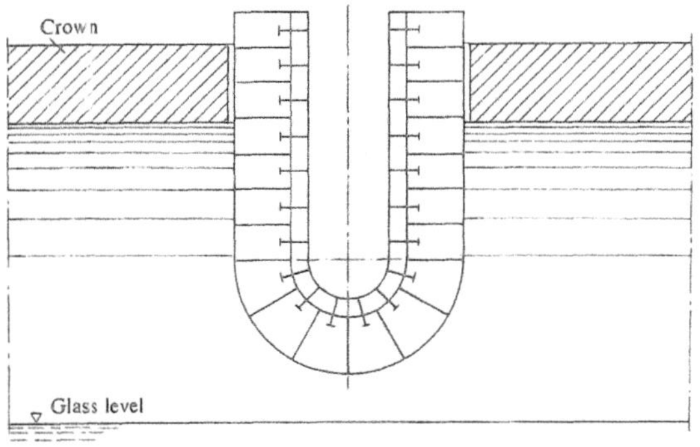

Figure 3.44. Separation by double-skinned suspended crown construction (Didier)

is one solution; alternatively a combination of a low set crown with suspended silica tiles can be used. However, a suspended crown is now only rarely used as it is expensive and involves high maintenance costs: the use of tiles, especially if made from good quality silica, is a better solution. To improve the sealing of joints the tiles can extend beyond the skewbacks (Figure 3.45). The tiles are held by threaded rods let into holes in their upper ends and they can thus be suspended in a steel frame construction; their distance from the glass surface can be adjusted by screws at the point of suspension. A disadvantage of this is its high cost.

3.2.6 Ports

The ports of a glass furnace must be of such a size and must be installed in such a way that, taking the construction of the furnace into account, they meet the requirements of the glass melting process and the supply of heat to the furnace. For side-fired furnaces this gives even heating over the width of the tank whereas longitudinal firing gives the highest degree of heating in the batch melting area. Differences of heating over the tank width which cannot be avoided with regeneratively fired furnaces are evened out by the periodic reversal of firing.

It is more difficult to achieve even heating along the length of end-fired tanks and problems arise in recuperative tanks from the absence of flame reversal which evens out differences in heating.

In cross-fired tanks the sum of the widths of ports on a side is 30–50% of the length of the melting end, a value determined largely by construction details such as bracing and access. Tanks for container glassware have three to five pairs of ports, flat glass tanks five to eight. The distance between the first pair of ports and the end wall (*a* in Figure 3.46) should be 1·2–1·8 m. If it is too small the flames will cause excessive batch dust and endanger the end wall by overheating, if it is too large the highest melting rate which could be

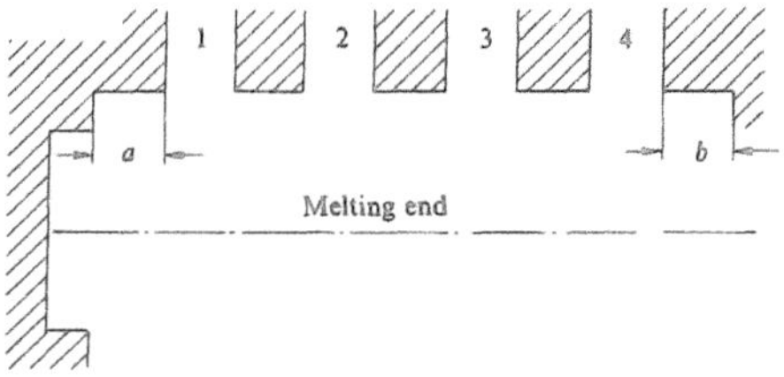

Figure 3.46. Distances between ports of cross-fired tank furnaces

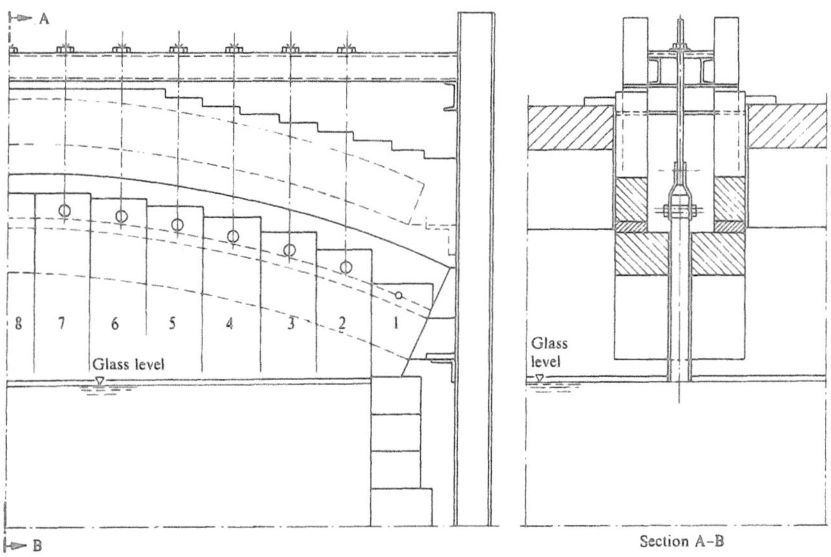

Figure 3.45. Curtain made of suspended blocks

attained will not be achieved. The last pair of ports should be 0·8–2 m from the bridge wall (dimension *b*). In end-fired furnaces the distance between port and sidewall should be about 0·3 m (*a* in Figure 3.47) and the length-to-width ratio should be 1·5–1·8 as it becomes difficult to heat the refining zone adequately above this value.

The port area required for one side of a regenerative cross-fired tank is usually stated as total port area divided by tank melting area, with values usually lying between 0·025 and 0·04;[13] modern container tanks aim for smaller values. Figure 3.48 shows this ratio expressed as a percentage for end-fired tanks with U shaped flames and cross-fired tanks for container glass; the latter are a little higher because of the higher flame velocities associated with end-fired tanks. For side-fired tanks the port area first increases with size of tank due to the increase in specific melting rate for a constant width of tank. Above about 50 m^2 the melting area is lengthened, leading to better fuel distribution and at the same time the increased distance from doghouse to throat enables the specific melting rate to be increased while maintaining operational control. The efficiency of the plant increases, leading to a reduction in total port area.

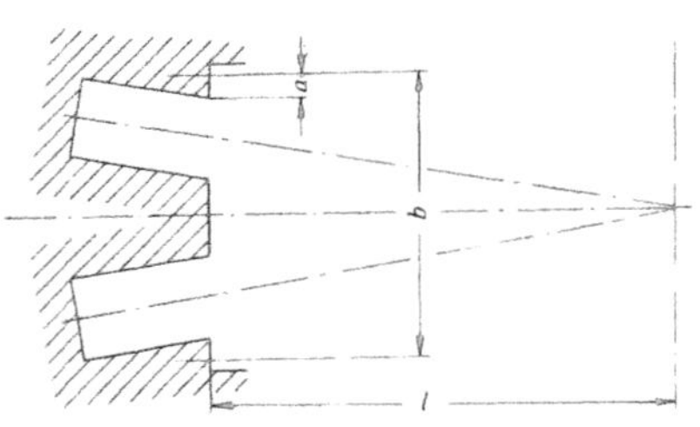

Figure 3.47. Port dimensions for U shaped flame tank furnaces

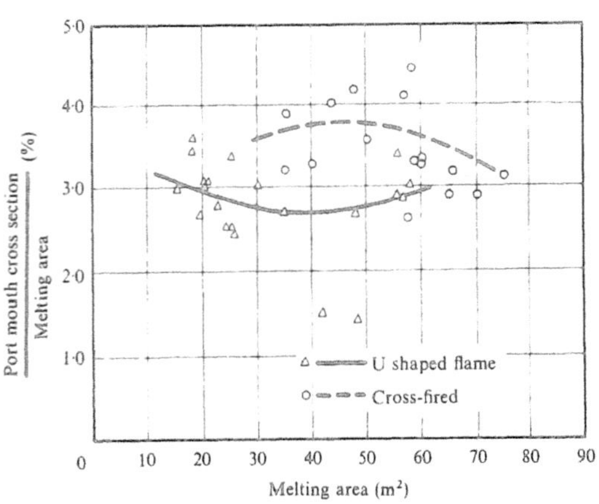

Figure 3.48. Relationship between port mouth cross section/melting area and melting area

End-fired tanks with U shaped flames behave in the opposite way: the curve first decreases as large tanks call for higher flame velocities. The increase is small with a length-to-width ratio of 1·2 to 1. After passing through a minimum at 40 m^2 the curve rises: increasing melting areas demand increases in fuel consumption and flame velocities reach values which cause batch dust problems in large tanks so that one is forced to reduce flame velocity and enlarge port areas. Velocities at the port mouth for cross-fired furnaces, calculated at their actual temperatures, are 8–12 m/s for air and 10–14 m/s for waste gases; these values are a little higher for U shaped flame tanks.

The design of ports depends on the type of fuel, the arrangement of fuel nozzles, burners, or atomisers, and the construction of the regenerators. The port is separated from the regenerator chamber by a vertical joint, an arrangement which permits separate free thermal expansion of tank and regenerator. Figure 3.49 gives an example of a port of an oil fired U shaped flame tank.[20] Inclination of port and oil burner must be designed carefully and more detailed consideration of this point is given in Section 6.3. The port construction for a large natural gas fired flat glass tank is shown in Figure 3.50: the port arch is joined to the furnace crown via a supporting arch.[21]

Both examples have vertical regenerators, which eliminates the problem of connection to the ports. Large flat glass tanks have small slag chambers next to the regenerator, which reduces the wear on the regenerator packing but demands that the ports be located on top of the regenerator; an example is given in Figure 3.51 where it may be seen that the

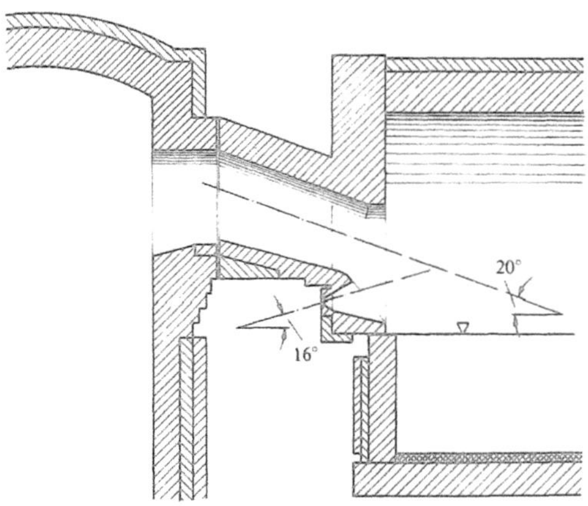

Figure 3.49. Port and burner of a U shaped flame tank furnace[20]

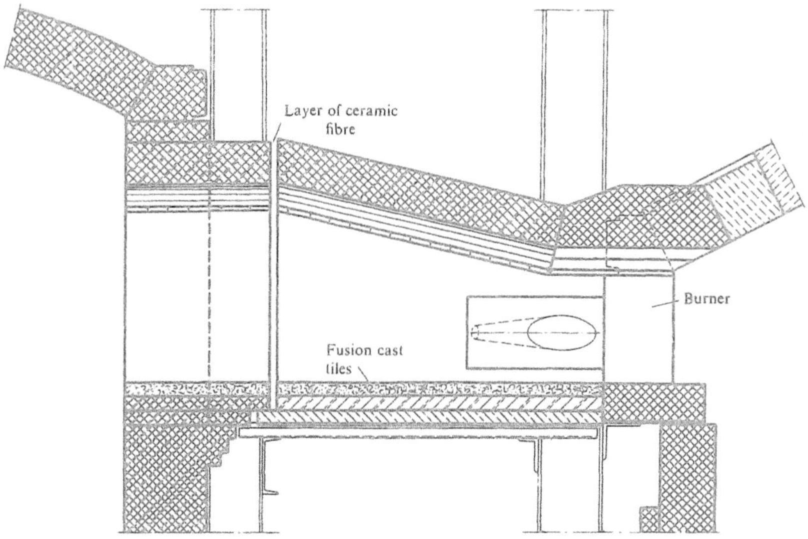

Figure 3.50. Port and burner of a cross-fired tank for flat glass[21]

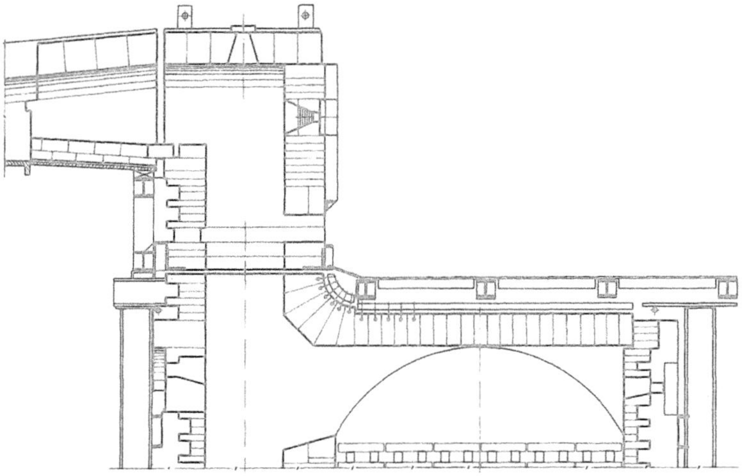

Figure 3.51. Port with slag pocket[22]

Figure 3.52. Sidewall of a cross-fired tank furnace showing root of flame and port sill

regenerator roof is suspended in the area where it turns from vertical to horizontal.

As to the positioning of oil burners, underport firing is now preferred (Figures 3.52 and 3.53(d)); the recessing of burner blocks creates more space for the flame (Figures 3.23, 3.24 and 3.52). Oil burners are mounted to leave a small annulus between burner and burner block. For natural gas and town gas the nozzles are usually located at the side as shown in Figure 3.50. Mixing is improved by locating the nozzles fairly low and by constructing baffles, or steps (Figure 3.54), which help to mix the waste gases with the fresh gas and assist in igniting it; they also contribute to the control of the carbonisation of natural gas (Figure 3.54). The baffle is easy to replace and it is even possible to instal a new design. Developments with natural gas are in

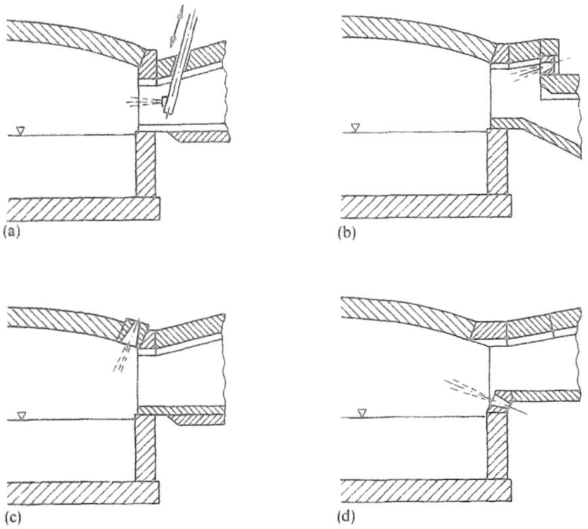

Figure 3.53. Alternative burner positions

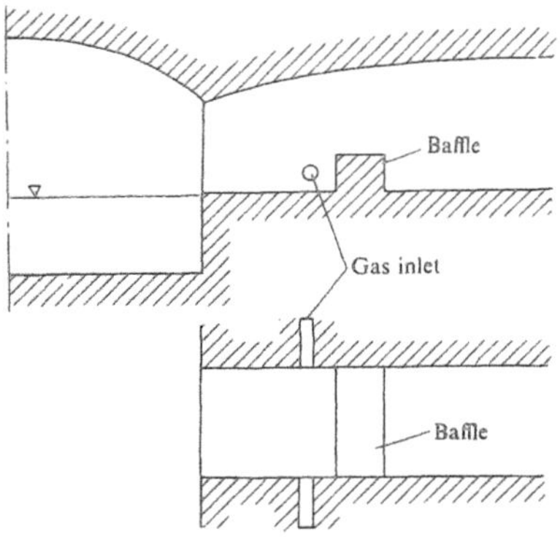

Figure 3.54. Port with baffle for natural gas (obsolete)

the direction of underport firing, similar to that for oil, and gas would be fed into the furnace at high pressure. The large refractory port with tongue, used in the past with producer gas, has disappeared now that this type of gas is no longer used.

3.3 Regenerators

The function of regenerators is to use the thermal energy of the waste gases to preheat the combustion air and thereby provide better heat transfer via high flame temperatures. This gives a considerable reduction in the specific energy consumption of the furnace.

When using fuels of low calorific value such as producer gas the preheating of air is essential to obtain the necessary temperatures for melting and refining; the producer gas itself is also preheated. For fuels of high calorific value such as natural gas or oil, the necessary temperatures can be reached without preheating; in this case the simple objective of preheating is to save energy. Depending on the extent of preheating, energy savings of 50–70% can be achieved.

3.3.1 Layout

Regenerators should be located next to the furnace in order to gain the maximum benefit from using the heat from the waste to achieve high preheat temperatures. Depending on the type of link to the furnace, one can have either tall regenerators or regenerators with the port located on top of the regenerator (Figure 3.55). As a result of improvements in refractories for regenerator checkers and the use of fuels of high calorific value the vertical regenerator is preferred.

Depending on the flow pattern through the checkers, there are vertical or horizontal, as well as single or multiple path constructions (Figure 3.56). The vertical chamber (Figure 3.57) is now normal

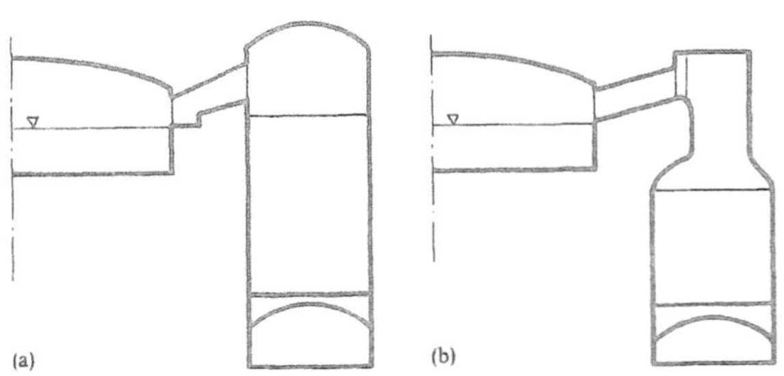

Figure 3.55. Connections between regenerator and melting chamber
(a) vertical regenerator, usually with a single chamber, but also multichambered (modern design)
(b) obsolete design with port placed on top of regenerator

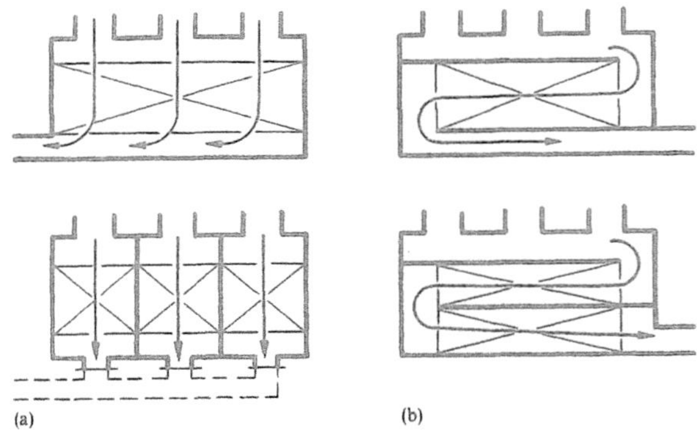

Figure 3.56. Different designs of regenerator differentiated by flow direction
(a) vertical (b) horizontal

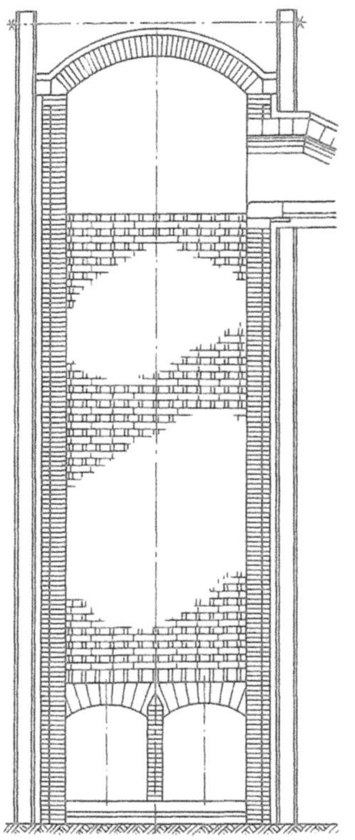

Figure 3.57. Designs of a modern vertical, vertically extended regenerator (Didier)

as it is the only construction which permits the installation of vertical checkers. Advantages previously claimed for the horizontal chambers, such as easier access for cleaning, have become obsolete by the use of improved refractories; horizontal chambers can only be justified for small furnaces. Vertical chambers with waste gas entry on one side and large collecting chambers give a more even flow through the checkers than those with ports placed on top of the regenerator.

3.3.2 Dimensions

The size of a regenerator is determined by the objective of obtaining a high air preheat temperature at an economic price. Accurate design calculations of preheat temperature and of temperature decrease during a reversal period have so far proved difficult but the use of computers with suitable programmes has improved the situation. It is now possible to establish the necessary data from elementary considerations of heat exchange without using calculations based on recuperators and applying them to regenerators (see Chapter 12). In simple cases dimensions can be fixed by using reference data (Table 3.2), the criteria being the melting area or the energy introduced in unit time in reference to one regenerator chamber.

The efficiency of a regenerator is defined as the ratio between the energy transferred to the combustion air and that carried into the regenerator by the waste gases. It seldom exceeds 60% which, among other factors, is related to the high waste gas enthalpy of modern fuels when compared to air. Care must be taken when comparing regenerator efficiencies since definitions of efficiency vary.[23] When calculating the area of checker heating surfaces one must assume that the energy transfer conditions between waste gas and checkers and between checkers and air are constant or, at least, comparable. The type of checker, the width of the ducts, and the checker material are contributary factors and the radiation from the waste gases depends on the fuel used. The energy relationships between these factors will be described in more detail in Chapter 12. Among recent work on this subject are papers by Delrieux,[24] Greschat,[25] and Krings.[26]

Table 3.2. *Nominal values and characteristics of regenerators for a single regenerator chamber*

Nominal value / *Characteristic*	*Unit*	*Data*
$\frac{\text{Checker heating surface (m}^2\text{)}}{\text{Melting area (m}^2\text{)}}$	–	25–45
$\frac{\text{Checker heating surface (m}^2\text{)}}{\text{Heat flow (W)}}$	$\frac{m^2}{W}$	$130 \times 10^6 - 260 \times 10^6$
$\frac{\text{Checker volume (m}^3\text{)}}{\text{Melting area (m}^2\text{)}}$	m	2–3·5

3.3.3 Gas flow through the regenerator

An even flow of gases is nearly always assumed when designing regenerators but in practice this assumption is not necessarily valid. The design of a regenerator provides some factors which can describe the gas flow through the checkers. A detailed description has been given by Trier,[27] a summary of which is given below.

3.3.3.1 Regenerators with smooth plain packing

There has been no experimental or theoretical study of the actual flow through smooth plain packing such as that used for glass furnaces. There are, however, papers dealing with flow through a system of ducts which resembles this type of checkerwork and the relationships discovered offer interesting conclusions when applied to flow through regenerators. The most important and comprehensive papers on this subject are by Reinhardt & Tollmien[29] and by Veit.[30] Figure 3.58 shows the type of flow pattern investigated.

Experiments have shown that the ratio $n\Delta f/b_u$ (where Δf is the cross section of a single passage, n is the number of passages, and b_u is the cross section of the flue through which the gas passes) has a strong influence on the flow distribution between individual passages.

If $n\Delta f/b_u < 1$, then the gas accelerates when it enters the passages and the change in flow direction in the regenerator flue causes only minor flow losses. In this case the gas tries to distribute itself uniformly between single passages. The distribution of flow between individual passages is governed only by the design of the regenerator top space and the resistance to flow in the passages. Whether the gas flows to the passages from the same or from opposite sides is irrelevant as more gas will always flow through the passages near the exit of the regenerator top space than at its blind end.

In order to achieve an even flow through branched passages in a construction involving a flue and a top space, the top space must be larger than the flue; Eck[28] recommends a cross sectional ratio of 2 but this cannot be achieved in regenerators with their intermittent periods of flow. The weakest link is the heat transfer during the air preheat and the objective must therefore be to obtain an even flow during this period. This leads to the provision of a generous volume in the regenerator top space above the checkers.

The flow conditions can be followed mathematically. Reinhardt & Tollmien define the unevenness of the flow, ω, by

$$\omega = \frac{v_1 - v_0}{v_0}, \tag{3.6}$$

where v_1 is the rate of flow in the passages near the junction with the regenerator top space and v_0 the rate of flow at the opposite ends of the passage. Theoretical considerations, confirmed by experiment, show that the relationship

$$\omega = \cosh \alpha_G - 1 \tag{3.7}$$

applies, where α_G is given by

$$\alpha_G = \frac{n\Delta f/b_o\sqrt{2}}{\sqrt{1-\zeta_a}}; \tag{3.8}$$

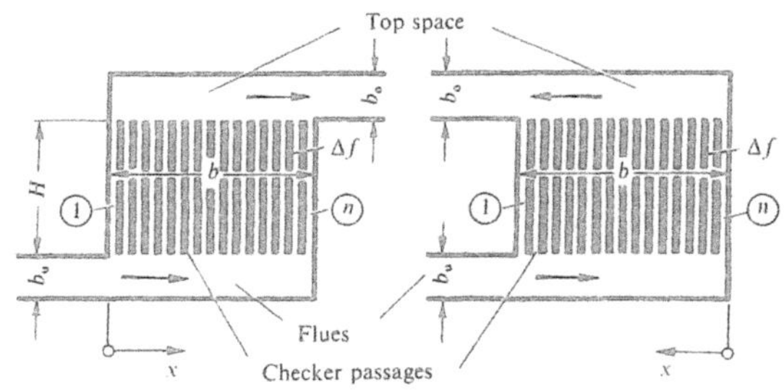

Figure 3.58. Schematic diagram of flows in regenerators[29]

here $n\Delta f$ is the total cross sectional area of all passages, b_o the cross section of the regenerator top space, and ζ_a the total resistance to flow of a single passage, which is assumed to be the same for all passages. ζ_a is defined by

$$\zeta_a = \frac{p_u - p_o}{v^2 \frac{\rho}{2g}} \tag{3.9}$$

where $p_u - p_o$ is the loss of pressure between regenerator flue and top space and v is the velocity of flow in the passages.

To calculate the flow profile across the width of the chamber (v) rather than the degree of unevenness, one may use the relationship

$$\frac{v}{v_m} = \frac{\alpha_G \cosh \frac{\alpha_G x}{b}}{\sinh \alpha_G}, \tag{3.10}$$

where v_m is the average velocity which is derived from the assumption of even flow through the chamber, b is the width of the chamber, and x is the running coordinate of the chamber width.

The relationships demonstrate the dominating influence of the relative width of the regenerator top space ($b_o/n\Delta f$) and the resistance to flow of the passages. This will be discussed in more detail later.

Experiments show that if $n\Delta f/b_u > 1$, the flow in the regenerator top space tends to bypass the first passages and to pass preferentially through the last ones; however this situation cannot be investigated mathematically.

Other investigations of flow distribution have been conducted by Kröll[31] who particularly investigated the influence of the entry into the passages. His investigations were aimed at relationships such as occur in drying equipment and he worked with Reynolds numbers of about 100 000, whereas in regenerators these are about 50 000. Despite this it is unlikely that a transfer of his basic conclusions to regenerators would be in error. His results are shown in Figure 3.59 and it should be noted that the vertical positioning of the regenerator duct above the passages (Figure 3.59(a)) leads to a significantly less even flow distribution through the passages than if the regenerator duct was situated to one side (Figure 3.59(b)). The regenerator duct above the passages is, however, very narrow and it is probable that a larger flue would also give a more even flow distribution. In any case, for ports located above the regenerators, the results demonstrate that the top space must be large enough and particularly that the exit velocities of the waste gases into the port must be kept

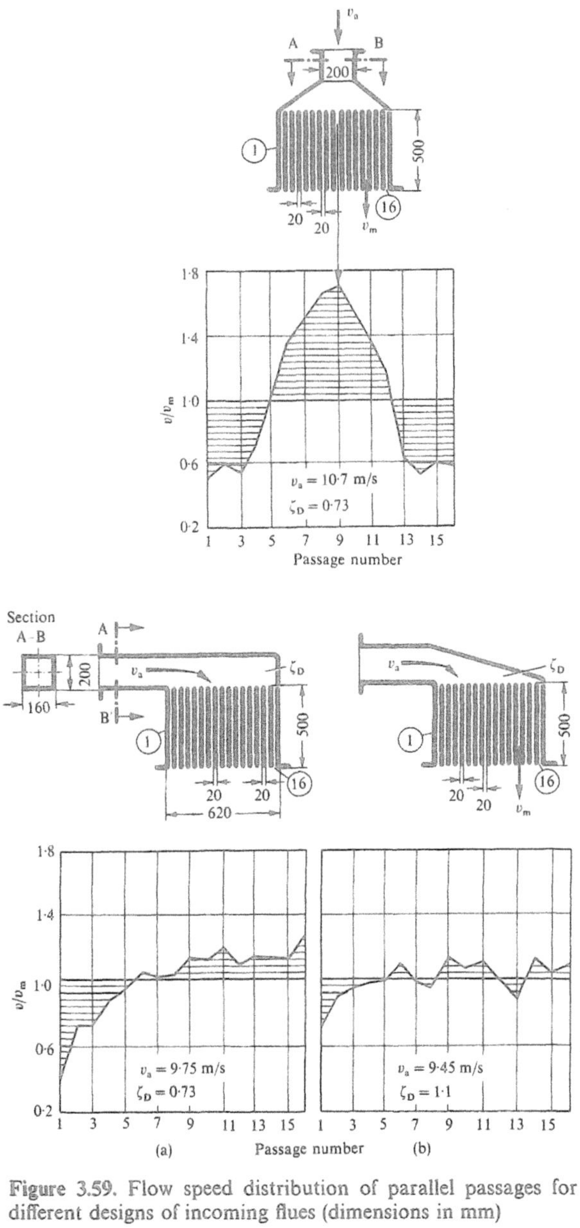

Figure 3.59. Flow speed distribution of parallel passages for different designs of incoming flues (dimensions in mm)

low. As far as flow distribution in the regenerator is concerned advantages are to be found in the design which extends the regenerator vertically to the port.

Kröll achieved significant improvements in the flow through individual passages by tapering the regenerator flue (Figure 3.59(c)). Reichardt & Tollmien showed that this tapering makes no contribution if $n\Delta f/b_u < 1$; if this value is greater than one, then there are advantages, a fact that has been confirmed by investigations by de Graaf mentioned below.

The flow through a hot blast furnace, which somewhat resembles regenerators with smooth plain packing, was studied by Hansen.[32] Among other factors he varied the height of the crown and the flow resistance of the passages, and his results also show that a high dome (i.e. a chamber in which the direction of flow is changed) favours uniform flow. Unfortunately, his results do not permit conclusions to be drawn about the correct design of regenerator chambers with smooth plain packing in which changes in flow direction occur. Despite this, his work also shows the dominating influence of flow resistance on the evenness of flow through the passages.

3.3.3.2 Regenerators with straight packing

The flow characteristics of pigeon hole checkers have so far hardly been investigated. Schleicher[33] carried out a study of flow in the regenerators of a Siemens–Martin furnace using a small model made of glass and demonstrated that air entering from the regenerator flue largely misses the first part of the checkers. Since, in his case, $n\Delta f/b_u$ was far above unity this result is not surprising.

De Graaf[34] also investigated the flow through the regenerator of a Siemens–Martin furnace, concentrating on temperature measurements and related deductions covering flow characteristics.

A mathematical treatment of the problem, such as that carried out by Reichardt & Tollmien for regenerators with smooth plain packing, has, so far, regrettably not appeared. The most significant difference between this packing and pigeon hole checkerwork as far as flow characteristics are concerned is that the gas is not confined to a particular flow direction in pigeon hole checkers but is free to move in any direction. A mathematical treatment of flow in pigeon hole checkers is possible provided that the problem is simplified by making suitable assumptions.[27]

3.3.3.3 Comparison between different regenerator packings

To elucidate the effect of different designs of regenerator (height of top space, width of chamber, type of packing, etc.) a study was made of the flow profile. Although the branched flows shown in Figure 3.58 and covered by Equation (3.10), differ from those in chambers of actual glass furnaces, this type of flow was investigated further because it was believed that the fundamental tendencies evident here would also be reflected in chambers of other designs. The results of these calculations are given in Figure 3.60 in which the ratio between the actual and the average velocities is plotted in relation to the reference plane divided by the chamber width. The parameters are the total width of the chamber, the height of the regenerator top space, and the type of checkers used: the height of the chamber is fixed at 5 m in all cases.

If one compares the flow distribution in a vertical chamber 6 m wide and with straight packing with one 4 m wide, then the conclusion is that the narrower chamber gives a significantly more even flow than the wider one, leading to the general conclusion that narrow chambers give a more even flow than wide chambers.

The next comparisons cover flows over straight packing, smooth plain packing, and staggered packing in a chamber 6 m wide; the evidence leads to the conclusion that staggered packing gives the most even flow and smooth straight packing the most uneven, the main cause being the different resistance to flow by the different types of checker. It is generally true that, with the same regenerator flue and top space, the greater the resistance to flow of a given type of checker the more even is the flow distribution; staggered packing offers the highest resistance and therefore gives the most even flow. It is commonly believed in industrial practice that the flow through smooth plain packing is more even than that through straight or staggered packing; this can indeed be so, but only when the resistance to flow of both types is the same. It is also necessary to remember that flow conditions through straight packing as distinct from smooth plain packing depend on the position of the regenerator cross section being studied. The flow is much more uneven just below the regenerator top space when compared to the regenerator flue. In the cases mentioned above, the centres of the regenerator chambers were compared.

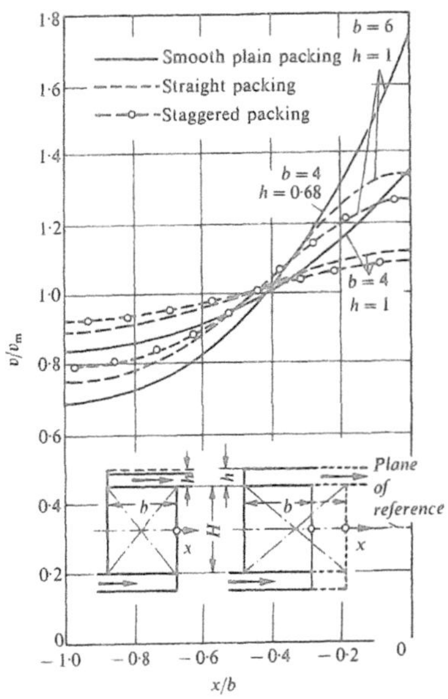

Figure 3.60

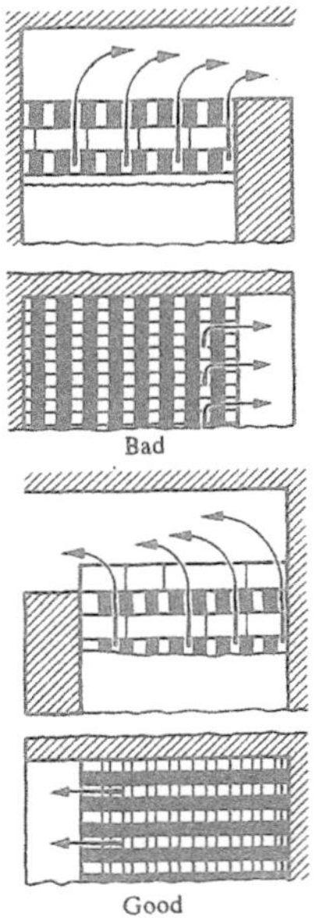

Figure 3.61

Figure 3.60. Standardised flow rate related to the distance x along the width of b of the chamber

Figure 3.61. Influence of packing on flow characteristics

As a general guide, the type of flow is governed by the height of the regenerator top space. When one compares the flow from top spaces 1 and 0·68 m high through a regenerator chamber 4 m wide, the results show that the lower top space gives a significantly more uneven flow. Experience indicates that it is better to reduce the height of the checkers rather than that of the top space as the benefit due to more even flow outweighs the loss arising from a reduced heat exchange surface. As a general guide, the cross section of the top space should equal that of the open cross section of the checkers; if it can be done, a regenerator top space of even larger cross section is desirable, especially if the checkers are rather open and have a low resistance to flow.

There is no information covering the correct dimensioning of return passages in regenerators with smooth plain packing and it is therefore recommended that they should be designed on the basis described above.

Careful placing of bricks in chambers with straight or staggered packing can give wide collecting channels as shown in Figure 3.61. If the top layer of bricks is placed so that their length lies in the direction of flow, then the height of the regenerator top space is, in effect, one brick width taller than if the bricks were positioned across the flow.

With this setting the gas flow turns in the second layer of bricks from the top, whereas the alternative setting allows the flow to turn only after leaving the top layer. This results in the flow velocity, and thereby the pressure drop, being lower so that the flow through the checkers is more even but without losing checker volume.

3.3.3.4 The influence of temperature distribution

The factors considered in Sections 3.3.3.1 and 3.3.3.2 assumed that the gas, on passing through the checkers, was not subject to temperature changes and the influence of buoyancy, a situation which does not occur in practice. It is worth considering whether these factors create particular effects which influence the type of flow.

Consideration of this point leads to the conclusion that in regenerator chambers which offer only small resistance to flow compared to the draught effect of the chimney (and this applies to all glass furnaces operating without fans) the flow distribution is indeed influenced by the draught. It follows that the flow will be more even when the hot gases flow downwards through the hot chamber and the cold air upwards. If the flow directions are reversed, as is often the case with multicompartmented vertical regenerators, then the flow, influenced by the draught, is more even than it would be in a cold regenerator. This result, at first sight remarkable, can be explained if one looks at Figure 3.62. Uneven flow causes one side of the chamber to become hotter than the other and this temperature difference gives rise to a convection current acting as a secondary flow superimposed on the main flow. The Figure shows that with hot gases from above and cold gases from below, these secondary flows are weakened, whereas they are augmented if the flows are the other way. Investigations carried out in Britain on hot and cold regenerators in which the hot gases flowed from top to bottom showed that the flow through a hot chamber was more even than through a cold one; this is in agreement with earlier conclusions.[34] In practice, the preferred flow arrangements are usually adopted and while vertical multichamber regenerators sometimes have their flow in the less favoured direction, they should be avoided if at all possible.

In horizontal chambers, especially those fitted with smooth plain packing, the draught effect of hot waste gases has highly undesirable effects in relation to the cold air. The hot gases flow preferentially through the upper parts of the checkers, while the cold air flows through the lower parts, thereby noticeably reducing the heat exchange.

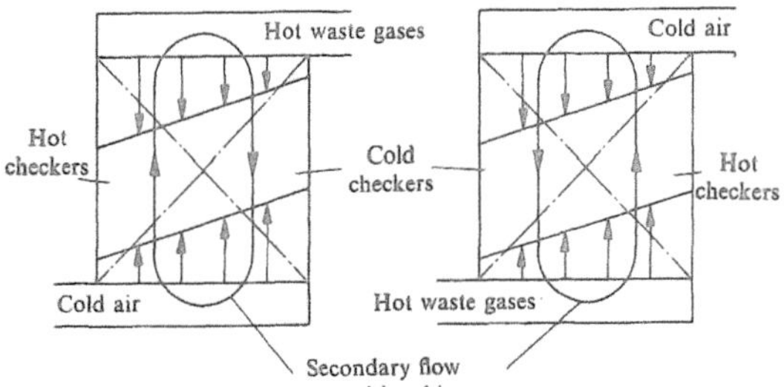

Figure 3.62. Influence of flow direction on the uniformity of flow through regenerators

3.3.4 Operating data for regenerators

Vertical regenerators will reach temperatures of 1400 to 1600°C at the top of the chamber during the passage of waste gases. The exit temperature of the waste gases depends on the size of the regenerator but is 300 to 600°C. The entry temperature of the air is about 100°C and a preheat temperature of 1100 to 1400°C can be assumed. The average temperature difference between waste gases and air is about 200 K. Depending on the radiation characteristics of the gases and the related improvement in heat exchange from the waste gases, the surface temperature of the bricks is not an average but is nearer to that of the waste gases. Values of 1/3 and 2/3 may be assumed so that the temperature difference between waste gas and checker brick surface is about 70 K and that between brick and air about 140 K. The temperature reduction during a cycle can be assumed to be 100 K. The reversal period is now normally 20 min, and while a longer period will reduce losses during reversal it will cause a greater temperature reduction and reduce the average preheat temperature. Temperature measurements in regenerator chambers follow the pattern shown in Figure 3.63.

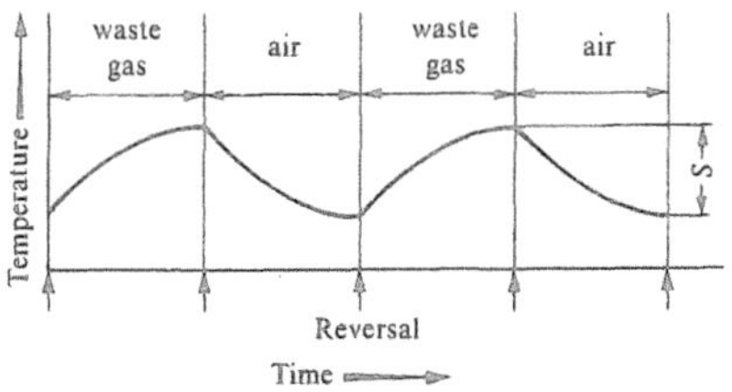

Figure 3.63. Schematic of temperature changes in a regenerator

For average conditions, flow velocities in the checkers are designed to be 1 to 3 m/s and pressure loss can be calculated from

$$\frac{\Delta p}{L} = \psi \frac{\rho w^2}{2d} \tag{3.11}$$

where the resistance, ψ, can be taken from Table 3.3, L is the length of passage through the checkers, ρ the average density of the gas, w the rate of flow, and d the hydraulic diameter of the flow channel (see Chapter 12).

Table 3.3. *Flow resistance constant, ψ, for different type of checker*

Type	*Reynolds number*	*$\psi \times 10^2$*
Smooth plain packing	$2\text{–}8 \times 10^3$	2
Basket weave packing	$2\text{–}8 \times 10^3$	6–8
Straight packing	$5 \times 10^2\text{–}4 \times 10^3$	12–15
Staggered packing	$5 \times 10^2\text{–}4 \times 10^3$	31–34

3.3.4.1 Types of checker packing

From the point of view of thermal efficiency, checkers should have a high heating surface per unit volume of chamber and good heat transfer between waste gases or air and bricks (Chapter 12); from the operational side the flow resistance should not be too large; nor should the performance of the refractories be forgotten. In practice only checker packings constructed from normal sized bricks have been found acceptable for glass furnaces and all attempts to use special shapes designed from considerations of heat technology have so far been unsuccessful.

The most commonly used checker packings are straight, staggered, basket weave, and combinations of basket weave and straight for vertical chambers and smooth plain packing for horizontal chambers. The most commonly used checker types are given in Figure 3.64.

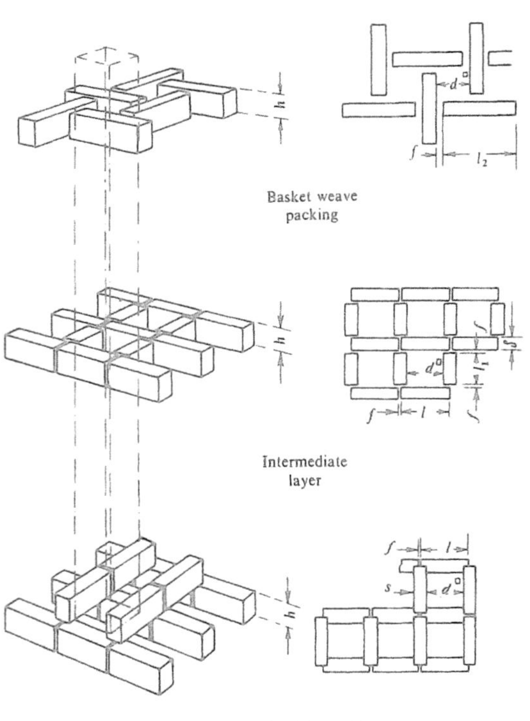

Figure 3.64. Construction of a combined checkerwork

3.3.4.1.1 Smooth plain packing

This is very stable and easy to clean; since there are no dead spots for flow between the bricks dust deposits are small. Disadvantages are the smaller heating area and high weight of bricks in relation to the regenerator chamber volume and the low heat exchange resulting from laminar flow in the ducts. Thinner bricks give improvements since a reduction in thickness of less than 4 cm increases the available heating surface.

3.3.4.1.2 Straight and staggered packing

For normal thicknesses of brick and chamber widths these offer the largest heating surface, and heat exchange is at its optimum when using staggered packing; straight packing gives slightly lower heat transfer values, but the risk of blockage is less.

In straight packing the distance between the rows of bricks dominates the heat transfer process and the height of the brick has only a minor influence. Heat transfer improves as the distance between bricks increases: model studies[36] have shown that it reaches a maximum when the ratio between the distance apart of the rows and the thickness of the brick is 1·3 to 1·5 and decreases again if these values are exceeded.

For many years straight packing has been preferred for the firebrick section of the chamber and it is also used sometimes for part of the basic brick section: the straight ducts permit cleaning in case of need.

3.3.4.1.3 Basket weave packing

With the advent of basic bricks basket weave packing was introduced because it gives a higher contact surface between bricks, is more suitable for the high thermal expansion basic bricks, and gives very stable checkers. Heat transfer in basket weave checkers is significantly higher than in checkers with smooth plain packing for only a small increase in flow losses. Heat transfer is not as good as with straight or staggered packing, but the ratio of pressure loss to heat transfer is smaller.[37] Basket weave packing uses a weight of bricks per m^3 of chamber similar to that used for smooth plain packing but the storage capacity is increased.

Mixed checkers

Regenerators of high performance tank furnaces use mixed packing as a result of using both magnesite and fireclay bricks. In a chamber of constant cross section the upper part is basket weave packing which, via an intermediate layer, is changed to straight packing (Figure 3.64). In parts of the uppermost layers 'roof tiles' on which the dust settles only with difficulty, are used. A combination of box blocks and straight packing is shown in Figure 3.65.

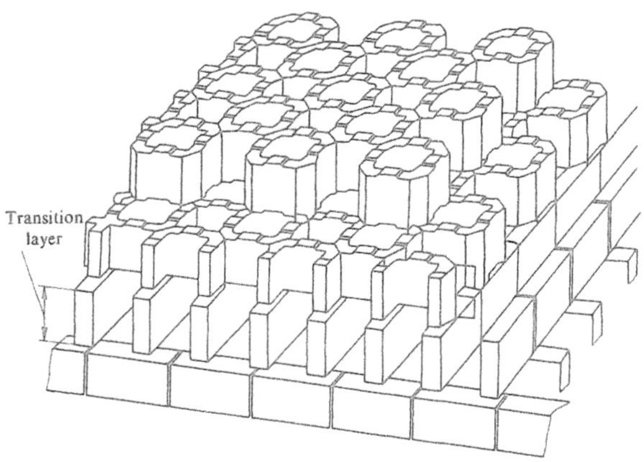

Figure 3.65. Checker packing with a combination of straight and box blocks

3.3.4.1.4 Brick sizes

Standard bricks and standardised square bricks are now always used for checkers since they are relatively cheap and available at short notice. Special shapes which provide higher surface areas for heating have not been successfully introduced because of the risk of blockage and their higher cost.

In the main, bricks 65 mm thick are used for checkers; the use of thicker bricks makes no sense as, with reversal periods of not more than 30 min, the core of the brick hardly plays a part in the heat exchange process and the degree of brick utilisation (Chapter 12) is reduced. Brick length is between 230 and 375 mm. For wide channels in straight packing, bricks $1\frac{1}{2}$ times the normal length are used and these are supported twice, with their total thickness on the layer below, which improves the structural strength of the checkers.

Standard brick height varies between 100 and 150 mm with a tolerance of only $\pm 0{\cdot}5$ mm to ensure checker stability.[38,39] Large deviations cause uneven surface contact and surface pressure and can, despite exact positioning of bricks, lead to collapse of the checkers.

3.3.4.2 Data for checkers

The open passages have considerable influence on the heat exchange process.[36] As they are reduced so the available heating surface and thereby the heat exchange capacity is increased; at the same time, however, the flow resistance increases as does the risk of blockage by batch dust or slag.

Blockages with firebrick checkers can be avoided if passages are constructed with clear widths of up to 300 mm. Basic bricks do not collect dust on their vertical surfaces and widths of 120–170 mm are normal, varying with furnace size. Widths of less than 100 mm are rare even with basic checkers.[38,39]

Other parameters for checkerwork are

1. the heating surface (m^2) related to the volume of the chamber (m^3), which is a measure of the utilisation of the chamber
2. the weight of checker bricks (t) related to the chamber volume (m^3), which is a measure of the storage capacity of the checkerwork
3. the free flow diameter as a percentage of, e.g. for vertical checkers, the cross section area; it is a measure of the flow resistance
4. the brick thickness, which governs the stability of the checkerwork
5. the equivalent brick thickness, which is given by

$$2 \times \frac{\text{brick volume/m}^3 \text{ of chamber volume}}{\text{heating area/m}^3 \text{ of chamber volume}};$$

this corresponds to the thickness of an infinitely large block which, as a parameter for calculations, gives the same conditions for heat exchange as the checker bricks actually used.

Checker dimensions are calculated from the equations given in Table 3.4. The specific tonnage of bricks required (t/m^3) is calculated from the volume and bulk density of the bricks as provided on the manufacturers' data sheets.

Table 3.4. *Formulae for checker characteristics; for symbols see Tables 3.5 and 3.6 (a = d + s = distance between centres)*

Parameters	*Units*	*Smooth plain packing; basket weave packing*	*Straight packing; staggered packing*	*Open basket weave packing*
$\dfrac{\text{Weight of checkwork (kg)}}{\text{Volume of chamber (m}^3)} \times 10^3$	$\dfrac{t}{m^3}$	$\left[1-\left(\dfrac{d}{a}\right)^2\right]\gamma$	$\dfrac{s}{a}\gamma$	$\left[1-\left(\dfrac{d}{a}\right)^2-\dfrac{2f\cdot s}{a^2}\right]\gamma$
$\dfrac{\text{Heating surface (m}^2)}{\text{Checker volume (m}^3)}$	$\dfrac{1}{m}$	$\dfrac{4d}{a^2}$	$\dfrac{2}{a}+\dfrac{2ds}{a^2h}$	$\dfrac{2(1-s)}{a^2}$
$\dfrac{\text{Cross section of free passages (m}^2)}{\text{Total flow cross section (m}^2)} \times 10^2$	%	$\dfrac{d^2}{a^2}$	$\dfrac{d^2}{a^2}$	$\dfrac{d^2}{a^2}$
Equivalent brick thickness	m	$s\left(1+\dfrac{s}{2d}\right)$	$s\dfrac{1}{1+\dfrac{sd}{ha}}$	$s+\dfrac{s^2}{2(d-f)}$
$\dfrac{\text{Volume of checker bricks (m}^3)}{\text{Checker volume (m}^3)}$	–	$1-\left(\dfrac{d}{a}\right)^2$	$\dfrac{s}{a}$	$1-\left(\dfrac{d}{a}\right)^2-\dfrac{2f\cdot s}{a^2}$

Table 3.5. *Data for horizontal and vertical checkerwork*

Smooth plain packing

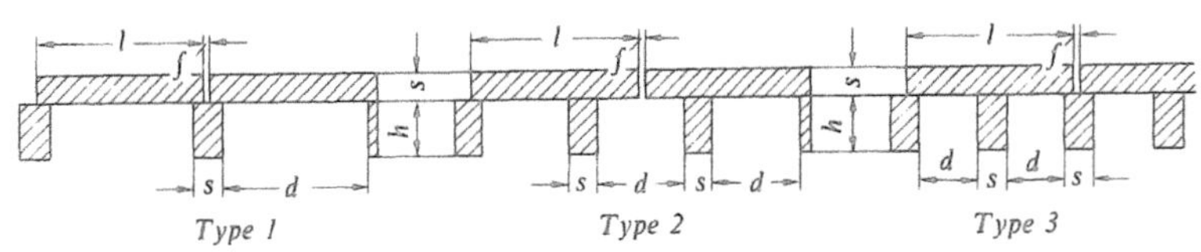

Passage width d × h (mm)	*Type*	*Brick dimensions (mm)*	*Relative heating surface (m^2/m^3)*	*Cross section of free passages (%)*	*Relative number of bricks (no/m^3)*	*Relative brick volume (m^3/m^3)*
102 × 123	2	205 × 123 × 65	14·3	40·2	298	0·595
122·5 × 123	3	375 × 123 × 65	13·9	42·8	190	0·57
172 × 115	1	230 × 115 × 65	13·5	47·2	314	0·54
185 × 123	1	250 × 123 × 65	13·0	48·0	260	0·52
172 × 230	1	230 × 115 × 65	11·5	56·6	279	0·48

Straight packing; basket weave packing

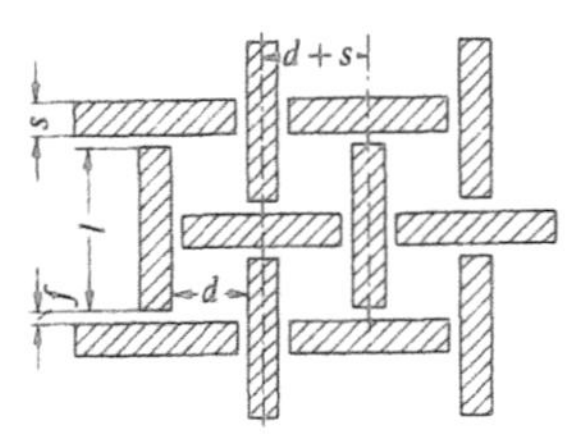

Passage width, d (mm)	*Brick dimensions (mm)*	*Pitch, d + s (mm)*	*Expansion joint, f (mm)*	*Relative heating surface (m^2/m^3)*	*Cross section of free passages (%)*	*Relative number of bricks (no/m^3)*	*Relative brick volume (m^3/m^3)*
122	300 × 150 × 65	187	9	14·0	42·6	194	0·57
145	345 × 115 × 65	210	11	13·1	47·6	201	0·52
160	375 × 123 × 65	225	12	12·6	50·8	163	0·49
218	250 × 123 × 65	283	7·5	10·9	55·4	205	0·41

Straight packing; cruciform packing

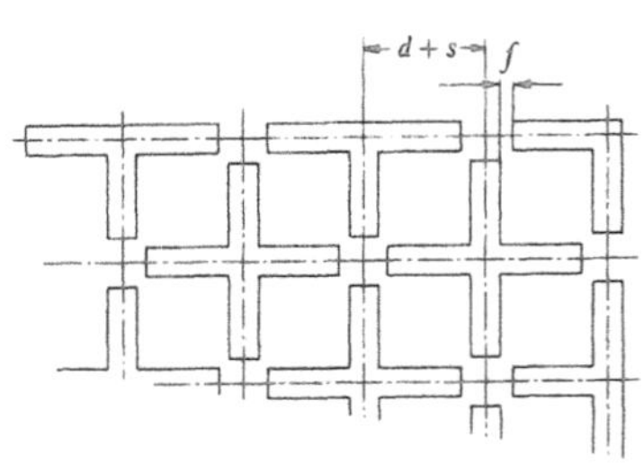

Passage width, d (mm)	Brick dimensions (mm)	Pitch d + s (mm)	Expansion joint, f (mm)	Relative heating surface (m²/m³)	Cross section of free passages (%)	Relative number of bricks (no/m³)	Relative brick volume (m³/m³)
140	333 × 280 × 40	180	20	15·12	60·49	46·34	0·321
170	333 × 330 × 40	210	5	15·42	65·53	45·35	0·317

Straight packing; box blocks packing (this may also be assembled staggered)

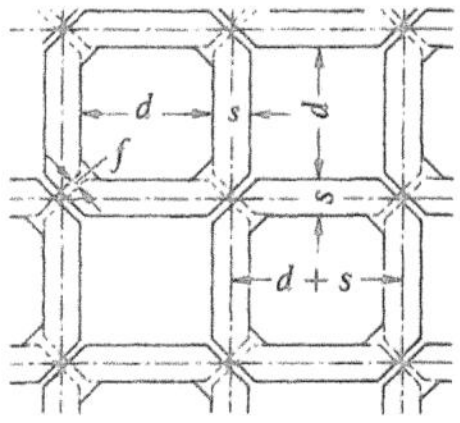

Passage width, d (mm)	Brick dimensions, s (mm)	Pitch d + s (mm)	Expansion joint, f (mm)	Relative heating surface (m²/m³)	Cross section of free passages (%)	Relative number of bricks (no/m³)	Relative brick volume (m³/m³)
140	40 × 150 high	180	6	15·97	0·578	102·9	0·395
160	40 × 150 high	200	6	14·94	0·625	83·3	0·361

Table 3.6. *Data for straight and staggered packing*

Straight packing

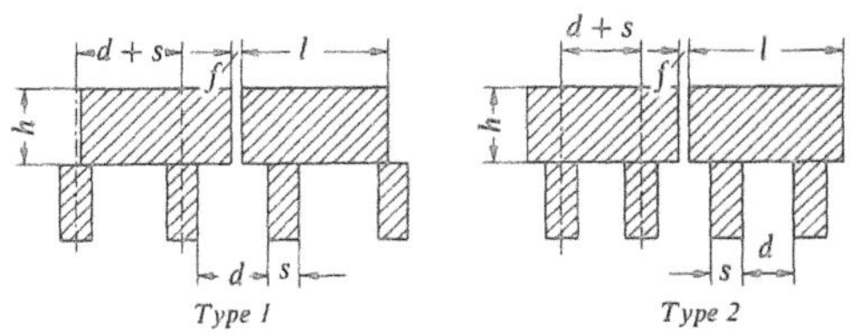

Passage width, d (mm)	Type	Brick dimensions (mm)	Pitch d + s (mm)	Expansion joint, f (mm)	(%)	Relative heating surface (m²/m³)	Cross section of free passages (%)	Relative number of bricks (no/m³)	Relative brick volume (m³/m³)
100	2	345 × 115 × 76	176	7	2·0	15·63	32·28	140	0·423
105	1	250 × 123 × 65	170	5	2·0	15·60	38·15	188	0·375
115	2	375 × 123 × 76	191	7	1·9	14·37	36·25	111	0·391
120	2	375 × 123 × 76	196	17	4·5	14·06	37·48	160	0·371
130	2	375 × 123 × 65	195	15	4·0	13·87	44·44	107	0·321
140	1	300 × 150 × 65	205	7·5	2·5	12·64	46·64	160	0·309
160	1	345 × 115 × 76	236	9	2·6	12·27	45·96	104	0·314
170	1	345 × 115 × 65	235	7·5	2·2	11·99	52·33	105	0·271
180	1	375 × 123 × 76	256	9	2·4	11·21	49·44	83	0·290
190	1	375 × 123 × 65	255	7·5	2·0	10·93	55·52	83	0·250

Diagonally staggered packing

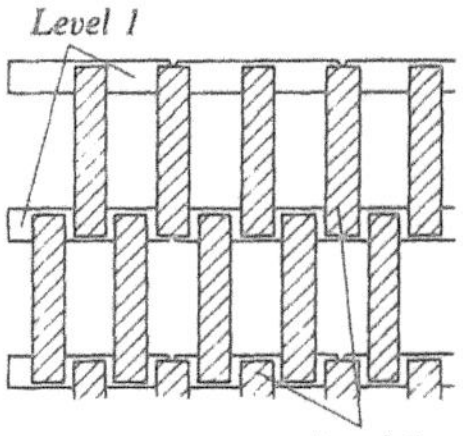

(Table 3.6 continued overleaf)

Passage width, d (mm)	*Brick dimensions (mm)*	*Pitch d + s (mm)*	*Expansion joint in Level 1 (mm)*	*Relative heating surface* (m^2/m^3)	*Cross section of free passages (%)*	*Relative number of bricks* (no/m^3) *Level 1*	*2*	*Relative brick volume* (m^3/m^3) *Level 1*	*2*
100	230 × 115 × 65	165	5	14·56	36·73	224	319	0·385	0·549
105	250 × 123 × 76	181	5	13·30	33·65	176	248	0·412	0·580
110	250 × 123 × 76	186	5	13·02	34·98	171	235	0·400	0·550
120	250 × 123 × 65	185	5	15·58	42·07	172	238	0·344	0·476
130	250 × 123 × 65	195	5	16·27	44·44	164	214	0·328	0·428
140	250 × 123 × 65	205	5	15·33	46·64	155·5	193·5	0·311	0·387
150	300 × 150 × 76	226	6	14·01	44·05	96·5	130·5	0·330	0·446
160	300 × 150 × 76	236	6	13·21	45·96	92	120	0·315	0·410
170	300 × 150 × 65	235	6	13·03	52·33	93	121	0·272	0·354
180	300 × 150 × 65	245	6	12·36	53·98	89	111	0·261	0·325
190	300 × 150 × 65	255	6	11·77	55·52	85·5	102·5	0·251	0·300

Tables 3.5 and 3.6 give the most important data for different types of checker packing.

3.3.5 Regenerator crowns and walls

Crowns of regenerator chambers are constructed as circular arches or barrel arches, in the main with a rise of 13·4% or 20%; barrel arches are more difficult to control and are found only rarely. A rise of 13·4% (which implies an included angle of 60°C at the arch centre) is preferred as it permits the use of standard arch bricks for all possible arch widths. An example of a regenerator crown and the upper part of the walls is shown in Figure 3.66. Special expansion bricks are not required if basic bricks are used but if silica is used, expansion bricks such as are used in the furnace crown may also be used here. Thin bricks which rest on corbels in the sidewalls are also used (Figure 3.67).

If basic bricks are used for wide thermally insulated arches subjected to high temperatures, then only refractory grades possessing low creep should be used. High creep leads to dangerous sagging of the apex of the arch.

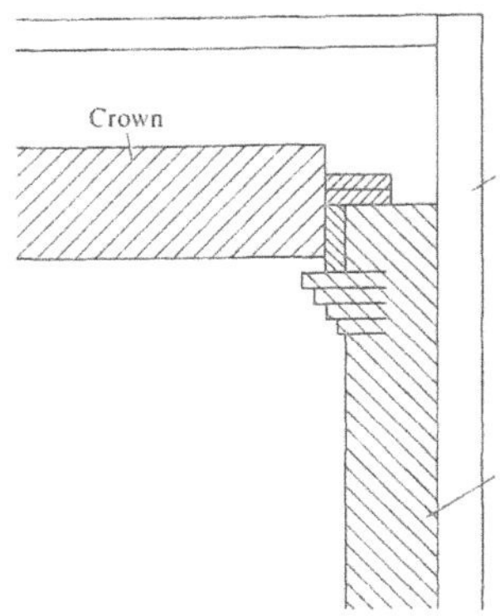

Figure 3.67. Sealing of regenerator arch

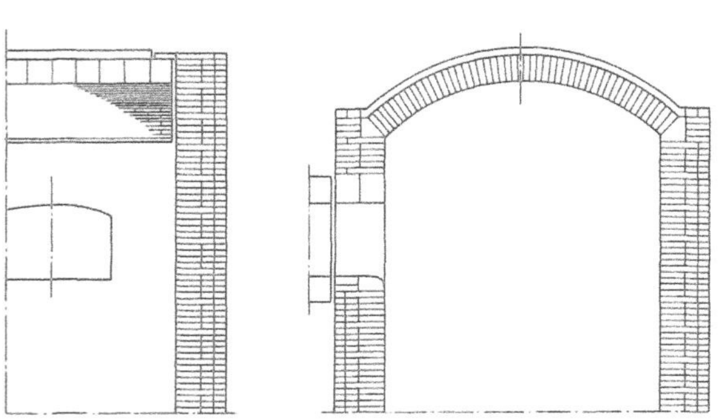

Figure 3.66. Regenerator arch with upper sections of wall (Didier)

For the chamber walls of modern vertical regenerators standard bricks should always be used for headers and stretchers (Figure 3.57). In line with the refractories used for the checkerwork, the upper section of the inner wall should be built of basic bricks and the lower section of high quality firebrick. Separating walls for end-fired tanks and the dividing walls of multicompartment regenerators for cross-fired furnaces are generally constructed of only one type of refractory.

3.3.6 Bearer arches and checkerwork support bricks

The checkerwork rests on bearer arches which separate the checker part of the chamber from the regenerator flue. The bearer arches are subject to high stress arising in part from the weight of checker bricks carried and in part from the temperature changes between waste gases and air to which they are subjected. It is important to ensure that chemical attack due to sulphate condensation is not added to these stresses; this can be avoided by establishing temperature conditions when designing the regenerator. As repairs are very difficult to carry out during the course of a campaign, it is essential that the refractories be selected and constructed very carefully. For chamber widths up to 2 m single arches are sufficient but above this double arches must be used. The use of standard size blocks is recommended. Examples of construction are given in Figures 3.68 and 3.69. Between arches the support ledge carrying the arches is covered with feather edge bricks.

The checkerwork cannot rest directly on bearer arches and therefore support bricks, whose length

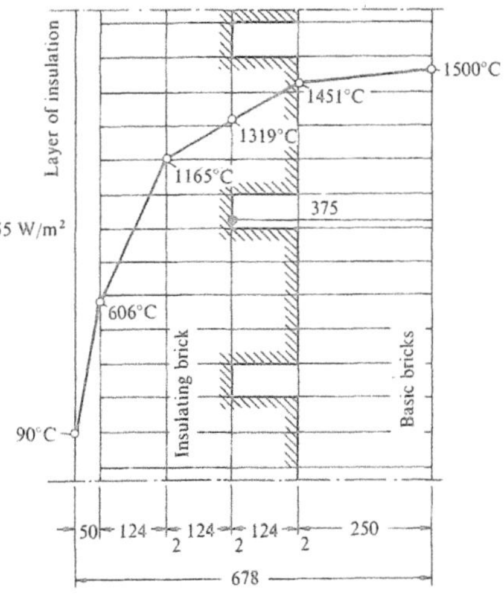

Figure 3.68. Example of regenerator wall construction (Didier)

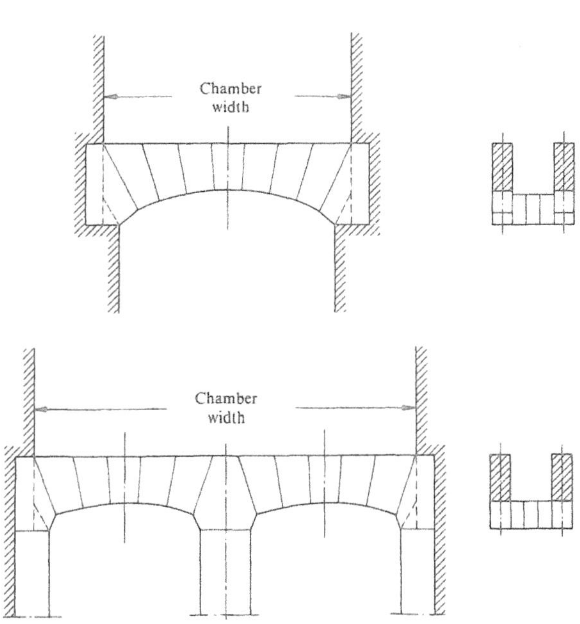

Figure 3.69. Construction of bearer arches (Didier)

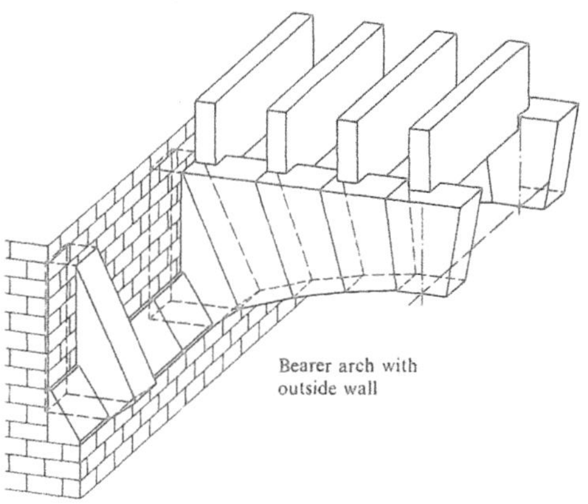

Figure 3.70. Bearer arches with checkerwork support bricks (Didier)

depends on the width of the flue, are necessary (Figure 3.70). The centre-to-centre distance between arches should be twice the width of the flue plus the thickness of the checker brick.

3.4 Recuperators

Small tanks and pot furnaces are fired with fixed flames and their combustion air is preheated by recuperators which may be of ceramic or metal; the trend is in favour of metal recuperators.

3.4.1 Ceramic recuperators

The main problem with ceramic recuperators is the gas-tight sealing of waste gas passages against air. The two-flow systems operate under a pressure differential of 3 to 5 mbar and if there is a substantial leak, the flue sucks the combustion air into the waste gases, satisfactory firing becomes impossible, and the furnace must be shut down. The design and construction of recuperators are therefore concentrated on maintaining a system with gas-tight joints with elements which have lowest possible thermal resistance.

Waste gas flues use box or tubular construction units of a hexagonal design such as those shown in Figures 3.71 and 3.72; the units are assembled one on top of the other so that their own weight ensures good sealing of the joints. A slightly different approach is the use of elliptical shapes (Figure 3.73) in which the air passes through elliptical channels while the waste gases pass round the outside of the shapes. The units described can give heating surfaces of 10 to 12 m^2/m^3 of recuperator chamber

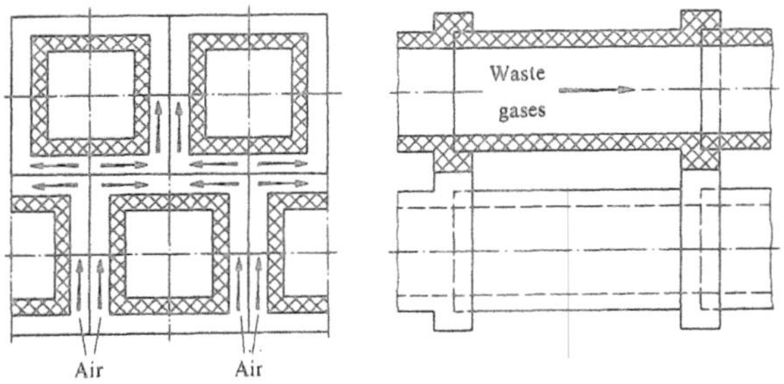

Figure 3.71. Box-shaped recuperator ceramic tubes

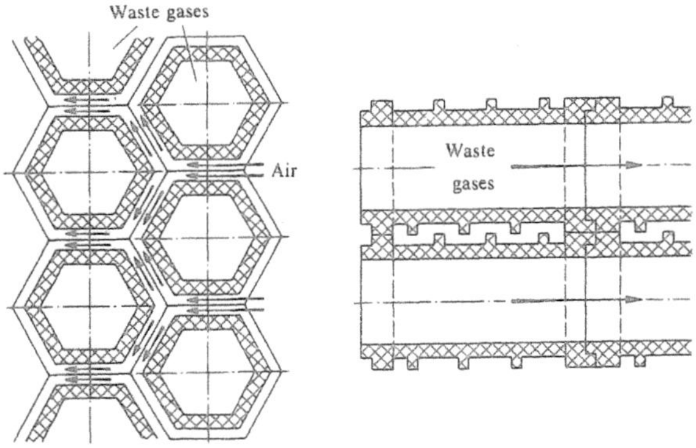

Figure 3.72. Tubular recuperator refractories

Figure 3.73. Elliptical recuperator elements

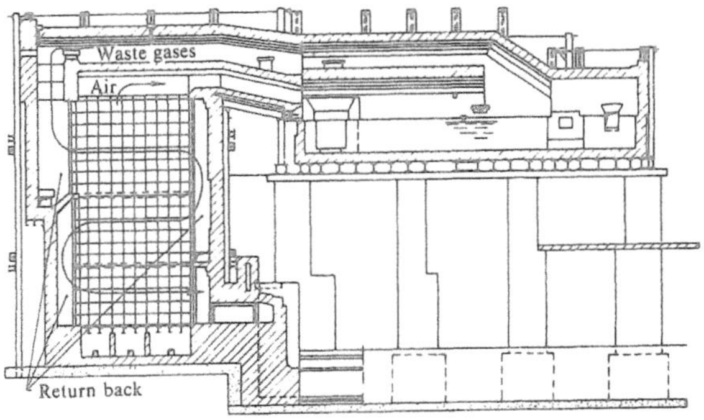

Figure 3.74. Tank furnace with double crown heated recuperatively[40]

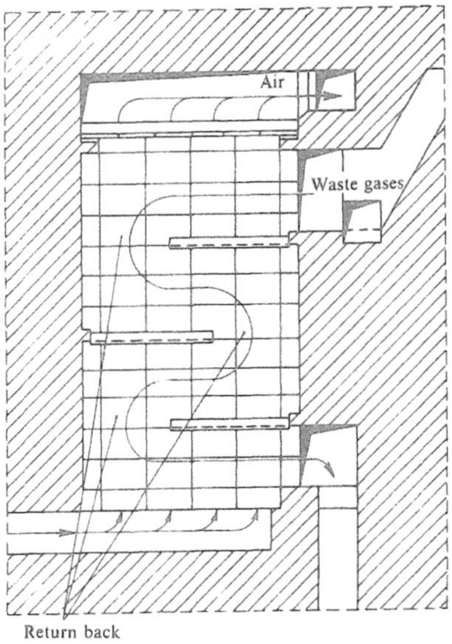

Figure 3.75. Recuperator with return backs within the checkerwork[40]

volume. The heat transfer coefficient depends, among other factors, on the waste gas temperature but values of 6 to 8 W/m^2 K are obtained.

The construction of a recuperator from hexagonal elements (Figure 3.72) used with a double crown tank is shown in Figure 3.74.[40] Waste gases take a serpentine passage horizontally around the elements while the air passes vertically upwards through the recuperator; the free flow cross section changes with the waste gas temperature. With tubular recuperator elements also the air flow can be on the inside and the waste gas flow on the outside, provided the tubes are installed one on top of the other. This arrangement avoids the exterior chambers necessary for reversing the flow direction and gives a compact construction (Figure 3.75), but it does offer the risk that dust will settle on the horizontal separating tiles, blocking the gas passages and leading to the termination of the furnace campaign. If at all possible, free return backs should be used.

Ceramic recuperators have two critical areas; the first is in the gas-tight sealing between the upper collecting channel for air (or for waste gases if they flow on the inside) and the layer immediately below it and the second is in the return backs. The air collecting channels used with intermittently operating furnaces such as pot furnaces and day tanks are difficult to seal since the structural elements move with changes in temperature.

Refractories for recuperator elements are made from dry-formed fireclay or, for the hot parts, from corundum-free sillimanite. Basic materials should, in principle, show advantages, but their use is prevented by the difficulty of manufacturing the complex elements needed for a recuperator.

3.4.2 Metal recuperators

These can be kept gas-tight much more easily than ceramic recuperators and are also easier to construct, but air preheat is limited by the temperature stability of the materials. For large furnaces requiring more than 10 000 m^3/h of air (at NTP), only tubular recuperators can be considered. The tubes are usually mounted vertically with a collecting ring top and bottom and the assembly of tubes is sited inside a steel casing lined with refractory material; parallel flow is preferred. The upper air manifold sits on the tubes while the lower manifold is free to move and allows the tubes to expand. Under these conditions preheat temperatures of 500°C can be obtained and the construction of such a radiation recuperator is shown in Figure 3.76.[41] If larger quantities of air are needed and if control is possible in two groups, as in unit melters, then a recuperator with two rings of tubes is used (Figure 3.77). Additional safety is provided by a refractory tube which projects half way up the waste gas flue.[42] The recuperator operates under counterflow conditions, giving air preheat temperatures of up to 700°C.

For smaller units such as day tanks and pot furnaces the double shell radiation recuperator has been successfully used (Figure 3.78). Waste gases and air usually flow in parallel, the waste gases moving slowly through the inner cylinder, the diameter of which is related to the rate of waste gas flow, while the air for preheating flows at high speed through the annulus between inner and outer cylinder. The inner cylinder is cast and presents a smooth surface to the waste gases, but has small ribs on the side of the air flow (Figure 3.79). Data for the design of recuperators, both ceramic and metal, are summarised in Table 3.7.

Alloys used for metal recuperators are sensitive to reducing atmospheres and the deposition of soot causes their embrittlement; metal recuperators must therefore be used only in conjunction with oxidising firing conditions.

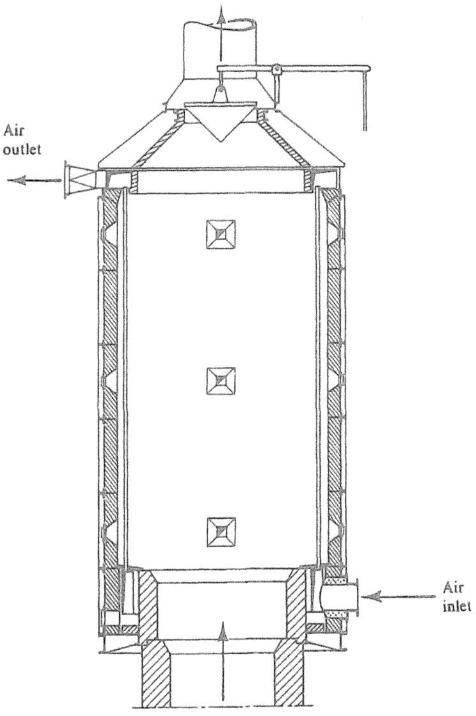

Figure 3.76. Radiation recuperator[41]

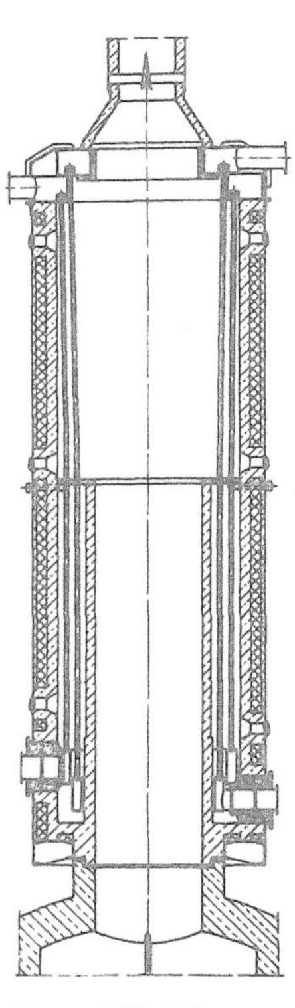

Figure 3.77. Inka recuperator with double wall and refractory supporting wall for counterflow operation[42]

Table 3.7. *Data for recuperators*

Type		*Permissible entry temperature of waste gases (°C)*	*Attainable air preheat temperature (°C)*	*Pressure loss in recuperator (mbar)*		*Gas velocities at NTP (m/s)*	
				Air	*Waste gases*	*Air*	*Waste gases*
Ceramic	Horizontal	1500	950	0·2–0·6	0·2–0·6	0·1–0·3	0·5–1
	Vertical	1500	<1150	0·2–0·6	0·2–0·6	0·2–0·5	0·3–0·6
Metal	Annulus	1300	500	13–45	0·1–0·2	10–20	1–2
	Tubular (Inka)	1500	700	13–45	0·1–0·2	10–20	1–2

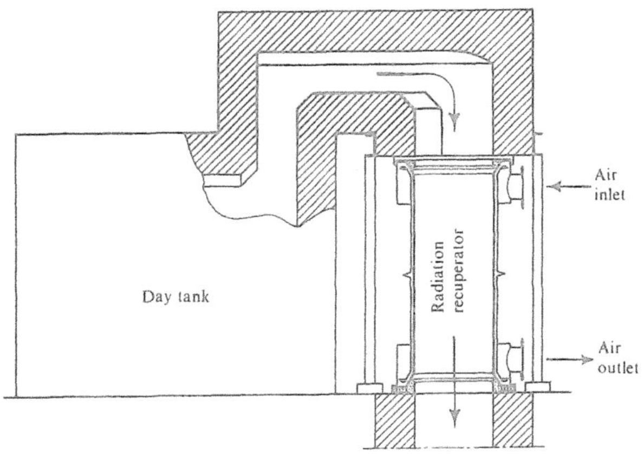

Figure 3.78. Day tank with radiation recuperator located to one side[41]

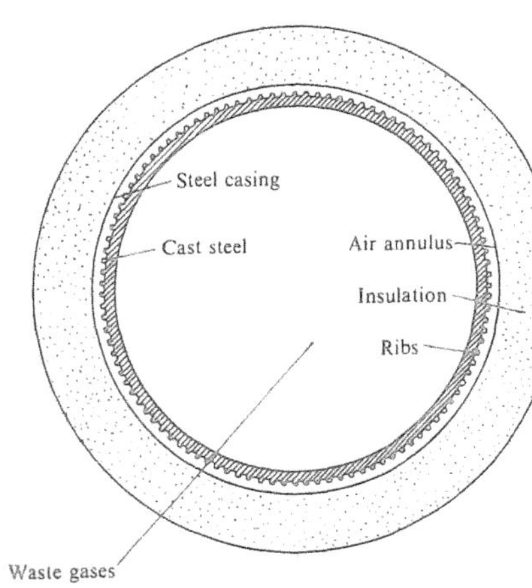

Figure 3.79. Cross section through an annular radiation recuperator[42]

3.5 Flues

The term flue is used for the system which includes the reversal system and the duct to the chimney; flues serve to remove waste gases from the heat recovery system and to supply combustion air to the furnace. Depending on space available and the layout of the flues it is possible to instal secondary regenerator chambers to work in conjunction with the regenerative firing system.

The flues between reversal valves and regenerator chambers must be arranged symmetrically to ensure steady furnace conditions. The cross section of flues should be such that the flow rates of gases are the same everywhere. The flow rates in the flues should not be far above those in the checkers since a nearly uniform distribution of waste gas and air flow is possible only under these conditions. Flues for sampling should be constructed with rounded or stepped corners; branch flues going off at right angles should be avoided since they cause loss of draught and uneven distribution of flow rates. The following average flow rates can be used for general guidance in the design of flues: below the bearer arch 2–3 m/s, leading to the reversal valve 3–5, inside the reversal valve 5–7, and from reversal valve to chimney 3–5.

Flues must be positioned above ground water level and must be air-tight under all conditions to prevent ingress of air. They should be quick and easy to instal and move in case the furnace is moved and may be made from bricks, premanufactured special shapes, or refractory materials of Class A III with a refractoriness under load value of 1380°C and upper limit of use between 1200 and 1300°C.

Thermal insulation of flues is not needed if they are placed underground. If the flues are crossed by roadways their loading and that of entry holes must be taken into account. If the flues are located above ground, they are usually insulated, especially if the remaining energy in the waste gases is used in secondary regenerators.

If flues are constructed of bricks the cost of materials is low but the cost of moving the flues is high; the many joints bring the risk of imperfect sealing if the flue is moved. The use of premanufactured units of up to 2 m^2 gives a higher material cost but a lower cost of moving. These units are made from suitable materials and are usually supplied unfired and are assembled in that condition; sections may be joined by cementing or in units with tongues and grooves which are subsequently joined together with cement. It is also possible to use ramming mixes and make the parts on site; this gives low material costs and also makes construction in several layers possible.

The layout of flues depends on the type of furnace and the local conditions. Figure 3.80 shows layouts for an end-fired furnace with connections to burners and flues on the same side in (a) and on opposite sides in (b). Normal layouts for cross-fired furnaces are given in Figure 3.81: (a) shows the flues joining the regenerator at its front end – this is simple to construct; individual dampers to influence the utilisation of the checkers can be installed if the connections are taken separately from the side of the regenerator, as shown in (b), and for a multichamber regenerator with this type of layout all the flues must eventually combine into one collecting flue situated externally; one waste gas flue, as shown in (c), is sufficient provided the combustion air is introduced in its own pipework, including all branch piping.

If the regenerator height is limited for any reason, or if waste heat recovery is to be increased, then secondary regenerators operating with horizontal flow are inserted into the flue. The use of firebricks is adequate in view of the low operating temperatures and their dimensions will be governed by the space available; Figure 3.82 shows an example using stretchers. The use of secondary regenerators requires widening of the flues. For end-fired furnaces the layout shown in Figure 3.80(a) is suitable provided the secondary regenerator is located under the furnace and the chimney flue under the refining or the working end. For cross-fired furnaces it is recommended that the secondary regenerators be connected from the side of the working end and that they be placed under the melting end. Easy access must be provided for the thorough

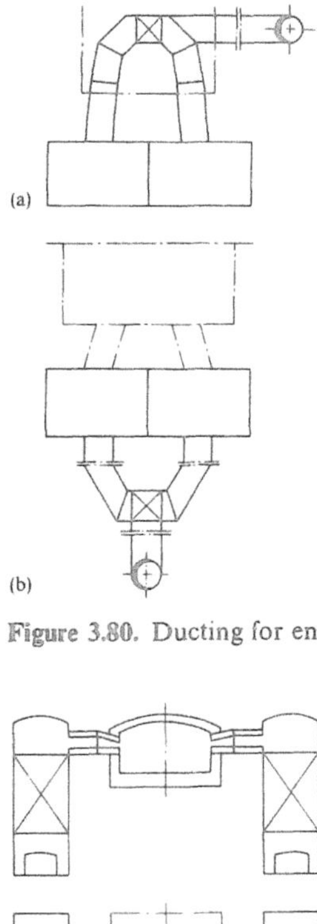

Figure 3.80. Ducting for end-fired tank furnaces

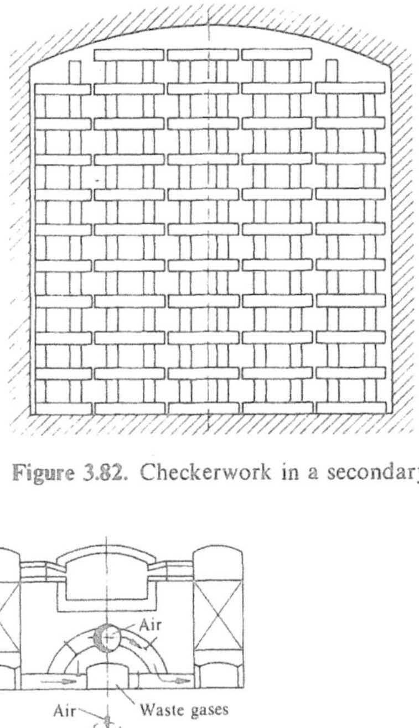

Figure 3.82. Checkerwork in a secondary regenerator[43]

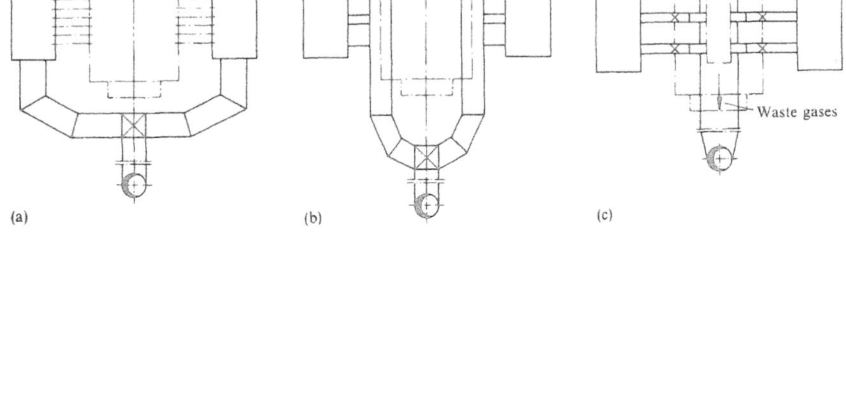

Figure 3.81. Ducting for cross-fired tank furnaces
(a) waste gas ducts coming from the regenerator end wall
(b) system of dual ducts
(c) waste gas collecting duct with separate air entry

cleaning of these regenerators as dust is readily deposited in them.

3.6 The reversal mechanism

Regenerative systems used with glass furnaces require mechanisms for the alternate opening and closing of flues;[43] these are referred to as reversal valves and a diagram of them and their associated equipment is shown in Figure 3.83.

The sizes of the reversal valves for waste gases and combustion air are in line with the relatively large cross sections of flues and pipes; however, standard fittings used generally in industry can be used for natural gas and oil. The reversal process itself is described in Section 5.3 and only the actual components are introduced here; they must function in accordance with a precise schedule at predetermined intervals.

The size of the installation and the optimising of the sequence and time of the reversal process calls for mechanical activation by auxiliary energy sources such as electric power, compressed air, or hydraulics via a central control system; in an emergency, manual operation is envisaged. The reversal drums used in former days, when producer gas was used as the source of heat, are now found only with small old fashioned tank furnaces.

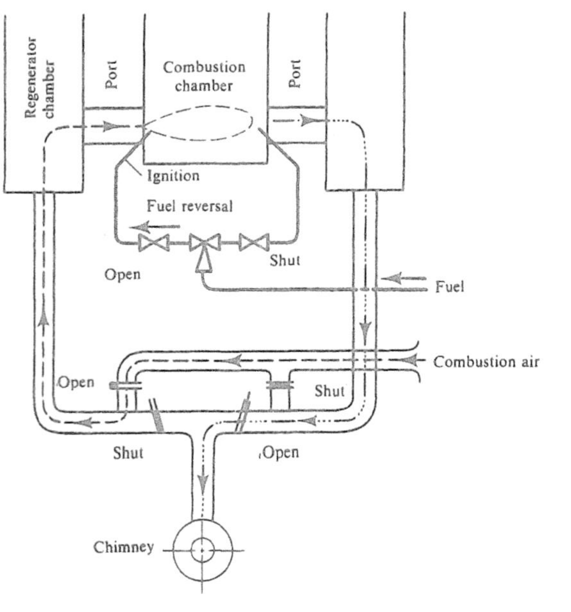

Figure 3.83. Reversal mechanism for oil firing (firing from the left). Further details are to be found in Table 5.12 and Figure 5.40

Occasionally one still finds Forter valves used as the reversal mechanism. A three-way valve, invented as early as 1900 and continuously improved since, it is a simple mechanism which in an improved version was provided with an emergency shut-off (Figure 3.84). It is sealed by water, just as is the reversal drum, but this causes corrosion in the sealing area. Further disadvantages are the high draught losses caused by the many changes in flow direction and the difficulties of carrying out repairs. On small furnaces the valves can be worked by hand; large furnaces call for drive motors.

An improvement derived from the Forter valve is the two-level damper, two of which are combined to form one waste gas/air reversal mechanism (Figure 3.85). Its great advantage is the saving in space when used in this way as the dampers can be sited conveniently and operated by a single mechanism. The mechanical coupling of the two dampers ensures positive reversal but there is a disadvantage in the movement of the dampers as no account can be taken of interference with or delays in particular movements. An improved design is shown in Figure 3.86; this layout of flues gives better flow. Modern installations are suitable for waste gas temperatures up to 800°C without water cooling.

Developments on modern large tanks have led to single valves which fulfill only one function, and this permits a certain latitude in the operating and switching of fuel, air, and waste gas routes. The amount of equipment increases but automation of the reversal becomes easier; however, if the automatic processes fail, higher demands are made on personnel. Draught losses are lower than with the two-level damper systems as the design makes flow easier. Problems of corrosion are also less as the

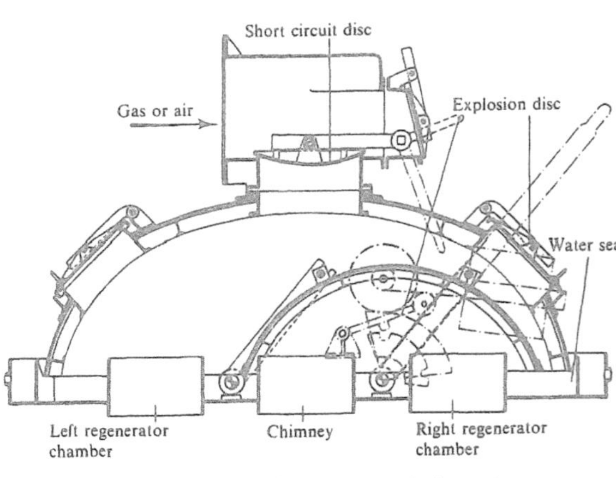

Figure 3.84. Forter valve (Zimmermann & Jansen)

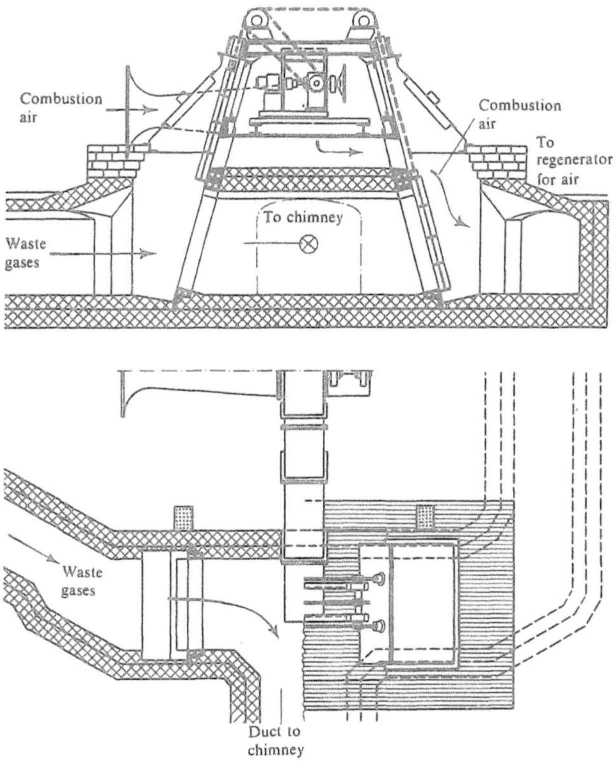

Figure 3.85. Two-level reversal system with dampers in the ducts opposite one another (Zimmermann & Jansen)

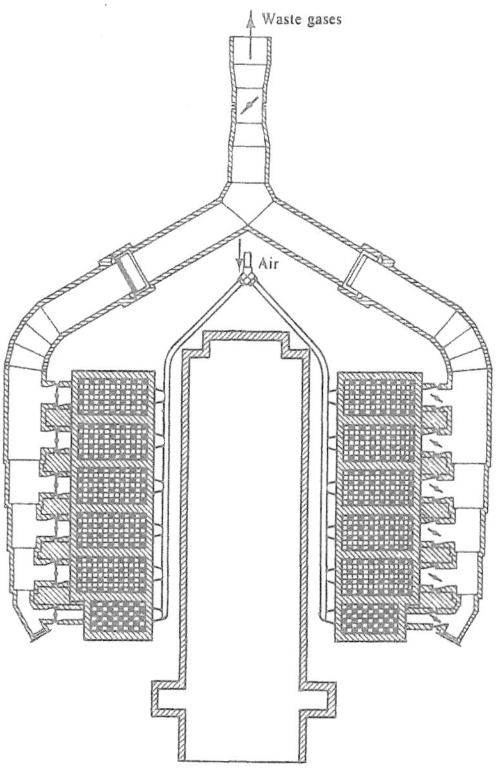

Figure 3.87. Butterfly valves for shutoff and control of waste gases from multichamber regenerators[43a]

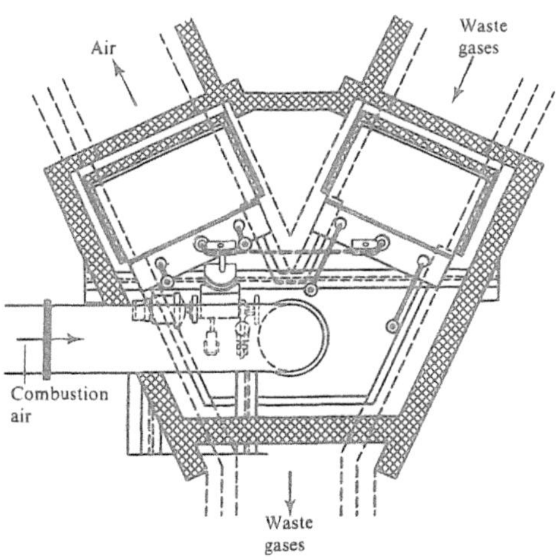

Figure 3.86. Two-level reversal system with waste gas ducts meeting at an acute angle[43]

Figure 3.88. Butterfly valve for shutting off waste gases[43]

temperature of the waste gas falls below the dew point for only comparatively short periods.

Butterfly valves for controlling waste gas flow are located in the chimney flue and, in the case of multichamber regenerators, downstream from each chamber (Figure 3.87). They serve only as regulating valves and have no function as part of the reversal mechanism. In terms of construction, they are tight fitting (Figure 3.88) and are used in flues of brickwork or brick-lined pipework. Their simple construction is an advantage but a disadvantage is that an absolutely gas-tight fit cannot be obtained.

3.7 Chimneys[44-46]

3.7.1 Sizing of a chimney

The draught and pressure conditions existing in the combustion space and waste gas regions of a regenerative tank furnace are given in Figure 3.89.

In addition to the hot chambers, draught is created by the chimney with or without being induced. In old times chimneys were sized to create the draught needed for a given furnace installation; nowadays the height of a chimney and the exit velocity of waste gases are governed by regulations covering emissions.

The natural draught of a chimney is the result of the difference in atmospheric pressure, Δp, between the internal pressure at the base of the chimney, p_i, and the external pressure, p_a. The hot waste gases in the chimney have a lower density, ρ_A, than the outside air, ρ_L. The relationship between these is

$$\Delta p = p_i - p_a = gh(\rho_A - \rho_L), \qquad (3.12)$$

the change in density as a function of chimney height, h, being ignored. Losses due to flow and to acceleration of the waste gases reduce the pressure difference by 10–15%. The rate of cooling of waste gases in a chimney can be taken as 2 degC/m although at low pulls it can be 5–10 degC/m. The exit velocity of gases is now normally 10–15 m/s.

Chimneys can be divided according to the waste gas temperatures to which they will be subjected; cold chimneys up to 100°C, low temperature chimneys 100–200°C, medium temperature chimneys 200–300°C, and high temperature chimneys above 300°C: glass furnaces usually require chimneys of the last two types. If fuels containing sulphur are used (e.g. oil), a chimney of acid-resisting materials is required and the temperature of the waste gases in the chimney should always be above the dewpoint of SO_3, 80–150°C. The condensation of SO_3 can lead to the formation of gypsum ($CaSO_4.2H_2O$) which, because of its insolubility, causes undesirable swelling deposits to form.

The exterior of chimneys can be made of brickwork or reinforced concrete, or a mixture of both. Brickwork is cheaper for heights up to 100 m but above this, reinforced concrete is preferred because of its lower weight, shorter time for construction, and less need for maintenance.

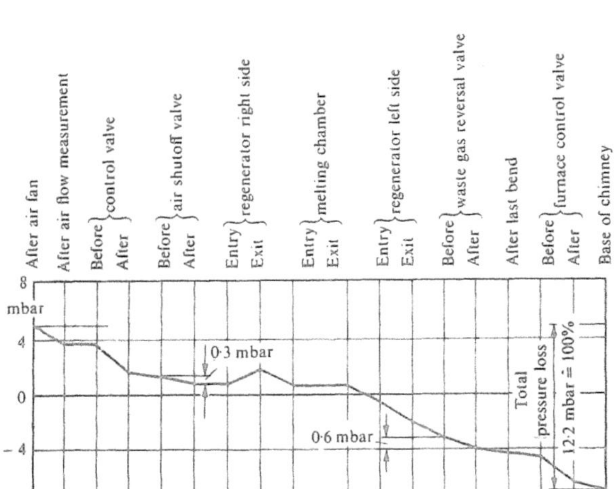

Figure 3.89. Pressure changes in a large sheet glass tank furnace[43]

3.7.2 Parts of a chimney

The foundation is a circular slab of reinforced concrete of the required load bearing strength and the chimney itself consists of a load-bearing part and a lining. The design limit of the compressive strength of bricks (350×10^5 N/m^2) must not be exceeded and for economic reasons the actual load should not be far below this value. Cracks due to temperature fluctuations can be avoided by including in the brickwork reinforced concrete rings with specially-shaped reinforcing steel parts. If the chimney is built of reinforced concrete, then a design using reinforced blocks is preferred to a monolithic construction because of its better corrosion resistance. Thermal stresses should be minimised by ventilating the load-bearing part of the chimney.

The load-bearing part of the chimney must be protected from attack by the waste gases by an acid-resisting, largely gas-tight tube through which the waste gases flow; this is called the lining. In brick chimneys this lining should extend to the chimney outlet and it must be taken to the outlet in a reinforced concrete chimney. The lining is usually made in sections; only in cases where the risk of acid attack or condensation is high is the lining made as one continuous tube. For use up to 200°C the lining can be made of densely fired clinker (hard brick); for higher temperatures or under conditions of strong condensation the lining should be built of acid-resistant firebrick. The anticipated thermal stresses must not exceed 16×10^5 N/m^2 anywhere and if this stress is likely to be exceeded, then a heat barrier must be included in the construction; Moler bricks have been used successfully and foamed glass is useful when there is a high rate of condensation but loose insulation is not recommended as it can lead to sagging.

The outlet of a chimney is always constructed of acid-resisting material: if the load-bearing part is of clinker (hard brick), then the joints are of synthetic resin cement; if of reinforced concrete, then acid-resisting paint is applied, or the outlet is constructed of acid-resisting firebricks or hard bricks. The height should be at least three times that of the diameter of the load bearing part of the chimney. The outlet itself is covered by special shapes of acid-resisting firebrick or stainless steel but expansion of the lining must not be restricted and ingress of water must be avoided.

3.7.3 Chimney equipment

The outside of the chimney must be provided with metal steps or a ladder with back hoops; near the top stainless steel or lead-covered steel must be used in line with the use of acid-resisting materials. The inside must also be provided with steps or means for mobile inspection (e.g. a suspended cage attached to a winding mechanism) as this, coupled to regular equipment checks, is now preferred. A two-level steel platform is attached at the top of the chimney and the distance between the two levels should not exceed 1·5 m.

Lightning conductors as required under applicable building regulations are obligatory; the steel ladders may sometimes be used instead of special lightning conductors.

In the vicinity of airports special markings, such as rings painted in white and red, and/or lights are required.

Authorities dealing with environmental protection call for openings to be constructed at various heights for taking samples and measuring smoke density.

Chimneys with ventilated interiors should have temperature sensors fitted near the cooling air exit so that the temperatures of both the cooling air and the waste gases may be measured.

Chimneys constructed with acid-resisting lining must be allowed to dry out before being taken into service so that the jointing cement hardens sufficiently. During this period the lining must be protected against frost and damp.

3.8 The supporting steelwork

Glass furnaces are large, sometimes very large, installations constructed from selected materials. The various parts of the construction are consolidated and held in place to take the often considerable loads by surrounding the furnace proper with extensive interlocking steelwork, generally referred to as bracing. The starting point of all bracing for glass furnaces is a grillage of I beams resting on piers and beams, the sizing of which is governed by the laws of static loading and regulations covering steel constructions. In the vicinity of a furnace all buildings and support structures must be surrounded by fireproof materials to a height of 1·5 m. No supply piping must lie in zones of possible glass breakout, nor must any oil, gas, or compressed air lines be located in the brickwork.

3.8.1 The bracing of the crown

Next to piers and grillage, the bracing of the crown is the most important part of the steelwork of every furnace and the forces involved here can be very considerable. Expansion during heat-up must be carefully controlled as the crown can only last through a furnace campaign when it is free of cracks and open joints. Apart from very small furnaces and some special furnaces, the crown must be supported and braced independently and must not be carried on sidewalls or ports; the replacement of worn parts below the crown must be possible without replacing the crown. Since the operating life of steelwork is very long, new developments take a long time to become established in practice.

3.8.1.1 Buckstays and their hinged supports

Buckstays are attached at their base in line with the width of the crown to beams carrying the grillage by pivoted supports (Figure 3.90). The skewbacks of the crown are located in bearers of channel or angle iron resting on welded or cast brackets attached to the buckstays. The thrust of the crown is taken by tie rods between opposite buckstays. A drawback of this design is that as a result of expansion during heat-up and the releasing of tie rods to take account of it, the buckstays tip outwards and it is difficult to avoid cracks forming in the crown. An improvement would be the use of an adjustable skewback bearer (Figure 3.91).[47] The

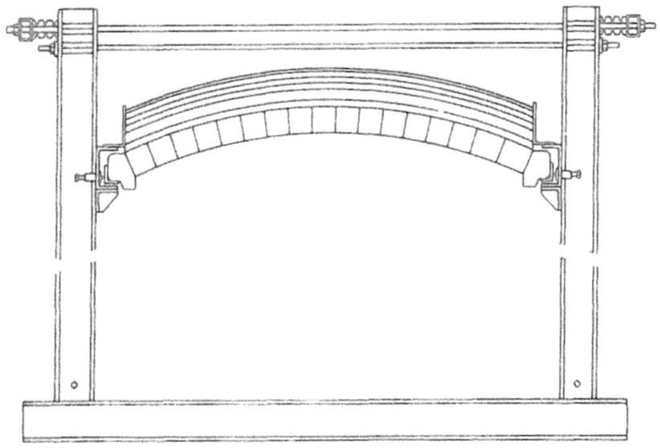

Figure 3.90. Crown bracing with hinged buckstays

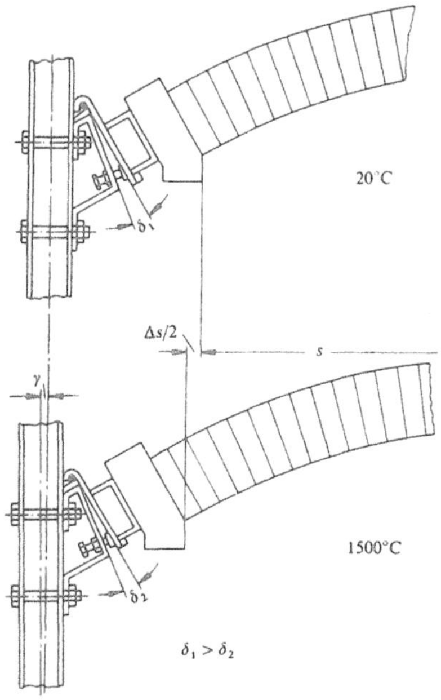

Figure 3.91. Suggested design for hinged bracing of the skewbacks[47]

problem with this design is that during the critical stages of the heat-up not only must the tie rod nuts be loosened to permit the stanchion to move through the angle γ, but the adjustable plates must also be corrected from δ to δ_2 to prevent joints in the crown from opening. For fusion cast crown blocks special skewback bearers are used so that compensation for the differences in width of each arch can be made independently[48] (see also Section 4.3.1). The tipping of buckstays and skewback bearers can be avoided if the buckstay support is placed in a shoe on roller bearings (Figure 3.92).

The distance between buckstays depends on the size and the number of burners; generally speaking, the aim is to have a distance not exceeding 2 to 3 m to avoid deformation of the profile of the skewback

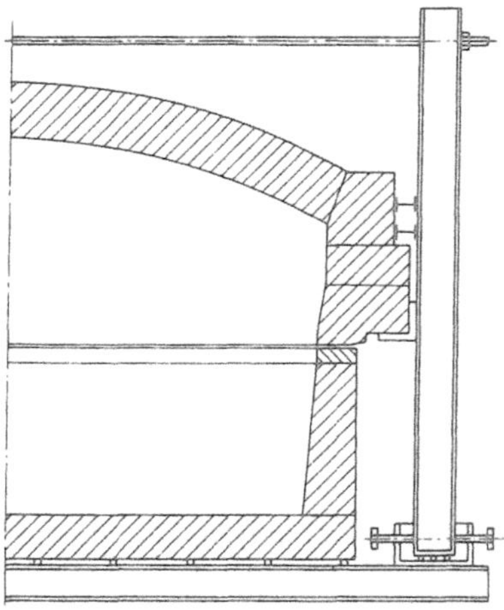

Figure 3.92. Buckstay with roller bearing and buckstay shoe

bearers. The buckstays themselves can be made from one or more joists of different cross section (Figure 3.93).[49] It has also been proved advantageous to locate the buckstays some distance from the furnace sidewalls to prevent them from overheating and to provide better access. In consequence special designs of the supports for skewback bearers are required to bridge the additional distance.

A significant improvement in the bracing of crowns resulted from the insertion of compression springs between the tie rod and the pendulum support. The springs prevent forces in the crown from increasing too rapidly even if the tie rods are not released at the correct time: disc springs have been found very suitable and up to 50 such springs can be formed into an assembly (Figure 3.94).

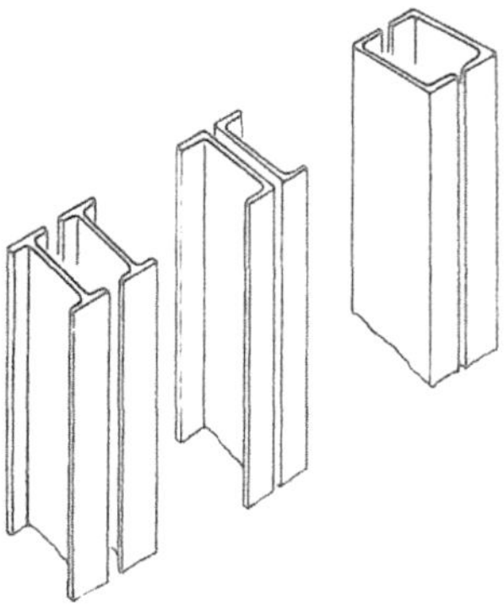

Figure 3.93. Examples of buckstay designs[49]

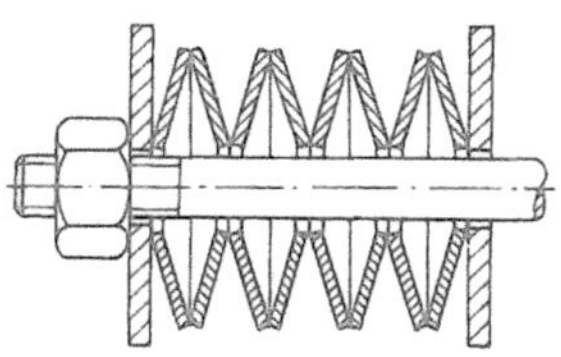

Figure 3.94. Set of disk springs

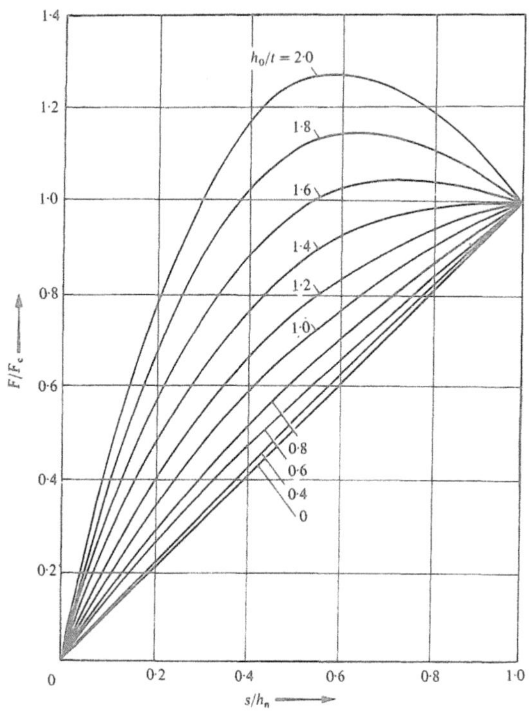

Figure 3.95. Load deflection data for disk springs (Bauer)

Table 3.8. *Dimensions and data for disk springs at 0·75 s/h_0 from DIN 2093*

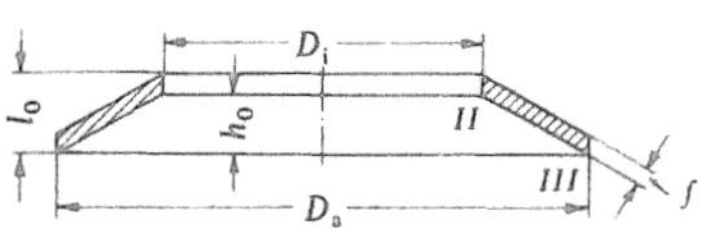

D_a	D_i	t	L_0	Force, F (N)	Deflections (mm)	h_0/t
112	57	3·0	6·90	10 700	2·95	1·3
112	57	4·0	7·20	18 300	2·40	0·8
112	57	6·0	8·50	45 000	1·88	0·42
125	64	3·5	8·05	15 900	3·40	1·3
125	64	5·0	8·50	31 000	2·65	0·7
125	64	8·0	10·60	88 000	1·95	0·33
140	72	3·8	8·7	20 200	3·70	1·4
140	72	5·0	9·0	28 500	3·00	0·8
140	72	8·0	11·2	87 000	2·40	0·4
160	82	4·3	9·9	25 600	4·20	1·3
160	82	6·0	10·5	42 000	3·40	0·75
160	82	10·0	13·5	140 000	2·60	0·35

Depending on dimensions and design, the springs can have linear or progressive characteristics (Figure 3.95), the latter being very suitable for the bracing of crowns as the pressure can be held almost constant within certain limits. Dimensions and data covering selected disk springs covered by DIN 2093 are assembled in Table 3.8. The required spring deflection can be obtained by design of the assembly of springs: single springs give constant force; by using two or three springs the forces can be doubled or trebled. Disk springs can be obtained in corrosion-resistant and heat-resistant steels.

The reader is referred to DIN 2092 for calculation of the forces and tensions. Under static loading the pressure on the upper edge of a disk spring when in the flattened condition should not exceed 3000 N/mm^2. The edges under risk of failure are at II and III (Table 3.8); for $h_o/t > 0{\cdot}8$ the maximum tension is at III while for $h_o/t < 0{\cdot}8$ it is at II. The tensile stresses at II and III for the flattened spring should not exceed 1600 N/mm^2.

3.8.1.2 Rigid crown bracing

The rigid crown bracing used originally for the Siemens–Martin furnace has in recent years been transferred to the glass furnace. This design allows precise control even of the very large crowns 12 m or more wide such as are used for flat glass tanks. The increasing and now almost universal use of insulated crowns makes high demands on the control of thermal expansion during heat-up. In the rigid construction the crown is suspended at the skewback bearers: one side serves as a fixed point and is supported against the buckstay while the other side is supported by a strong adjustable spring and serves as a moving bearing (Figure 3.96). Disk springs are used in preference to coil springs, which are less suitable because of the high pressures encountered.

Details of a skewback bearer construction for a suspended crown with disk springs are shown in Figure 3.97. The springs are only used during heat-up and afterwards fixed bolts are adjusted to hold

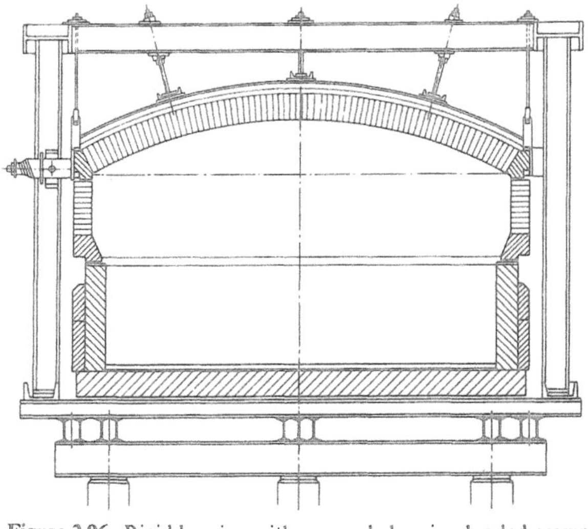

Figure 3.96. Rigid bracing with suspended spring-loaded crown

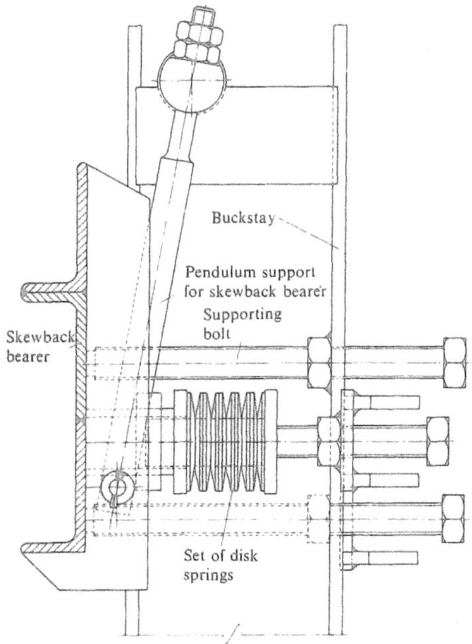

Figure 3.97. Skewback bearer for suspended crown with disk spring bracing (Sorg)

the skewback bearer in position, and the thrust of the disk springs is removed. This avoids fatigue of the springs and makes the use of high temperature resistant and therefore expensive springs unnecessary. A modified design of suspended crown is sometimes used for electric furnaces.

A variation of the rigid crown bracing is the use of rigid buckstays with sliding supports for the skewback bearers and furnace sidewalls. Steel rods are used as rollers between the load bearing refractory and the support to give easy movement. The horizontal movement of the crown is taken up by pressure bolts located one above the other and any temporary tilting of the skewback bearers as a result of differential yielding of those bolts can be tolerated. The gap between sidewall and skewbacks is chosen such that it completely closes during heat-up; a slight lifting of the skewback bearers is acceptable. Spring-loaded pressure bolts are not needed for crowns up to 8 m wide. A disadvantage of this design for side-fired furnaces is that the pressure bolts for the skewbacks are difficult to adjust because of their location between the burners. If the heat-up is controlled correctly it is possible to avoid tilting of the skewback bearers and formation of cracks in the crown with this design. Adjustable pressure bolts, with or without assemblies of disk springs, take up the movement of the crown.

The positioning of the sidewalls and of the pressure bolts holding the tank blocks is not influenced by changes to the crown during heat-up; undesirable tilting of the sidewall supports is also avoided (Figure 3.98).

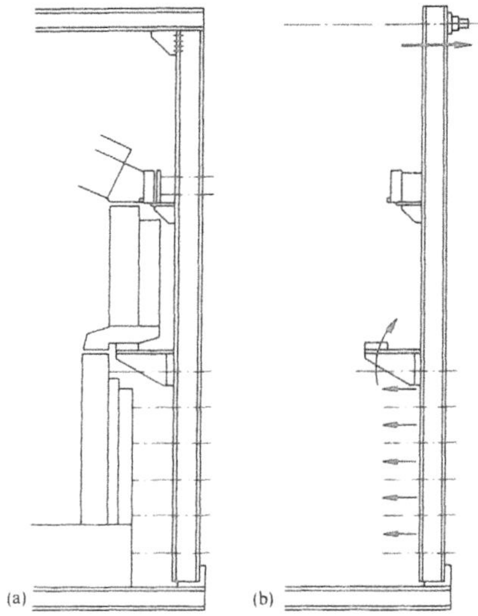

Figure 3.98. Types of bracing
(a) rigid frame bracing, no tilting
(b) buckstay with tilting

3.8.2 Bracing steelwork for walls

3.8.2.1 Superstructure sidewalls

The sidewall steelwork consists essentially of a support structure for the tuckstone attached to the buckstays carrying the crown; special steelwork or angle iron is used. Good sealing and thermal insulation between tuckstone and steelwork are important (Figure 3.99). In large tank furnaces,

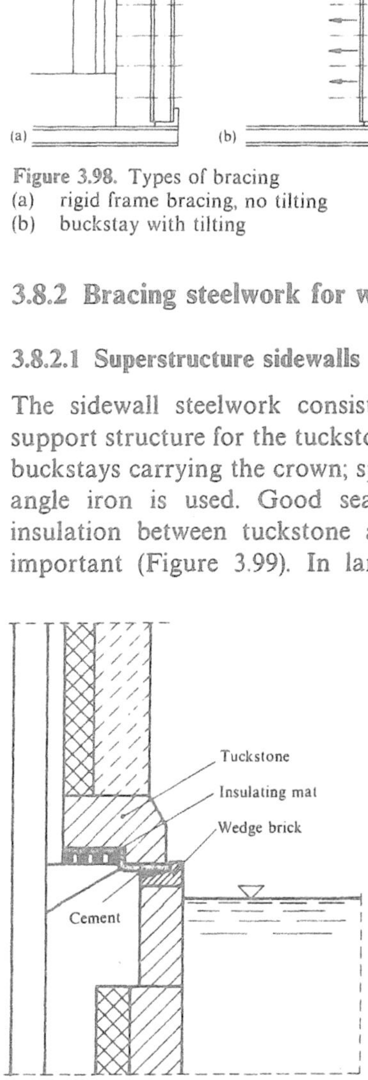

Figure 3.99. Supporting steelwork and sealing in the area of the tuckstone[1]

joints can open on the inside as a result of longitudinal expansion.

3.8.2.2 Tank blocks

The walls containing the glass are made of blocks, now mainly soldier blocks, which must be heated up to give tight joints and must be supported to prevent the hydrostatic pressure of the liquid glass from pushing the blocks outwards. Threaded rods attached to the buckstays usually secure the tank blocks in position via angle or U iron (Figure 3.100) and any adjustments are made by nuts on the threaded section of the rod. Furnaces with electric boosting or all-electric furnaces require electrical insulation between tank blocks and retaining steelwork. For example, square-edged firebrick held in channel iron may be used (see the insert in Figure 3.100).

Rapid changing and adjustment of retaining bolts are facilitated by fixing tubes to the buckstays.

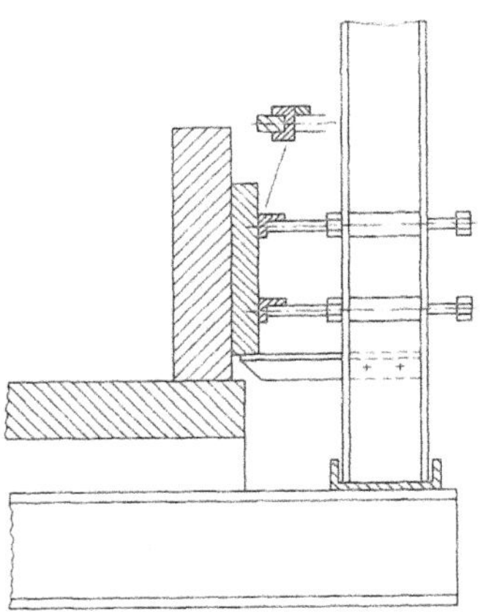

Figure 3.100. Example of tank block bracing

3.8.2.3 Steelwork for end and gable walls

This consists of stanchions similar to the buckstays used for the crown fastened together by horizontal steelwork. The stanchions are connected between gable and end walls by tie rods so that the expansion joints of the crown and the joints between crown and vertical walls may be closed tight after heat-up. The tight closing of these joints has a major influence on the life of the furnace.

3.8.2.4 Steelwork securing the doghouse

The need for sound construction of this steelwork increased in importance as doghouses became wider (Figure 3.101). If possible, this steelwork should be separate from that for the crown, so that each can be adjusted separately. The shear forces emanating from the crown can be considerable and a correspondingly sound construction of doghouse steelwork is therefore required.

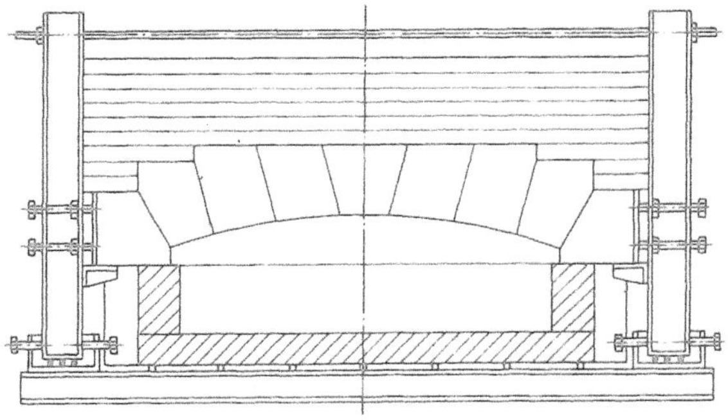

Figure 3.101. Bracing around the doghouse

3.8.3 Bracing of ports

From the point of view of construction, the ports constitute the link between melting tank and regenerator chamber. During heat-up both parts of the structure expand independently and the steelwork for the ports must therefore be constructed in a manner which permits independent movement; the steelwork carrying the port neck is usually on rollers (Figure 3.102). The end of the port where it connects with the regenerator, which is normally unsupported, is carried by steelwork; plates and piers or, better, horizontally placed steel sections and supports are used. The steelwork must leave sufficient access to permit renewal of the ring of

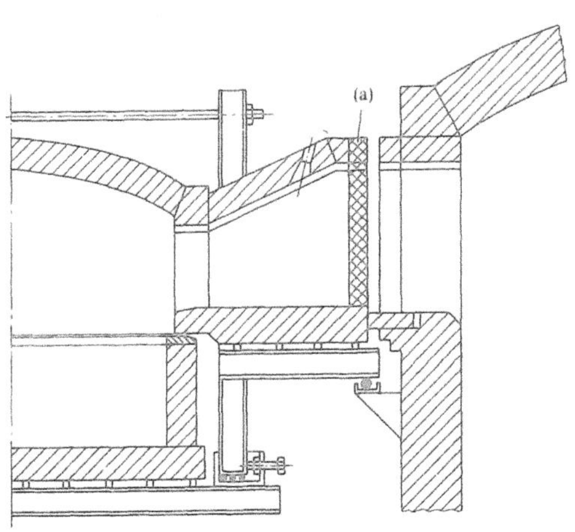

Figure 3.102. Connection between regenerator and port neck with supporting steelwork on rollers and (a) layer of replaceable halfbricks

connecting bricks and for cooling of tank blocks. The remaining steelwork of the port consists of tie rods and vertical stanchions to support the port arch and to make the port structure gas-tight, the stanchions being steel sections 160–400 mm across and the tie rods 20–64 mm in diameter. The port arch is constructed like a small crown with vertical stanchions and corresponding tie rods.

3.8.4 Regenerator bracing

The steelwork of the regenerators stands independently, the large vertical walls being held by strong stanchions connected horizontally with steel sections or tie rods. The steelwork for the arch is usually simpler than that for the melter crown and is placed on the sidewalls, not carried independently. Horizontal expansion is taken up by special skewbacks connected to the stanchions.

An example of the steelwork of a small cross-fired tank furnace with regenerator is given in Figure 3.103.[50]

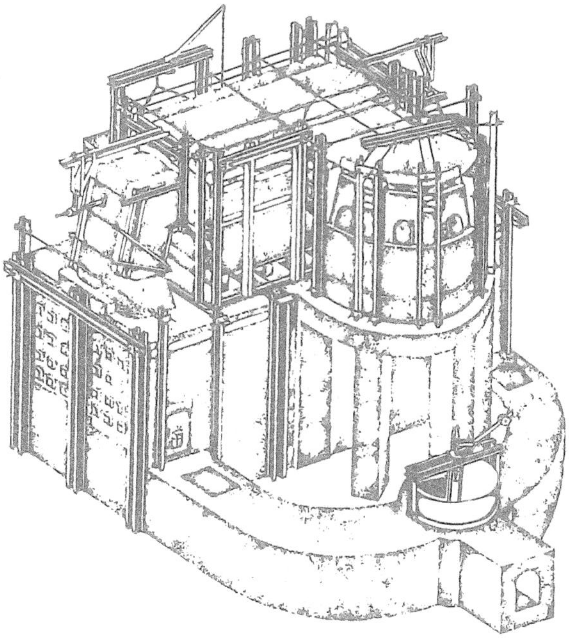

Figure 3.103. Bracing of a small regeneratively heated tank[50]

Table 3.9. *Tie rod and buckstay bolts, metric ISO threads to DIN 13,T1 and DIN 934,T1 (mm)*

Bolts					*Nuts*	
Nominal diameter, d	*Pitch, P*	*Major diameter, d_2*	*Minor diameter, d_3*	*Strain diameter, A_s* (mm^2)*	*Thickness, t (mm)*	*Width across flats, s (mm)*
10	1·5	9·026	8·160	58·0	8	17
12	1·75	10·863	9·853	84·3	10	19
16	2	12·701	11·546	115	13	24
20	2·5	18·376	16·933	245	16	30
24	3	22·051	20·319	353	19	36
30	3·5	27·727	25·706	561	24	46
36	4	33·402	31·093	817	29	55
42	4·5	39·077	36·479	1120	34	65
48	5	44·752	41·866	1470	38	75
56	5·5	52·428	49·252	2030	45	85
64	6	60·103	56·639	2680	51	95

*The definition of A_s is given in DIN 13 T28

$$A_s = \frac{\pi}{4}\left(\frac{d_2 + d_3}{2}\right)$$ where d_2 and d_3 are nominal values

Table 3.10. *Types of steel according to DIN 17100 with guaranteed values of mechanical and technical properties (density 7·85 kg/dm³)*

Type of steel		*Upper yield point* (N/mm^2) for material thicknesses (mm)*					
Reference	*Material number*	*≦16*	*>16≦40*	*>40≦63*	*>63≦80*	*>80≦100*	*>100*
St 37-2	1.0037	235	225	215	205	195	As agreed
St 44-2	1.0044	275	265	255	245	235	
St 52-3	1.0570	355	345	335	325	315	

*The yield point is that stress at which, with increasing extension, the tension remains constant for the first time or decreases. The upper yield point is the highest stress before the first decrease in tension with increasing extension (DIN 50145)

3.8.5 Calculations

Only indications covering steelwork calculations can be given here. It is necessary to differentiate between those parts of the steelwork which are under a permanent fixed load, such as the buckstays carrying the crown and members carrying the weight of other parts of the furnace, and those which hold the framework together. These, the tie rods, serve only to consolidate the structure during heat-up and cooling and to ensure the continuing absence of gas leaks through the brickwork.

Load-bearing forces can be predicted relatively easily. However, this is not so for tie rods or other members fulfilling a similar function and their strength is chosen on the basis of experience at a level as high as the load-bearing steelwork. Calculations for the load-bearing steelwork are made on the basis of ordinary mechanics such as used in high rise and machine construction. It is not necessary to assume lower requirements for the construction of a glass furnace or to make any specific adjustments or safety factors. If the buckstays are extended into the foundations or are associated with carrying an unsupported crown, they must be checked for buckling; buckstays of suspended crowns must be checked for pressure and bending.

The supports for skewback bearers and tuckstones must be checked for buckling, the supporting steelwork itself for bending, and the bolts for the possibility of stripping. Dimensions and extension limits taken from DIN 13, DIN 934, and DIN 17100 have been assembled in Tables 3.9 and 3.10.

The permitted extension limits are a fraction of the extension limits of the materials used for construction and the factors of safety used are between 3 and 4 (Table 3.10). Reserves for scaling, flame corrosion, etc. are not made when calculating for tension but are allowed for when choosing the cross sections. For steelwork of correct design the influence of temperature need not be taken into account as 200–300°C will not be exceeded. Should these temperatures be exceeded in any section, then the permitted strength values should be reduced accordingly.

4. Selection and use of refractories in glass tank furnaces

4.1 Selection criteria

The primary objective of a glass making plant is to manufacture glass of the required quality and in the required quantity at the lowest attainable cost. In meeting these objectives, the selection and correct use of refractories is of major importance.

The requirements for the glass are dictated not only by the type of glass but also by the section of the furnace being considered. In the melting zone the formation of small stones or other reaction products can be tolerated since the subsequent homogenisation process ensures the dissolution of these stones and their uniform distribution in the molten glass. If this were to happen in the working end it would create a considerable problem since no homogenisation can follow and the contaminating material would appear as faults in the glassware.

The different mechanical, physical, and chemical conditions existing in each part of a furnace require matching by the properties of refractories in order to obtain a reasonable furnace life; another factor is the behaviour of one refractory in contact with another.(1)

A further factor is price, not only relative to the refractory as a material but also to the influence of shape, i.e. standard shapes, possibly available ex stock, as against special shapes.

Generally speaking, the selection of refractories for the different parts of a furnace should aim at achieving uniform wear so that the furnace as a whole achieves an adequate life. If rapid wear in particular locations cannot be avoided (e.g. at the flux line), then facilities for repair must be taken into consideration.

4.2 Compositions and properties of the more important refractories(1–3)

4.2.1 Clay blocks, firebrick

For glass furnaces, refractories based on clay have lost much of their former importance, but are still used in less critical positions. They are made from new and burnt clay and depending on their alumina content, are classified as low, medium, or high alumina clays. Table 4.1 gives the most important chemical and mineralogical data. Low porosity is an objective, but this reduces the resistance to thermal shock. Data covering thermal expansion and thermal conductivity are given later in Table 5.10.

4.2.2 Sillimanite and mullite

Sillimanite and mullite have a higher refractoriness under load than clay. Such dense and fine-grained materials show a low tendency to create gas bubbles in glass and are therefore suitable for feeder channels, rotating cylinders, plungers, and dies and rings at outflow positions. They are now only used rarely for tank blocks. The properties of different sillimanite blocks can be seen in Table 4.2.

4.2.3 Zircon

The chemical composition of this material approximates to that of zirconium silicate, i.e. 67% ZrO_2, 33% SiO_2, and its refractoriness under load is even higher than that of sillimanite. Its resistance to thermal shock depends to a considerable extent on density and grain structure. This material is especially suitable in areas of high attack in the superstructure of glass furnaces but for use as tank blocks it is suitable only for low alkali glasses. Properties are given in Table 4.3.

Table 4.1. *General properties of fireclay blocks (in the main taken from Stahl-Eisen-Werkstoffblatt 910–64)*

Type	Al_2O_3 (%)	TiO_2 (%)	Fe_2O_3 (%)	$CaO + MgO$ (%)	$K_2O + Na_2O$ (%)	Seger cone	Density (g/cm³)	Bulk density (g/cm³)	Total porosity (%)	Cold crushing strength (N/mm²)	Refractoriness under load t_a(°C)	t_e(°C)	Thermal shock resistance	Approximate upper temperature limit of application without chemical effects (°C)
High alumina fireclay	42–46	~ 1·5	0·6–1·5	< 0·6	0·4–1·2	> 34	~ 2·74	2·25–2·4	12–18	35–85	≮ 1450			1450
AO p		1·0–3·5		About 0·5–1·0			2·55–2·75	p, 1·8–2·0; h, 1·9–2·1; t, 2·0–2·2	26–30	10	1450	About 150–200 degC above t_a	As delivered, types AO–AI are better than AII–AIII; also dry formed are better than half dry or plastic formed types	
AO h	> 42		< 3·0		< 2·5	> 34			22–27	15	1430			1400
AO t									19–25	25	1450			
AIS p									26–30	13	1370			
AIS h	40–42		< 2·5		< 2·8	> 33/34			22–27	20	1400			1350–1400
AIS t									19–25	30	1420			
AI p									26–30	13	1350			
AI h	37–40		< 2·5		< 3·0	> 33			22–27	20	1380			1300–1350
AI t									19–25	30	1400			
AII p									25–30	13	1330			
AII h	33–37		< 2·5		< 3·0	> 32			22–26	20	1350			1250–1300
AII t									19–25	30	1370			
AIII p									23–30	15	1300			
AIII h	30–33		< 3·0		< 3·0	> 30			21–25	25	1320			1200–1250
AIII t									19–25	30	1340			
High silica fireclay	22–30					> 26			< 30	> 10	1250			1100

Table 4.2. *Properties of sillimanite refractories*

		Ceramically bonded sillimanite Great Britain	German Federal Republic	German Federal Republic	USA	USA	*Khasi sillimanite India*
Chemical composition (wt%)							
Al_2O_3	56·4	59·3	64·5	65·5	59·5	59·9	
Fe_2O_3	0·96	0·80	1·0	0·71	0·81	0·92	
TiO_2	1·12	1·13	0·5	1·13	1·19	1·30	
SiO_2	39·5	38·1	33·4	31·9	37·6	35·7	
CaO	0·28	0·08	–	trace	0·32	0·34	
MgO	0·88	0·17	0·1	0·26	0·18	trace	
Na_2O	0·26	0·19	0·1	} 0·46	0·40	trace	
K_2O	0·51	0·24	0·4			0·07	
Loss on ignition	–	–	–	–	–	2·13	
Physical properties							
Density	g/cm^3	3·04	3·05	3·22	3·25	3·05	3·16
Bulk density	g/cm^3	2·36	2·34	2·58	2·32	2·35	3·14
Total porosity	vol%	22·4	23·2	19·6	28·6	23·0	0·6
Apparent porosity	vol%	–	20·8	16·2	21·8	21·5	0·4
Permeability	$\frac{10^3\ cm^3\ cm}{cm^2\ s\ cm\ WG}$	20	6·6	15	10	–	–
Refractoriness under load t_a	°C	–	1620	1650	1570	1580	1700
t_e	°C	1710	1720	1720	1770	1720	1700
Cold compressive strength	$\times 10^5\ N/m^2$	–	600	700	–	550	–
Thermal expansion coefficient 20–1500°C	$\times 10^6\ K^{-1}$	–	5·5	6	6·6	5·4	5·4

Table 4.3. *The properties of zircon used in glass melting furnaces*

		Normal zircon blocks	*Dense zircon blocks*
Chemical composition (wt%)			
ZrO_2		65·1 –66·2	64·4
SiO_2		34·4 –32·6	33·4
TiO_2		0·05– 0·25	1·07
Al_2O_3		0·4 – 0·7	0·40
Fe_2O_3		0·05– 0·25	0·05
CaO + MgO		0·03– 0·40	0·11
$Na_2O + K_2O$		trace– 0·11	–
Physical properties			
Density	g/cm^3	4·56– 4·63	4·56
Bulk density	g/cm^3	3·61– 3·67	4·05
Total porosity	vol%	20·2 –22·0	11·2
Apparent porosity	vol%	16 –20	–
Permeability	$\frac{10^{-3}\ cm^3\ cm}{cm^2\ s\ cm\ WG}$	1 – 3	0·5
Refractoriness under load t_a	°C	1730	1730
t_e	°C	1800	1800
Compressive strength at room temperature	$\times 10^5\ N/m^2$	5·5	5·5
Thermal expansion coefficient 20–1500°C	$\times 10^6\ K^{-1}$	1000–1500	3000
Thermal conductivity at 1100°C	W/Km	1·5 – 2·3	3·5

4.2.4 Fusion cast refractories[4]

Apart from the void area, these blocks have a very dense structure of interlocking fibrous crystals, which makes them highly resistant to attack by molten glass, and are now of major importance for glass furnaces. The early corundum–mullite blocks have been superseded by corundum–zircon (AZS), α and β alumina, and, more recently, corundum–chrome blocks; refractories containing 31% and 41% ZrO_2 are now extensively used as glass contact blocks and in parts of the superstructure. Data are presented in Table 4.4. The high proportions of glassy phase which are occasionally found in these blocks, their high thermal conductivity (Figure 4.1), and their thermal expansion should be noted. AZS blocks show an expansion anomaly between 900 and 1100°C which is caused by a

Table 4.4. *Chemical and mineralogical composition and physical properties of fusion cast refractories*

	Corundum–zircon					*Al_2O_3–Cr_2O_3–ZrO_2*	*Alumina*		
	Trade name *ER 1681*	*ER 1682*[(a)]	*Unicor 501*	*Unicor I*	*ER 1711*	*ER 2161*	*JARGAL M*	*JARGAL H*	*ER 5312*[(a)]
Chemical composition (wt%)									
ZrO_2	32·5	32·5	33·8	40·0	41·0	26·0	–	–	–
Al_2O_3	50·6	50·6	49·2	45·6	45·5	31·5	95·0	94·5	87·5
SiO_2	15·6	15·6	15·2	13·0	12·2	13·0	1·2	0·1	8·0
Na_2O	1·1	1·1	1·5	1·1	1·0	} 3·5	3·5	5·2	4·5
CaO, Fe_2O_3, TiO_2, MgO	0·2	0·2	0·12	0·14	0·3		–	–	–
Fe_2O_3	–	–	0·14	0·14	–	–	0·05	0·05	–
Other	–	–	–	–	–	–	0·25	0·15	–
Cr_2O_3	–	–	–	–	–	26·0	–	–	–
Mineralogical composition (wt%)									
Baddeleyite	32·0	32·0			41·0	25·0	–	–	–
Corundum	47·0	47·0			42·0	–	45·0	2·0	–
Chrome–corundum	–	–			–	56·0	–	–	–
β-alumina	–	–	No information		–	–	53·0	97·5	–
β'''-alumina	–	–			–	–	–	–	Main phase
Glassy phase	21·0	21·0			17·0	19·0	2·0	0·5	–
Physical properties									
Density $(g/cm^3)^{(b)}$	3·84	3·84	3·77	3·9	4·09	4·11	3·54	3·26	3·37
Apparent density $(g/cm^3)^{(b)}$	3·72	–	–	–	4·0	4·0	3·40	–	–
Bulk density $(g/cm^3)^{(c)}$	3·50	3·30	3·49	3·5	3·70	–	3·14	2·80	2·90

(a) only suitable for regenerator checkers
(b) density of powder
(c) valid for normal refractories; refractories fusion cast by special methods have higher values

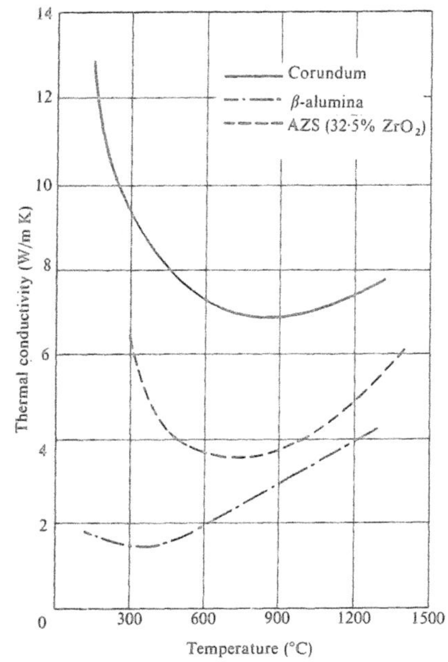

Figure 4.1. Thermal conductivities of fusion cast tank blocks as a function of temperature

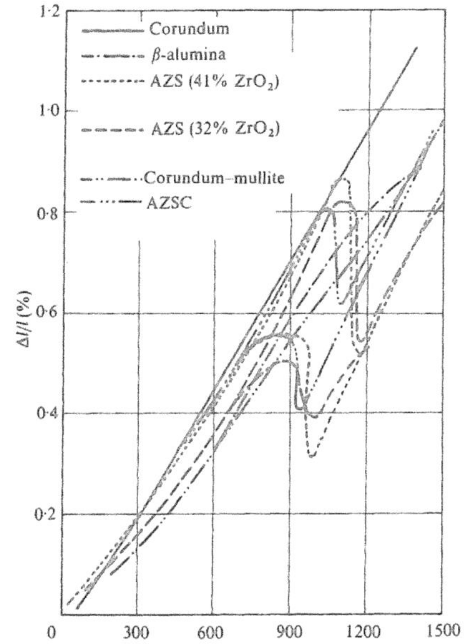

Figure 4.2. Coefficient of thermal expansion of fusion cast refractories as a function of temperature

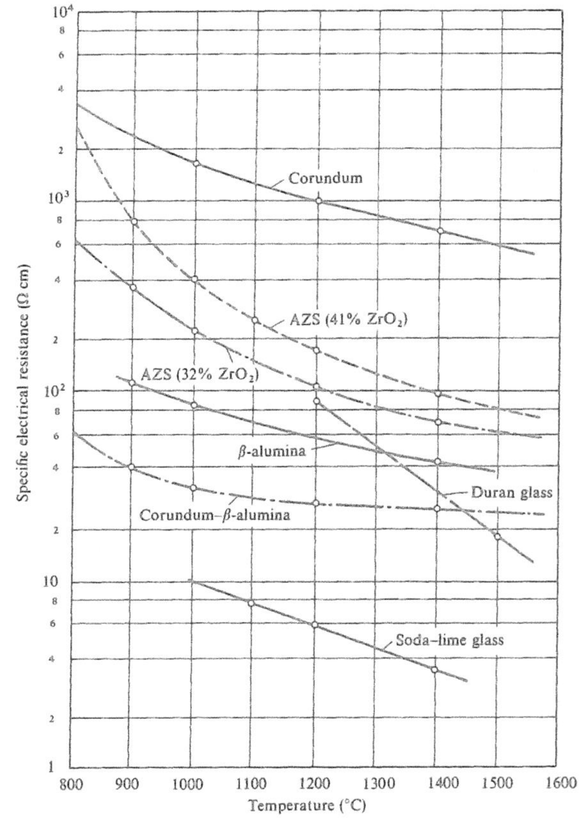

Figure 4.3. Specific electrical resistance of different tank block refractories as a function of temperature[4]

reversible change in the crystal structure of the ZrO_2 from monoclinic into tetragonal; the exact behaviour depends on the ZrO_2 content itself (Figure 4.2).

In electrically heated furnaces the electrical conductivity of tank blocks is important. Some fusion cast blocks have a relatively low electrical resistance (Figure 4.3) so that, when melting glasses of low alkali content possessing a high electrical resistance, it is necessary to use ceramically bonded zircon blocks as electrode blocks and, in some cases, for all glass contact blocks.

4.2.4.1 Ramming materials used with fusion cast blocks

Ramming materials have become very useful for improving the sealing of joints, for forming a separating layer for hot and cold repair work, and, especially during construction, for forming separating layers and special shapes. Special ramming materials have been developed for use with fusion cast refractories; they match their properties and can prevent penetration and upward drilling by glass in furnaces of high throughput operating at high temperatures. Table 4.5 gives the properties and applications of a range of such materials. It is important to follow the manufacturers' instructions for their use.

Table 4.5. *Ramming materials, cements, etc. used in contact with fusion cast refractories*

	Group and mineralogical composition after firing						
	Corundum–mullite–zirconia	*Corondum–zircon*	*Corundum*	*Corundum*	*Corundum–zirconia*	*Corundum–zirconia*	*Zircon*
Trade name	ERSOL 06L	ER 66	ER 67	ER 67H	ERPLAST 121	ERPLAST 221	ERPLAST 271
Type of material	Castable	Cement	Cement	Cement	Cement	Mortar	Mortar
Chemical composition (wt%)							
Al_2O_3	48	35·8	88·2	93	77	77	0·4
ZrO_2	30	34	–	–	15	15	60
SiO_2	20	25	7	–	1	1	31
Fe_2O_3	} 2	0·5	0·3	0·1	0·1	0·1	0·1
P_2O_5		–	–	2·4	2·4	2·4	4·5
Loss on ignition		4·7	4·5	4·5	4·5	4·5	4·0
Loss (%) at							
1300°C	–	1·3	1·2	1·2	1·2	0·5	0·2
1500°C	–	–	–	2·0	2·0	1·2	1·8
Expansion 0–1000°C (%)	0·5	No binding	No binding	0·7	0·7	0·56	0·8
Areas of application	Setting and sealing of paving blocks	Setting and repair of paving blocks in melting zone	Setting of paving blocks in working end and, for flat glass, in melting area zone	Sealing between fusion cast blocks and insulation	Tank and superstructure: sealing between fusion cast blocks and insulation	Hot and cold repairs of fusion cast refractories, ramming material under paving blocks	Hot and cold repairs of superstructure; separating layer between silica and fusion cast refractories

4.2.5 Silica

Despite the higher temperatures at which furnaces now operate, silica refractories have not lost their importance for making the crown. Their suitability for use is governed by their porosity and by any residual quartz content which can lead to dangerous aftergrowth of the bricks (Figure 4.4). The various forms of silica and the temperature ranges within which they exist are given in Figure 4.5.

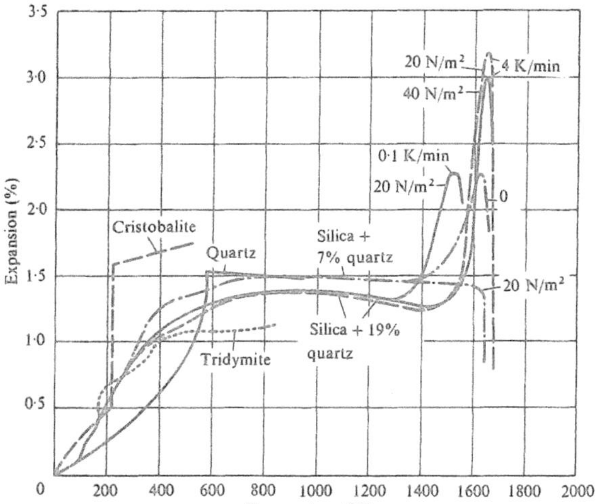

Figure 4.4. Thermal expansion of silica minerals and silica refractories of different thermal histories[7,8]

4.2.6 Basic refractories[6]

Basic refractories have a high resistance to attack by alkali and are now used as checker bricks; these refractories are available in many qualities and data covering them are given in Table 4.6.

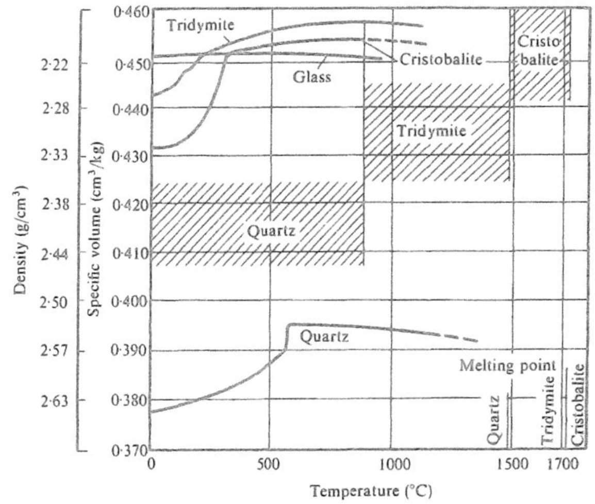

Figure 4.5. Densities and specific volumes of some forms of silica and their regions of stability as a function of temperature[5]

4.3 Refractories, their selection and use in particular cases

4.3.1 Tank furnace crowns

The refractories in tank furnace crowns, especially wide crowns, are subject to high compression. Temperatures on the hot face reach 1600°C or more and, except with some special glasses, the atmosphere contains substantial amounts of alkali. The crown must be stable and therefore the refractories must not creep over many years of use; silica brick has become the preferred material because of its properties and price.

Silica bricks consist mainly of SiO_2 with calcium hydroxide as a binder. Their main constituents are

Table 4.6. *Chemical and physical data and properties relevant to the use of basic refractories*

		Magnesite	*Magnesite–chrome*	*Chrome–magnesite*	*Chromite*	*Forsterite*	*Sintered dolomite*	*Limestone*
Chemical composition (wt%)								
SiO_2		0·2–6	< 5	< 5	< 7	< 40	< 1·5	0·3–1
Al_2O_3		0·1–5	< 11	< 22	< 28	< 8	< 1	0·3–2
Fe_2O_3		0·2–6	5–12	7–15	9–18	< 8	≲ 1	0·3
CaO		0·8–3·9	< 3	< 2·5	< 2	< 1·5	< 61	96–97·5
MgO		85–98	55–80	25–55	< 25	50–60	< 36	0·2–1
Cr_2O_3		–	6–20	15–25	> 25	< 1·5	–	–
Physical properties								
Density	(g/m^3)	3·5–3·8	3·7–3·85	3·8–4·0	3·8–4·1	3·3–3·4	3·38	3·30
Bulk density	(g/m^3)	2·85–3·1	2·8–3·1	2·8–3·2	3·0–3·3	2·6–2·85	2·6–2·9	2·87
Apparent porosity	(vol%)	13–18	16–23	16–23	17–23	15–23	14–25	13
Permeability		0·5–12 (20)	3–35	3–35				
Comprehensive strength, cold	(N/mm^2)	40–115	> 20	> 15	> 30	> 25	30–80	60
Young's modulus	($\times 10^4$ N/mm^2)	7–14	1–6	1–6				
Refractoriness under load	(°C)	1560–1700	> 1500	> 1500	> 1500	> 1600	> 1700	> 1700
Thermal shock resistance		Poor to medium	Poor to medium	Poor to medium	Poor	Medium	Good	Good
Maximum temperature of application	(°C)	1450–1850	1600–1750	1600–1750	–	1600	1500–1700	–

cristobalite, tridymite, and, usually, residues of quartz; the quantities of pseudowollastonite and vitreous phase present depend on the amount of fluxing material and the firing and cooling conditions. The forms of SiO_2 mentioned above have separate temperature ranges of stability and different physical properties, as shown in Figure 4.5.

Although the expansion anomaly at about 600°C has been known for a long time and must be carefully controlled during heat-up, the phenomenon of aftergrowth above 1500°C has only become known and understood more recently as this temperature has been reached and exceeded in operation.[7,8] The critical parameter governing the aftergrowth of silica refractories is the residual quartz content which, in turn, is a function of the choice of raw materials, the manufacturing conditions, and, above all, the firing time and temperature (see Figure 4.4).

This aftergrowth, particularly if crown bricks differ in behaviour between one another, presents considerable problems since it affects both the stability of the crown and its function as a gas-tight structure. Areas of crown which are not gas-tight permit sting-out of flue gases containing alkali so that alkali sulphate, which is invariably present, condenses in the cooler portions of the silica bricks and then attacks them; this can rapidly lead to substantial damage to the crown. Silica bricks for glass tanks must therefore be of well converted silica with a low and uniformly constant residual quartz content of not more than 9%; badly converted bricks are denser and can be sorted and rejected by measurement. Also important is the refractoriness under load; values of 1640°C or, in more demanding cases, 1680°C must be obtained. Compared to these properties, porosity and permeability to gases are of minor importance.

After long periods of use, silica bricks in the hottest parts of the furnace form a clear zone of slagging. At lower temperatures, e.g. in the batch melting area, slagging occurs at the surface and causes accelerated wear of bricks. At higher temperatures, such as those found in the refining zone, the brick is outwardly clean and the slagging zone lies slightly inside the brick.[9,10] The relationship can be seen in Figure 4.6.

Alkali condensation at low temperatures occasionally produces a phenomenon referred to as tridymite frost, a loose surface layer of well formed crystals of tridymite. The tendency to form frost increases with the porosity of the bricks, which allows the rapid

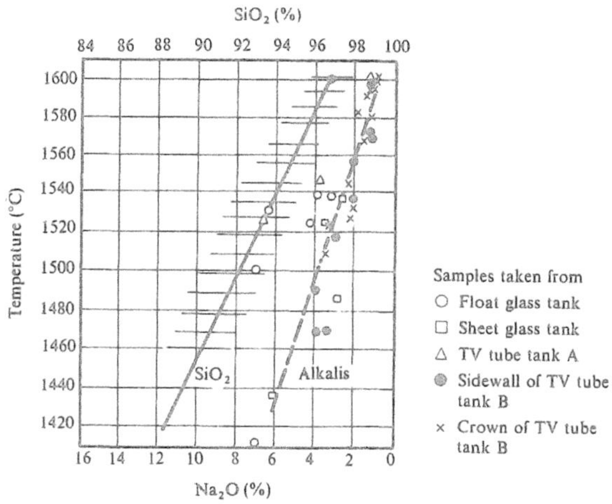

Figure 4.6. Relationship between temperature and Na_2O and SiO_2 contents of the hot face of silica bricks[9]

migration of vitreous siliceous material into the interior of the brick.

When constructing a crown, special attention must be paid to expansion joints. Past designs called for 14 mm for every metre run, corresponding to the maximum expansion of silica bricks of 1·4%, and expansion joints of 50–80 mm were provided every 4 to 6 m; more recently, similar joints for the same length of run have been reduced to 30–50 mm. Longitudinal compression acting on the crown is preferred to a joint not fully closed and ensures that the joints across the crown are closed and become gas-tight. The sealing of expansion joints is made easier by the use of special joint bricks (Figure 4.7). So that their closing is not in

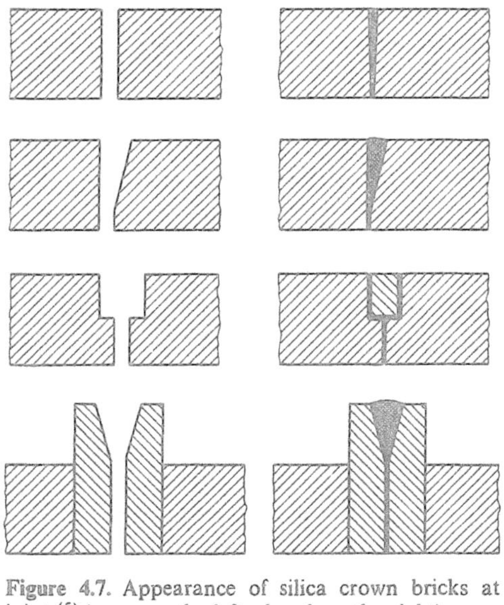

Figure 4.7. Appearance of silica crown bricks at expansion joints[5] (open on the left, closed on the right)

any way obstructed, these joints often extend past the skewbacks into the sidewalls. Within the sidewalls they sometimes taper down to zero but in large tanks melting container or flat glass they extend right down past the tuckstones. In some plants the surfaces of the skewback bearers are ground smooth.

Silica bricks in a crown are jointed using a matching cement of the same chemical composition as the brick. During heat-up and the first days of operation the mineralogical processes, which were completed in the bricks during their firing process, take place in the cement. Additives are used to increase the capacity of the cement for retaining water and thus delay the migration of water into the dry bricks; this helps in the setting of dry silica crown bricks without causing cement drops to form. To avoid mistakes the cement is often coloured green by the suppliers. A thickness of 250 mm is sufficient for bricks for dense crowns as the rate of wear is small but 350 mm is preferred for large container glass tanks while for crowns of large flat glass tanks, bricks of up to 450 mm thickness are used.

Silica crowns are now normally provided with thermal insulation but for reasons of safety this is not added if the temperature of the melting end exceeds 1600°C or if, when firing with strongly radiating flames, the distance between glass surface and crown is only marginally sufficient. Insulation is carried out after heat-up and after sealing any joints or cracks that may have appeared. Heating up with the insulation in place is rare[12] and is only recommended for small tanks when the heat-up is under perfect control. To apply the insulation the crown is first smoothed over with a layer of silica cement followed by a first course of light weight silica bricks flat on their sides. These bricks should be set using silica cement in the joints to prevent dust, which settles on the crown and which contains fluxes, from reaching the crown proper from above. A second course of light weight silica bricks and cement is laid so that the joints of the first layer are covered. A third layer is formed of good silica or Moler insulation with cement and this is followed by a final layer of insulating cement. The placing of insulation should be done in stages avoiding, if at all possible, leaving parts of the crown insulated and parts uninsulated as this prevents a uniform increase in temperature on the outer surface. Large crowns are best insulated by starting from both sides simultaneously. Insulation of an old crown at a later date is risky and cannot be recommended because the areas into which materials have migrated will be moved and the bricks will drip and become weakened.

If the operating temperatures of a furnace exceed 1600°C or when glasses with low alkali and high lime contents are melted, the life of silica is reduced to an extent that makes it necessary to look for alternative materials. Trials using top quality fusion cast AZS have been encouraging.[13] Since the high density of AZS would give a very heavy crown, specially shaped bricks are used. The design of such a crown is shown in Figure 4.8.

Expansion joints are placed between adjoining arches. Arch blocks and skewbacks are set without cement on a supporting framework. A special method of assembly had to be developed to deal with the problem that arches will be of different lengths because the arch blocks cannot be pressed together except with special tools. This was achieved by permitting the skewbacks to rotate along a horizontal axis, which ensures that all joints close during heat-up and enables insulation to be placed on the crown cold. A final verdict on the performance of this type of crown is still awaited; drawbacks are the high weight and the tendency to creep.

Silica bricks cannot be used in the construction of tanks melting sodium silicate (water glass) because of the extremely high alkali concentrations; crowns made of (corundum-free) sillimanite or chrome magnesite have performed better. Low iron, chrome-free special magnesite bricks have in some cases been successful in tanks melting soda–lime and lead glasses. Chrome–magnesite has also been used successfully in tanks melting glasses for insulation fibres, which contain B_2O_3 and fluoride and have a high alkali content. When designing crowns using basic bricks it is important to take into account their great weight and the fact that their expansion characteristics are different from those of silica. Tanks melting alkali-free glasses with high lime contents (such as E glass) for glass fibre production have crowns constructed of mullite. Tanks melting chemically resistant glasses at high temperature use zircon bricks in the crown.

4.3.2 The crown of the working and conditioning ends

Silica bricks are used almost exclusively here also. The first rule is that the crown must be gas-tight.

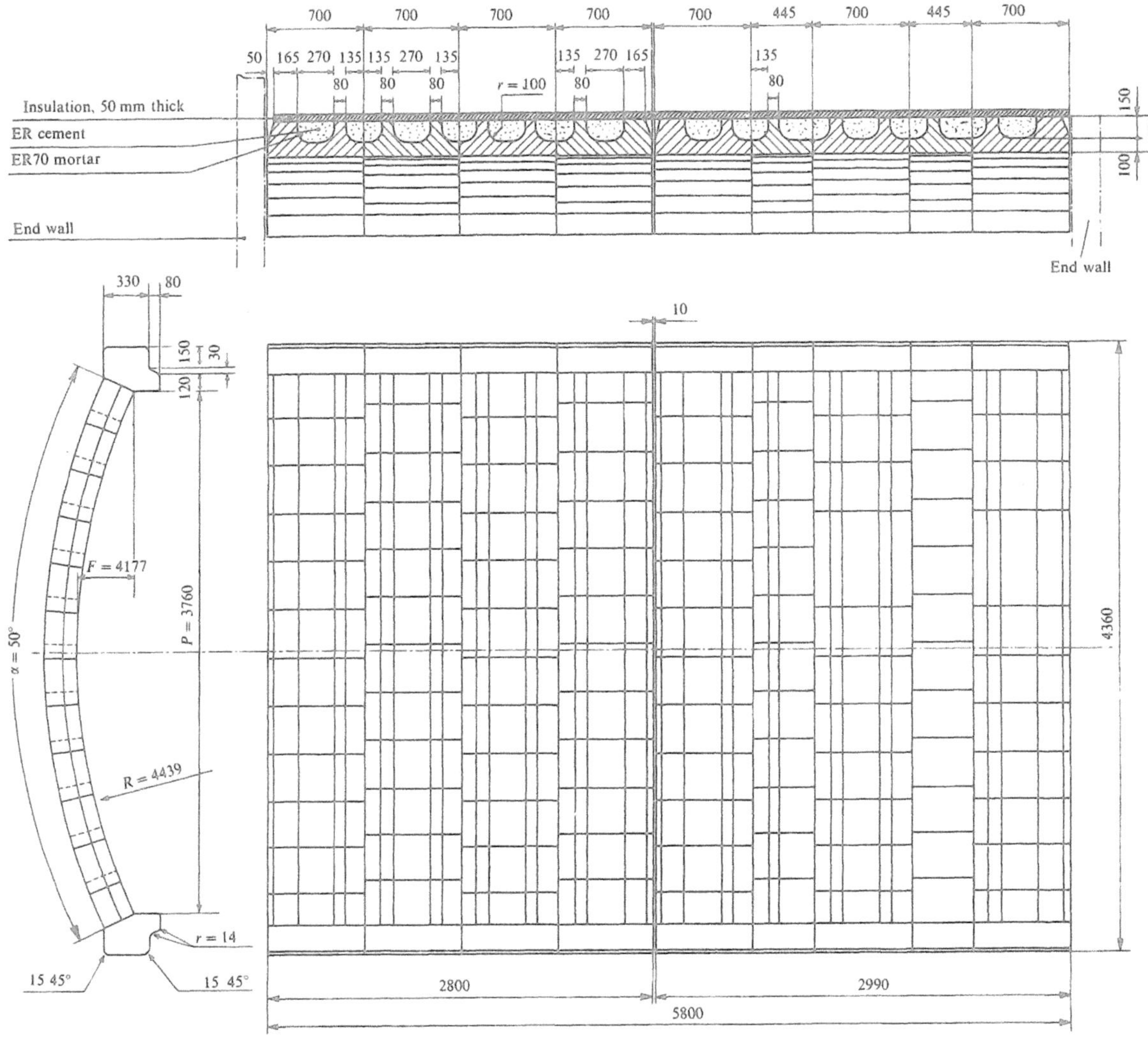

Figure 4.8. Construction of a crown using special AZS fusion cast refractory shapes (dimensions in mm)[13]

Since temperatures are lower, aftergrowth and residual quartz content are of less significance.

4.3.3 Melting end sidewalls

When selecting refractories for sidewall and port construction it is as well to remember that the majority of faults in the glass originating in the superstructure are in fact caused directly or indirectly, by the sidewall and port construction. A careful selection of refractories, including their compatibility, is therefore of major importance.

In the late 1940s the superstructure of glass tank furnaces was almost exclusively built of silica. As operating temperatures increased with the advent of oil and natural gas firing the sidewalls and, particularly, the port mouth entry arch became the weak parts of the installation.

This applies especially to the first and second pair of ports which are subject to severe conditions due to temperature, batch dust, condensation, and erosion. Changing to sillimanite did not produce any significant improvement. The then widespread use of AZS as tank blocks suggested that this refractory might also be used in the superstructure and it gave a significant improvement over silica and sillimanite. The port mouth entry arch held its shape for many years, which helped in controlling the firing of tanks with U shaped flames. A disadvantage was that it exuded a vitreous phase which reacted when it came into contact with silica slag originating from the crown. The initial impression that this exudation was caused by a combination of high temperature and gravity was soon disproved[14] and it was shown that the AZS blocks then available containing 15–20% glassy phase were contaminated with carbon or carbides, sulphides, and nitrides. When placed in an oxidising atmosphere O_2 would diffuse into the block and react there with these materials to give CO_2, SO_2, and N_2 which under pressure would push the glassy phase out of the block to appear as bubbles on its surface. This problem and the glass faults it caused were significantly reduced by producing the blocks under oxidising conditions and the use of raw materials of greater purity.

The problem of siliceous material flowing over AZS blocks and the glass faults that this produced has been reduced by the construction of a drip course.[15] A further contribution to the easing of this problem was that the increase in operating temperatures together with the insulation of silica crowns stopped the flow of slag in the refining section because no materials condensed on the crown. Slag formed in the melting zone is of less significance since it is absorbed into the glass by homogenisation in the refining end.

As AZS blocks improved in quality their use spread into the area of port mouths and large sections of the sidewalls. Although much reduced, the glassy phase in AZS blocks still reacts with silica bricks with which it comes into contact (Figure 4.9). Not all reactions between these two

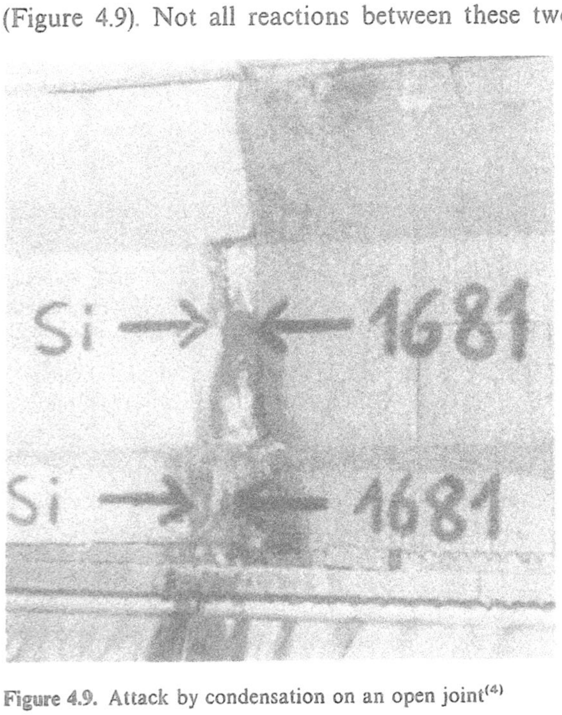

Figure 4.9. Attack by condensation on an open joint[4]

materials originate in the glassy phase: different rates of expansion give rise to improperly sealed joints and condensation of materials carried in the flue gases is then the main cause of damage. To avoid this reaction zircon bricks of low porosity are used as a neutral layer between AZS and silica and it is sometimes sufficient to make a joint with zircon cement instead of bricks (Figure 4.10).

AZS containing 31% ZrO_2 or, for high duty applications, 41% ZrO_2 is now used in container and flat glass tanks for tuckstones, burner mouths, sidewalls between burners, burner sills, parts of the port neck, the sidewalls above and adjacent to the doghouse, the wall between melting and working ends, and, for single chamber furnaces, the throat block and shadow wall. Silica is also used in Float tanks. Port sills and tuckstones are sometimes constructed of void-free bricks. Only AZS should be used in the hot sections of burner ports as periodic temperature fluctuations in the range 1000–1200°C eventually lead to cracking and spalling due to its anomalous expansion (Figure 4.2).

No ideal material has yet been found for arches above the doghouse and burner blocks of regeneratively heated tanks although special designs using AZS make it possible for these parts to last the campaign. Burner blocks are also made of high alumina, high duty refractories of fused mullite.

High thermal conductivity and sensitivity to temperature changes, especially in the region 1000–1200°C, together with the refractoriness of this material have made it necessary for fusion cast refractories to be well insulated. This can be done for the superstructure sidewalls, gable wall, and above the doghouse by using insulation bricks, light weight refractories, and ceramic fibre boards as these have no structural function. Careful calculations must be made to ensure that the temperature limits of any one material are not exceeded; the bracing steelwork must also be taken into consideration so that no overheating or bending can occur. If additional fibrous insulation such as

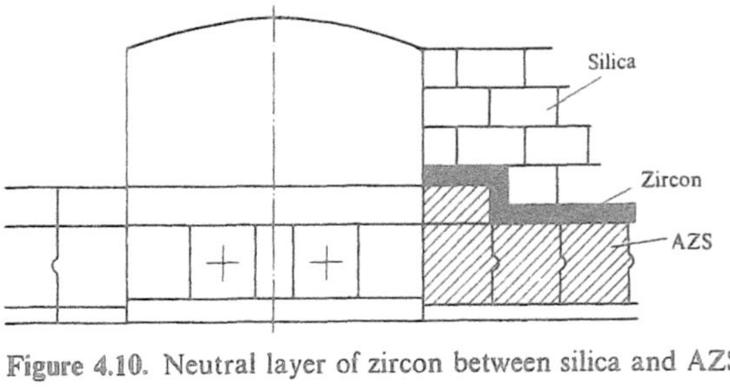

Figure 4.10. Neutral layer of zircon between silica and AZS

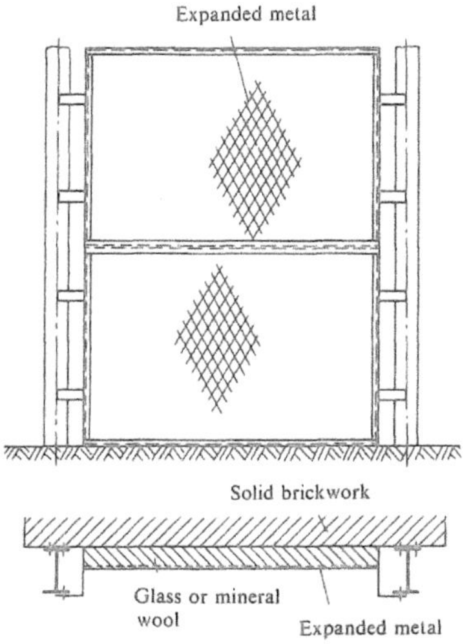

Figure 4.11. Frame with expanded metal for attaching additional insulation

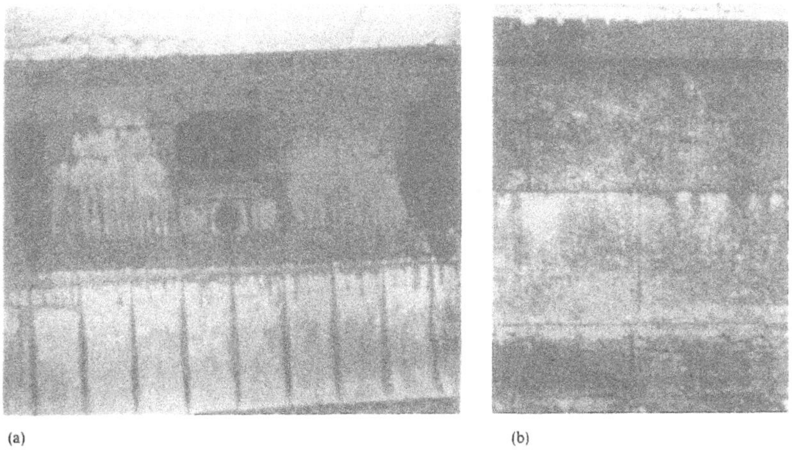

Figure 4.12. Cracks in an AZS superstructure sidewall[4]
(a) soda–lime glass tank
(b) borosilicate glass tank

glass or ceramic fibre is used, the best arrangement is to contain this in metal frames with a covering of expanded metal (Figure 4.11).

Under conditions of good thermal insulation and high temperature the surface of AZS develops a crazed surface (Figure 4.12). It is caused by crystal growth on the surface and can lead to spalling in the form of cup-shaped pieces of refractory.[4]

4.3.4 Sidewalls of working and conditioning ends

The sidewalls of the working ends of container glass tanks are still constructed of silica. In tanks for high quality flat glass fusion cast β-corundum bricks are used: these possess a very high thermal shock resistance and contain only very small amounts of vitreous material ($<0{\cdot}5\%$) so that exudation of the glassy phase and contamination of the glass is avoided.

4.3.5 The melting end

This is constructed almost exclusively from fusion cast blocks because these have the highest resistance to attack by molten glass. The major advantage of fusion cast blocks over ceramically bonded blocks is their intercrystalline bonding and the close packing of their crystals so that they do not show any open porosity. The rate of corrosion, as for ceramic blocks, is strongly temperature dependent and the composition of the blocks is of secondary importance as is shown in Figure 4.13 for blocks in contact with soda–lime glass. The behaviour of different fusion cast materials in contact with a variety of glasses is shown in Table 4.7.[4] The dominating influence on the corrosion of

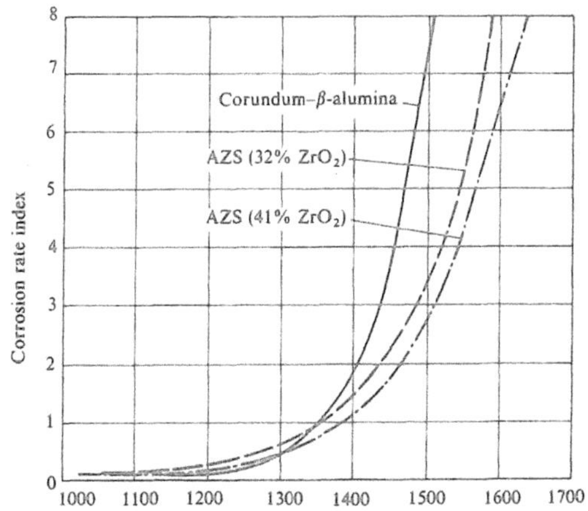

Figure 4.13. Corrosion of various fusion cast refractories used as tank blocks in a tank melting soda–lime glass as a function of temperature[16]

Table 4.7. *Corrosion tests and stone formation for four types of refractory with various glasses*[17]

	Soda–lime glass	E glass	Pyrex, Duran	Neutral glass	PbO glass	Technical glass	TV glass
	Test temperature (°C)						
	1550	1580	1570	1560	1450	1430	1560
Attack (mm)							
AZS ($32ZrO_2$)	9	15	0–0·1	0·5	0–0·1	2·2	1
AZS ($41ZrO_2$)	7·5	8	0	0·3	0	1·4	0·8
Zircon	45	5	0	1	–	4	2·5
Corundum–alumina	22·5	18·8	0·15	1·2	0·4	3·7	2
Stone formation							
AZS ($32ZrO_2$)	0–1	4–5	3	2–3	0	2	1–2
AZS ($41ZrO_2$)	0–1	2–3	2	2	0	0	1
Zircon	5	1–2	4	4	0	–	–
Corundum–alumina	0	–	0	0	0	0	0

these refractories, and this applies to both fusion cast and ceramically bonded types, is the preferential migration of alkali out of the molten glass into the vitreous phase of the refractory, leading to a reduction in its viscosity; the equilibrium between phases in the refractory is disturbed and results in the dissolution of the refractory itself. The diffusion of alkali causes the alkali concentration in the refractory to exceed that in the molten glass, i.e. the alkali migrates against the concentration gradient.[17]

Refractories containing very little glassy phase are attacked by molten glass, especially if they have a high open porosity, and migration into the refractory interior proceeds via wetting.

The presence of corundum and baddeleyite, which are formed in AZS blocks during the cooling process after pouring, is an important factor in their dissolution. The aim is to achieve the formation of a sieve-like structure by the simultaneous crystallisation of Al_2O_3 and ZrO_2. Crystals of corundum are thereby covered and interspersed with small round precipitations of baddeleyite which protect the corundum crystals from rapid dissolution.

In addition to the corrosion of refractories by molten glass already mentioned, particularly strong attack occurs at the flux line and in the areas of horizontal joints. This attack cannot be explained solely on the basis of the high temperatures involved and its cause is ascribed to convection at the interphase surfaces, which will be discussed further in Section 6.9. The flux line attack by molten glass on high alumina blocks, e.g. AZS, is considerably greater than on high silica blocks. From the point of view of glass composition, both soda–lime and borosilicate glasses containing less than 6% alkali oxide cause similar corrosion at the flux line under similar conditions.

As far as the tank proper is concerned the strongest attack is on the doghouse corner blocks, the region around the throat (for container glass tanks), and the waist of flat glass tanks. For highly aggressive glasses such as alkali borates and alkali-free glasses with high lime contents, blocks of chrome–corundum–spinel are used but these can only be supplied in small sizes and discolour the glass very slightly.[18]

Figure 4.14 shows the fusion cast chrome–corundum–spinel throat of a furnace melting a glass of special composition. The throat had a life of 28 months while under similar conditions an AZS throat cover block had to be reinforced after nine months, thereby proving that the resistance of chrome blocks is much better.

As they are manufactured from a melt, fusion cast blocks are inhomogeneous and this is shown most forcefully by the appearance of voids. Improvements to pouring techniques has led to a reduction in the size of the voids or has enabled them to be moved to less critical parts; in extreme cases they can be sawn off to give void-free refractories. Void-orientated blocks, with the font located at the outside bottom end, are now preferred for glass contact. Void-free blocks should be used when assembling throats and weirs, for carrying electrodes or bubblers, and for use as backing-up tiles, e.g. such as those placed to back up severely worn flux line blocks.

The soldier blocks constituting the tank proper are now thermally insulated with high quality ceramically bonded insulation bricks, often in two layers.[20] The hot face would be of 38% alumina, and the outside a suitable light weight fireclay insulation brick.

The tank bottom is constructed of high alumina (38%) fireclay, usually in two courses placed direct-

(a)

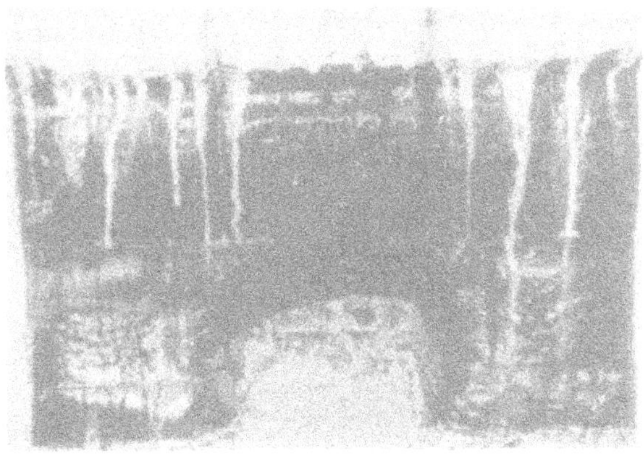

(b)

Figure 4.14. Chrome–corundum–spinel throat of a tank for a special glass[18]
(a) when new (b) after 28 months' operation

ly onto the supporting steelwork of the tank. Placed above these are two courses of void-free fusion cast paving tiles set with special cement. Insulation is usually placed under the fireclay to reduce heat losses downwards. Well insulated tank bottoms should have their vertical joints sealed with special AZS cements (see Section 3.1.2.2). Thick AZS bottoms have been used for tanks melting lead glass, as precipated lead can cause strong local corrosion.

For special glasses massive floors of rammed zircon have been used instead of tiles.

4.3.6 Working and conditioning ends

The corrosion resistance of working end blocks is of less importance due to the lower temperatures prevailing. The most important objectives are to obtain and maintain the quality of the glass and to avoid the formation of seeds, stones, and cords. AZS blocks are normally used in this zone in container glass tanks but when high quality is demanded, corundum–β-alumina may be used. Flat glass tanks use the latter since AZS blocks create sealing problems.

Insulation of the working and conditioning ends presents no problems as suitable bricks are readily available.

The bottoms of the working and refining ends can also be covered with AZS tiles set in cement while for high quality requirements, the use of corundum–β-alumina should be considered. Use of the latter will avoid all problems arising from the dissolution and accumulation of ZrO_2 on the tank bottom.

4.3.7 Port necks

The port necks of glass tanks are subject to severe thermal stresses arising from high temperatures and temperature changes due to the reversal of the regenerative heating system and to attack by batch dust, which is unavoidable with oil firing. The high vanadium content of oil ash also has adverse effects on many refractories. Care must be taken to avoid contamination of the melting end by port refractories.

For tanks with low pull, the port necks may be constructed of silica but for tanks with high pull and waste gas temperatures of above 1500°C special expensive refractories are required. The chemically bonded refractories covered in steel sheet used frequently in the past and which, set without cement, formed a gas-tight construction due to the formation of magnesium ferrite, are now used only rarely. For intermediate conditions magnesite bricks fired at high temperature are recommended. In addition high alumina, sintered mullite, sintered alumina, and sintered clay with zircon, all fired at high temperature, have been found satisfactory. Fusion cast AZS is not recommended as the temperature changes linked to reversal cause scaling due to the crystal structure changes described earlier.

4.3.8 Regenerator chambers

Satisfactory furnace operation can only be achieved if the regenerator chambers are in good condition. Apart from the flux line and, for container glass tanks, the throat, the life of a furnace is determined by the life of the regenerator and in particular its checkers. When selecting the refractories both the thermal and the corrosion properties of the materials available must be considered and a careful choice must be made.

The performance required of the checkerwork of a tank melting soda–lime glass varies widely within the regenerator; in addition, the operating conditions of the furnace and the fuel used play major roles. The high alumina firebrick checkers used in the past have, except for some special cases, been replaced by basic materials.[23,24]

The best criterion to use for selection is the waste gas temperature. In regions where the waste gas temperature exceeds 1500°C the checkerwork is exposed to strong attack by CaO, SiO_2, and Al_2O_3 batch dust. Low-iron, directly bonded (MgO–MgO) magnesite bricks of 94% MgO content and fired at high temperature should be used for tanks fired by natural gas.[25] High firing temperatures and the resultant direct bonding anticipates the grain growth which would otherwise take place over many years and changes in checker structure can thereby be avoided.

For oil fired furnaces the directly bonded bricks are often not suitable because the residual dicalcium silicate is destroyed by a combination of vanadium in the ash from the oil and large amounts of batch dust carried over. The bricks recommended in this case are low-iron magnesite bonded by forsterite and fired at high temperature. Up to 2% Ni has been found in oil ash and this can also cause problems. Nickel, as NiO, forms mixed crystals with MgO and if nickel compounds find their way into the melting batch and if sulphides are formed in the batch as intermediate compounds, then there is the risk of the much feared NiS appearing in the glass. The use of chrome–magnesite bricks in the upper layers of hot regenerators has not been very successful as batch dust and heavy metal oxides have led to their crumbling and the preheated air then carries particles of chromite into the molten glass causing defects.

In the temperature region above 1500°C, the sidewalls and arch can be constructed of magnesite bonded with silicates, although the wall opposite the port neck (where batch particles can impinge) should be made of bricks with direct bonding. Since these bricks are subject to cooling from the outside, the conditions under which they operate are somewhat less onerous than those for checker bricks. The creep of magnesite bricks becomes noticeable in large regenerator crowns over 4 m wide subject to high temperature and in this case it is best to construct the crown of magnesite fired at high temperature or even of silica.

In the regions where the waste gas temperature lies between 1500 and 1300°C reactions between batch dust and silicate binder must be assumed to occur. If long campaigns are the target then, similar to the example given above, low-iron magnesite bonded by silicates should be replaced by the directly bonded variety fired at high temperature. Directly bonded chrome–magnesite fired at high temperature, or low-iron magnesite, can also be considered for arches and sidewalls subjected to these temperature conditions.[25–27]

Contact reactions between checkerwork and chamber walls are avoided by using basic bricks or tiles, which are indifferent to contact effects, on the wall hot face (Figure 4.15). Consolidation of the wall is improved by using longer bricks to effect a tie into the backing of firebricks. Attention needs to be paid to the fact that the thermal expansion of basic bricks is higher than that of firebricks; in the vertical direction, expansion joints can be ignored for heights up to 10 m while along the length, pieces of cardboard 2–2·5 mm thick are inserted every 250 mm.

Batch dust hardly reaches those regions where the waste gas temperature is 1300–1100°C and here also the vaporised constituents in the waste gases (alkalis) cannot react with the refractories as the waste gas temperature is too high. This section can be constructed of low-iron magnesite of 90–93% MgO content.

Alkalis are condensed where the waste gas temperatures fall to 1100–600°C.[25–30] There are sections in which the reversal mechanism causes several corrosion mechanisms to occur. During the

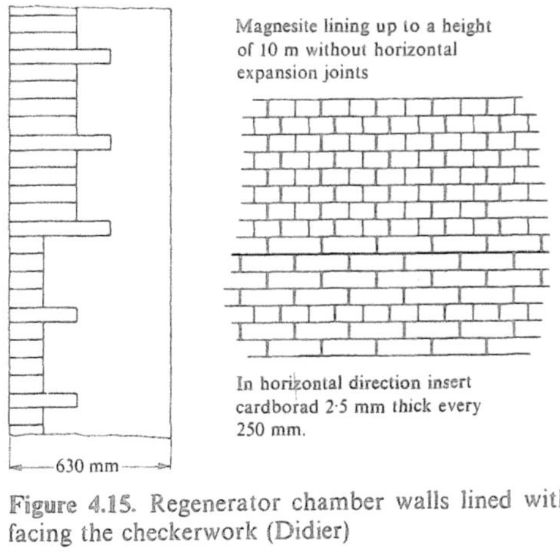

Figure 4.15. Regenerator chamber walls lined with magnesite facing the checkerwork (Didier)

waste gas interval the zone of condensation moves downstream while condensates evaporate upstream as the temperature increases. The condensate remains liquid downstream but solidifies after reversal when it makes contact with incoming cold air. The condensate consists chiefly of Na_2SO_4 but with excess SO_3 and in the presence of V_2O_5 from oil firing, $MgSO_4$ is formed. If only Na_2SO_4 is expected to be formed, then basic bricks with a high CaO content in the binding phase can be used; if attack by SO_3 and vanadium has to be taken into consideration, then forsterite or basic bricks containing chrome should be used and they should, if possible, be directly bonded and fired at high temperature.

In vertical regenerators the lower parts and the bearer arches are under high load due to the weight of the checkerwork they carry. In this section high quality firebrick, low in fluxes and with an Al_2O_3 content of about 44%, are used as they have some resistance to Na_2SO_4 slag. At higher temperatures even this quality of firebrick is not good enough and flaking occurs as a result of the formation of felspar.

If the tank is oil fired (i.e. the bricks are subjected to attack by SO_3), then the use of chrome–magnesite fired at high temperature is recommended; fused mullite–corundum bricks containing 90% Al_2O_3, fired at high temperature, and of low porosity may also be used. If natural gas is the fuel (i.e. no SO_3 attack), then pure, direct bonded magnesite fired at high temperature can be used for the checkerwork, with sulphate-resisting fused mullite–corundum containing 90% Al_2O_3 for the bearer arches.

The melting of lead, borosilicate, and opal glasses as well as glasses for fibres gives rise to high vaporisation losses which strongly attack basic refractories[34] and this has to be borne in mind when constructing checkerwork for special glasses. Silica bricks were used in the past but are no longer adequate in view of the longer campaigns expected and higher operating temperatures employed. The use of basic bricks has brought certain advantages in the melting of lead glasses and other advances resulted from the use of fusion cast AZS.[35] Special bricks, of cruciform shape, were developed to extend checker life and to improve heat regeneration; because of the high thermal conductivity of AZS, these are only 40 mm thick (Figure 4.16).

Figure 4.16. Cruciform refractories made of ER1690, 170 mm across and 250 mm high[35]

The trial use of zircon–mullite bricks for borosilicate glasses was successful.

4.3.9 Flues

Firebricks containing 32–42% Al_2O_3, selected according to the temperatures at which they will operate, may be used in the flues and special shapes, pressed and cast from suitable materials, are used.

Details concerning the construction of chimneys are given in Section 3.7.

5. Operation of glass tank furnaces

5.1 The heating process

5.1.1 Temperature distribution

A glass tank furnace is primarily a piece of process equipment and only secondarily a furnace for the heating of a specific material. The equipment must be constructed and operated so that the separate stages of glass manufacture follow one another in the correct sequence and take place in that part of the furnace allocated for them.

The different zones (see Figure 5.1(a)) are (I) preheat, (II) melting and reaction, (III) refining, and (IV) conditioning of working zone. Temperatures of 1000–1300°C are adequate for the melting and reaction processes of the batch constituents while 1400–1550°C and sometimes higher is required for good homogenisation and refining. Temperatures in the conditioning or working zone are governed by the manufacturing processes employed.

In zone I the batch (usually supplied at ambient temperature) is subjected to heating by flames from above and by the bath of molten glass from below. Its poor thermal conductivity (0·12–0·23 W/mK) creates steep temperature gradients in the batch. When the 'melting temperature' of the batch is reached, the temperature increase ceases and the temperature then remains constant, especially on the surface of the batch, despite intense heating due to the fact that the primary melt runs off the batch or drains into it (zone II).

The refining process is a particularly important part of the glass melting process: gases must be removed and the molten glass homogenised and, apart from temperature, sufficient time (about 1 hour) is required. In the refining zone the glass surface should be clear so that the energy required for completing the melting of the batch from below and replacing the heat losses through the tank blocks can reach it (Figure 5.1(b)).

A dwell time which is long enough so that no foam or batch residues can reach the refining zone or even the conditioning zone is ensured by creating a thermal barrier in the refining zone; this is sometimes combined with a mechanical barrier, e.g. a row of bubblers. An alternative barrier between melting and refining zones is a weir, a device which has been particularly successful for furnaces with a melting area of less than 150 m^2 (Figure 5.1(c)).

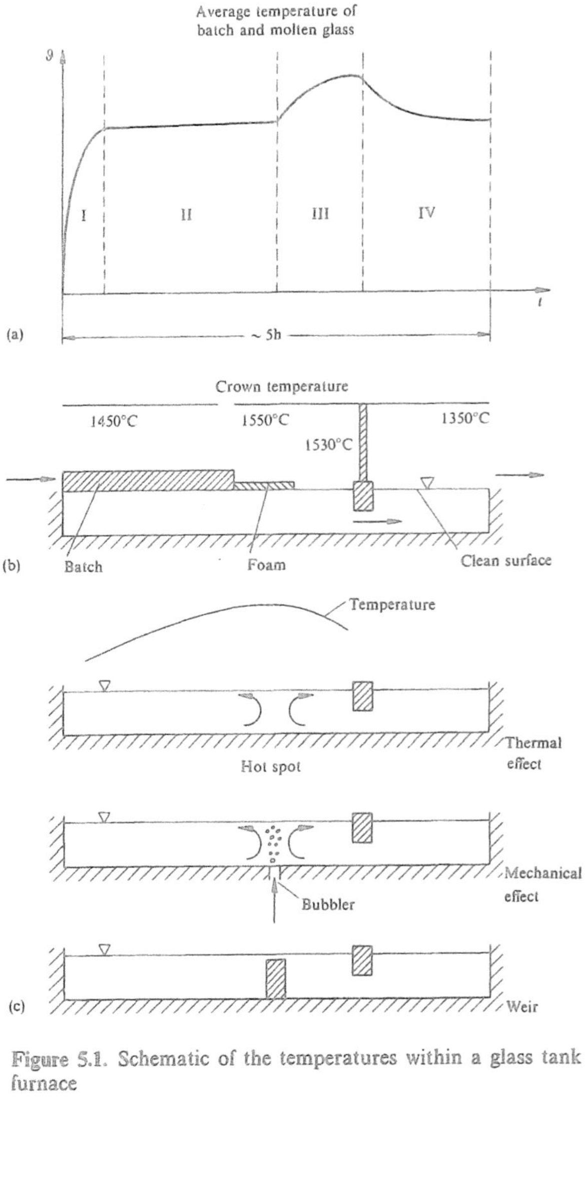

Figure 5.1. Schematic of the temperatures within a glass tank furnace

5.1.2 Heat flux density and fuel distribution[1,2]

Several parameters are used in considering the firing of furnaces; the combustion space loading, the melting area loading, and the heat flux density. The combustion space loading is defined as the energy supplied via the fuel per unit volume of combustion space, the heating area loading as the energy supplied per unit area of melting surface. Table 5.1 gives these values for glass furnaces compared to other furnaces and boilers; glass furnaces are clearly not in a leading position despite their high operating temperatures.

The heat flux density between the superstructure of a flame fired furnace and the glass surface varies widely between locations and depends on flame distribution and the different properties of the glass surface, i.e. whether there is melting batch, foam, or a clear glass surface: indicative figures are given in Table 5.2. Due to the high temperatures employed, heat transfer is mainly by radiation (about 95%) and the rest by convection. A description of heat flux density by means of a heat transfer coefficient is not suitable since the superstructure temperatures vary from place to place and temperature differences can therefore only be given approximately. In the melting zone a temperature difference of 250 degC can be expected, which corresponds to a heat transfer coefficient (radiation) of about 600 W/m^2K for currently known heat flux densities.

In cross-fired furnaces the fuel distribution is arranged so that the glass surface in the refining zone is clear; the optimum situation is that the glass surface is one third covered with batch, one third covered with foam, and one third clear. To achieve this the combustion space loading of the zone covered with batch must be much higher than that of the refining zone. Examples of fuel distribution for container and flat glass tanks are given in Figure 5.2: the large differences between different plants are significant as, generally speaking, the proportion of fuel used over the batch zone increases with increasing pull. The heat flux density along the length of a U shaped flame fired tank can only be influenced by the length-to-width ratio of the tank and the length of the flame. When the tank is in operation, it is only possible to adjust the flame length, which makes it difficult to obtain optimum results in cases where the pull varies widely. The flame return should be two-thirds down the length of the melting end and if normal combustion is assumed, then a combustion space loading such as that shown in Figure 5.3 will result: it corresponds to the local loadings for cross-fired furnaces or the fuel supply distribution.

Table 5.1. *Indicative thermal data for different types of furnaces*[1]

	Combustion space (kW/m³)	*Heating area (W/m²)*
Glass furnace	60–160	90–200 × 10^3
Siemens–Martin furnace	120–350	290–470 × 10^3
Power plant boiler	120–350	Up to 150 × 10^3
Large water tube boiler	580–1520	-

Table 5.2. *Indicative heat flux density in glass melting furnaces* (kW/m^2)[1]

Flame to batch	120–230
Flame to foam cover	50–70
Flame to bath of molten glass (normal)	70–80
Flame to bath of molten glass (cool) dusty seed	80–90

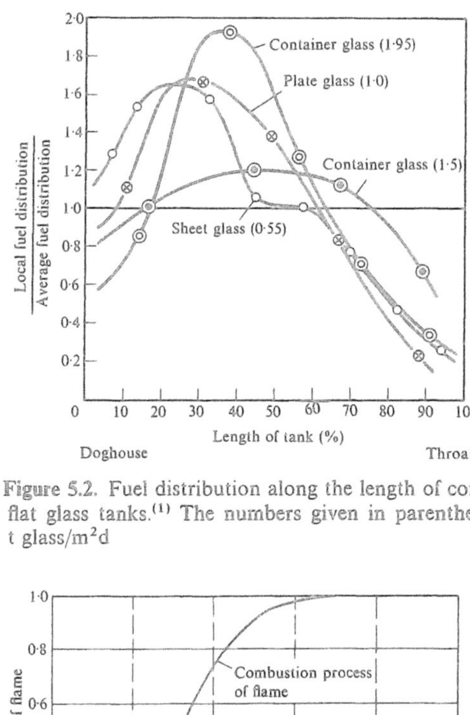

Figure 5.2. Fuel distribution along the length of container and flat glass tanks.[1] The numbers given in parentheses refer to t glass/m²d

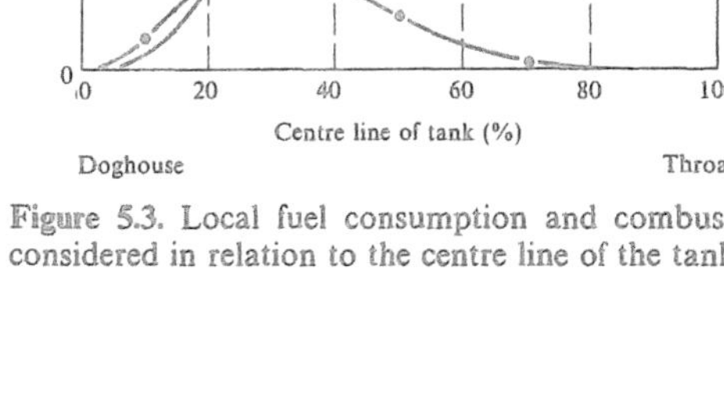

Figure 5.3. Local fuel consumption and combustion process considered in relation to the centre line of the tank

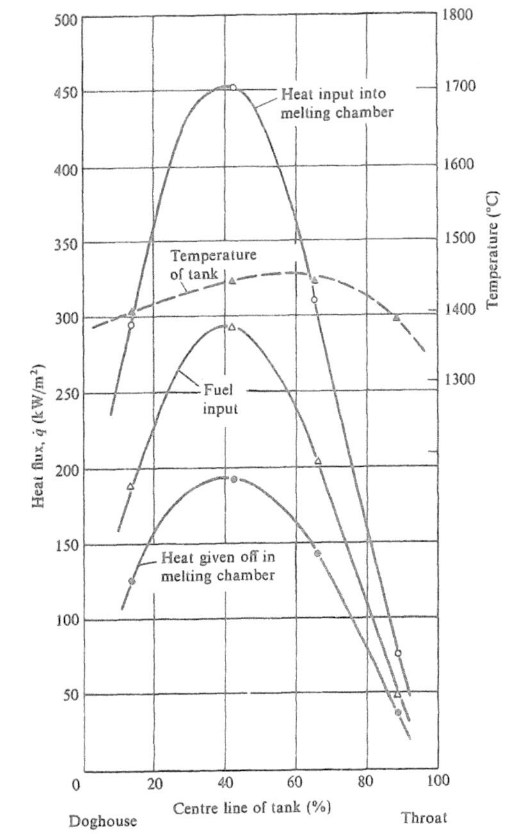

Figure 5.4. Distribution of heat flux along the centre line of a tank[1]

The local distribution of specific heat flux and the resultant tank temperatures along the centre line of the crown are given in Figure 5.4; these are based on data for a clear container glass tank fired with heavy fuel oil.

5.1.3 Efficiencies

5.1.3.1 Fuel efficiency

Different parameters have been suggested for judging the fuel efficiency, the best known being by Heiligenstaedt.[3] If the energy supplied by the fuel into the upper part of the furnace is $\dot{E}_B$, the enthalpy of combustion air and burning gases is $\dot{H}_R$, the enthalpy of the waste gases leaving the furnace is $\dot{H}_A$, the heat lost through sidewalls and crown is $\dot{Q}_S$, and the heat transferred to the molten glass is $\dot{Q}_G$, then the fuel efficiency, η_f is as given in Table 5.3. Strictly speaking, η_f should not be considered as the fuel efficiency but the furnace efficiency since it contains not only the efficiency with which heat is transferred in the combustion chamber, but also the efficiency of the regenerator or recuperator used in conjunction with it. These are parts of the furnace but not of the firing system, and η_f is therefore better described as furnace efficiency.

Gailhbaud[4] suggested that the fuel efficiency should be related not to the energy supplied by the fuel but to the total energy, $\dot{E}_B + \dot{H}'_R$, supplied to the upper part of the furnace; this is called by him 'facteur d'utilisation' and the term combustion space efficiency, η'_f is suggested for this definition. For a consideration of fuel efficiency, η'_f appears to be more suitable than η_f and a definition of η'_f is given in Table 5.3. A disadvantage in the use of η'_f is that the energy created in the upper part of the furnace is treated as a total and no differentiation is made between energy that is transferred to the material to be heated, i.e. the glass, and the energy that is lost through crown and sidewalls. The definition of the efficiency with which energy is supplied to the glass itself is here defined by η''_f (see Table 5.3). This definition has the disadvantage that $\dot{Q}_S$ is difficult to measure and one has to rely on estimates. There are a series of other definitions which cannot be mentioned for lack of space. However, to avoid misunderstanding, the urgent need to arrive at a generally agreed definition of efficiency is clear.

Table 5.4 gives fuel efficiency data for a number

Table 5.3. *Efficiencies for judging the performance of glass tank furnaces*

	Definition	*Diagram of furnace*
Furnace efficiency	$\eta_f = \frac{\dot{E}_B + \dot{H}_R - \dot{H}_A}{\dot{E}_B} = \frac{\dot{Q}_G + \dot{Q}_S}{\dot{E}_B}$	
Combustion space efficiency	$\eta'_f = \frac{\dot{E}_B + \dot{H}_R - \dot{H}_A}{\dot{E}_B + \dot{H}_R} = \frac{\dot{Q}_G + \dot{Q}_S}{\dot{E}_B + \dot{H}_R}$	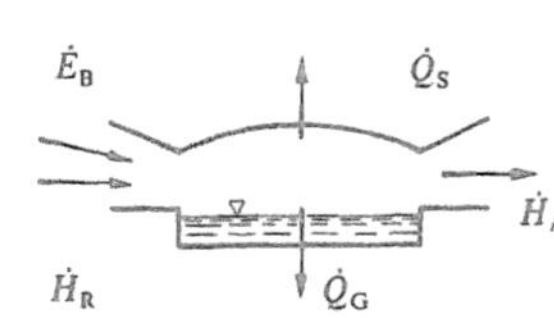
Useful heat	$\eta''_f = \frac{(\dot{E}_B + \dot{H}_R) - (\dot{H}_A + \dot{Q}_S)}{\dot{E}_B + \dot{H}_R} = \frac{\dot{Q}_G}{\dot{E}_B + \dot{H}_R}$	
Fuel utilisation	$\eta_u = \frac{\dot{Q}_G}{\dot{E}_B}$	

Table 5.4 *Thermal data from some old container glass tank furnaces*

		Tank I	II	III	IV	V
		Type of glass Clear	Clear	Clear	Amber	Green
		Fuel used Coke oven gas	Producer gas	Oil	Coke oven gas	Producer gas
Specific fuel consumption	(kJ/kg glass)	7953	8540	7245	11552	7517
Pull	(t glass/m²d)	1·41	1·35	2·24	1·26	1·23
Useful heat	(%)	21·1	25·1	32·44	18·85	26·9
Melting area	(m²)	65·7	67·2	68	48·3	63·7
Furnace efficiency (η_f)	(%)	66·9	76·45	73·1	64·45	64·53
Combustion space efficiency (η_f')	(%)	44·2	43·2	46·1	37·9	43·3

of furnaces, all container glass tanks, taken from measurements carried out by the Hüttentechnische Vereinigung der Deutschen Glasindustrie. In order to ensure reliability of the figures only data from tanks which were properly controlled and whose data were accurately recorded were used. If the specific fuel consumption of each furnace is compared to its fuel efficiency (Heiligenstaedt's η_f), then results show that the tank with the highest η_f (II) does not also give the lowest specific energy consumption. Tank II shows a fuel efficiency of 76·45% but a specific fuel consumption of 8540 kJ/kg glass while Tank V, where the fuel efficiency is 64·53%, has a specific fuel consumption of 7517 kJ/kg glass. The application of fuel efficiency as a criterion of energy consumption is therefore only of limited use.

A comparison of 'useful heat' stated as a percentage with η_f shows a better correlation, even if the differences are larger (see Tanks II, III, and IV in Table 5.4), a situation which cannot be explained on the basis of their respective fuel efficiencies.

If η_f' is calculated for each tank the specific fuel consumption shows a better correlation with η_f' than with η_f. Tank IV with its relatively high specific fuel consumption of 11 552 kJ/kg glass has an η_f' of only 37·9%. However, even using this factor there are deviations, as is shown by the differences between Tanks II and V.

Figure 5.5 shows efficiencies η_f and η_f' and useful heat against specific fuel consumption: an increase in fuel consumption should give a steady decrease in efficiencies and useful heat but as Figure 5.5 shows, this is only approximately true for η_f' while η_f and useful heat show even greater fluctuations. This suggests that η_f' is more suitable than η_f for assessing the efficiency of a plant. The efficiency related to η_f' of the combustion space of modern container glass tanks has now risen to values of 50% and above as a result of improved control over firing, better insulation, and higher preheat temperatures.[5]

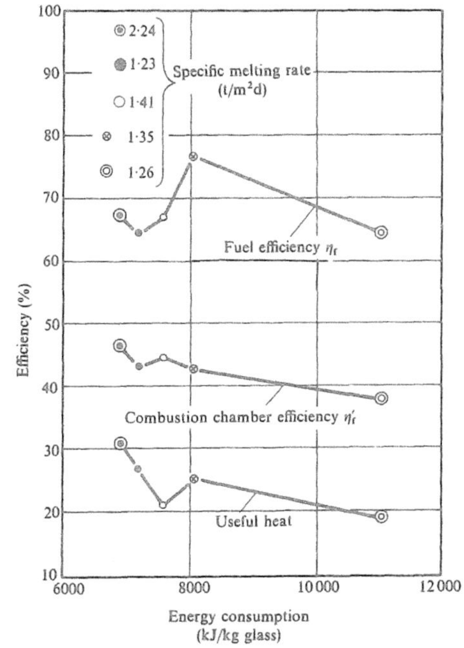

Figure 5.5. Efficiencies of glass tanks as a function of the specific energy consumption[2]

Figure 5.6 shows the fuel efficiency η_f for different air preheat temperatures and waste gas temperatures for air factors $\lambda = 1{\cdot}0$ (thin lines) and $\lambda = 1{\cdot}14$ (thick lines). Account was taken of the fact that, of the combustion air, only part was preheated in the regenerator (to 1350°C) and the rest was introduced cold into the furnace in the form of atomising air and air induced through the doghouse, open joints, etc.[5]

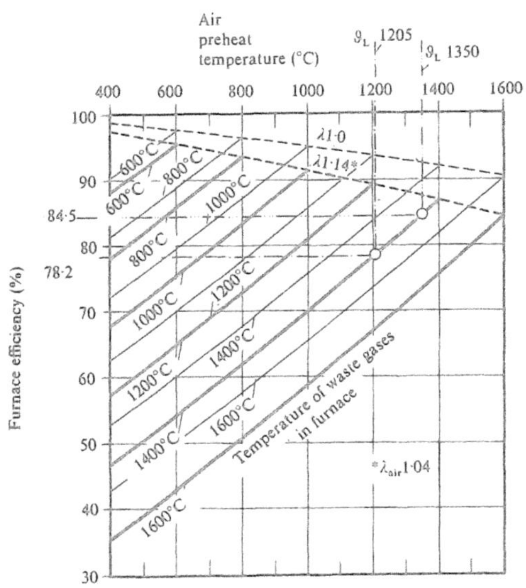

Figure 5.6. Furnace efficiency as a function of air preheat temperature[5]

5.1.3.2 The theoretical heat requirement of molten glasses

The different definitions of fuel efficiency use the quantity of energy transferred to the glass as a reference parameter, i.e. for a glass tank it is $\dot{Q}_G$. Since the molten glass is itself subject to heat losses through walls and via convection, which can vary from case to case, it seems necessary to consider more accurate means for establishing the energy required for glass melting itself. The literature uses the concept of 'useful heat' or, better, 'theoretical energy requirement';[6] this is understood to be the enthalpies of the reactions involved in the formation of glass, of the molten glass, and of the gases originating in the batch (Table 5.5). The delivery temperature must be fixed in order to state the enthalpies and a temperature range of 20–1500°C is generally assumed. Although this can vary from case to case the differences are usually ignored since the actual temperatures can only be determined with difficulty and, for open tanks, are also difficult to define.

The theoretical energy requirement is expressed as a percentage and is often described as the thermal efficiency of a glass furnace. However, this definition is unsatisfactory from a theoretical point of view, since this efficiency does not show a limit.

The determination of the enthalpies of molten glass and batch gases is possible in theory but impossible in practice: when making comparisons between efficiencies this particular point must not be overlooked.

The theoretical energy requirement of different glasses depends on the batch formulation and, in practice, changing proportions of cullet must be allowed for.[7]

5.1.4 Heat consumption

Experience has shown that the heat consumption of glass tanks can be given approximately by

$$\dot{q} = \dot{q}_a + sb \tag{5.1}$$

where $\dot{q}$ is the heat supplied per unit time and melting area (e.g. W/m^2) and s is the pull per unit area (e.g. kg glass/m^{2}d). The quantities $\dot{q}_a$ and b are virtually constants, and depend on the design and the operating parameters of the furnace. The term $\dot{q}_a$ comprises the heat losses of the furnace independent of pull and is often referred to as 'heat consumption when empty'; this is a somewhat loose definition since an empty furnace has a totally different temperature distribution and, therefore, a different fuel consumption from a filled furnace continuously supplied with batch. The 'heat consumption when empty' is therefore more a parameter for purpose of calculation than a physically identifiable magnitude. The term b depends on the efficiency of the energy transfer in the furnace – flame formation and efficiency of the regenerator or recuperator. Any improvement in the heat transfer process or the efficiency of heat exchange reduces the value of b. Within narrow limits b is independent of s because, in general, the heat transfer process and the efficiency of heat exchange worsen

Table 5.5. *Theoretical heat requirement of different glasses*[6]

	Soda–lime glass (container glass)		*Laboratory glass (about 8% B_2O_3)*		*Lead crystal glass (about 19% PbO)*	
	(kcal/kg)	*(kJ/kg)*	*(kcal/kg)*	*(kJ/kg)*	*(kcal/kg)*	*(kJ/kg)*
Heat of reaction	116	487	98	412	96	403
Enthalpy $\Delta\vartheta$ (20–1500°C)	449	1886	405	1701	403	1693
Batch gases $\Delta\vartheta$ (20–1500°C)	71	289	33	138	39	164
Theoretical total heat requirement	636	2671	536	2251	538	2260

with increasing pull. A plot of $\dot{q}$ against s, as shown in Figure 5.7, gives almost horizontal lines turning only slightly upwards. For a new, well insulated, correctly set container glass tank $\dot{q}_a = 2000$ MJ/m^{2}d and $b = 4$ MJ/kg glass. The specific fuel consumption, $\dot{q}/s$ (kJ/kg glass), is often plotted against s, the pull to give the relationship

$$s(\dot{q}/s - b) = \dot{q}_a \qquad (5.2)$$

which is an equal sided hyperbola shifted upwards by the distance b. Corresponding theoretical considerations (Figure 5.8 and Section 7.3) tend to agree well with practical results (Figure 5.9). In practice the scatter of actual results is very large since they include a multitude of different influences including that of different glass quality requirements. In general, the specific fuel consumption

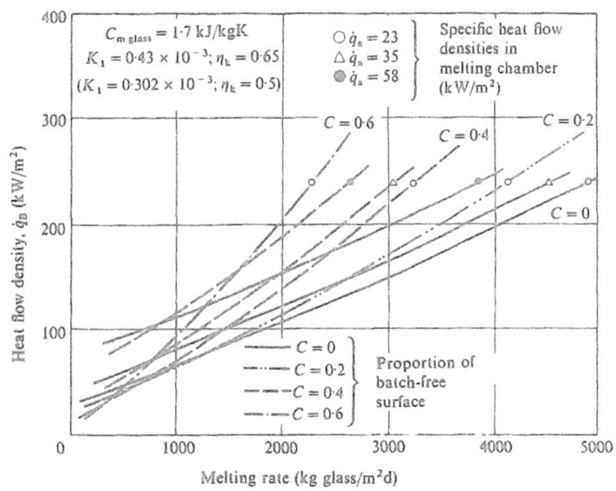

Figure 5.7. Specific heat flow as a function of melting rate (for the parameters chosen see Section 7.2)

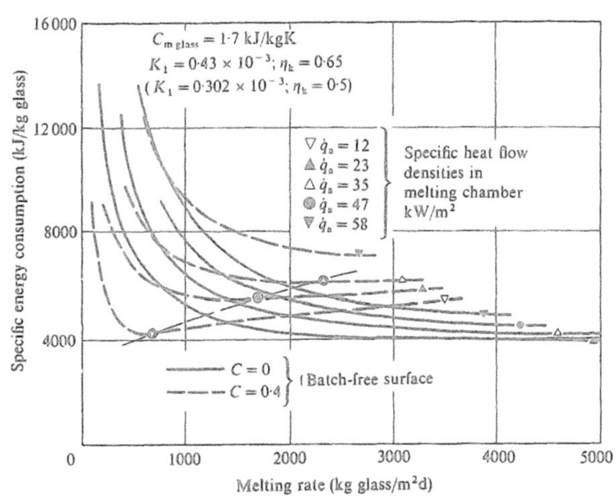

Figure 5.8. Specific energy consumption as a function of the specific melting rate (for the parameters chosen see Section 7.2)

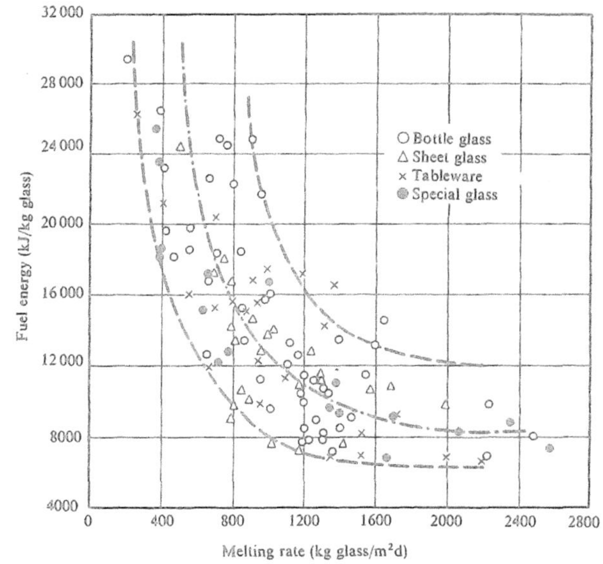

Figure 5.9. Fuel energy introduced as a function of the specific melting area (see also Section 7.2)

decreases with increasing pull, but passes through a minimum beyond which it again increases. The tank is then, from a thermal point of view, pulled to excess. Whether it is possible to reach the minimum specific fuel consumption depends on the design and operation of the tank. If the pull can be increased beyond the minimum point, then factors governing the thermal performance of the tank, such as flame formation and heat recovery, must be improved. On the other hand, if the minimum point cannot be reached then other technical factors must be improved.

5.1.4.1 The influence of furnace age and ambient temperature on fuel consumption

Experience has taught us that the fuel consumption of glass tank furnaces increases with age. The causes are many: wear of superstructure refractories and attack by molten glass on tank blocks reduce their thickness and increases the heat losses; wear of the throat cover block reduces the effectiveness of the throat as a barrier and permits a return current to be formed; wear of the port mouth entry arch reduces control over firing; and materials deposited in the checkers reduce the preheat temperature.

The use of high quality refractories has reduced the increase in fuel consumption considerably. Literature covering the rate of increase is scarce; for container glass tanks the increase is about 0·1–0·2% per month and for flat glass tanks about 0·3–0·4% (Figure 5.10). The major causes are now ascribed to deposits formed in the regenerator and an increasing proportion of induced air leaking through joints etc. into the furnace. In addition, partial loss of insulation in areas of higher refractory wear adds to this increase. Furnaces with large regenerator chambers show a smaller rate of increase than those with small chambers.

Some furnaces show a remarkable annual cycle of fuel consumption, with increases of 10% or more in winter (Figure 5.10); the ambient air temperature affects both the heat losses and the preheat temperature of the combustion air.

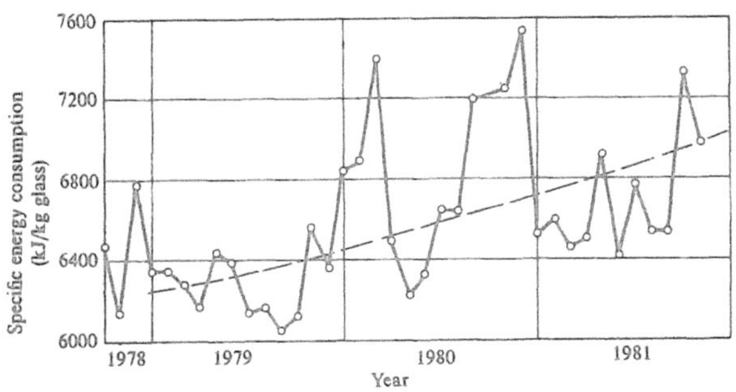

Figure 5.10. Specific energy consumption with time (average monthly figure) of a large Float glass tank

5.1.4.2 The influence of excess air and induced air on fuel consumption

Complete combustion of the fuel fed into a furnace requires a specific minimum of excess air (see Section 11.1.1), which depends on the burner construction and the firing conditions in the combustion space. Every increase in excess air reduces the flame temperature (Figures 11.2 and 11.3) and increases the quantity of waste gases, reducing the heat transfer and increasing the losses in waste gases (Figure 5.6). Fuel and costs can be saved by minimising excess air.

Apart from excess air, any other air introduced cold has an adverse effect on the fuel efficiency of the furnace. This cold air may be introduced as atomising or cooling air, or it may be induced involuntarily and will include air which leaks into the preheat system (whether regenerator or recuperator). Air which leaks into the exhaust side of the waste gas system can be ignored as it is of minor importance.

An indication of the excess air above that required for the combustion of the fuel, the 'air factor', is given by a measurement of the oxygen content of the waste gases between furnace superstructure and regenerator checkers; it can be controlled by the use of suitable sensors (see Section 5.2.5). The air factor alone cannot give the heat loss caused by the excess air; it is also necessary to know the proportion of preheated air and the change in this proportion as the amount of excess air is changed.

An accurate determination of the relationship between air factor, possible saving in fuel consumption, and amount of preheated air is analytically impossible if, as is necessary, the changes in waste gas temperature and heat regeneration are also taken into account. A computer programme covering heat regeneration, combustion, and heat transfer conditions enables such interrelationships to be determined. Results of the calculations are assembled in Figure 5.11. The starting point is the assumption that 10% of the total air with an air factor of 1·245 enters the furnace cold; this must not be overlooked when considering the flows of waste gases and air. The regenerator is designed so that, with a waste gas temperature of 1440°C and an entry temperature of 110°C, an air preheat temperature of about 1200°C is obtained. The evidence leads to the conclusion that a change in the proportion of preheated air has an effect as large as a change in the air factor. The amount of cold air is significant in relation to the amount of fuel consumed. A reduction in the air factor or an increase in the proportion of preheated air of 1% causes a reduction in fuel consumption of about 0·5%.

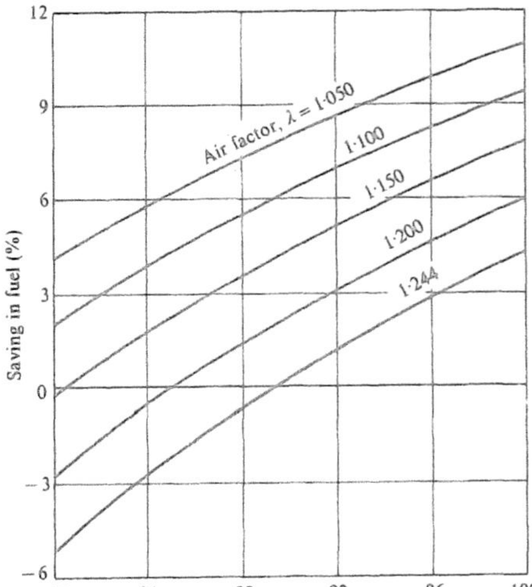

Figure 5.11. Nomogram covering the influence of cold air on fuel consumption

5.1.5 Heat balances[8]

There are many situations when an investigation into the energy consumption of glass tank furnaces is desirable and can provide valuable information for improving the design and method of operation or for making comparisons between plants.

5.1.5.1 Subdivision for balances

The construction of a balance requires clear boundaries to be established between the system to be studied and its surroundings, sometimes referred to as drawing a balance boundary. When this has been drawn it should be noted that, for a balance, all energy flowing in either direction across this boundary must be measured and only if they can be measured with sufficient accuracy is there any point in constructing a balance. In addition, the energy flows must have a parallel value in practice; for example, if the superstructure heat losses are taken overall, then they must include the losses of both the melting and the working ends. Since the losses from the working end, as distinct from those from the melting end, are technically not losses at all, especially for furnaces of high pull, a balance constructed on this basis will be unrealistic in relation to actual conditions. Open furnaces, such as are used in the manufacture of flat glass, present substantial problems in the separation of melting and conditioning zones as far as heat balances are concerned. However, two compartment furnaces with throats give similar problems as it is difficult to define the temperature of the glass and the effect of the return glass flow through the throat if this takes place.

5.1.5.2 Terms used for the construction of heat balances

The first law of thermodynamics is used to define the energy contained in the chemical compounds constituting the fuel and the thermal energy flows. Other forms of energy can usually be ignored as they are of minor importance.

The chemical energy in the fuel is determined by

$$\dot{Q}_c = H_u \dot{M} \qquad (5.3)$$

where H_u is the net calorific value and $\dot{M}$ the mass flow rate. The enthalpies of the various media flowing in the furnace, particularly those of the air, the gases, the waste gases, and the molten glass, to name only the major ones, can be calculated from the well known relationship

$$\dot{H} = c_p \vartheta \dot{M}, \qquad (5.4)$$

where c_p is the average specific heat and ϑ the temperature (°C).

Tanks with electric boosting have part of their energy supplied as electric current.

Other important heat flows are the heat losses of furnace walls via conduction which may be calculated from

$$\dot{Q}_{Lt} = \lambda A \frac{d\vartheta}{dx} \qquad (5.5)$$

where λ is the average thermal conductivity, $d\vartheta/dx$ the local temperature gradient, and A the area.

The heat radiated through openings is calculated from

$$\dot{Q}_R = C_s A \left[\left(\frac{T_1}{100} \right)^4 - \left(\frac{T_2}{100} \right)^4 \right] \qquad (5.6)$$

where C_s is the black body radiation, for which a value of 5·67 W/m^2K^4 may be used.

5.1.5.3 Reference units

When constructing a heat balance, the various flows of energy, and in particular the thermal energy, need to be stated in consistent units. What units are used depends largely on the use to which the results of a heat balance will be put (Figure 5.12).

5.1.5.3.1 Heat introduced stated as a percentage

Usually the heat introduced into the furnace in unit time is taken as a reference parameter and equated to 100%. This normally includes the enthalpies of the combustion air, the fuel, and the batch. Since these energies are usually only about 0·5–3% of the chemical energy provided by the fuel, and since the enthalpies are not of major importance in the operation of a furnace, it is usual to take only the chemical energy of the fuel as the reference datum.

The choice of the fuel energy with or without the enthalpies of air, gas, and batch as data has the great advantage that the energy available from current normal fuels such as oil, natural gas, or town gas, can be determined with accuracy. In addition, a heat balance constructed in this way allows the energy to be divided into meaningful parts which can be related to furnace design. A disadvantage is that different furnaces can only be compared under certain conditions. For example, if

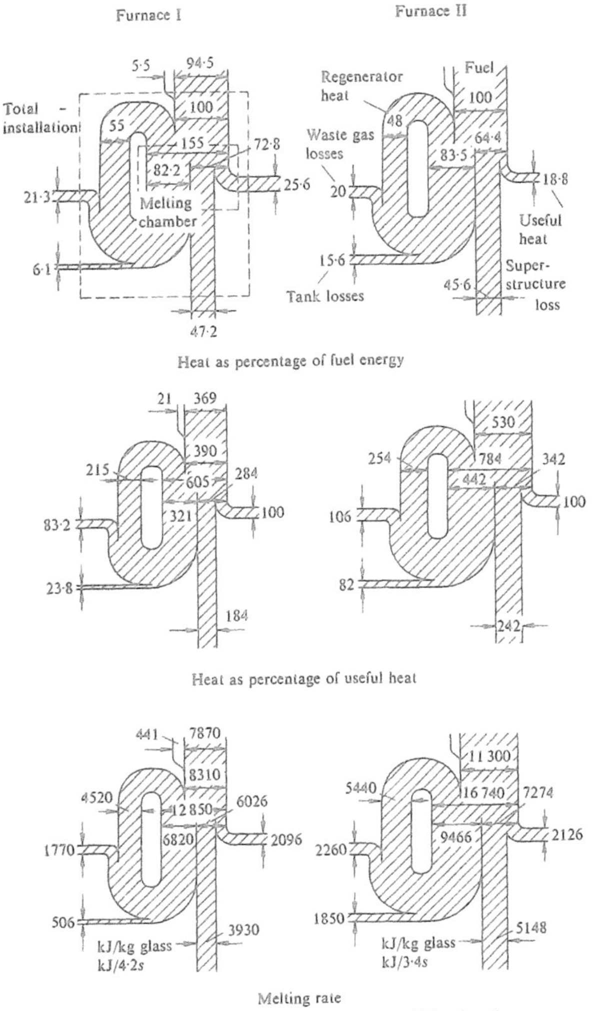

Figure 5.12. Energy balances for two furnaces based on different reference parameters[8]

there are two furnaces with different fuel consumptions but the same heat transfer in a specific part of the furnace, e.g. the working end, then the energy losses here, expressed as a percentage, will be different. If the heat balances of the two furnaces are compared, it will not be possible to decide whether the difference between the two parts is due to different fuel consumptions or different heat losses. Heat balances expressed as percentages are therefore not conclusive.

5.1.5.3.2 Useful heat in the glass stated as a percentage

The enthalpy of the glass as a result of the melting process, which is often referred to as 'useful heat', can be set at 100%, similar to the procedure above for fuel energy. This has the advantage that the heat energy properly required for obtaining, say, 1 kg of molten glass can be readily deduced. A disadvantage is that the useful heat is more a function of the working processes employed than the design of the furnace. A machine making flat glass may call for glass at 950°C, whereas a container glass tank, depending on the processes to be used, will have to supply glass at 1300°C. The problem of working in percentages applies here just as it does for fuel energy as described above. Furthermore, it is still impossible to determine the useful heat of glasses accurately; the uncertainty surrounding this reference parameter brings uncertainty to all the others.

5.1.5.3.3 Heat balance using unit weight of glass as a reference parameter

It would appear sensible to refer to the quantities in a heat balance involving molten glass not in percentages but in terms of the weight of glass produced, i.e. in kJ/kg glass; this has the advantage that the energy required to melt glass stands out clearly in the heat balance. Comparisons between different furnaces may then be made much more easily than with balances based on percentages; the function of the furnace is, after all, to supply glass at a specified working temperature, not just to use up energy. The difficulty of defining useful heat becomes irrelevant or affects other heat quantities only as far as residual heat (losses from the superstructure) is concerned.

A disadvantage of this approach is the different densities of glasses so that furnaces melting heavy lead glass, for example, will appear in a comparatively favourable light. This suggests that a unit volume of glass (say, 1 dm^3) not the unit weight should be the reference quantity; this argument is strengthened by the fact that glassware is made with reference to a certain volume, not weight, of glass. However, since the densities of the glasses used in large scale production (containers, flat glass) are nearly the same, and since weight is easier to determine than volume, on balance it is better to use 1 kg as the reference quantity.

5.1.5.3.4 Unit of time

It is also possible to interlink the different heat quantities by reference to a specified time interval, thus giving flow rates. The disadvantage is that a comparison of different balances becomes more difficult because different plants require different quantities of energy and different heat centres are only comparable quantitatively in absolute terms. In a way, a balance based on 1 kg glass can also be constructed on the basis of a unit of time, i.e. the time necessary for melting 1 kg of glass.

5.1.5.3.5 Examples

Heat balances of modern container glass tanks have been published by Schulte-Kellinghaus[5] who chose as reference quantities the heat consumed expressed as a percentage and the quantities of heat per kg of molten glass (Figure 5.13).

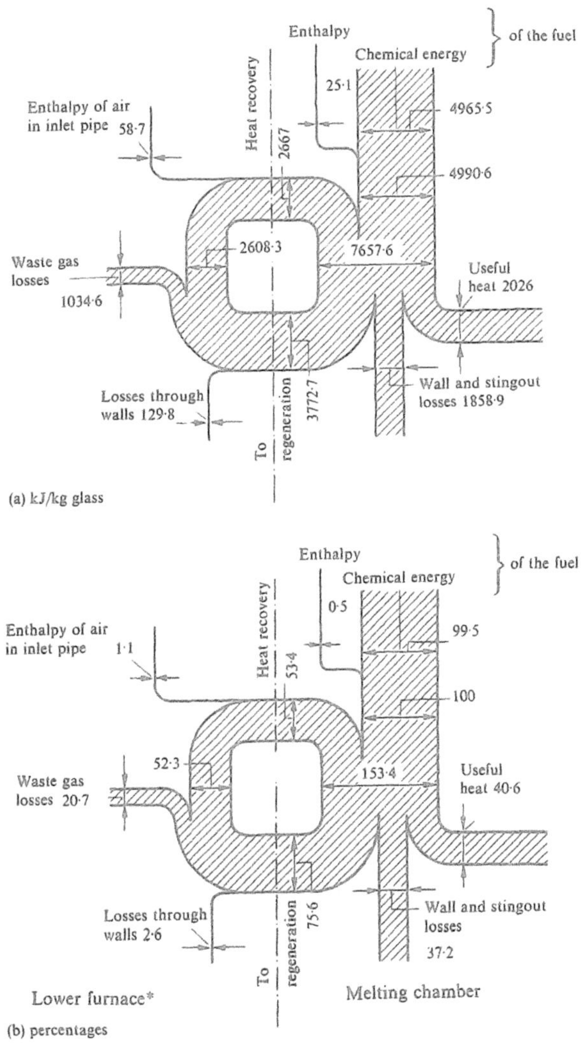

Figure 5.13. Energy balance of a cross-fired tank furnace of 100 m^2 melting area (clear glass, heated with oil grade S, $s = 3$ t/m^{2}d, $K = 4{\cdot}2$ m^3/m^2, air preheat temperature 1400°C, waste gas temperature beyond reversal valve 350°C)
*for definition see Figure 2.1

5.1.5.4 Multi boundary balances

If a single boundary heat balance cannot provide results of the required accuracy then it is advisable to construct a multi boundary balance. However, the measurement techniques are very demanding so their construction can only be justified in exceptional circumstances.[8]

Multi boundary balances can be represented clearly in a diagrammatic form as Figure 5.14 shows. The individual secondary boundaries (regenerators, melting end, working end) are best presented diagonally from upper left to lower right of the diagram so that it is then possible, even for complicated representations, to follow the convention by which inward energy flows from above or below and outward energy flows to the sides, heat

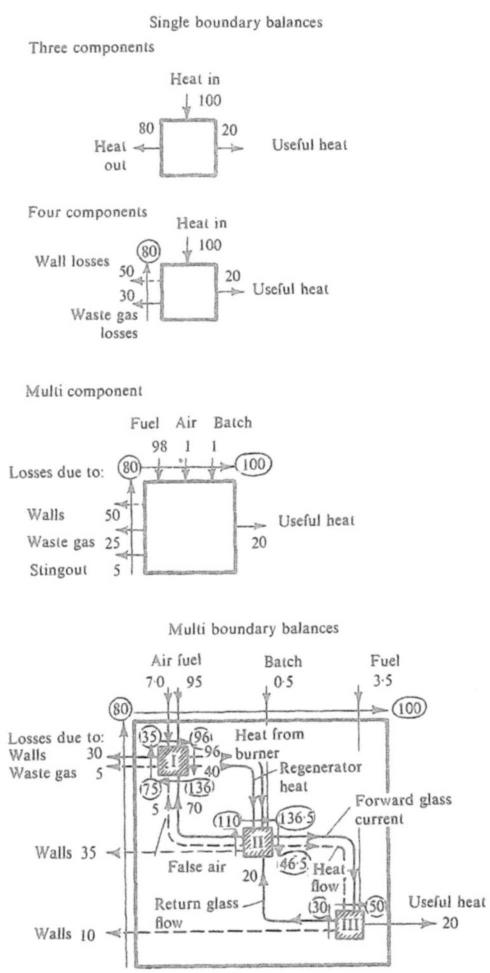

I = Lower furnace (for definition see Figure 2.1)
II = Melting chamber
III = Working end

Figure 5.14. Construction of energy balances as simple linear diagrams[8]

losses to the left and useful heat to the right. Flow directions are clearly indicated by arrows. The representation of a heat balance in this form may at first sight appear unusual, but has many advantages.

5.1.5.5 Exergy balances[9,10]

The use of the first law of thermodynamics in the evaluation of any technical process involving heat, and this includes glass melting furnaces, has the major disadvantage that the presentation of the value of the heat energy and its further use in such equipment as waste heat boilers and steam driven electric generators is distorted. The drawing up of an exergy balance is recommended to overcome this point, but has found little acceptance in practice so far.

Exergy is defined as that form of energy which can be changed into any other form of energy and thereby into mechanical work. For energies which cannot so be changed the term 'anergy' has been suggested. There are forms of energy which consist wholly of either exergy or anergy and there are also forms of energy which consist of both types.

Exergy balances, analogous to energy (heat) balances, can be drawn up. Calculations of exergy flows must include entropy and enthalpy. Exergy has the same dimensions as energy and is calculated from the enthalpy, $\dot{H}_1$, of the carrier of the energy flow in whatever form it is available; the enthalpy of the same energy carrier when it is exhausted to a point at which it is in equilibrium with its surroundings, $\dot{H}_0$, is calculated from the difference in entropy flow $(S_1 - S_0)$ between the conditions 1 and 0:

$$\dot{E}_{1,0} = \dot{H}_1 - \dot{H}_0 - T_0(\dot{S}_1 - \dot{S}_0). \qquad (5.7)$$

The exergy of a quantity of heat, Q, which is removed from a system at a temperature, T, is given for constant temperatures by

$$E_Q = Q\frac{T - T_0}{T}, \qquad (5.8)$$

where T and T_0 are the thermodynamic temperatures.

Figure 5.15 illustrates the energy and exergy flows of a glass tank furnace presented opposite one another, the reference parameter being the useful heat of the glass. The differences between the energy and exergy balances are essentially in the areas of combustion losses and waste gas losses.

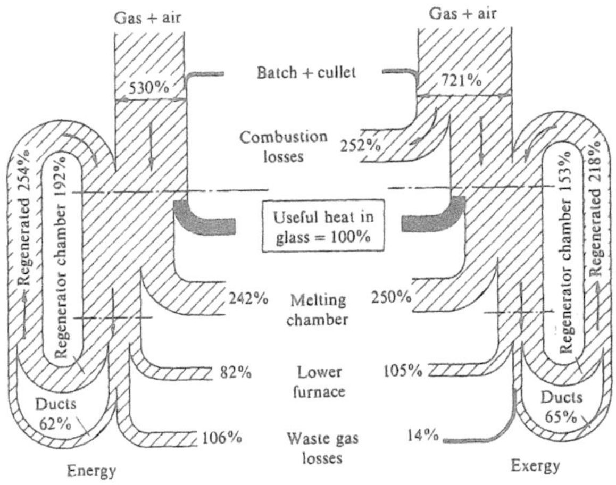

Figure 5.15. Energy and exergy flow diagram of a glass tank furnace[9]

5.1.6 Thermal insulation and cooling

5.1.6.1 Thermal insulation

The consumption of thermal energy can be reduced by providing the glass furnace with good insulation. Reservations about insulation and its avoidance in the past were due to limitations to the temperature to which the furnace superstructure could be subjected. Better refractories and improvements in the design and operation of furnaces have significantly enlarged the possibilities of insulation.

However, more insulation also creates more risks. The ability to observe refractories operating under critical conditions is more limited and a rapidly deteriorating situation may not be recognised in time; defects in the glass or, in extreme cases, a breakout of glass or collapse of the crown may result. Thorough insulation assumes sound control of the heat up and operation of the furnace. Insulation should not be applied without a full consideration of the consequences. The temperature of each layer of insulation must be calculated and the proper materials selected.

Not all parts of the furnace can be insulated. The flux line and throat of a tank must be left uninsulated and after a period of operation, or even from the beginning, they will have to be cooled to extend furnace life, a contradictory procedure from the point of view of the furnace. The attack at the flux line and on the throat cover block is so great that a balanced wear of different parts of a furnace can only be achieved by partial cooling.

In certain parts of the furnace and under certain operating conditions insulation must be avoided or limited for operational reasons. After refining, the glass temperature must be reduced as quickly as possible to that required for working; circumstances, especially the pull, will dictate whether the heat losses from the working end, whether it is provided with insulation or not, are sufficient to remove this heat. The conditioning zones of flat glass tanks have to be provided with openings which permit heat losses by radiation in order to achieve the necessary reduction in temperature; on the other hand, container glass tanks operating at high pull have to have their working ends insulated and supplied with additional heat.

Day tanks and pot furnaces present special cases because excessive insulation can make the periodic operation of such furnaces difficult.

These factors confirm that the provision of insulation must be carefully considered; each case must be studied in depth to assess the effects aimed for and likely to be achieved, points which are made particularly strongly by Boggum.[15]

5.1.6.1.1 Calculation of thermal insulation

In terms of space glass furnaces are complicated structures, which makes design calculations for insulation difficult. Essentially, however, we are dealing with the tank walls, i.e. the sidewall blocks containing the glass, the superstructure sidewalls, the crown, the regenerators, etc., which are almost one dimensional structures and which can therefore be treated as such for design calculations.

For a flat wall consisting of several layers with a hot face temperature, ϑ_h, and an ambient temperature, ϑ_k, the temperature gradients under stable conditions and for n layers will be like the example

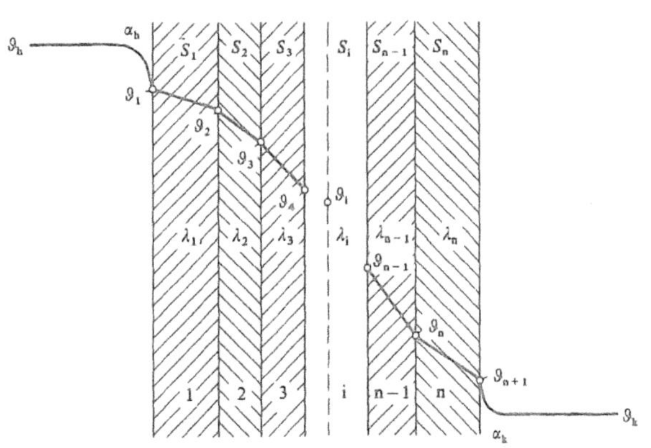

Figure 5.16. Temperature profile of a wall consisting of parallel layers

shown in Figure 5.16. The heat flow, $\dot{q}$, at right angles to the hot face may be written

$$\dot{q} = \frac{\vartheta_h - \vartheta_k}{\dfrac{1}{\alpha_h} + \dfrac{1}{\alpha_k} + \dfrac{s_1}{\lambda_1} + \dfrac{s_2}{\lambda_2} + \ldots + \dfrac{s_n}{\lambda_n}} \tag{5.9}$$

where α_h and α_k are the heat transfer coefficients of the hot and cold surfaces respectively and S is thickness and λ the thermal conductivity of each layer.

The selection of suitable materials depends on the intermediate temperatures, ϑ_2, ϑ_3, etc., and these are governed by the equations

$$\vartheta_h - \vartheta_1 = \frac{\dot{q}}{\alpha_h},$$

$$\vartheta_1 - \vartheta_2 = \dot{q}\frac{s_1}{\lambda_1},$$

$$\vartheta_2 - \vartheta_3 = \dot{q}\frac{s_2}{\lambda_2},$$

$$\vartheta_i - \vartheta_{i+1} = \dot{q}\frac{s_i}{\lambda_i}.$$

The intermediate temperature, ϑ_i, may be calculated from

$$\vartheta_i = \vartheta_h - (\vartheta_h - \vartheta_k) \times \frac{\dfrac{1}{\alpha_h} + \dfrac{s_1}{\lambda_1} + \dfrac{s_2}{\lambda_2} + \ldots + \dfrac{s_{i-1}}{\lambda_{i-1}}}{\dfrac{1}{\alpha_h} + \dfrac{1}{\alpha_k} + \dfrac{s_1}{\lambda_1} + \dfrac{s_2}{\lambda_2} + \ldots + \dfrac{s_n}{\lambda_n}}. \tag{5.10}$$

The above equations assume good (ideal) thermal contact at the interfaces.

Data governing dimensions are known with some accuracy and to a certain extent this also applies to the thermal conductivity of each insulating layer. But special consideration and calculations must be applied to the heat transfer coefficients, α_h and α_k.

Due to the high temperatures involved α_h is many times larger than α_k and consequently its influence on $\dot{q}$ and ϑ_i is small. Accurate calculations are difficult and literature references of doubtful value: Brüning[17] uses $\alpha_h = 232\ \mathrm{W/m^2K}$ for the tank walls (glass contact refractories) and Hrma[17] $815\ \mathrm{W/m^2K}$ for the flux line regions. It is certain that α_h for the tank walls is strongly dependent on glass depth and colour. Differences in results could also be related to methods of measurement. The value of α_h for the superstructure can be taken as 500–1000 $\mathrm{W/m^2K}$ (see Sections 6.3 and 7.33).

Calculation of α_k involves separate calculations of the heat transfers by convection, $\alpha_{k,c}$, and radiation, $\alpha_{k,r}$. When compared to α_h, α_k is very much smaller and therefore has a more persistent influence on temperature distribution and heat flux density in the wall.

It is possible to derive $\alpha_{k,c}$ from a study of similar heat transfer situations involving free, laminar, and turbulent flow on vertical walls of height h at constant surface temperature where the following relationship applies:[18]

$$Nu = 0{\cdot}517(Gr\,Pr)^{1/4} \qquad (5.11)$$

where Nu is the dimensionless number for heat transfer (the Nusselt number), Gr is the Grashof number, and Pr is the Prandtl number. The Nusselt number is calculated from $\alpha l/\lambda$, where α is the heat transfer value (in this case $\alpha_{k,c}$), l the height of the wall, and λ the thermal conductivity of the surrounding air at a temperature of $(\vartheta_{n+1} + \vartheta_k)/2$. The Grashof number is given by

$$Gr = \frac{ghl^2 \dfrac{\rho_\infty - \rho_0}{\rho_0}}{\nu^2}$$

where g is the acceleration due to gravity, h the width of the wall, ρ_0 the air density at temperature ϑ_{n+1}, ρ_∞ the density of the surrounding air, and ν the kinematic viscosity of the air at $(\vartheta_{n+1} + \vartheta_k)/2$.

The Prandtl number is given by

$$Pr = \frac{\nu}{a} = \frac{\eta c}{\lambda},$$

and for air under normal ambient pressure is only slightly dependent on temperature, so that $Pr = 0{\cdot}67$ can be assumed.

Equation (5.11) is plotted in Figure 5.17. When used in conjunction with the data given in Table 5.6 for air, Gr and Pr can be readily determined and Figure 5.17 then provides Nu, from which $\alpha_{k,c}$ may be calculated; $\alpha_{k,c}$ is usually between 5 and 10 W/m²K.

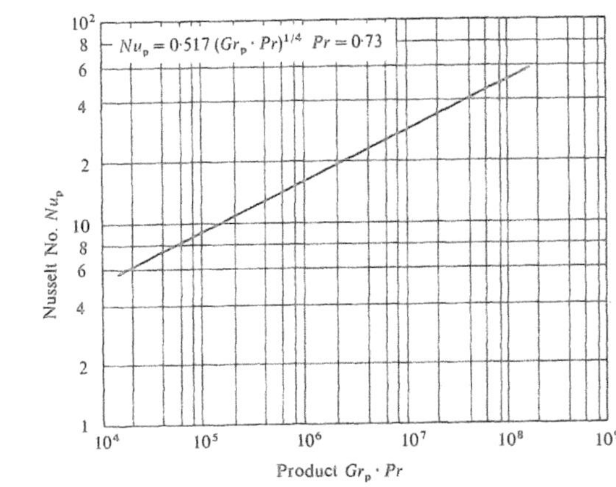

Figure 5.17. Heat transfer by convection for free laminar and turbulent flow[18]

Table 5.6. *Properties of dry air at a pressure of 1 bar*[18]

ϑ temperature
ρ density
C_p specific heat capacity at constant pressure
β coefficient of thermal expansion
λ thermal conductivity
η dynamic viscosity
ν kinetic viscosity
α thermal conductivity
Pr Prandtl number

ϑ (°C)	ρ (kg/m³)	c_p (kJ/kg K)	β (× 10⁻³/K)	λ (× 10⁻³W/m K)	η (× 10⁻⁶kg/m s)	ν (× 10⁻⁶m²/s)	α (× 10⁻⁶m²/s)	Pr
0	1·2754	1·006	3·671	24·54	17·10	13·41	19·1	0·70
20	1·1881	1·007	3·419	26·03	17·98	15·13	21·8	0·70
40	1·1120	1·008	3·200	27·49	18·81	16·92	24·5	0·69
60	1·0452	1·009	3·007	28·94	19·73	18·88	27·4	0·69
80	0·9859	1·011	2·836	30·38	20·73	21·02	30·5	0·69
100	0·9329	1·012	2·684	31·81	21·60	23·15	33·7	0·69
120	0·8854	1·014	2·547	33·23	22·43	25·33	37·0	0·68
140	0·8425	1·017	2·423	34·66	23·19	27·53	40·5	0·68
160	0·8036	1·020	2·311	36·07	24·01	29·88	44·0	0·68
180	0·7681	1·023	2·209	37·49	24·91	32·34	47·7	0·68
200	0·7356	1·026	2·115	38·91	25·70	34·94	51·6	0·68
250	0·6653	1·035	1·912	42·43	27·40	41·18	61·6	0·67
300	0·6072	1·046	1·745	45·91	29·20	48·09	72·3	0·67
350	0·5585	1·057	1·605	49·31	30·90	55·33	83·5	0·66
400	0·5170	1·069	1·485	52·57	32·55	62·95	95·1	0·66
450	0·4813	1·081	1·383	55·64	34·00	70·64	107	0·66
500	0·4502	1·093	1·293	58·48	35·50	78·86	119	0·66
600	0·3986	1·116	1·145	63·5	38·30	96·08	143	0·67
700	0·3577	1·137	1·027	67·8	40·87	114·3	166	0·69
800	0·3243	1·155	0·932	71·3	43·32	133·6	190	0·70
900	0·2967	1·171	0·852	74·3	45·65	153·9	214	0·72
1000	0·2734	1·185	0·786	76·8	47·88	175·1	237	0·74

Little is known about the effect of position on heat transfer. Schack[19] based his results on old measurements of Rosin and states that, for horizontal surfaces, the upper and lower surfaces have heat transfer figures which are 6% higher and 6% lower respectively than a vertical surface, i.e. $\alpha_{k,c}$ (horizontal, upper surface) = $1{\cdot}06\,\alpha_{k,c}$ and $\alpha_{k,c}$ (horizontal, lower surface) = $0{\cdot}94\,\alpha_{k,c}$.

The radiation component, $\alpha_{k,r}$, may be calculated from the well known laws of radiation

$$\alpha_{k,r} = \frac{C_s(T_{n+1}^4 - T_c^4)}{\left(\dfrac{1}{\varepsilon_{n+1}} + \dfrac{1}{\varepsilon_c} - 1\right)(T_{n+1} - T_c)} \qquad (5.12)$$

where $C_s = 5{\cdot}67\ \mathrm{W/m^2K^4}$ is the emissivity of a black body and ε_{n+1} and ε_c the emissivities of the body involved in the exchange of radiation. Parts of a furnace radiating freely into their surroundings can be assumed to have $\varepsilon_c = 1$ and this gives

$$\alpha_{k,r} = \varepsilon_{n+1} C_s \frac{T_{n+1}^4 - T_c^4}{T_{n+1} - T_c}.$$

Often T_c^4 can be ignored when compared to T_{n+1}^4 and this gives

$$\alpha'_{k,r} = \varepsilon_{n+1} C_s \frac{T_{n+1}^4}{\vartheta_{n+1}}.$$

Table 5.7. *Radiation of a black body*

Temperature (°C)	*(K)*	*Radiation flux density (kcal/m²h)*	*(W/m²)*
0	273	276	320·93
50	323	540	627·9
100	373	960	1 020·2
150	423	1 587	1 845·3
200	473	2 483	2 887
250	523	3 711	4 315·1
300	573	5 349	6 219·8
350	623	7 473	8 689·6
400	673	10 175	11 831
450	723	13 550	15 755
500	773	17 715	20 599
550	823	22 755	26 459
600	873	28 810	33 500
650	923	36 000	41 860
700	973	44 450	51 686
750	1023	54 340	63 186
800	1073	65 820	76 535
850	1123	78 900	91 745
900	1173	93 900	109 187
950	1223	110 950	129 013
1000	1273	130 230	151 431
1050	1323	152 030	176 780
1100	1373	176 300	205 002
1150	1423	203 370	236 479
1200	1473	233 500	271 514
1250	1523	266 900	310 351
1300	1573	303 600	353 026
1350	1623	344 100	400 119
1400	1673	388 600	451 864
1450	1723	437 200	508 376
1500	1773	490 100	569 883
1550	1823	547 900	637 098
1600	1873	610 300	709 657

Table 5.8. *Emissivity of refractories*[21]

Refractory	*Temperature (°C)*	*Emissivity*
Silica (98% SiO_2)	1000	0·62
	1200	0·54
	1400	0·49
	1500	0·46
Clay (38% SiO_2, 58% Al_2O_3, 0·5% Fe_2O_3)	1000	0·61
	1200	0·52
	1400	0·47
	1500	0·45

Table 5.9. *Emissivity of some metals*

Metal	*Surface condition*	*Temperature (°C)*	*Emissivity*
Iron	Heated to redness	20	0·61
	Rolled	20	0·66
	Clean polished	20	0·29
Copper	Polished	20	0·04
	Stained	23	0·048
	Oxidised black	25	0·78
Aluminium	Polished	23	0·052
	Raw	28	0·07
Zinc	Galvanised steel sheet	28	0·23

Table 5.7 gives the heat flux of a black body.

The emissivities of clay, silica, and various metals are given in Tables 5.8 and 5.9. Unfortunately, similar data for insulation materials are not available.

The determinations of α are followed by determinations of the thermal conductivities, λ, of each layer which should then be entered in Equations (5.9) and (5.10). Values of λ for insulating materials are given in Table 5.10.

5.1.6.1.2 Installation of thermal insulation

5.1.6.1.2.1 Glass contact refractories

Recent experience has shown that, except in the immediate vicinity of the flux line, tank blocks can be insulated provided void-orientated fusion cast refractories in the form of soldier blocks are used and these have been ground flat to guarantee tight vertical joints.[22] Insulation is applied in two layers, the hotter side a tank block refractory and the outside a lightweight clay quality;[15] Figure 5.18 gives an example of an insulated sidewall by which heat losses can be reduced from about 12 500 to 3000 W/m². The thickness of fusion cast soldier blocks has recently been reduced to 200 mm. The

Table 5.10. *Thermal conductivity of refractory insulating materials*[20]

	Temperature limit of application (°C)	*Thermal conductivity (W/mK) at 300°C*	*at 700°C (600°C)*	*at 1000°C (900°C)*
Insulation bricks				
Siliceous insulation bricks, fired	1250	–	0·3	0·39
Siliceous insulation bricks, chemically bonded	1200	0·26	0·34	0·34
Fireclay insulation bricks, cement bonded	1100–1250	–	0·23–0·27	–
Forsterite insulation bricks	1500	–	0·4–0·56	–
ZrO_2 insulation bricks – oxidising conditions	1800–2400	–	–	0·4
– reducing conditions	1600	–	–	–
Unformed insulation materials				
Fireclay light weight cement used in load bearing application	1200	0·4	(0·42)	(0·47)
Fireclay light weight cement used in load bearing application, as poured	1100	0·22	(0·28)	(0·40)
Light weight concrete with organic material removed during firing	800–1600	0·16	(0·18)	(0·31)
Bubble alumina light weight concrete	1700	0·8	(0·58)	(0·67)
Refractory light weight cement for spraying	1250	0·30	(0·33)	(0·36)
Vermiculite tiles	< 1000	–	(0·1)	–

uninsulated section is 200 mm high for clear glass, and 150 mm high for coloured glass.

Good face-to-face contact between soldier blocks and insulation layers is important and it is best to use a special cement. Vertical joints of tank blocks

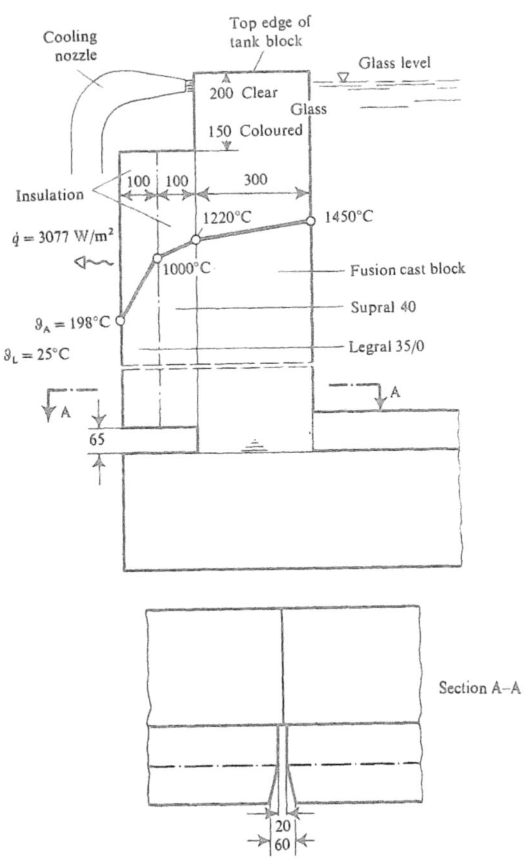

Figure 5.18. Heat losses from an insulated sidewall of a glass tank; temperature at flux line 1450°C[15] (dimensions in mm)

are either left exposed or, if insulation is applied in two layers, are left exposed at the top and covered with offset joints at the bottom.

Insulation of tank sidewall blocks does not increase their rate of wear but gives a more even wear. The insulation of working and sidewalls is best done by applying clay tiles 150 mm thick.

Insulation of tank bottoms is similar to that of the sidewalls. The risk of a breakout with its disastrous consequences is avoided by using a multilayer construction such as that shown in Figure 5.19, especially for clear glass; with this structure, heat losses can be reduced from 5100 to 1900 W/m². Of special importance is the installation of fusion cast paving blocks laid either on a bed of cement or, better, on a layer of special ramming compound of the same thickness. This prevents the glass from penetrating under the tiles and subsequent accelerated metal attack. A major influence on temperatures and heat losses is the hot face temperature of the floor and this in turn is affected by whether clear or coloured glass is being melted.

Superstructure sidewalls are insulated like the tank blocks of the working end, by a single layer of lightweight firebrick. The wall must be gas-tight and free from open joints which would cause alkali condensation leading to erosion and glass defects. The heat losses can be reduced from about 11 000 to 2300 W/m².

5.1.6.1.2.2 Crown and sidewalls

Due to its large area, the crown will give a significant saving in energy when insulated. A

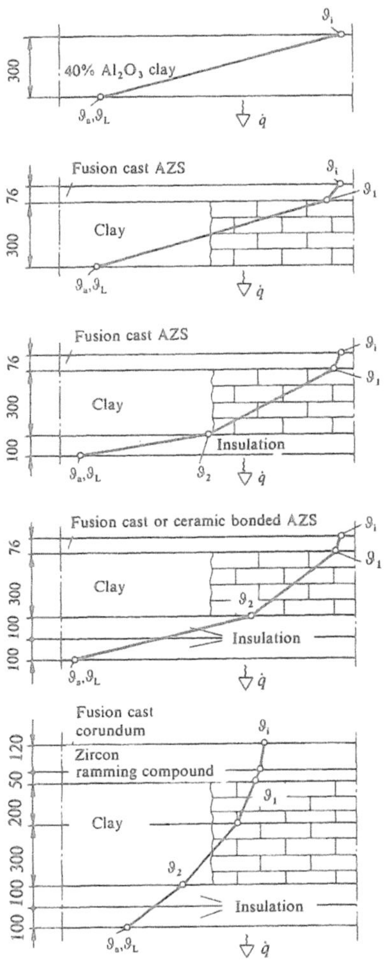

Temperature (°C) ϑ_i	ϑ_1	ϑ_2	ϑ_a	ϑ_L	Heat losses (W/m^2)	($kcal/m^2h$)
1400			260	20	5171	4446
1400	1339		252	20	4931	4240
1400	1366	770	166	20	2761	2374
1400	1377	968	142	20	1980	1614
1200	1127	685	156	45	1195	1030

Figure 5.19. Heat losses through glass tank bottoms[15]

precondition for insulation is that the crown must have been constructed of well converted silica bricks of accurate dimensions and low deformation under pressure. The pressure between faces should not exceed 0·2 N/mm². Silica crowns are now usually 375 mm thick and their load carrying ability is calculated on the basis that two-thirds of the outer thickness can be load bearing.

Of special importance is the need to make the crown gas-tight, to carry out the heat up correctly, and to cover the crown with slurry after heat up. Insulation consists of one or two layers of light weight silica bricks laid on top of a special cement. The upper insulation layer must be covered with a joint-free sealing material to prevent the ingress of batch dust; this material should be strong enough to bear the weight of a person walking on it. Insulation mats with aluminium foil on the outside are now used, the foil preventing the ingress of batch dust. Insulation of crowns can reduce heat losses from 4800 to 2300 W/m².

Heat losses through such structures as port necks and regenerator crowns are considerable due to the high thermal conductivity of the basic bricks from which they are built. If the hot face temperature is about 1550°C then the loss will be about 15 000 W/m² but insulation using light weight firebrick 64 mm thick will reduce this to 5800 W/m². Losses from regenerator arches provided with 100 mm of insulation are reduced to about 3000 W/m².

The large surface of regenerator chambers can be insulated with building bricks or by the application of insulating concrete.

5.1.6.2 Cooling of tank blocks

The wear of tank blocks is very uneven, the attack being strongest at the flux line and the throat, and premature failure of these parts is prevented by cooling. The benefit of cooling arises from the reduction of the hot face temperature and thereby of the rate of corrosion. Figure 4.13 showed that the corrosion rate of even fusion cast blocks rises strongly at temperatures above 1400°C.

5.1.6.2.1 The effect of cooling

In practice, cooling can only be achieved by raising the heat transfer coefficient on the outside of the block. Equation (5.9) gives the heat flux density for an uninsulated block as

$$\dot{q} = \frac{\vartheta_h - \vartheta_k}{\dfrac{1}{\alpha_h} + \dfrac{1}{\alpha_k} + \dfrac{s_1}{\lambda_1}}.$$

If one recalls that α_h is many times greater than α_k, then, for $\vartheta_1 \approx \vartheta_h$,

$$\vartheta_1 \approx \dot{q}\left(\frac{1}{\alpha_k} + \frac{s_1}{\lambda_1}\right) + \vartheta_k.$$

For fusion cast blocks 200–300 mm thick s_1/λ_1 is several times larger than $1/\alpha_k$, which means that a temperature reduction via α_k is only possible to a limited extent. Measurements have shown that the reduction is only about 10 degC.[24] Only if the wall thickness is reduced to 100 mm and less is the cooling effect more noticeable.

Daudans[25] has drawn attention to the fact that the corrosion of tank blocks is a function not only of temperature but also of the glass currents near the blocks. These currents are mainly convective and are encouraged by cooling the blocks; increased cooling can therefore lead to a reduction in the hot face temperature and a consequent increase in the convection currents and corrosion of blocks. Daudans believes that there is a definite maximum in the convection currents at 1500–1550°C. It follows that the rate of corrosion of tank blocks is not constant. If the effect of convection currents is included, then the corrosion rate increases with decreasing block thickness until, at about 100 mm or less, it decreases significantly.

These considerations do not apply to the throat since the convection currents in this area are not influenced by cooling.

5.1.6.2.2 Design of a cooling system

In the majority of cases cooling is by air, nozzles connected to a supply system being placed at the height of the flux line along the sides of the melting end so that the air is blown against the blocks (Figure 5.18). The heat transfer coefficient which may be achieved with such a system will vary widely as one is considering flow against a baffle, an arrangement which is difficult to calculate in advance: optimum cooling is achieved by blowing air at right angles against the tank block[16] but this setting also gives the greatest pressure loss.[23] The location of nozzles and joints must be such that no cooling air can enter the interior of the furnace as this would negate the effect of the cooling air, quite apart from causing an increase in the uncontrolled air induced into the combustion process. Examples of throat cooling are given in Figures 3.14 and 3.15.

The effect of air cooling on heat transfer can be calculated to an order of magnitude for flows along a flat wall by using known equations.[18] Using the theory of similarity for heat transfer, functions have been developed similar to those for heat transfer by convection which link the dimensionless Nusselt, Grashof, and Prandtl numbers with one another. For the laminar component

$$Nu_{\mathrm{lam}} = 0{\cdot}664\sqrt{Re}\,\sqrt[3]{Pr}, \tag{5.13}$$

and for the turbulent component

$$Nu_{\mathrm{turb}} = \frac{0{\cdot}037\,Re^{0{\cdot}8}\,Pr}{1 + 2{\cdot}443\,Re^{-1}(Pr^{2/3} - 1)}. \tag{5.14}$$

The relevant length is the length of the flat surface on which the air impinges; in the case of tank block cooling this is the distance between the upper edge of the tank block and the line where thermal insulation begins (Figure 5.18).

Since we must assume that both laminar and turbulent zones of flow will occur, Equations (5.13) and (5.14) can be combined. For a simple evaluation, the function $Nu = f(Re, Pr)$ is shown diagrammatically in Figure 5.20 against the relevant Prandtl number: the Prandtl number for air is 0·7. The heat transfer coefficients attainable with air cooling are about 200 W/m^{2}K and the exit flow rates from the nozzles are about 30 m/s in these instances. There is little point in going above this flow rate since the cooling effect increases only slowly while the energy requirement increases exponentially. The quantity of cooling air required is 0·3–0·5 m^3/s at a pressure in front of the nozzles of 0·005–0·01 bar. The layout and distribution of cooling air ducting is best left to be decided according to local conditions. Figure 5.21 shows an old air cooling system for a tank: the preference nowadays is for the ducting to be located between the regenerator chambers and the tank. The power required for the air cooling fan is, on the average, about 0·6 kW for each metre of tank block.

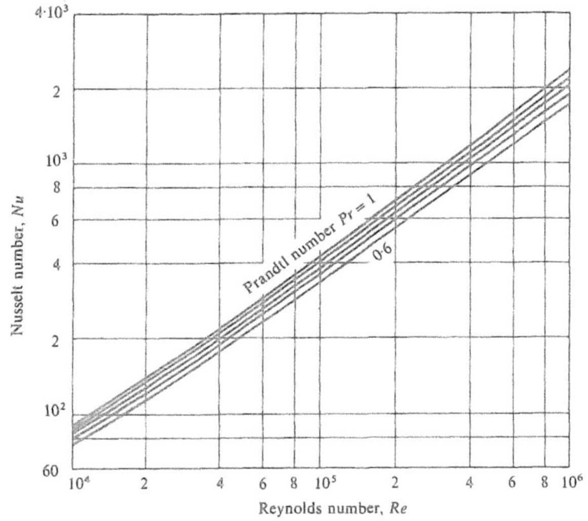

Figure 5.20. Heat transfer on a wall cooled along its length[18]

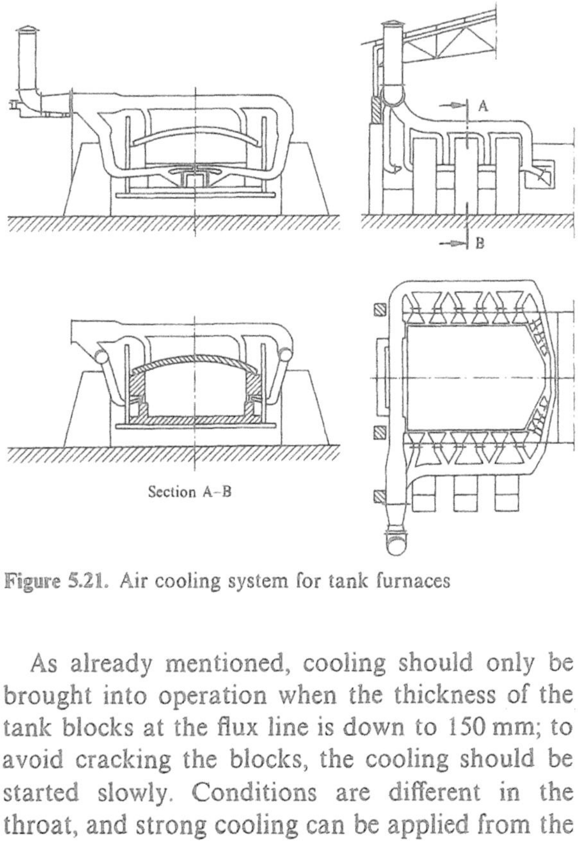

Figure 5.21. Air cooling system for tank furnaces

As already mentioned, cooling should only be brought into operation when the thickness of the tank blocks at the flux line is down to 150 mm; to avoid cracking the blocks, the cooling should be started slowly. Conditions are different in the throat, and strong cooling can be applied from the beginning.

As an alternative to air cooling, water cooling deserves a mention. Although calling for less energy and creating less noise, the investment costs are higher and the effect achieved depends strongly on the system. Cooling by water boxes, even if they are well fitted and installed using a contact cement, is not as effective as air cooling[16] and should only be used if a glass breakout appears imminent. Direct cooling of blocks by water is very effective but the refractories easily craze and crumble, and water running down from the cooled blocks could well cause damage in other areas. In all cases softened water should be used supplied, whenever possible, from a recirculating system.

The best method of cooling, whether by air or by water, is useless if it fails and the failure is not immediately noticed. Controls are essential to ensure that a failure of the cooling system is noticed immediately and that corrective action is taken at once. When considering power failure and the installation of a standby electric generator, the cooling of tank blocks should have a high priority.

5.2 Measurement and control

5.2.1 Temperature

Temperature plays the dominant role in the glass melting operation. The processes of melting, refining, and working are all primarily temperature-dependent and satisfactory performance and life of refractories can only be achieved within certain temperature limits. In practice, temperatures in glass furnaces are measured by thermocouples and radiation pyrometers, the latter being used only in special cases or as control instruments measuring a part of the radiation.

Temperature measurement by thermocouples depends on the Seebeck effect: if two wires of different metals or alloys are joined at each end, and each joined end is exposed to a different temperature, then a small direct current voltage can be measured at a suitable point in the circuit, its magnitude depending on the two metals or alloys and the temperature difference between the ends. If one temperature is known, then the voltage obtained is a measurement of the other; the voltage is usually measured at or near the cold junction. The junction exposed to the temperature to be measured is referred to as the hot junction, the one for comparison as the cold junction.

The thermocouples mainly used in the glass industry are of iron/constantan, nickel–chrome/nickel–aluminium, platinum–rhodium/platinum, and platinum–rhodium 30/platinum–rhodium 6 (International Code B); these are suitable for continuous use at 700, 1000, 1400, and 1600°C respectively. The curves relating temperature to thermocouple voltage have all been standardised in DIN 43710 (with the exception of the Code B type) and the temperature–voltage changes for the most common thermocouples are given in Figure 5.22 and Table 5.11.

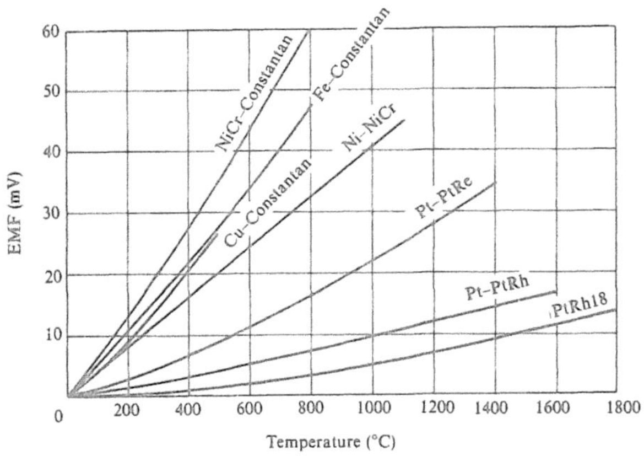

Figure 5.22. Temperature/voltage curves of the more common thermocouples[12]

Protection against mechanical damage and aggressive media is provided by protective sheaths made of metal or ceramics. From the head of the sheath, where the thermocouple ends in clamps, compensating cables lead to the cold junction. The compensating leads are insulated and are suitable established conductors which match the voltage curve of the relevant thermocouple up to 200°C but are constructed of cheaper materials. A schematic drawing of a thermocouple circuit is given in Figure 5.23.

Compensating methods are now used for the measurement of temperature by thermo-electric voltage; these overcome the influence of changes in resistance which would otherwise affect the measurement itself. The thermo-electric voltage to be measured, U_x, is opposed by a comparison voltage, U_v, which is varied until a galvanometer in the circuit reads zero current; the voltage to be measured then equals the comparison voltage (Figure 5.24).

Furnace temperatures are measured by installing

Table 5.11. *Operating data and limits of accuracy of thermocouples according to DIN 43710*

Type	+ − Cu/Constantan	+ − Fe/Constantan		+ − NiCr/Constantan	+ − NiCr/NiAl		+ − PtRh/Pt		+ − Pt.30Rh/Pt.6Rh	
Colour code	Brown	Blue		Black	Green		White		Grey	
International code	V	J		E	K		S		B	
Temperature of measurement (°C)	*EMF and permitted errors (mV)*									
−200	−5·70	−8·15	±0·5	–	–	–	–	–	–	–
−100	−3·40	−4·6	±0·4	–	–	–	–	–	–	–
0	0	0	–	0	0	–	0	–	0	–
100	4·25	5·37		6·21	4·04		0·64		–	
200	9·20	10·95		13·38	8·14		1·44		0·16	
300	14·89	16·55	±0·4	20·95	12·24	±0·3	2·32		–	
400	20·99	22·15		28·74	16·38		3·26		0·79	
500	27·40	27·84		36·75	20·64		4·22		–	
600	34·30	33·66		44·98	24·94		5·23		1·80	
700	–	(39·72)		53·31	29·15	±0·4	6·27		–	
800	–	(46·23)	±0·8	(61·45)	33·27		7·34		3·16	
900	–	(53·15)		(69·25)	37·32		8·45		–	
1000	–	–	–	–	(41·32)		9·60	±0·05	4·84	±0·10
1100	–	–	–	–	(45·22)	±0·6	10·77		5·79	
1200	–	–	–	–	(49·02)		11·97		6·81	
1300	–	–	–	–	–	–	13·17		7·89	
1400	–	–	–	–	–	–	(14·38)		9·00	
1500	–	–	–	–	–	–	(15·58)		10·13	
1600	–	–	–	–	–	–	(16·76)		(11·26)	
1700	–	–	–	–	–	–	–	–	(12·39)	
1800	–	–	–	–	–	–	–	–	(13·52)	

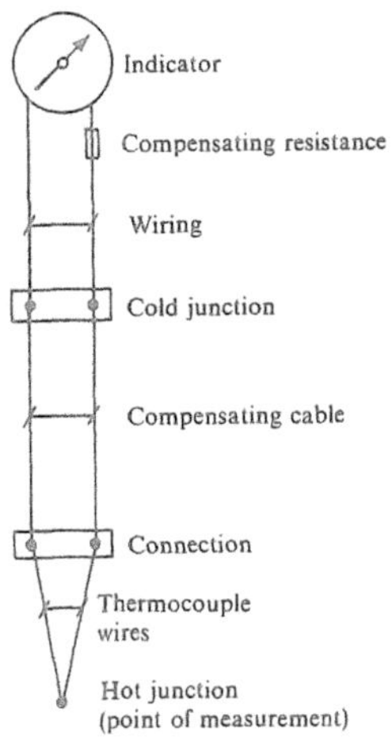

Figure 5.23. Schematic of thermocouple connections and circuit[54]

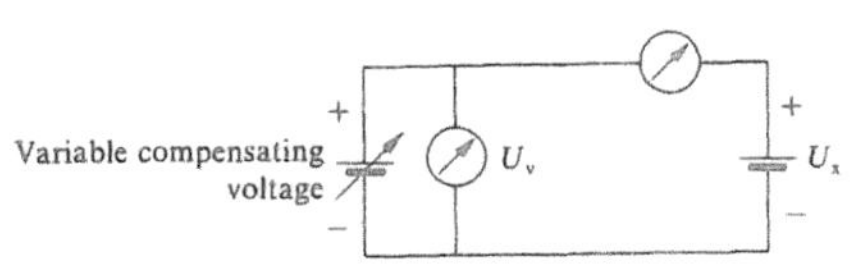

Figure 5.24. Principle of compensating measurements

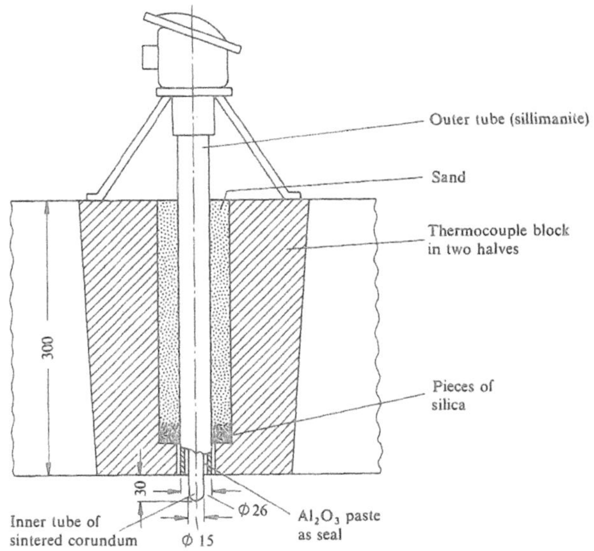

Figure 5.25. Installation of a thermocouple for operation at high temperatures (obsolete)[12] (dimensions in mm)

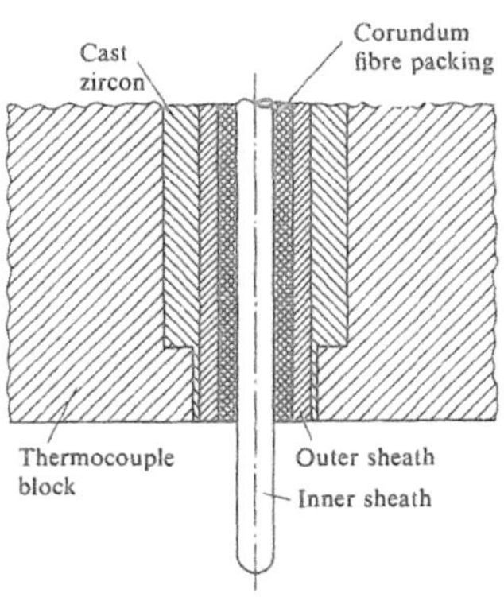

Figure 5.26. Installation of thermocouple with sealing using corundum fibres and zircon casting material

thermocouples along the centre line in the melting and refining zones and in the working end. Thermocouples appropriate to the temperature conditions are chosen and those of the Code B type have proved successful as their ageing effect at high temperature is less than that of normal Pt/PtRh thermocouples. In addition, the thermocouple voltage below 200°C is very small so that, if high accuracy is not required, compensating leads need not be used. Thermocouples must be protected against hydrogen, alkali vapours, and ash from oil firing by an impervious protective sheath; the best sheaths have proved to be of sintered alumina. Operating lives of more than one year can be obtained but the low thermal shock resistance of the material should be noted. When choosing the location of a thermocouple the following criteria should be considered: accuracy of temperature measurement, sensitivity, ability to cross check the measurement, the age of the thermocouple, and effect that its insertion may have on the crown.

Experience has shown that condensation of alkalis on thermocouple sheaths should be avoided, since this strongly attacks both sheath and crown. Such condensation can be avoided by use of a gas-tight installation (Figures 5.25 and 5.26).

The brick or block through which the thermocouple is inserted into the furnace should be made of silica or, for furnaces operating at very high temperatures, dense zircon. The protecting sheath should be of sintered corundum. The annulus between block and outer sheath used to be filled with pieces of silica brick as shown in Figure 5.25, but zircon pieces are now used instead. The annulus between inner and outer sheath is filled by winding high temperature resistant corundum fibre on the inner sheath. This construction (Figure 5.26) allows thermocouples to be changed during operation, provides a virtually gas-tight seal, and protects the inner sheath from sudden temperature changes. A platinum cover is sometimes used on the inner sheath in special circumstances. Accurate temperature measurement and sensitivity is assured if the sheath is inserted 3–5 cm into the furnace chamber.

If reliable temperature comparison and long operating life is sought, then the use of a 'dead end' thermocouple block protruding 2·5 cm into the furnace chamber provides a proven solution (Figure 5.27).

Internal diffusion processes cause even the best thermocouples to show a reduction in voltage with time and checks at regular weekly intervals are therefore necessary; in most cases such checks are carried out using a radiation pyrometer. Since it is comparatively easy to focus the pyrometer on the closed end of a dead end block and since the operating life of thermocouples is thought to be of increasing importance, the installation as shown in Figure 5.27 is gaining in favour.

Temperature measurements in the working end are less of a problem than in the melting and refining zones due to the low temperatures employed but the principles are the same as for the melting end. Each temperature measurement is the result of many influences and gives only approximate information about temperature conditions in the molten glass and it is for this reason that the measurement of temperature in the glass has become a valuable additional tool.[12,33] The strong attack by molten glass on alumina makes it necessary to use platinum as a protecting sheath.[33] If it is necessary to obtain only comparative readings, insertion of the thermocouple into a dead end refractory block may also be used here. A thermocouple assembly suitable for insertion into molten glass is shown in Figure 5.28 but in special circumstances a water-cooled housing

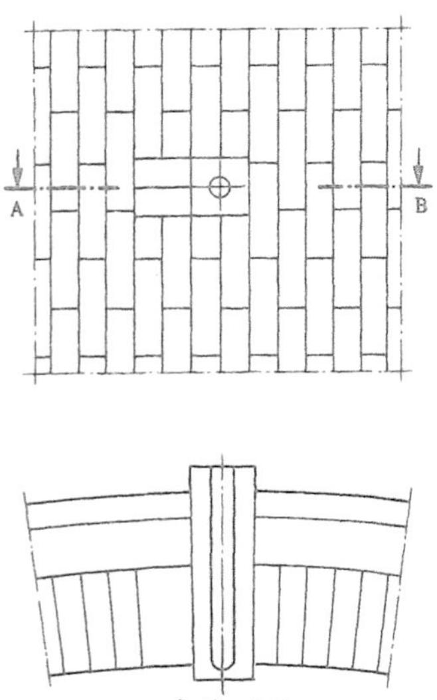

Figure 5.27. Installation of a thermocouple block in the crown (Didier)

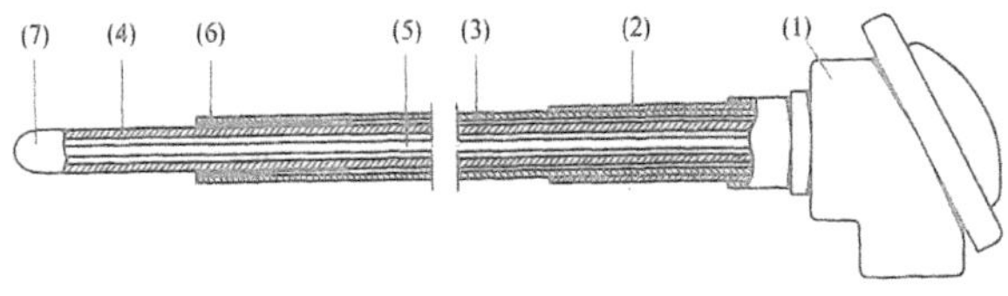

Figure 5.28. Platinum sheath for thermocouples immersed in molten glass[12]
(1) connecting head
(2) supporting tube (steel)
(3) & (4) alumina tube
(5) thermocouple wire insulating beads
(6) cement
(7) platinum sheath

which allows the thermocouple to be changed during the campaign is used (Figure 5.29). Temperature measurements in the glass give information on the formation of and changes in the flow of glass. Preferably, as in the crown, the measurement points should be located along the longitudinal axis of the furnace.

For the stable operation of the furnace, the temperatures in the regenerators are also measured by thermocouples. Since the large variations in temperature within the regenerator top space permit only comparative readings to be taken, it is normal practice to measure temperature at the base of the chamber. The temperature range with its maximum of 700°C permits the use of cheap and robust Ni–Cr/Ni–Al thermocouple wires.

Automatic temperature control is used only rarely for glass furnaces, which may seem surprising in view of the importance attached to temperature for their correct operation. The main reason is that temperature measurements under the crown provide only indicative readings which themselves are the result of many influences difficult to relate to one another. It has therefore been found best to control the most important factors such as batch feeding and furnace pressure by providing these with their own control circuits, and to control the quantities of fuel and combustion air required. Only in special circumstances has the supply of fuel been coupled to temperature measurement and an example is shown in Figure 5.30.[13] An experiment in controlling a U shaped flame tank via 'average

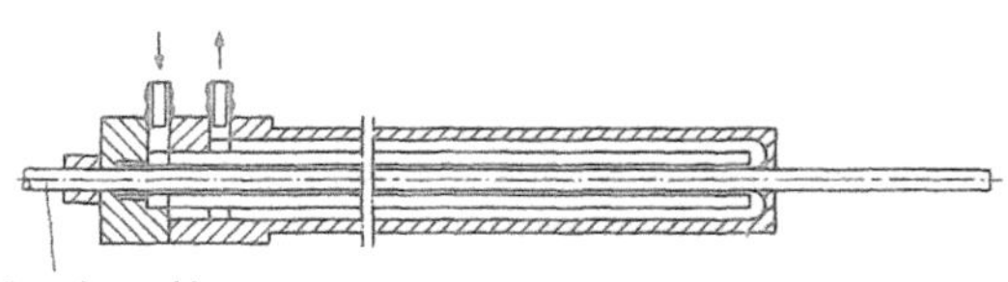

Figure 5.29. Water-cooled thermocouple with holder

temperature values' using an analogue computer has been described by Thies.[14]

The left hand side of Figure 5.30 shows the fuel control circuit. The quantity of fuel is measured by means of a diaphragm meter. The measurement of the flow rate of the oil is, in fact, a volume flow measurement through a standardised length and the temperature of the oil must therefore be constant as, otherwise, the necessary accuracy of measurement cannot be obtained. The signal converter 2 converts the pressure measurement into an electric current and at the same time the square root relationship between pressure and flow rate is converted into a linear function so that the signal current becomes proportional to the oil flow. The controller 3 compares actual to nominal values and its output signal activates the motor 4 which operates valve 5. The control circuit is linked to the temperature controller 8 which receives a signal from the sensor 6 and whose output signal is sent to the fuel controller 3.

The advantages over working with fixed rates of fuel consumption are automatic compensation for any changes in pull, reduced wear on the furnace, improved temperature stability under the crown, and the possibility of operating at higher temperatures. Maximum efficiency calls not only for automatic fuel control but also for control of the fuel/combustion air ratio.

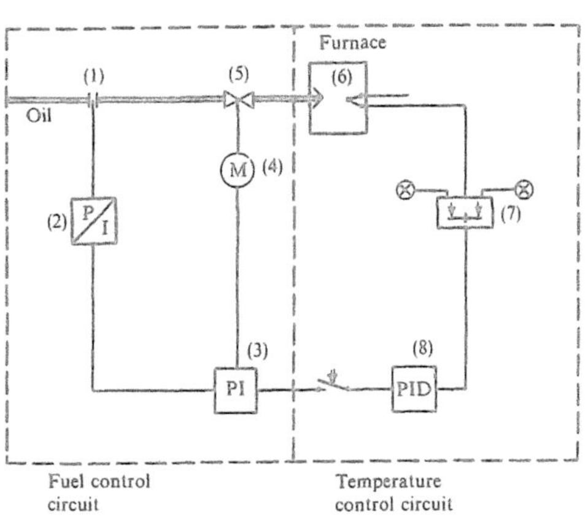

Figure 5.30. Schematic of control circuit for tank furnace controlled by temperature[13]
(1) oil flow meter
(2) signal converter
(3) pressure controller
(4) motor and regulator
(5) control valve
(6) thermocouple
(7) limit signal
(8) controller

5.2.2 Furnace pressure

Accurate measurement and control of furnace pressure is considered to be very important in the operation of glass furnaces. It is a precondition for the satisfactory operation of other control loops such as temperature and glass level. The size of the furnace installation and the existence of unavoidable openings emphasise the importance of maintaining a constant furnace pressure.

The decision on where to locate the pressure sensor is of major importance for the success of the plant; the melting end crown is preferred for cross-fired container glass tanks, in large furnaces the location is often between the fourth and fifth ports, while U shaped flame tanks have theirs half way between end wall and the end of the flame. Sometimes pressure is also measured in the working end. Unit melters usually have their sensor in the waste gas flue. There appears to be no general rule and every case needs to be decided on its merits.

The large tanks used in the flat glass industry have their sensors located in the conditioning zone as the pressure conditions near the drawing machines appear to have a major bearing on the quality of the sheet produced. Thus, pressure fluctuations in the melting zone cannot be avoided.

In many cases the place of measurement is along the centre line of the crown but there are also examples where the place of measurement is at the glass level. Pressure differences between left and right sides of regeneratively fired tanks are avoided by using two interconnected sensors. However, in these interconnected systems some gas will flow into the pipes, which suggests that batch dust and condensed matter may eventually cause blockages. This can be avoided if pressure is measured at the centre of the crown, but here the reading is influenced by the buoyancy of the gases in the furnace and must be adjusted for this effect. The pressure drop between crown and glass level can be taken as constant. Pressure measurements on furnaces with a constant flame, e.g. unit melters, are best taken in the sidewall.

The sampling tube should be of a suitable ceramic or heat resisting steel connected via a flexible pipe to an iron pipe connected to the pressure meter. A parallel compensating pipe is needed when there are large differences in height and temperature.

Diaphragm instruments have been found suitable for the measurement of pressure and the 'submerged bell' instruments are now largely superseded and are used only rarely. A diaphragm pressure instrument with pneumatic signal amplifier is shown in Figure 5.31,[54] the signal amplifier being placed in front of the pressure controller to achieve the desired accuracy. The time constant is reliably small so that the controller can function purely as an integral instrument without fluctuations being set up. An accuracy of control of ± 2 μbar has been stated. Lastly, reference should be made to the purely electrical Teleperm control instrument (Figure 5.32). Its function is based on a comparison between pressure and the current which flows through the coil of a permanent magnet; the current measured (50 mA) is led to indicator, recorder, and controller which are purely electrical instruments: even the drive motor is electrically actuated. The diaphragm of 200 cm^2 provides a range of measurement of 1 mbar with an error of less than 1%; the sensitivity is stated to be 2 μbar.

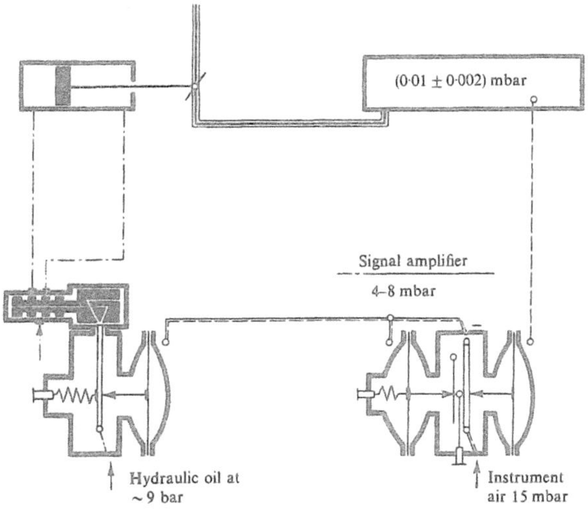

Figure 5.31. Furnace atmosphere pressure control system for high sensitivity[54]

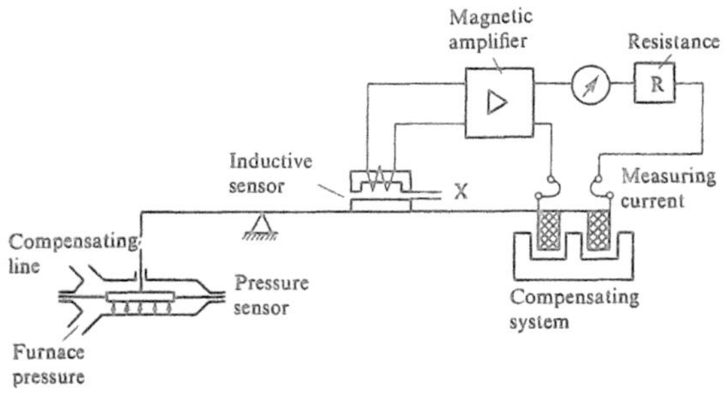

Figure 5.32. Teleperm signal transformer for furnace pressure control[12]

5.2.3 Volume and fuel/air mixture

Fuel and combustion air, including the necessary excess air, must be introduced into fuel-fired glass furnaces in the correct proportion. The maintenance of this correct proportion calls for continuous measurement of the flow of fuel and air; and on these depend the fuel consumption and the melting process.

There are many methods of measuring flow, the most important being those which measure pressure differences by means of orifice plates or nozzles. The principles of the measurement are illustrated in Figure 5.33 and details are given in DIN 1952 which gives the following basic equations for calculating the flow through pipes in which a standard orifice plate has been inserted:

$$\dot{V} = \alpha \varepsilon m A_D \sqrt{2\Delta p/\rho_1},$$

$$\dot{M} = \alpha \varepsilon m A_D \sqrt{2\Delta p \rho_1}.$$

where $\dot{V}$ is the volume flow
$\dot{M}$ is the mass flow
α is the flow, dependent on m, Re, and geometry
ε is the expansion factor (unity for incompressible fluids, less than unity for compressible fluids)
m $= (d/D)^2$, the orifice/pipe cross section ratio in which d is the diameter of the orifice in the orifice plate
A_D is the cross section of a pipe of diameter D
Δp $= p_1 - p_2$, is the pressure drop across the orifice plate as indicated in Figure 5.33
ρ_1 is the density of fluid being measured in front of the orifice plate

Flow rates of gases are best measured using standard orifice plates as these give highly accurate results. Flow rates of oil can be measured using quarter circle or double tapered orifice plates which give good results even for small Reynolds numbers.

There are several options available for mixture control; one controller may be used for the direct quantitative control of fuel and air or, in cascade control, one controller may be used to control the leading flow rate (e.g. the fuel) which, in turn, adjusts the set point of the controller for the second flow rate (e.g. the air).

The first method of close control of fuel and combustion air must be used if good control over

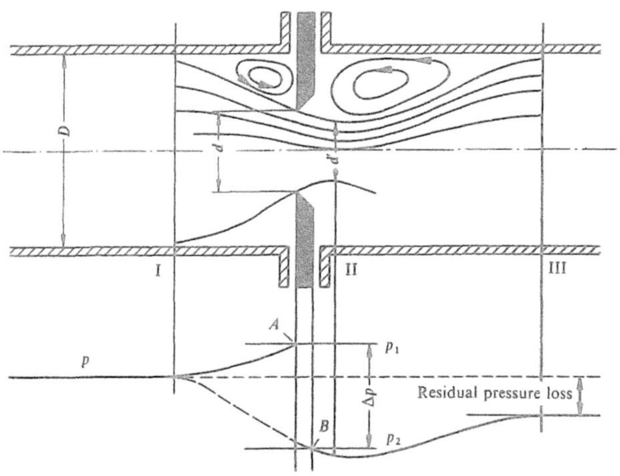

Figure 5.33. Diagram of flow and pressure changes near orifice plates

the whole operating spectrum is required. A good and not very expensive system is a mixture control where the leading material is controlled and the second material follows it via a proportional controller. An example of such a system incorporating ratio control with potentiometric setting and a signal converter is shown in Figure 5.34. This method is very stable and allows corrections to be made for air pressure, temperature, calorific value, etc.; however, these additional facilities are very expensive and can only be justified in special cases. For small and average sized tanks one control of total fuel and combustion air is sufficient; for large tanks the quantities of fuel to individual ports require controlling. If sectional regenerators are used then separate mixture control is needed for each section. If highly accurate control is required for individual ports or groups of ports using oil as fuel when subject to large variations in pull, then the range of a controlling valve (maximum 1:4) is no longer sufficient and a metering pump with a range of 1:10 will be required to solve the problem. The output tolerance for a turn-down ratio of 1:100 is about ± 1%. The metered flow alters linearly with the stroke setting, which can provide a pneumatic or electric signal enabling every flow rate to be recorded.

5.2.4 Glass level[12]

Accurate measurement and control of glass level is very important, especially in container glass tanks where the constancy of gob weight depends on it. Glass level measurement is now always combined with level control, with the controller activating the batch charger.

Since the objective is a constant level in the feeder, the level sensor should preferably be located in the forehearth; for furnaces with more than one forehearth, one is selected as representative. The best level probes are electric coupled to mechanical or pneumatic drive and Figure 5.35 shows a probe and system with pneumatic drive. The electrode is attached to a diaphragm working against a compression spring; air pressure and the distance between electrodes are, within the working range, related linearly so that pressure can be used as a measurement of glass level. The controller can be set for periodic operation. Accuracy of measurement, usually given in mm, is ± 0·1–0·2 mm.[55,56]

The probe electrode consists of an alumina tube with a platinum end. During installation care must be taken that the tube does not make contact with the cover blocks of the forehearth since the current

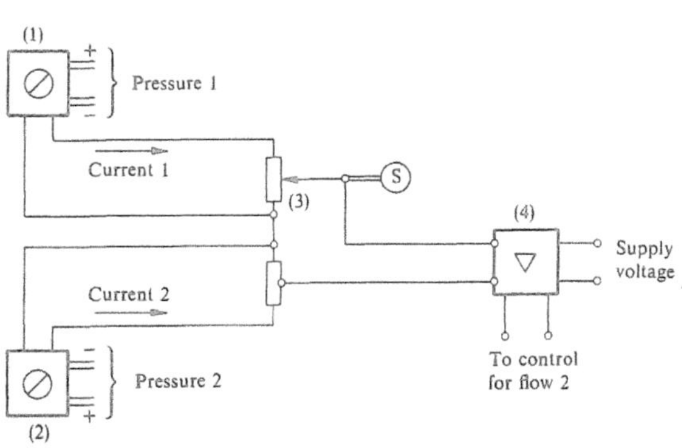

Figure 5.34. Electric mixture control with signal converter[12]
(1) signal converter for flow 1
(2) signal converter for flow 2
(3) potentiometer with proportional setting
(4) control amplifier (including return connections)

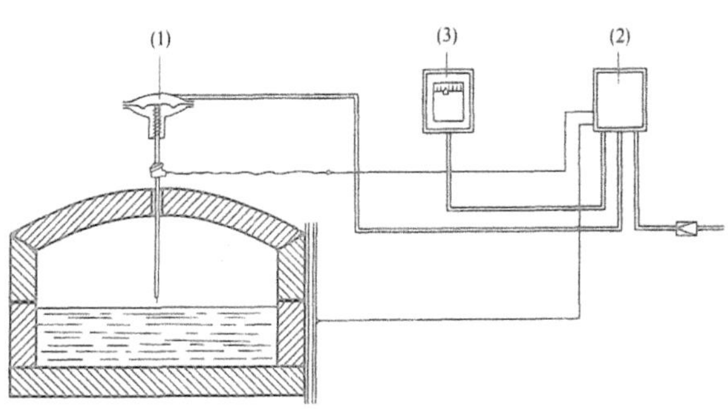

Figure 5.35. Schematic of a pneumatic method of measuring glass level[12]
(1) diaphragm sensor
(2) electropneumatic controller
(3) universal pressure recorder suitable for 1·2–2 bar calibrated in mm glass depth

created by contact can be enough to operate the controls. Careful electrical insulation from the furnace steelwork is also necessary. The contact current between electrode and molten glass should be chosen sufficiently large, an adjustable alternating current voltage of 24–60 V usually being sufficient. Systems with fixed electrodes have also been suggested.

Pneumatic probes are used particularly in flat glass tanks, back pressure probes being especially suitable; their operation is shown in Figure 5.36. Compressed air enters the probe from above and divides to cool the two temperature-dependent resistances heated by alternating current. The air leaving the nozzle impinges on the glass surface, whereas the air from the compensating tube (set by means of an adjustable damper) escapes into the furnace; the compensating tube ensures that changes in furnace pressure have no effect on level measurement. If the glass level changes, it causes a change in back pressure at the nozzle which, in turn, affects the flow rate, the cooling of the resistance, and, via an electrical bridge circuit, the output voltage. This is linked via a zero amplifier to a positioning motor so that the original probe position is restored. The probe position is used as an indication of glass level and is transferred to a recorder and controller; the accuracy is reported to be $\pm 0{\cdot}1$ mm. The great advantage of the system is that it does not make contact with the glass. Other methods employ radiation or optical sensing of the glass level but are only used in special cases.

Controlling the glass level in a tank is relatively simple and it has been demonstrated that an on/off control of the batch charger is adequate for such control. If the control system is as shown in Figure 5.35, then a diaphragm switch receives the controlling pressure from the output of the controller at the particular moment when the pressure corresponds to the glass level. This is compared to the opposing pressure in a second chamber of the switch, namely the nominal pressure. Depending on the pressure difference, the diaphragm separating the actual pressure chamber from the nominal pressure chamber moves and actuates a contact which, if the glass level is too low, causes the batch charger to be switched on via a power limiter. If the control involves long periods of inactivity then the insertion of a pneumatic two-term amplifier (proportional and derivative) into the signal line permits the generation of a differential impulse of adjustable duration.

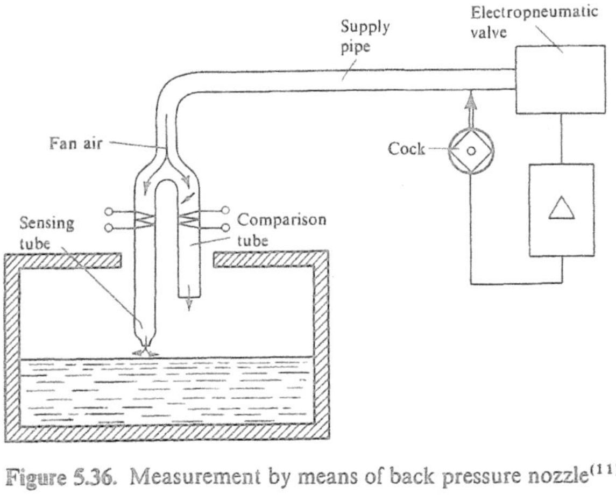

Figure 5.36. Measurement by means of back pressure nozzle[11]

If more than one batch charger is used they can be operated in parallel or alternately with different operating periods, depending on the evenness or otherwise of the batch blanket. A precondition for glass level control is the trouble-free uniform supply of batch from the storage bunker to the batch charger. This requirement is often inadequately provided for.

5.2.5 Composition of the waste gases

The objective of waste gas analysis is the maintenance of the specific furnace atmosphere required for melting the glass, fuel economy, and control over emissions into the atmosphere. The first two are covered in most cases by controlling the O_2 content of the waste gases, the last by control of SO_2.

Measurements of O_2 content should be taken in the regenerator top space to avoid false sampling arising from leaks through the brickwork. Established methods of measurement are based on magnetic oxygen analysers. In actual practice, problems of measurement are not connected with the analytical equipment but with methods of sampling. Water-cooled, largely self-cleaning systems have been shown to be the best. Regeneratively fired tanks require two sampling probes collecting waste gases from the right and left sides alternately. Furnaces with fixed cross firing may need separate sampling probes for the melting and refining zones.

Zirconium dioxide probes, which operate as solid state electrolytes, have been developed for measurements of O_2; their reaction time is very short and they are suitable for measuring small concentrations of O_2. They are also suitable for controlling the O_2 content of molten glass and molten metal (e.g. the tin bath of a Float glass plant).[28]

Equipment for measuring CO and CO_2 has not been adopted in connection with glass melting as the measurements are influenced by the waste gases themselves. Local government licensing authorities prescribe the use of SO_2 recording instruments upstream from the chimney in order to monitor chimney emissions; sampling devices and suitable instruments are available. Instruments for measuring the density of the smoke are also required sometimes and would be located in the chimney itself (see Chapter 13).

5.2.6 Instrumentation and control installations[29-32]

Having discussed individual measurement and control problems arising on glass furnaces, this section will briefly describe a comprehensive control installation by way of example.

Figure 5.37 is a schematic drawing of the controls used on a container glass tank; the separate control loops may readily be identified. It should be noted that the flow of oil to each burner is controlled after flow measurement as this ensures that, should a burner nozzle become restricted (due to coking), the necessary quantity of oil will still pass. Evidence confirms that this procedure reduces the incidence of atomisers coking up, probably because the forced flow dislodges the coke. The flow of total combustion air is controlled in proportion to the oil flow, furnace pressure and glass level are also controlled. Temperature control of feeders is indicated.

Figure 5.38 shows the instrumentation and control equipment of an oil fired unit melter with feeder. The furnace is divided into two firing zones, the melting zone and the refining zone, each with its own control loop. For gas and oil fired furnaces operating with preheated air from a recuperator each zone has its own temperature controller, a fuel/air proportional controller, and a control to safeguard the recuperator. A correction for temperature must be made if the air flow is measured in the hot parts of the supply ducting and if the supply is unstable it may be necessary to instal pressure regulators. Furnaces fired with heavy oil and using proportional burners have only one temperature controller per zone and adjustments originating from this controller operate a single mechanism to which all burners are linked; the fuel/air ratio is therefore predetermined for each burner. Although originally designed to operate with combustion air taken at ambient temperature, virtually all unit melters now operate with preheated air and thereby achieve substantially improved specific fuel consumptions.

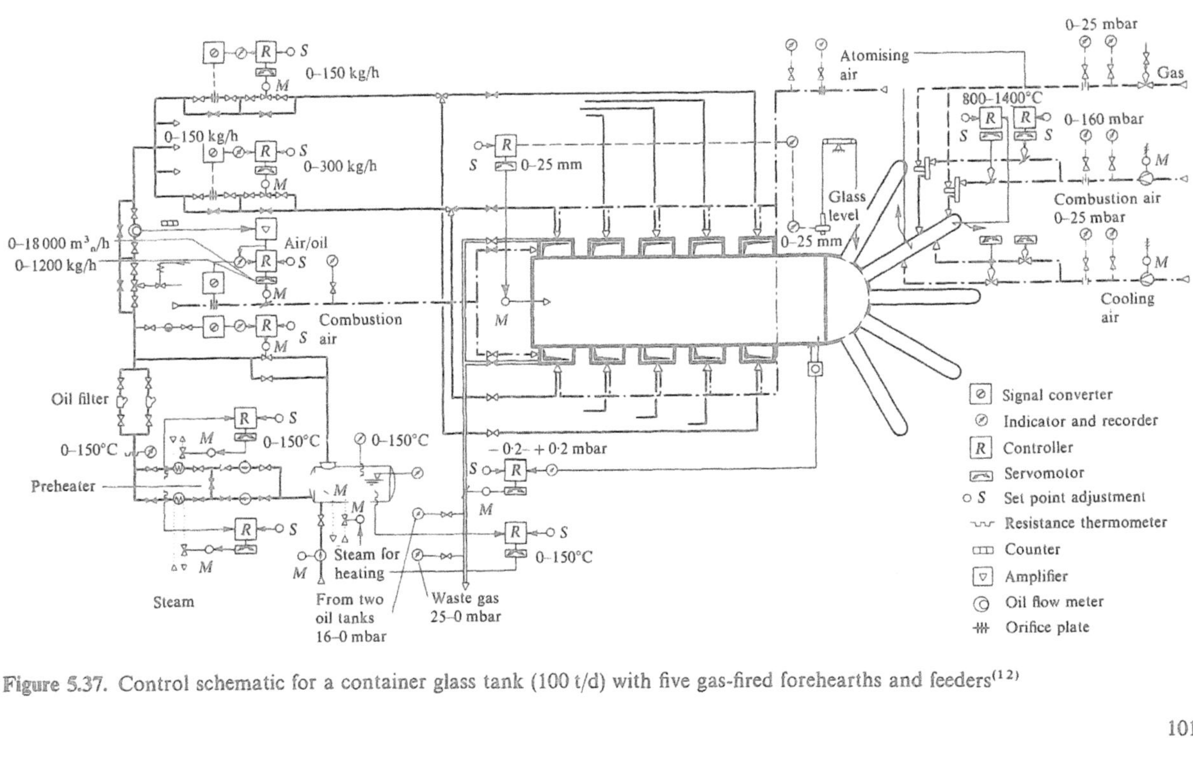

Figure 5.37. Control schematic for a container glass tank (100 t/d) with five gas-fired forehearths and feeders[12]

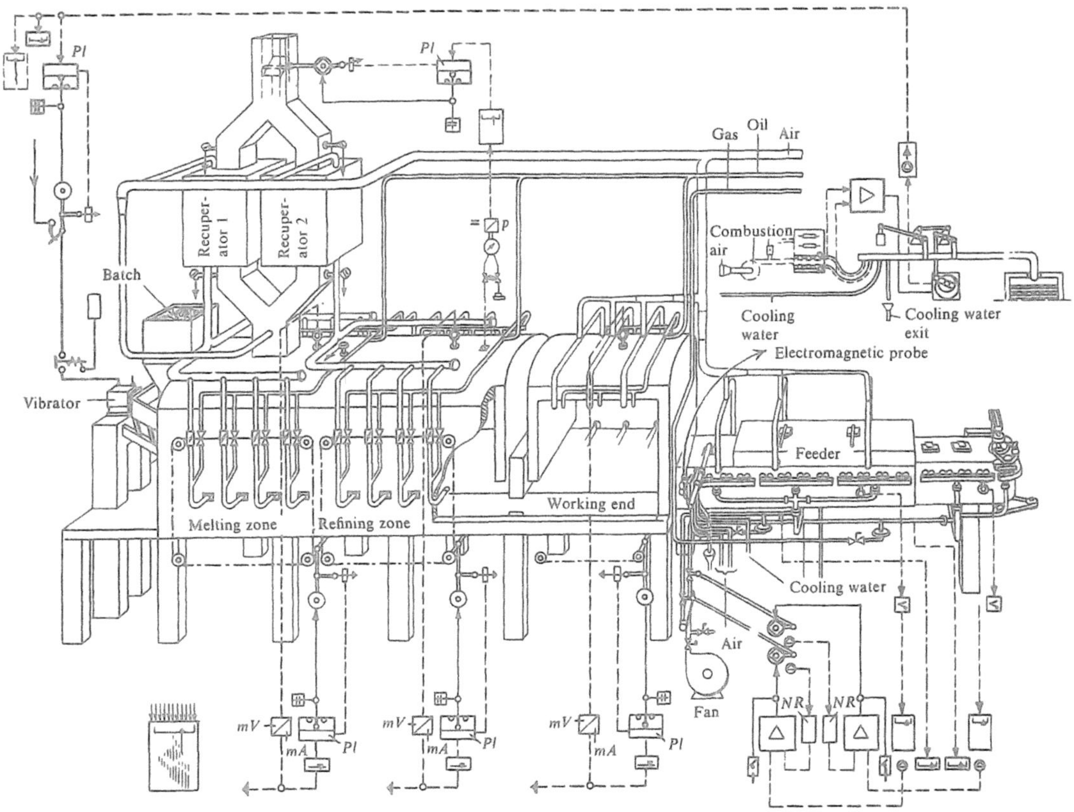

Figure 5.38. Control schematic for an oil fired unit melter with feeder[29]

5.3 Reversal[34]

Regeneratively heated furnaces require reversal of firing from one side to the other according to a predetermined time interval or other criteria such as temperature changes etc. The process must be planned so that it is carried out in the correct sequence and individual adjustments follow one another at correct time intervals. Fundamentally, the whole reversal process is a timed sequence of controlled opening and closing of valves or dampers.

5.3.1 Reversal as a periodic process

Figure 5.39 uses an oil fired tank to show the most important reversal mechanisms. This simplified schematic shows the section of a furnace with its pair of burners and relevant fuel, air, and waste gas reversal equipment. Normal open/closed or three-way valves are used in the fuel supply lines in positions (1), (2), and (3), and suitable closure equipment in positions (4), (5), (6), and (7) are used for combustion air and waste gases.

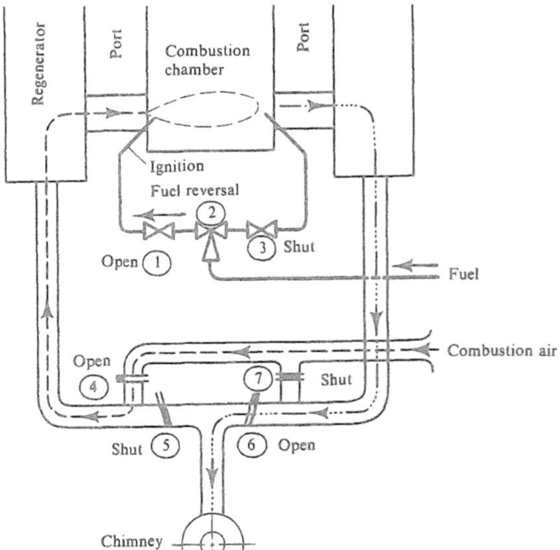

Figure 5.39. The setting of reversal mechanisms for oil firing[34]

These individual mechanisms must be moved from open to closed and vice versa in a specific sequence, the timing of which is governed by safety requirements designed to prevent dangerous conditions from arising; e.g. combustion taking place where it should not in places such as regenerator chambers, or explosions taking place in the waste gas ducts or at the outlet of the chimney. The basic law of reversal states: first cut off the fuel and then reverse the flow of waste gases and combustion air to flush the (former) waste gas chamber with air until the combustion air reaches the port; only then should the fuel valve on this side of the furnace be opened.

Safety must govern the whole process, including the selection of individual items of equipment which together form the reversal system. The timing of the process must also be controlled, either by suitable instructions to personnel or by programming an automatic installation which controls the process as a whole.

The planning of a new reversal system or the checking of an existing one, including all aspects of safety, is best done by constructing a table which lays out timing for all the phases of the process (e.g. as shown in Table 5.12). Every mechanism involved in the reversal process must be entered and, for practical reasons, each mechanism should be marked with a specific sign as shown in column 4 of Table 5.12: whether the mechanism operates on the right or left side or on both sides simultaneously should also be noted. Additional factors can refer to the number of the burner or the number of specific devices in the installation.

If closing valves for oil (position 1, identification B) are required, one for each side, then for five pairs of burners the factor is 5, and so on; since the reversal mechanisms operate simultaneously for all pairs of burners, this factor is of no importance when considering time. It should be noted that Table 5.12 includes mechanisms which play no part in the reversal process but are influenced by it, e.g. the furnace pressure control valve. Column 9 gives the times necessary for the opening and closing of each valve. A diagram covering the whole installation can be drawn up based on this Table.

Figure 5.40 is an example of a time phased diagram based on the data of Table 5.12, but actual layout varies between companies and plants. The X axis is best used for time, with vertical lines placed every 1 or ½ second while the Y axis is used for a list of all equipment and mechanisms required for the reversal. Signal contacts, on/off light indicators, alarms, etc. indicating the functioning of certain processes of the controllers, e.g. furnace pressure control, flow control, fuel pressure control, etc. can all be added in so far as they are influenced by or are associated with the reversal process.

By shading the relevant field parallel to the X axis for each piece of equipment the open and shut periods of each may be seen clearly. If the scale is large enough, the times needed for opening and closing of valves and dampers can also be included by tapering off the field to a point. Black fields indicate a device that is closed or out of use and white fields a device that is open or in use, while vertically shaded fields may be used to indicate, for example, an intermittent light contact.

Table 5.12. *Example covering the installation of reversal equipment with operating times for use with time-phased diagram*

Position according to Figures 5.39 and 5.40	*Medium*	*Equipment*	*Reference*	*Number of units of equipment* *Left side*	*Right side*	*Total*	*Total number*	*Operating time (s)*
1	Oil	Closing valve	B	1 per port	1 per port	–	10	1·0
2	Oil	Three-way valve	B	–	–	1 per pair of ports	5	1·0
3	Oil	Ringline valve		1 per port	1 per port		10	1·0
		Bypass valve		–	–	1 per pair of ports	1	1·0
		Control valve for quantity	R	–	–	1 per pair of ports	5	–
		Control valve for pressure		–	–	1 per pair of ports	1	–
4	Atomising air	Closing valve with fixed position for purging	Z	1 per port	1 per port	–	10	1·0
5	Combustion air	Closing clapper	L	1	1	–	2	3·0
		Control clapper		–	–	1	1	–
6	Waste gas	Closing damper	A	1	1	–	2	6·0
		Furnace pressure control clapper	R	–	–	1	1	–
7	Control (general)	Warning signal/alarm	S	–	–	1	1	$t_1 \doteq t_5$

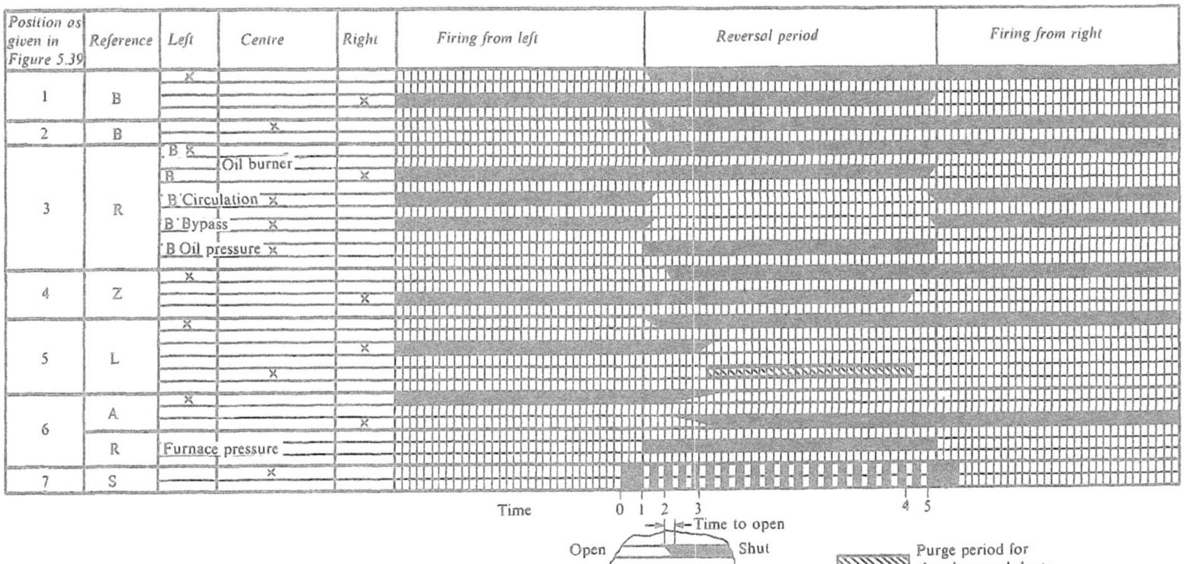

Figure 5.40. Time phased diagram of reversal equipment for oil firing[34]

Using time as the basis, the reversal begins with the closing of the fuel supply valve while the combustion air continues for a few seconds to complete the combustion of any fuel remaining in the furnace. The waste gas dampers are then reversed, followed by the opening of the combustion air supply on the other side: at this moment all flow in waste gas ducts, regenerators, and furnace comes to a halt. After confirming that the relevant combustion air supply valve is in the open position, the process of purging waste gas ducts, regenerator chambers, and furnace begins. The purge time, t_s, can be determined with sufficient accuracy from

$$t_s = \frac{\sum V_s}{\dot{V}_A}$$

where $\sum V_s$ is the sum of all volumes to be purged and $\dot{V}_A$ is the waste gas flow rate under normal operating conditions. The purge time must not be cut too fine, although the above Equation generally gives purge times which are too long and the best method is to adjust t_s empirically after commissioning the furnace; the point to be confirmed is that combustion air is available at the port when the fuel supply is opened. On the other hand, the reversal process should not be extended unnecessarily as the long interruption to firing causes heat losses and a reduction in furnace temperature which can lead to faulty melting. Reversal times of small furnaces are 25–30 s and those of large furnaces up to 40 s; reversals take place every 20–30 min. Long periods between reversals will give rise to large temperature fluctuations under the crown and a reduction in the thermal efficiencies of the regenerators while short periods will increase the losses caused by reversals: optimum conditions cannot be calculated but must be found empirically.

When the furnace has been in operation for some time, deposits which increase the flow resistance appear in the checkerwork of the regenerators. By occasionally extending the period between reversals to one hour or more the temperature of the downstream (waste gas) chamber can be raised so that these deposits can be partially melted down, thereby reducing the flow resistance of the chamber, but care must be exercised to avoid overheating of the checkerwork.

5.3.2 Control of the reversal process

Manual operation of all the reversal stages can only be done on small furnaces and, in view of labour costs, is only practical in special cases. Automation at first consisted of the electrical or pneumatic actuation of each device in sequence from a control room, with auxiliary contacts on a loop or series of switches being actuated at the end of one sequence to permit the operation of the next stage: the next development was the provision of sequence switching for all stages. The reliability of safety switches is particularly important since they are required to operate under adverse conditions; one improvement to these came from the use of reed contacts which are protected from the ambient atmosphere by being hermetically sealed in an inert gas atmosphere and a further improvement is the use of controllers without contacts using semiconductors in place of a combination of relays.[35]

In special cases, the reversal process can be coupled to the control of a glass furnace[37] or it may be controlled by regenerator temperature or difference in temperatures.[38] In periodically operating furnaces, e.g. pot furnaces, these procedures are of importance; in other cases reversal is always based on time.[39,40]

If a plant has several furnaces and, particularly, if they have a common chimney, the reversals must be planned in relation to one another. The operation and supervision is best done from a control panel with mimic diagram and illuminated switches. Manual override or auxiliary motors with power from a standby electricity supply must be provided for emergencies.

5.4 The heating and cooling of glass furnaces

The large size and high operating temperatures of glass furnaces as well as the expansion and sensitivity to thermal shock of the refractories used call for a process of heating from ambient to operating temperature for which a time and temperature schedule must be chosen with care.[41,42]

5.4.1 The expansion of refractories

Successful heating up depends on a knowledge of the expansion of the refractories used, the term expansion including not only the purely thermal expansion but also the changes in dimensions resulting from changes in crystal structure. Figure 5.41 gives the average values of the thermal expansion of various refractories. Silica expands considerably

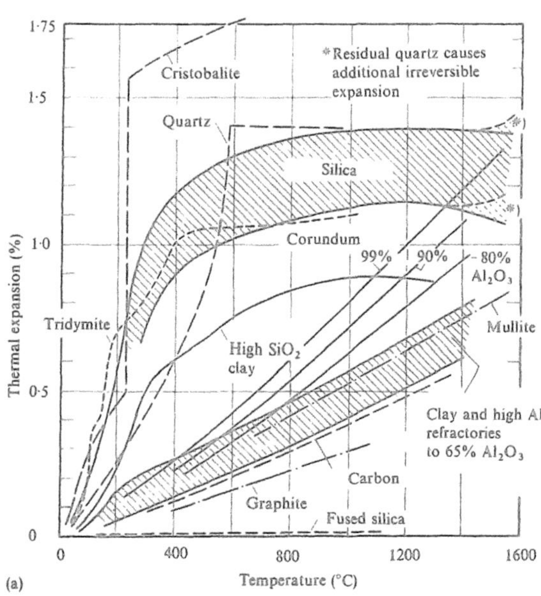

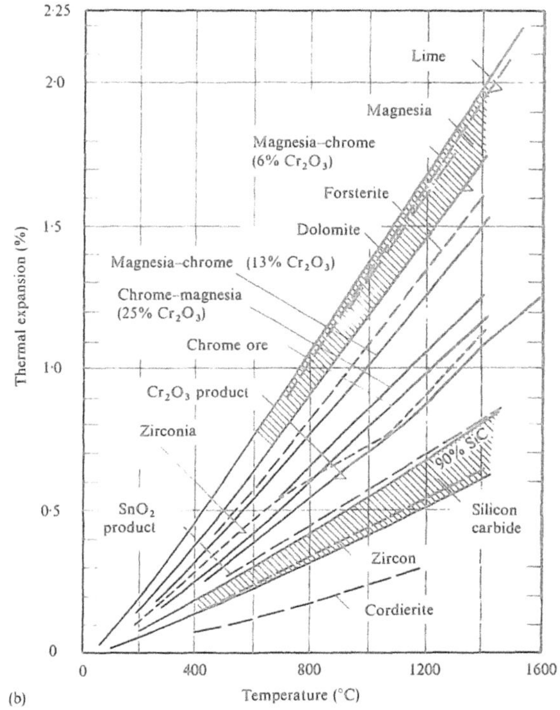

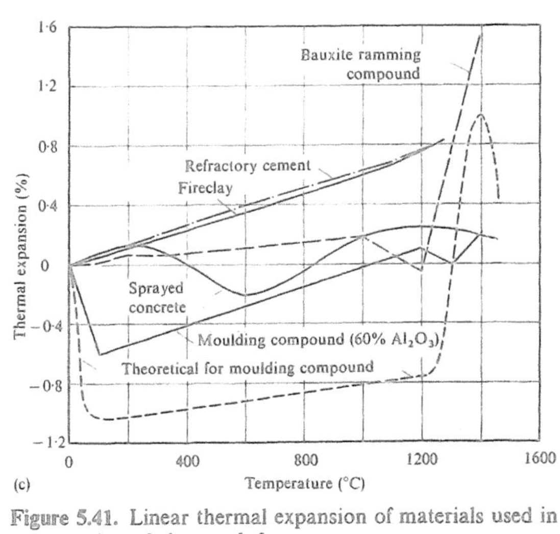

Figure 5.41. Linear thermal expansion of materials used in the construction of glass tank furnaces
(a) & (b) different refractory parts
(c) unformed refractory materials

between 100 and 650°C, whereas magnesite shows the greatest expansion above 1000°C; the curves for fusion cast AZS show the well known erratic expansion above 1100°C; the expansion of fireclay depends very much on composition, high silica clays expanding more than those with a higher alumina content. Accurate data are available from refractory manufacturers but for large expensive installations it pays to determine the expansion characteristics in your own laboratory.

5.4.2 Temperature gradients and thermal stresses in refractories

During heat up and in the course of operation, temperature gradients which are responsible for the setting up of thermal stresses and for causing deformation are established in the furnace refractories; a knowledge of the temperature gradients is therefore very important and Figure 5.42 shows the temperature gradients set up in the silica crown of a large flat glass tank. It should be noted that with a proper heating up schedule and with good control of the heating up process the temperature gradients set up in the refractories are practically linear, which means that the local gradient stays constant. This is an important point to remember for, leaving aside stresses from external forces, the refractory is not subjected to thermal stresses as long as the temperature change is linear.[43]

In practice it is important to know the deformation of the refractory since this leads to the opening of joints and pressure along edges.[44] According to Dämmer[46] the temperature changes shown in Figure 5.43 cause the dimensional

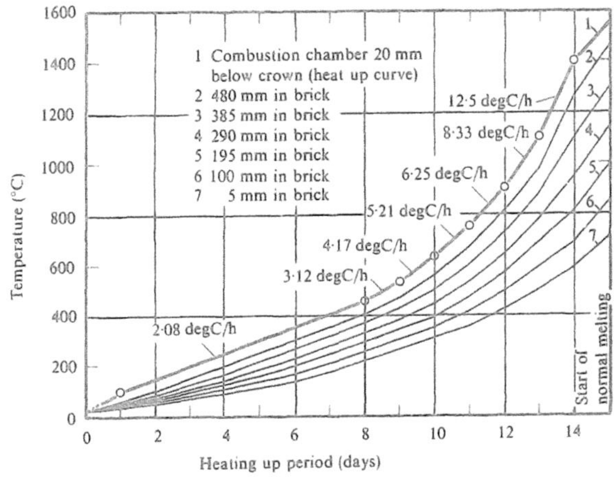

Figure 5.43. Temperature changes within a crown brick during heating up [46]

changes shown in Figure 5.46; this deformation must be taken into account during heating up so that the tie rods may be released in such a way that the opening of joints is avoided.

Tank blocks are subjected to similar temperature gradients. Dimensional changes in AZS are difficult to follow quantitatively due to the expansion anomaly and their inhomogeneous structure; in addition, there are internal stresses originating from the manufacture of the block. Observation of soldier blocks has shown that cracks are most likely to appear in the upper uninsulated part of the block (Figure 5.44) and in the blocks around the batch melting zone and the doghouse; temperature fluctuations associated with batch feeding could be a contributary factor.

A critical phase is the switching on of block cooling after heating up. It should not be started suddenly on full throttle and a gradual step-by-step

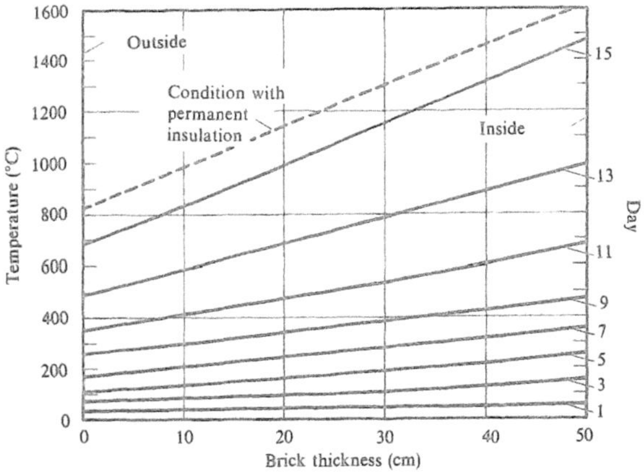

Figure 5.42. Temperature within a silica crown brick as a function of its thickness[46]

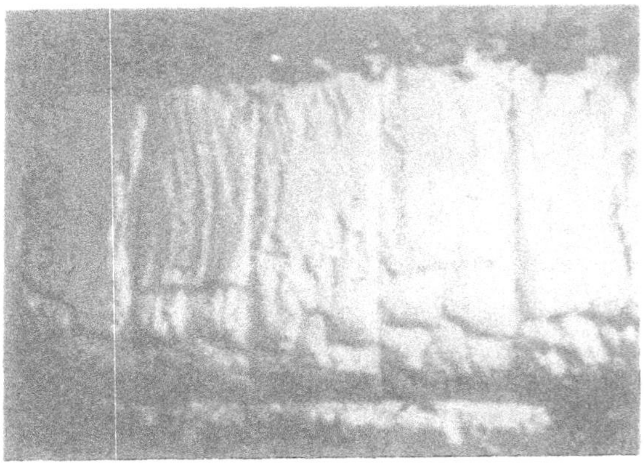

Figure 5.44. Cracks caused during heating up in the upper uninsulated parts of soldier blocks[47]

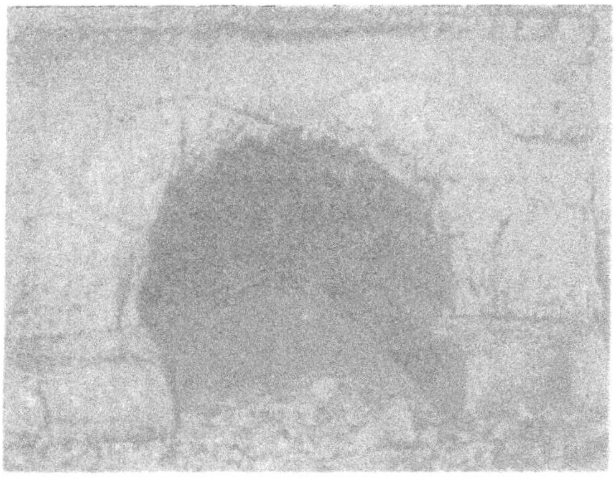

Figure 5.45. Horizontal joint in a tank as starting point of accelerated corrosion

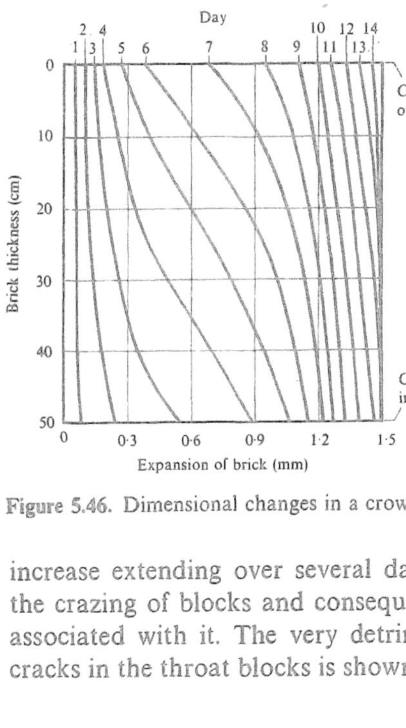

Temperature on inside of crown	
Day	(°C)
1	53
2	103
3	153
4	204
5	257
6	306
7	358
8	407
9	472
10	567
11	686
12	824
13	997
14	1296

Figure 5.46. Dimensional changes in a crown brick[48]

increase extending over several days will prevent the crazing of blocks and consequential problems associated with it. The very detrimental effect of cracks in the throat blocks is shown in Figure 5.45.

5.4.3 The heating-up curve

Successful heating up calls for a curve which is designed on the basis of the size and construction of the furnace and the refractories used in its construction. On the one hand the heating should save time and energy and on the other hand damage arising from it should be avoided.

Calculations by Sonntag[43] and Hargreaves[43a] based on thermal stresses set up in refractories during heating up and stresses caused by departure from linear temperature changes have shown that very short heating up periods of 48 h are possible. However, due to imperfections in the heating up process (local temperature peaks) and pressure on brick edges arising from expansion of the walls, even small tanks need 3 to 4 days; this gives a heating up curve which, in the temperature interval critical for silica (up to 650°C) is limited to 9 degC/h and above 650°C to 30 degC/h. But even these values are not reached in practice and nowadays one can assume rates of 2–5 degC/h up to 650°C and 10–12 degC/h between 800 and 1250°C, with the last figures applying to larger furnaces. A proven heating up curve for a Float glass tank (Figure 5.47) has been given by Dämmer[46] and modern container glass tanks follow a very similar curve.

Since the heating up process is governed by the expansion of the refractories and since this expansion, in certain temperature ranges, is not proportional to the increase in temperature, it is possible to base the heating up curve on the permissible expansion rate. The controlling value is that of the refractory with the highest rate of expansion at the relevant temperature and the literature[41] gives a value of 0·18%/d.

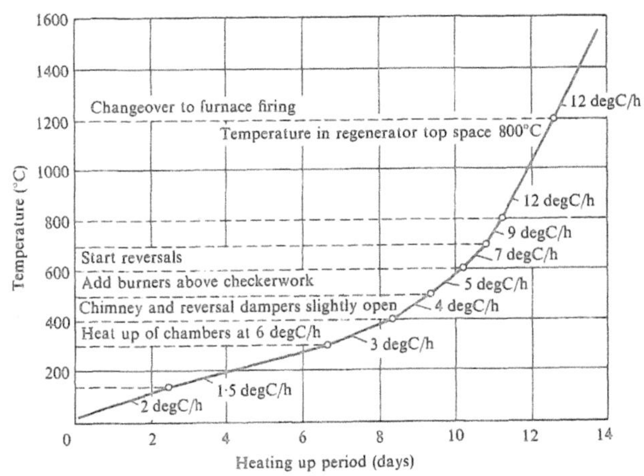

Figure 5.47. Heating up curve for a flat glass tank[46]

5.4.4 The heating-up process

5.4.4.1 Location of burners

There are two methods for carrying out the process and these govern the positioning of burners; they are heating from the bottom of the regenerator chambers and heating from the melting chamber. The first has the advantages of a positive pressure in the furnace chamber, the buffering effect of the checkerwork itself, the high degree of preheating in the regenerators, and the fact that most of the equipment and installations needed are available for use, e.g. combustion air and cooling air, and pipework under the furnace is easy to install or is in existence. Advantages of heating up from the melting chamber are better control over temperature distribution in the furnace, the ability to reach higher temperature in the melting chamber, and easier supervision of the burners.

With the introduction of high velocity burners operating with a large excess of air, both methods are usually combined; heating is started in the base of the regenerators and when the temperature in the melting chamber reaches 600°C, burners are installed there.

If a large excess of air is used in the process then it has been found advantageous to heat up the regenerators and the melting chamber separately. If a new chimney has been constructed, the heating up should be controlled by the chimney constructor. The furnace and chimney must be treated as a unit and the heating up planned so that the chimney has reached operating temperature when the melting chamber burners are lit.

Figure 5.48 shows the arrangement of burners in a flat glass tank; the process starts at the base of the regenerators and continues in the melting chamber. The time/temperature schedule and the points at which individual groups of burners are installed were given previously in Figure 5.47

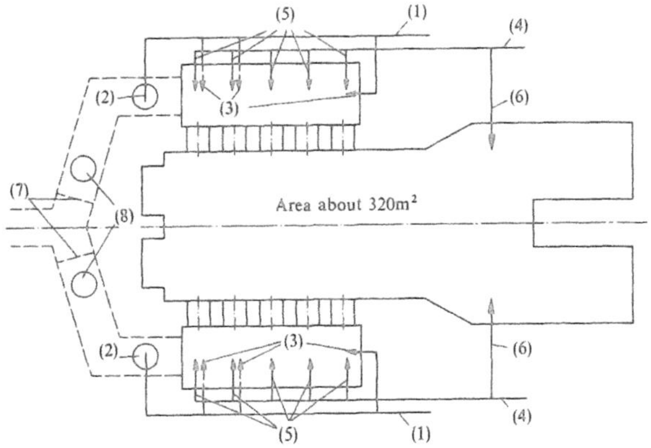

Figure 5.48. Arrangement of heating up burners[46]
(1) basement supply line
(2) forked burner in duct
(3) burner lance under bearer arches
(4) furnace floor supply line
(5) burner lance above checkerwork
(6) burner lance through floater opening
(7) reversal dampers
(8) air intake (consumption about 115 000 m^3/h)

5.4.4.2 Burners

Simple lance or fork burners can be used in the regenerator flues (Figures 5.49 and 5.50), the air required for combustion and dilution being supplied by a fan blowing into the flue. A shadow wall beyond the burners serves to mix the gases and to ensure firing. Safety requires that the burners are equipped with flame failure safety devices. Heating up of the melting chamber itself is best carried out by portable high-velocity burners with air mixers, an example of which is shown in Figure 5.51. Whenever possible the heating up is done by gas, even if the furnace is subsequently fired by oil, since gas is easier to mix and control.

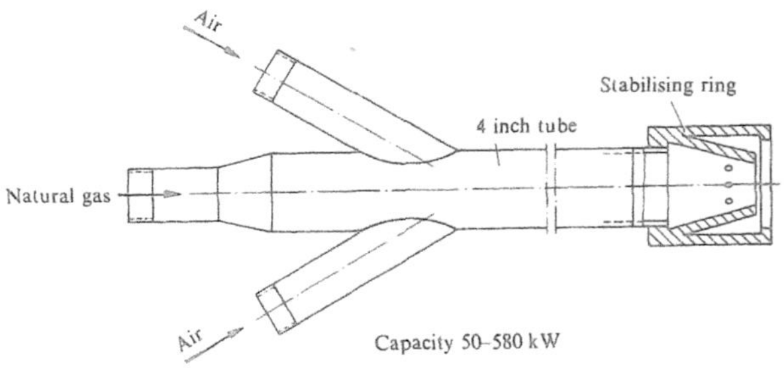

Figure 5.49. Heating up burner for natural gas[46]

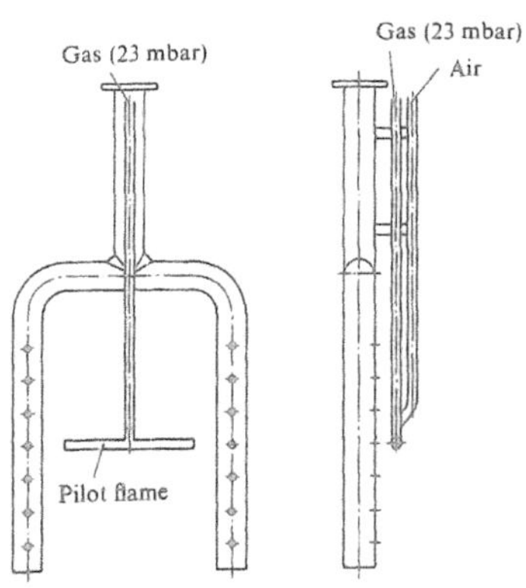

Figure 5.50. Heating up burner for coke oven gas (forked burner)[46]

5.4.4.3 Heating up

Burners for starting the heating up in the regenerators are placed in the regenerator flues, the chimney damper is closed, and the reversal dampers set in the mid position. The initial air flow is about four to five times that required under normal glass melting conditions. If the heating capacity of the burners in the flues is not sufficient to follow the temperature curve, then additional burners are

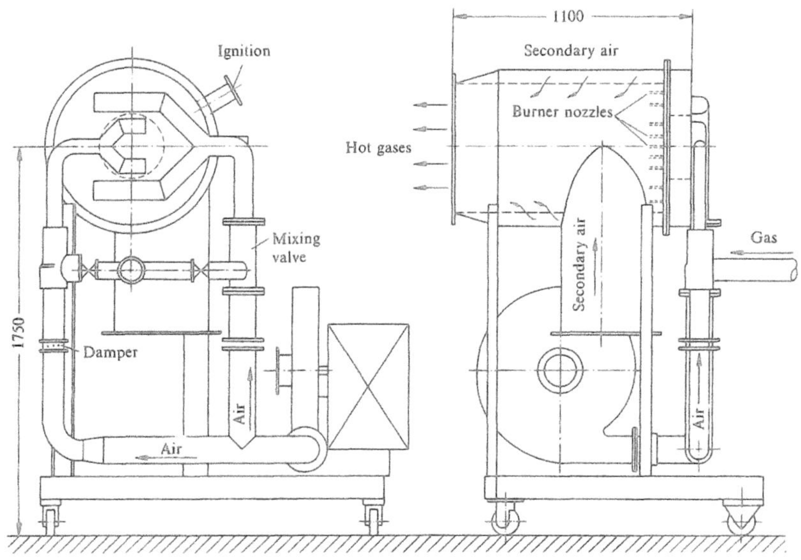

Figure 5.51. Heating up equipment (Küppersbusch system)[46] (dimensions in mm)

inserted under the bearer arches one by one and finally in the regenerator chambers themselves.

Additional separate heating is required in the working end or the conditioning zone. When the temperature in the melting chamber reaches about 600°C the chimney damper is opened slightly and the reversal process started in order to condition the regenerators to the normal operating conditions they will subsequently encounter. Initially, reversals are arranged to take place every 5–10 min so that large temperature fluctuations are avoided, and the period is gradually extended until, shortly after lighting the burners for glass melting, the normal reversal timing is established. The burners for glass melting are put into operation when the melting chamber reaches 1200°C and dangerous temperature jumps are avoided by not lighting all the burners simultaneously but gradually, one at a time. The heating of the working end should be carried out so that the temperature difference which will eventually exist when the furnace is in normal operation is gradually established between melting and working ends. The heat flow in a melting chamber shortly before the changeover to firing the glass melting burners is given by Dämmer[48] as 45–70 kW/m^2, with the lower values being normal for large flat glass tanks. It is often claimed that one advantage of starting the heating up in the regenerators is that the bricks forming the checkerwork pass rapidly through that temperature range in which they can be destroyed by the effect of steam (hydration).

Direct heating up of melting chambers is used for small furnaces and recuperatively heated furnaces, the waste gas ducts and recuperator being heated up from the base at the same time as the melting chamber; hot waste gases are dissipated through openings in the roof and at a later stage the chimney damper is opened and the waste gases drawn downwards into the chimney. When the melting chamber reaches 800–900°C the normal burners are lit one at a time and the melting chamber is slowly brought to operating temperature. It is claimed that the advantage of carrying out the process from the melting chamber is better control of temperature distribution which may be achieved by better placing of the burners. Hydration of the bricks constituting the checkerwork only terminates when the melting chamber temperature reaches 200°C.

5.4.4.4 Control of the thermal expansion of refractories

Special attention must be paid to the crown due to the expansion anomaly of silica, the forces involved, and the need to achieve a gas-tight crown. The expansion anomaly even of well converted silica causes arching of the tapered sides of the bricks (Figure 5.46) but this arching can be reduced by placing insulation on the crown; suitable insulation is provided by a blanket 2–3 cm thick of aluminium silicate fibre with 10 cm of mineral fibre mat on top.

If the temperature curve is calculated on the basis of the deformation of silica bricks and is designed to avoid any opening of joints on the hot face, then a large crown should follow curve 1 of Figure 5.52 which gives changes in the rise of the crown:[46] the large negative 'rise' of about 250 mm at 200°C should be noted. Since the calculations include factors of unknown reliability and since there is the danger of the crown collapsing if such a negative 'rise' were permitted, practical experience dictates acceptance of a much smaller reduction in height. An earlier technique for permitting the crown to rise is shown in curve 3, from which it can be seen that this procedure was not the optimum either. Experience has shown that a small reduction in the height of the crown, as shown in curve 2, gives satisfactory results.

Good control over the heating up of the crown requires suitable bracing: spring loading of the tie rod helps to hold and control the expansion (Figure 3.97) while tilting of the buckstay and, thereby, of the skewback angle can be avoided by using buckstays which can be adjusted at their base.

Large and correspondingly heavy crowns are best held suspended in rigid and fixed steelwork (Figure 5.53), with expansion being controlled by strong compression springs. The spring assemblies have rounded surfaces in contact with the suspended skewback angles (Figure 5.54) so that the latter can roll freely over them. For reasons of weight crowns of this type are divided into sections, each 4 m in length.

Carefully designed bracing operating in the direction of the centre line of the furnace is important for the life of the end wall and, if applicable, for that of a burner located in this wall. The jacking bolts require checking and may have to be loosened during heating up if the expansion joints close

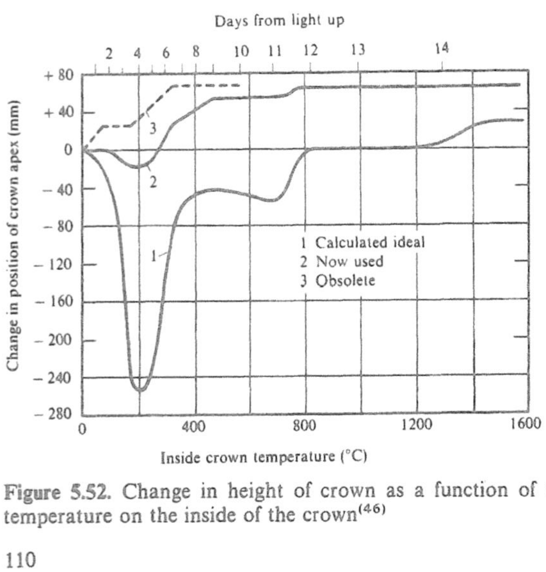

Figure 5.52. Change in height of crown as a function of the temperature on the inside of the crown[46]

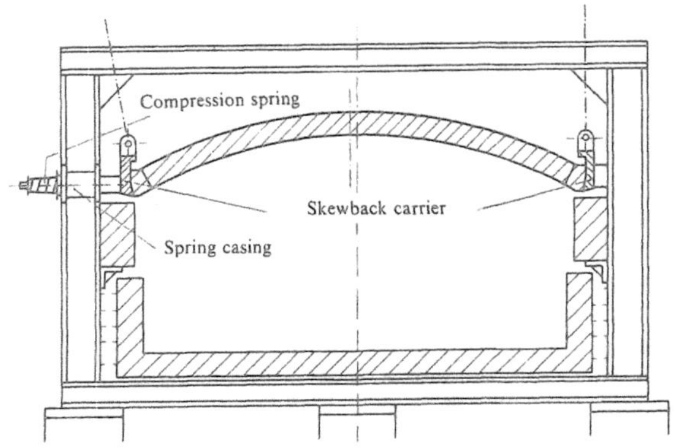

Figure 5.53. Rigid bracing of furnace with suspended crown[46]

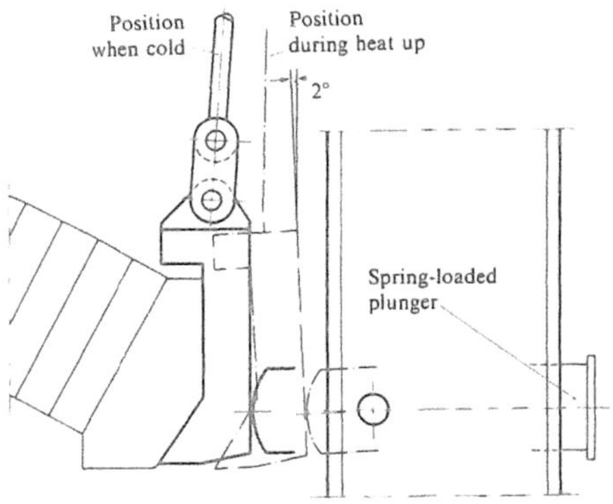

Figure 5.54. Crown suspension with pivoting skewback bearer[48]

prematurely. At the beginning of the heating up the expansion joints should be covered by loose bricks.

At the end of the process the temporary insulation on the crown needs to be removed and replaced by permanent insulation; this should be done over a small area at a time in order to avoid chilling of the crown and so that the temporary and permanent insulation in place will provide some protection against heat for the personnel. Careful cleaning by vacuum cleaners should be followed by the sealing of any remaining joints with a slurry of quartz sand and pulverised quartz or by the application of a dry mixture of these materials.

The interior of the melting chamber, including the floor joints, should have been cleaned before starting the process, with all foreign bodies which could prevent the closing of joints removed. The bottoms of flat glass tanks are covered with sheet glass after cleaning so that nothing can fall into the joints. Paving blocks of fireclay and fusion cast materials are positioned so that they can expand freely in all directions: a large joint, which is covered with paving blocks, is left between paving and tank sidewalls and every 10–15 cm large joints are left across the width of the tank bottom to make sure that no paving blocks lift during the heating up.

The blocks in the working end should not be allowed to move in view of their positioning with respect to machinery, and suitable construction and bracing enable the sidewalls to expand towards the melting end (Figure 5.55). Courses of tank blocks and soldier blocks are positioned either for individual expansion or, alternatively, every fifth vertical joint is left open to an extent that takes into account the relative expansion of the floor laid without cement.

The placing of courses of blocks must be such as to ensure that there are no continuous vertical joints otherwise there is a risk that the blocks could stick together during heating up, causing corners to break off. The off-setting of vertical joints must not be allowed to give rise to open horizontal joints.

Hexagonal blocks are often used at tank corners so that the expansion of the sidewall blocks will push the corner block out from its original position without causing open joints to form. Turning of the block can be prevented by the use of a dowel peg (Figure 5.56).

The superstructure sidewalls of the melting and refining zones are nearly always supported separately, being separated from crown, tank blocks, and burner ports by joints. The port mouth lies in the sidewall but does not extend into the crown as in earlier times. The joints are packed with high temperature insulation wool before heating up and

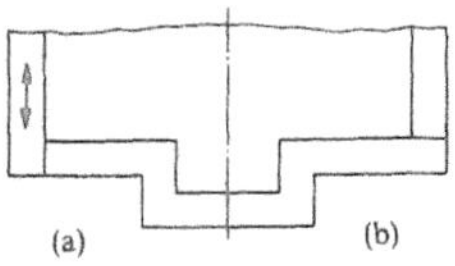

Figure 5.55. Expansion of tank sidewall blocks
(a) unhindered expansion
(b) obstructed expansion

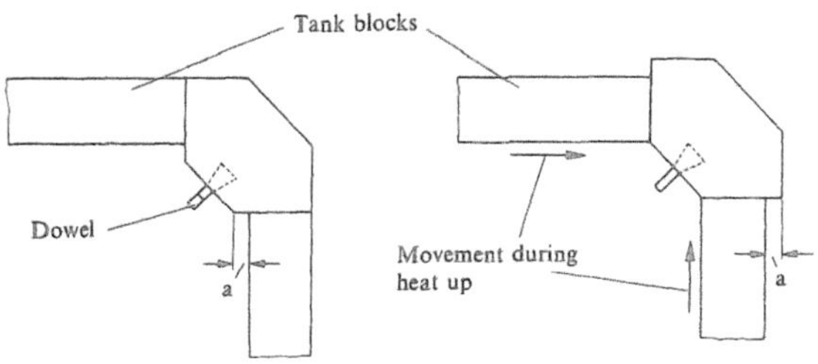

Figure 5.56. Dowel for controlling movement at corners[46]

larger openings are covered with ceramic fibre matting.

Regenerator crowns are treated very much like furnace crowns, with temporary insulation also being applied here. If the upper parts of the chamber are constructed of silica and the lower parts of firebrick then the expansions of both parts are very different, the silica expanding more than the firebrick. Stepped cracks can be avoided by providing a sliding zone and separate bracing for the firebrick section as has been pointed out by Greschat.[42] If the regenerator walls are built of magnesite then the transverse expansion can be absorbed in individual joints between bricks while in the direction of the length of the furnace each layer has cardboard 2–2·5 mm thick inserted every 250 mm. In the vertical direction expansion joints can be omitted (Figure 4.15).

Special steps must be taken at connections to port necks, details of which have been given by Greschat.[42]

5.4.4.5 Charging the furnace after heating up

Old literature recommended that tanks, especially large tanks, should be filled with cullet before heating up, the sidewalls being kept clear so that a controlled heating up was possible. It was soon pointed out that this procedure put great stress on the tank bottom because at the start of the process the cullet acts as insulation and when it melts, the bottom is rapidly raised to a high temperature, creating uncontrolled expansion which causes cracks to appear in the refractory. With large tanks it occasionally happened that the cullet, after it had melted in the melting end, flooded into the colder conditioning zone and caused the bottom blocks there to expand at such a rate that the bottom bracing could not be released quickly enough and had to be burnt through.

The problems described above have led to cullet being charged into the heated furnace by a centrifugal cullet thrower. The design of such a machine with a trajectory of between 4 and 23 m is shown in Figure 5.57. Such a machine should only be used to charge coarse cullet in order to avoid throwing small pieces of glass against the superstructure where they adhere and start premature corrosion.

The heating up of tank furnaces without cullet or batch can cause the bottom joints to open when the bottom is subsequently cooled during charging.

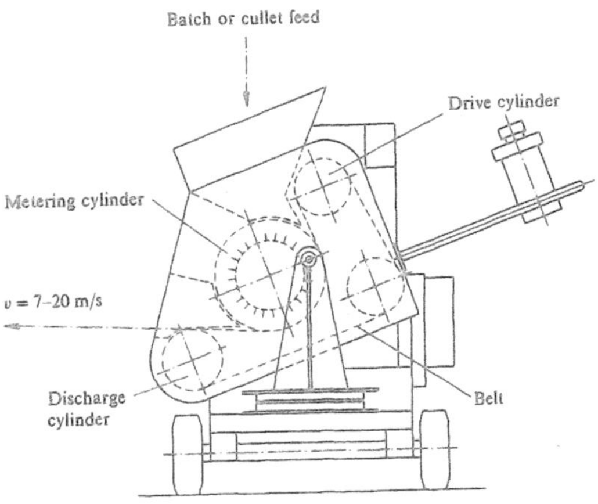

Figure 5.57. Schematic of a cullet thrower[46]

The use of intermediate layers of refractory which are plastic in this temperature range not only prevents these dangerous joints from appearing but also the corrosion which could subsequently take place beneath the bottom paving blocks (see also Section 3.1.2.2). Charging should only start when the bottom has reached its operating temperature.

5.4.4.6 Instrumentation and control during heat-up

All furnace instrumentation and controls should be ready for use before heating up commences. Additional thermocouples need to be inserted in the tank bottom, the regenerators, and the flues; pressure gauges in the gas lines to the melting chamber, regenerators, and flues are also useful. The heating up burners should be provided with all the controls which will ensure that gas, air, waste gas temperature, and waste gas volume will be set and controlled accurately.

5.4.5 The cooling down of glass furnaces

If a furnace is cooled down for a partial repair or because it will be used again later for production, then the cooling down must be carried out in a controlled manner. If enough time is available the furnace should be sealed and allowed to cool down slowly by itself although, depending on the size and condition of the furnace, this process may take too long or may be too rapid. A temperature/time schedule similar to that used for heating up can be drawn up and it can be followed by introducing air or by the use of auxiliary burners. Adjustment of the bracing must be done in conjunction with the cooling process.

Refractories which have been heated up and cooled down with care can certainly be used again, provided they have not been attacked too badly by molten glass.

5.4.6 The draining of tank furnaces

If a tank furnace were to be cooled without first draining off the glass, a solid block of glass would be formed which could only be removed by hard labour and at a high cost; this is best avoided by draining off the glass through a water jet, which causes it to granulate into an easily manageable form. The firing of furnaces is continued during draining, especially in the case of furnaces melting dark glasses containing iron, until the highly viscous layer of glass near the bottom has been melted.

Drain holes about 8 × 8 cm are provided in a number of places near the tank bottom but especially at the lowest position. They are located in the soldier blocks when the furnace is constructed and are sealed with a suitable plug which is pushed inwards for draining.

As insulation began to be applied to tank bottoms and as furnace operating temperatures increased, the drain holes became a problem in that they were the cause of corrosion by upward drilling and the provision of drain holes was abandoned in favour of equipment, recently developed, for drilling holes through hot refractory blocks. Nowadays the blocks are drilled just before tapping is to take place; drilling at this time has the advantage that the placing of drain holes can be decided in relation to the condition of the furnace and that a number of holes can be drilled one on top of the other for easier control of draining.[49]

The stream of glass is allowed to flow along a drain and is then caught in a jet of water at a pressure of 6–8 bar flowing at a rate of about 70 m^3/h, cooling the glass into a granular form which is easy to handle. The channel between the drain hole and the water jet is made of fireclay and has to be preheated so that the glass flows freely; past the water jet, steel ducts are used. The stream of glass is controlled and can be stopped by plungers which can be pushed against the tank block hydraulically or mechanically. A drainage system can also be provided and a gate of heat-resisting steel can be used for flow control. A facility for draining a tank is shown schematically in Figure 5.58; it can handle up to 30 t/d of glass.

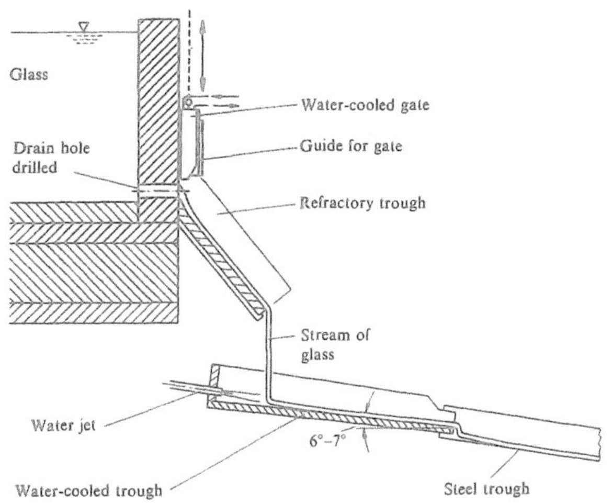

Figure 5.58. Schematic arrangement for the draining of a glass tank furnace

Some tanks are fitted with holes in the bottom for technical reasons and these can also be used for draining the tank; Figure 5.59 gives an example of such a bottom drain. It also is important that the flow here can be controlled, and this is achieved either by a sliding gate or via electrically heated floor nozzles (see also Section 6.5.3.1.6). A water jet is again used to granulate the glass.

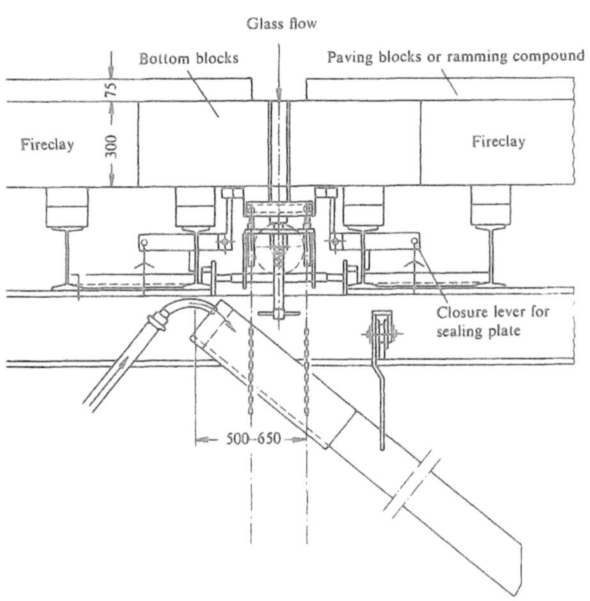

Figure 5.59. Draining of a glass tank through the bottom (dimensions in mm)

5.5 Hot repairs

If during the course of operation a glass tank developes serious structural faults such as damage to tank blocks, crown, or ports, which put the continuation of the campaign into question, then it must be decided whether the faults can be repaired as part of a hot repair or whether a cold repair is necessary. Since a cold repair implies cooling down the furnace, it follows that the shut-down will be of some length; if possible, therefore, hot repairs are preferred. Depending on the nature of the damage, production can either continue as normal, or the furnace temperature may have to be reduced, or even the glass level lowered.

5.5.1 Increasing the cooling

If the life of the furnace only needs to be extended for a relatively short period, perhaps to postpone a cold repair for a few months, and if the damage is largely local and becomes evident on the outside only when the temperature is raised, then increased cooling by air or water can often help.

Portable blowers of various sizes which are used for air cooling provide air at the critical location through flexible pipes. Fans which will give 200–30 000 m^3/h of air and pipes of up to 200 mm in diameter are necessary. Connecting this cooling to a central cooling system is usually not advisable since the available pressure is too low.

If the locations to be cooled are flat and easily accessible, then cooling by water boxes is possible. Boxes 100–150 mm deep made of steel plate 3–4 mm thick are used as they do not easily distort and thus cause poor heat transfer. Heat transfer can be increased by using a chemically setting cement containing SiC between box and refractory.[50] The box needs to be pressed firmly against the location to be cooled and flow control of the cooling water also needs to be provided. Cooling water manifolds

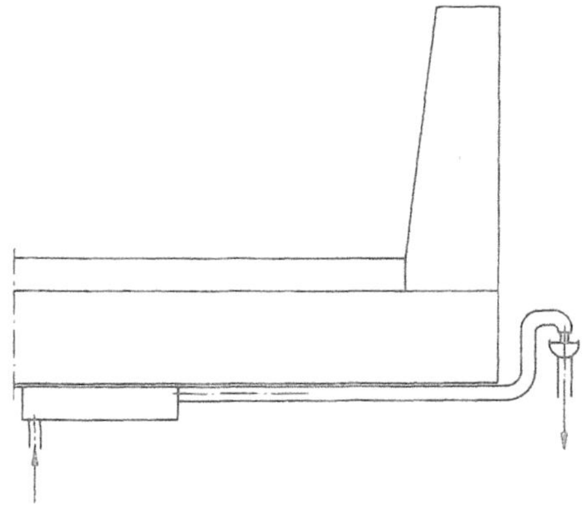

Figure 5.60. Water cooling box located against the tank bottom

made of pipes of square cross section can also be used. Water must flow upwards from below to ensure that any steam formed is washed out of the box and the temperature increase of the water should be limited to 20 degC. Figure 5.60 shows a water box cooling a bottom tank block.

5.5.2 Refractories for hot repairs

If the damage is so advanced that local cooling is not sufficient, then the area can be repaired by removing the old refractory and replacing it by suitable new refractories. To carry out this work, it is sometimes necessary to reduce furnace temperature to about 800°C and cover the opening created with suitable sheeting or curtains. A suitable replacement refractory is bonded quartz because its low thermal expansion is linked to high thermal shock resistance or, alternatively, if good resistance to corrosion is required, synthetic mullite is a possibility. The behaviour of refractories for hot repairs is summarised in Table 5.13. Bonded quartz should be used without cement and expansion joints need not be provided, but when using mullite an expansion of 0·5% must be allowed for; the use of a thin cement is recommended. Levelling off against the old brickwork is best achieved with zircon patching materials because of their good adhesion.

Hot repair refractories can be purchased either in the fired or hydraulically bonded forms. Quartz blocks can also be made on site using quartz ramming material, which is very useful if complicated shapes are needed.

Parts which are known from the beginning of the campaign to need replacing during operation, e.g. burner blocks, are held in place with high temperature fibre mats. The replacement of these components is therefore a relatively simple matter.

5.5.3 Refractory compounds for hot repairs

Casting, ramming, and spray compounds are used in hot repairs. Casting and ramming compounds provide temporary support until chemical or hydraulic bonding has taken place. Their advantage is that holes of very irregular shape can be closed and it is also possible to repair inaccessible parts but, for hot repairs, these materials are more likely to contaminate the molten glass than conventional refractories.

Damage to silica refractories is best dealt with by using quartz with water glass as a binder. If neutral conditions vis-a-vis fusion cast materials and silica must be met (port areas, sidewalls), then zircon ramming compounds, which set chemically at temperatures as low as 150–200°C, offer advantages.

A flat piece of hardwood saturated with water is recommended as a temporary support for hydraulically setting quartz compounds; it should fit the contours of the area to be repaired. This area is cleared of all old ramming compounds, residues of refractories and cement, and corroded silica bricks (Figure 5.61). The template, supported by wires of heat-resisting steel, is lowered into the opening so that the wires are pressed against the existing silica refractories (Figure 5.62) and the prepared quartz cement is rammed home to the full depth required. Within a short time the wooden support has burnt away, by which time the hydraulic bonding of the ramming compound should have progressed sufficiently to take the weight of the piece itself. The support wires can be pulled out when the wood has burnt away and the ramming compound has set adequately (Figure 5.63).

With chemically setting zircon compounds it is necessary to heat the edge zones rapidly in order to achieve the necessary rigidity; the use of steel plates may prove to be more appropriate but only if they can be removed after the compound has set. If the

Table 5.13. *Properties of refractories for hot repairs*

Material	*Behaviour when in contact with another material*	*Thermal shock resistance*	*Workability*	*Temperature limit of application (°C)*	*Durability*
Bonded fused silica	Like silica	Very good	Very good	1600	Like silica
Zircon	Neutral	Medium to good	Medium	1650	Function of glass composition
Zircon–cristobalite	Neutral	Good	Medium	1650	–
Synthetic mullite; sintered mullite directly bonded	Good, due to low proportion of glassy phase	Good	Poor	1700	Good
Fused mullite directly bonded	Good	Very good	Poor	1750	Good

Figure 5.61. Damaged area of crown[50]

Figure 5.62. Supporting the wooden template by wires of heat resisting steel[50]

Figure 5.63. Closing the hole using quartz ramming compound[50]

crown is damaged over a wide area, zircon compounds should be applied over the whole width of the surface starting from the skewbacks. A larger area of damage can be covered with a preheated tile of an appropriate size and the edges sealed with ramming compound (Figure 5.64). The repair of a damaged port arch using this technique is seen in Figure 5.65; the size of the hole made it necessary to close it in two steps and the zircon ramming compound was supported by an angled steel plate.

Hot repairs are also suitable for worn eyes of pot furnaces. A proven method is to lower a wooden box and fill the worn areas with ramming compound (Figure 5.66).

Figure 5.64. Positioning of a prefired tile

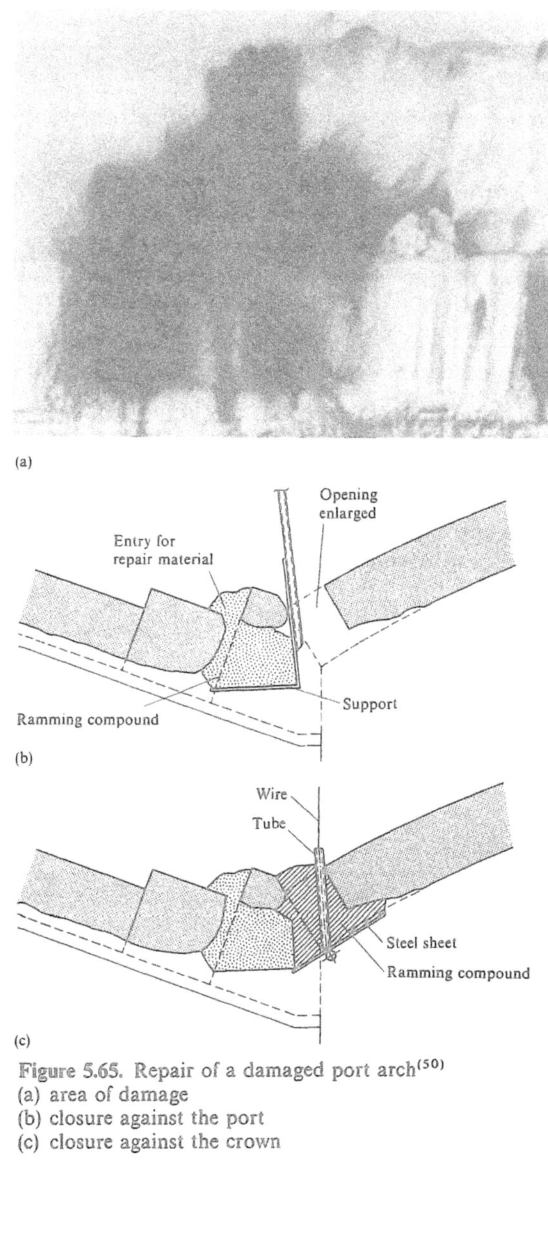

Figure 5.65. Repair of a damaged port arch[50]
(a) area of damage
(b) closure against the port
(c) closure against the crown

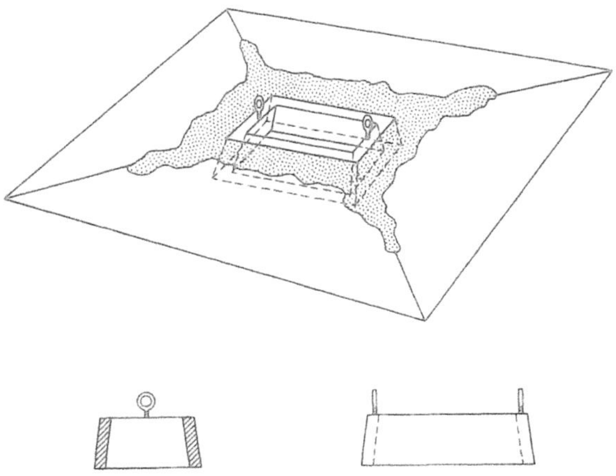

Figure 5.66. Repair of an eye using a wooden box[50]

5.5.4 Gunning

Although hot repairs with refractories and ramming compounds can be carried out by glass works personnel themselves, gunning requires specialised equipment available only to manufacturers of refractories. This method should be considered when extensive damage, which is not accessible from the outside, has taken place on the inside of the furnace. Even with careful working, about 30–50% of the sprayed material bounces off the target and falls into the glass; despite skimming, temporary contamination of the glass cannot be avoided.

Quartz compounds using water glass, phosphate, and boric acid binders are employed. The compounds are mixed with the binder in the machine and the mixture is led through hoses to the gunning lance, for which $\frac{3}{4}$ inch water pipes without cooling jackets have been used successfully.

The compound is applied in layers and good adhesion, especially of the first layer, is important. The addition of extra binder produces a rough surface onto which the compound may be sprayed; on vertical surfaces the layers must be relatively thin (20 mm) but on sloping or horizontal surfaces they may be thicker. After each layer has been applied 15–30 min must elapse to permit the compound to set and harden.

Figure 5.67 shows an example of a successful repair of a worn port mouth using the gunning technique; the repair was carried out using a curved lance inserted from below through the gap between the port sill and the upper surface of the tank block.

Gunning can also be used to repair suspended curtain walls on flat glass tanks; damage caused by faulty cooling, e.g. spalled refractories, can be repaired with quartz compounds (Figure 5.68). In the same way the separating walls of regenerator

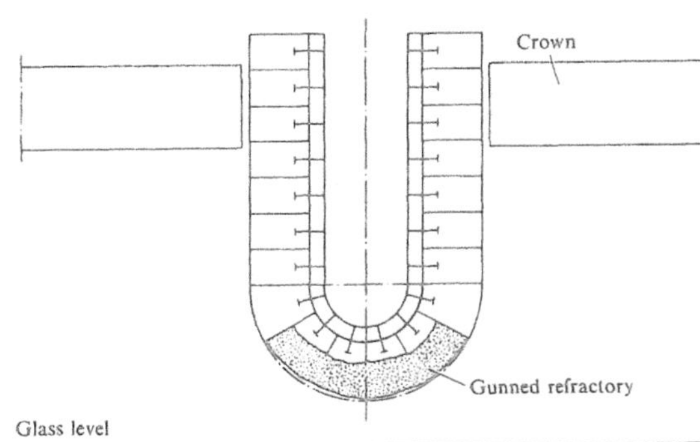

Figure 5.68. Gunned shadow wall of a flat glass tank[50]

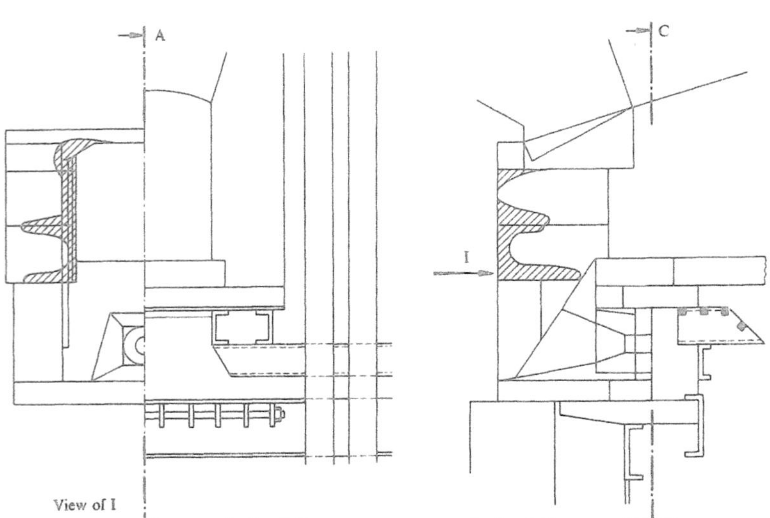

Figure 5.67. Port repair using a gunning technique, showing damage and direction of spray[50]

chambers can also be repaired but compounds suitable for this location must be used.

The welding of patching compounds to the structure under repair using oxygen (thermite) is still in its infancy. Silicon grains can be converted into molten silica using oxygen but as the temperatures arising in this process are above 2000°C, care is necessary when using it on glass furnaces.

5.5.5 Hot repairs using fusion cast refractories

Strictly speaking, this belongs in Section 5.5.2 but as fusion cast refractories are highly sensitive to temperature changes, special procedures have to be adopted and this justifies their being treated separately.

The most frequent hot repair involving fusion cast refractories is the backing up of the upper third of the tank blocks with tiles to counter the continuing corrosion at the flux line. The upper course of insulation is removed and void-free tiles are placed on the lower course; alternatively, the tiles can be placed on steel supports provided for this purpose. The layer of insulation can only be considered strong enough to carry the tiles if it has been constructed as a sandwich with an inner core of fireclay (Figure 5.69). The tiles are installed when wear has proceeded to an extent that, despite normal cooling, the outer surface glows medium red. Unnecessarily early installation of backing up tiles will interfere with the cooling of the other blocks and thereby accelerate the rate of corrosion. If the tiles are installed too late, there is the risk that the heads of the soldier blocks will break off, exposing the tiles to intense heat radiated from the body of the furnace; this will cause the surface of the tile to heat up at a rate which sets up high thermal stresses and causes the tiles to crack. The temperature shock can be reduced by applying, before installation, several layers of water glass to the side which will become the hot face. After installation, cooling should be increased slowly, not started at full strength.

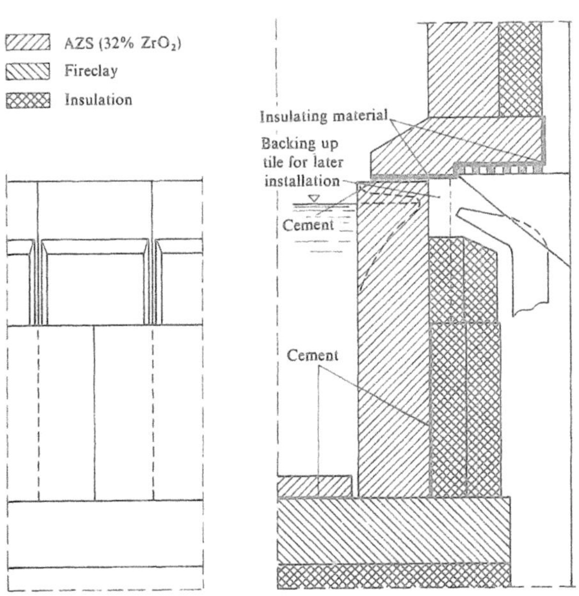

Figure 5.69. Insulation of tank sidewalls and bottom[52]

The sensitivity of fusion cast blocks to thermal shock demands careful design of joints between tuckstones and soldier blocks. It is possible to leave a relatively large gap between side block and tuckstone, close it with a wedge shaped fillet which prevents the refractory from falling into the tank, and seal the gap with gunned refractory as indicated in Figure 5.70. Backing up tiles must be placed so that their top edges are in contact with the bottom of the tuckstones. A different solution, illustrating how a narrow gap may be filled with gunned cement, is shown on the left hand side of Figure 5.70. The tuckstone has a step which extends beyond the backing up tile, which makes access easier as well as giving a good seal between tile and tuckstone. The introduction of cement into a narrow joint is difficult and for this reason a combination of both versions, i.e. the use of wedge shaped fillet and stepped tuckstone, could well be the best method. Solving the problems of sealing has become much easier since the introduction of high temperature fibrous mats.

Special care must be taken when carrying out any repairs to the throat. Heavy corrosion usually appears on the melting end side of the cover block

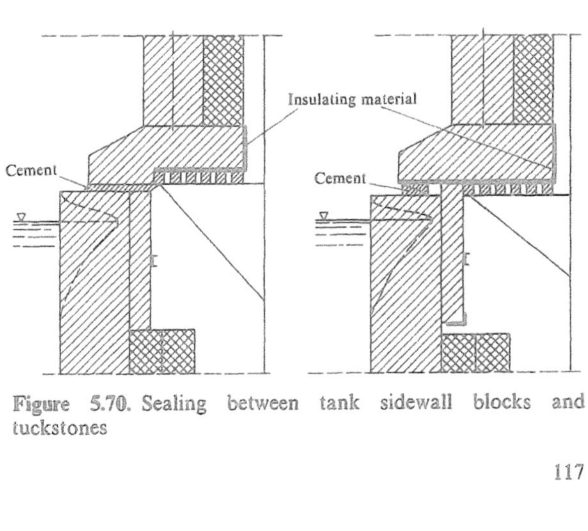

Figure 5.70. Sealing between tank sidewall blocks and tuckstones

and the placing of tiles on top of the cover block in good time will prevent a breakout. It is best to use pairs of wedge shaped tiles as shown in Figure 5.71; by forcing the tiles against one another they can be wedged tight so that they cannot lift off. The tiles must be without voids and to minimise the risk of cracking they should not be too thick.

An alternative procedure is to freeze the throat temporarily so that the damaged blocks can be completely replaced. Throat cooling must be adjusted to suit the new conditions and should be increased slowly to keep thermal stresses in the refractories to a minimum.

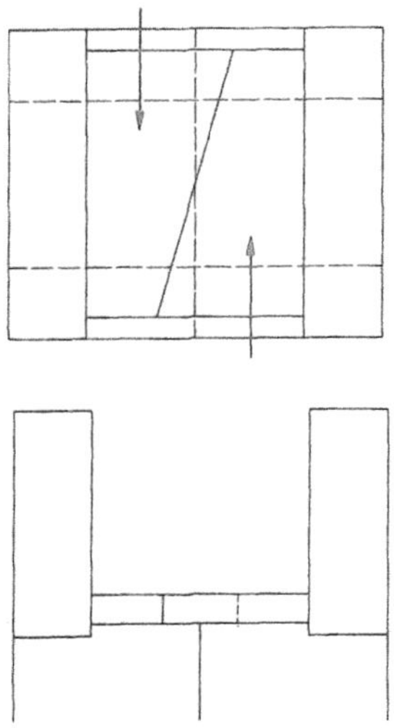

Figure 5.71. Covering the throat cover blocks[52]

5.5.6 Hot repairs to regenerators

Unfortunately it frequently happens that, in a long campaign, parts of the checkerwork collapse causing restrictions to the distribution of combustion air to individual ports or preventing the furnace pressure from being maintained due to insufficient pull from the chimney. If this situation arises and a cold repair cannot be considered, then the regenerator must be replaced in whole or in part as a hot repair.

Suitable working conditions for removing faulty checkerwork and replacing it with new bricks can be created in a regenerator by lowering the temperature in the melting chamber, keeping the flame constant over longer periods, and closing off parts with ceramic fibre sheeting or temporary walls. For a while the tank must be operated with cold combustion air, which is possible when combined with a reduction in the melting chamber temperature. If for technical or production reasons the melting chamber temperature cannot be reduced or can only be reduced slightly, then oxygen may be introduced or even the heating up burners must be brought into operation.[53]

If the regenerator is in sections, interference with furnace operation can be minimised or even totally avoided.

6 Tank furnaces – their behaviour in operation

6.1 Flow of combustion gases in the combustion space

6.1.1 Cross-fired tank furnaces

Direct measurements of gas flow in the melting chamber are difficult and costly. The literature therefore only gives data based on measurements using models and results covering the residence time of combustion gases.

Detailed measurements on an experimental section of an oil fired tank have been performed by Ullrich,[1] the main interest being the position of the oil burner; if this is located above the air flow (Figure 6.1(a)), then a strong broadening of the circulating flow above the air flow results. The flow levels just above glass level are relatively high, which can lead to high evaporation losses from the molten glass as well as local reduction of the glass itself. If the burner is situated below the port (Figure 6.1(b)) then the circulating flow under the crown is somewhat less and the air flow from the port is prevented from sinking onto the glass surface by the atomised jet; the cooling effect is thereby avoided and the flow rate above the glass is also lower. Investigations showed that the positioning of the burner above the air called for very accurate setting in relation to the burner axis. Even small deviations led to undesirable flow conditions, causing both overheating and local reducing effects on the bath of molten glass. The positioning of burners under the air stream is less critical and the system of burner jet plus combustion air stream is very stable.

Information covering the flow between two pairs of ports is derived from investigations into the residence time of combustion gases.[2] The flames from two adjacent ports give rise to two opposing turbulent flows which cause an exchange of combustion gases of the order of 15–25% (Figure 6.2). Quantitative proof was achieved by injecting a tracer material in one port and analysing samples in the adjacent port opposite; the concentration was about one order of magnitude lower than in the main stream. The minima and maxima appearing at the tail ends of the curve in Figure 6.3 indicate that more or less stable rotating waves exist be-

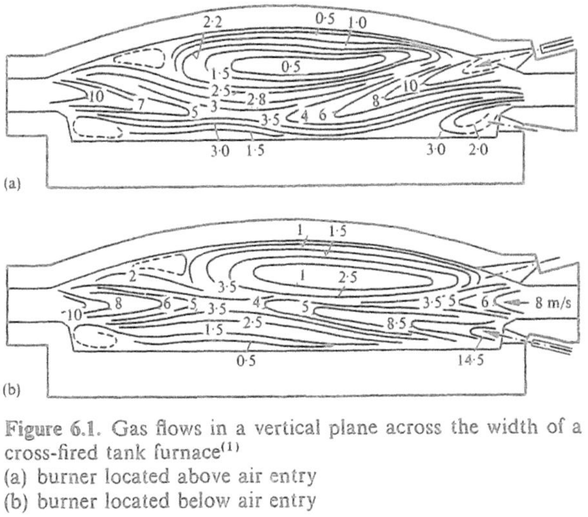

Figure 6.1. Gas flows in a vertical plane across the width of a cross-fired tank furnace[1]
(a) burner located above air entry
(b) burner located below air entry

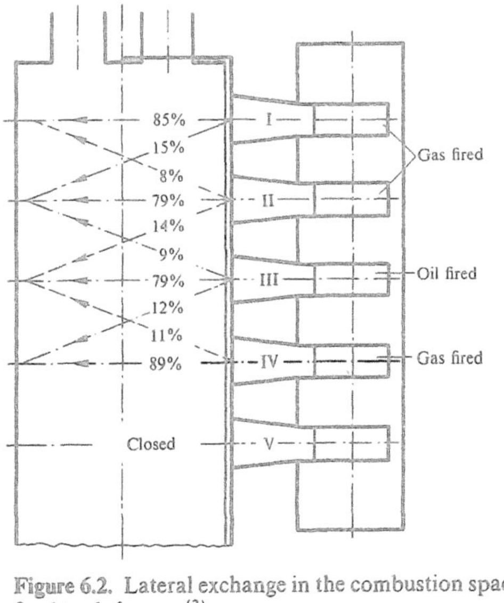

Figure 6.2. Lateral exchange in the combustion space of a cross-fired tank furnace[2]

tween individual ports through which material exchange takes place.

The crosswise exchange is strongly dependent on the momentum of the flame as is shown in Figure 6.4 which gives the residence time in a tank furnace in which one port was fired with oil and the adjacent one with gas. The proportion of direct flow, i.e. that flow which moves in the plane of a pair of ports without crosswise exchange, increases with the momentum of the flame. The result is that gas flames, with their lower momentum, cross the combustion space more uniformly whereas oil flames, with their higher momentum, draw in larger quantities of waste gas and air. The flame crosses the combustion space quickly but the remaining air and the waste gases mix with the flame only slowly on their way to the exit port.

Figure 6.4 gives the standardised residence time

$$E(\tau) = \frac{c(\tau)}{\int_0^\infty c(\tau)\,d\tau}$$

for oil and gas firing as a function of dimensionless time $\tau = t/t_{th}$. Standardisation was necessary to make gas and oil firing results comparable despite their very different mass throughputs. The residence time $E(t)$ is defined by

$$E(t) = \frac{c(t)\dot{V}}{n_{ind}} \quad (6.1)$$

where $c(t)$ is the measured change of concentration of the tracer material with time, $\dot{V}$ the volume stream entering the system, and n_{ind} the quantity of tracer material added at time t_0. For standardisation a selected $E_0 = \int_0^\infty c(t)\,dt$ and the theoretical average residence time t_{th} was used. With V_s as the total volume of the system,

$$t_{th} = \frac{V_s}{\dot{V}}. \quad (6.2)$$

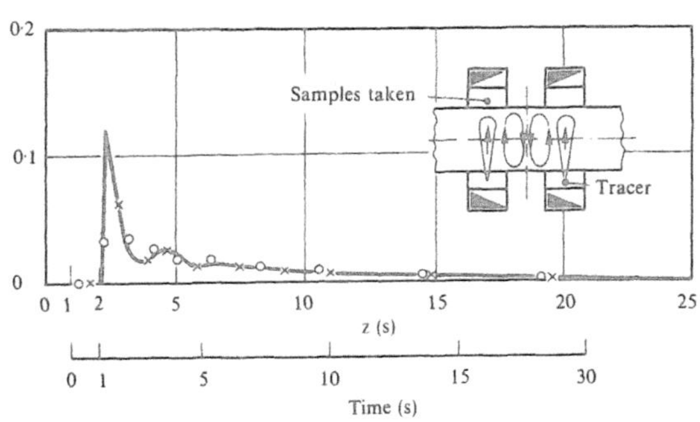

Figure 6.3. Gas flow along a cross fired tank furnace as a function of time[2]

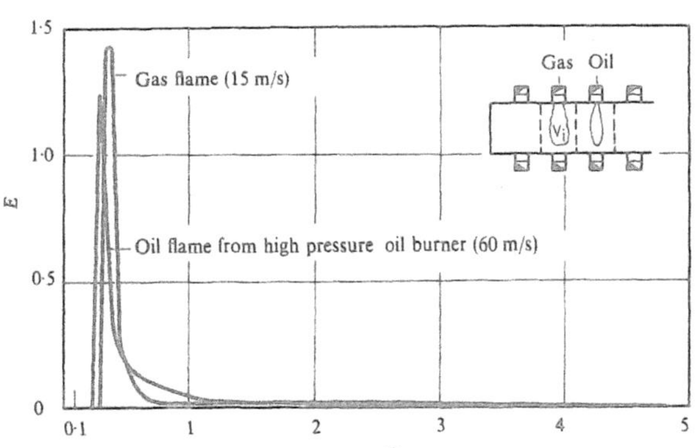

Figure 6.4. Normalised residence time distribution of an oil and a gas flame[2]

6.1.2 U shaped flame tank furnaces

The flow relationships in a U shaped flame tank are more complicated than those in a cross-fired tank. Results of measurements on models are shown in Figure 6.5 in which the currents are shown in three horizontal planes at different heights inside the combustion space; several rotating vortices may be seen and the whole picture of these currents must be envisaged in three dimensions.

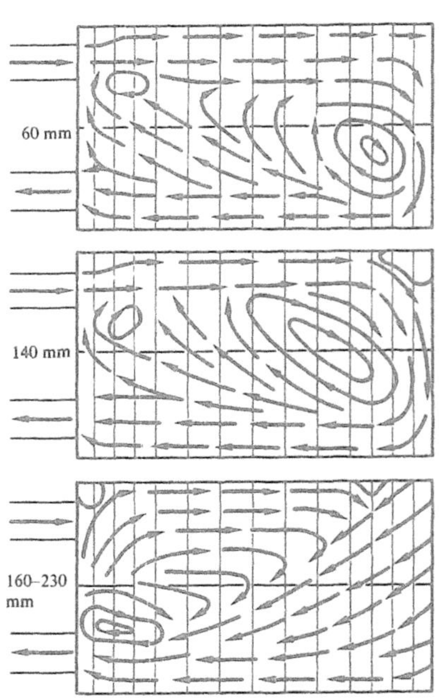

Figure 6.5. Gas flow in U shaped flame tanks in three horizontal planes at different heights[3]

If cold air is induced from a doghouse placed along a sidewall, then the current which enters is no longer drawn against the sidewall but goes almost straight into the tank.

If sting-out losses occur, then the mass and momentum flow of the flame jet are reduced and the entrainment of surrounding material diminished. The inversion point of the flame takes place further downstream on a tank without sting-out losses and the main vortex almost reaches the breast wall.

Crowns which slope downwards to the flame return zone give rise to unstable flow. The reduced volume of the combustion space in the area of the inversion point creates a positive pressure which disturbs the stabilising negative pressure between flame and sidewall and leads to a fluttering of the flame jet with all its consequent disadvantages.[3]

Turning the port axis by 10° towards the centre line of the tank only creates stable flow conditions for larger distances between ports if $b/B > 0{\cdot}7$, where B is the width of the tank and b the distance between the centre lines of the ports. The flame jet can now develop without interference from the sidewall and cover the melting zone adequately.

Many U shaped flame tanks tend to have unstable flows even if the crown does not slope downwards. The flame jet moves to and fro in the combustion space between a flow parallel to the sidewall and a flow at an angle to the centre line of the tank. The residence times measured under these conditions using a tracer technique are given in Figure 6.6 in terms of concentration changes of the tracer with time. The normal distribution shows a strong maximum with only a very weak secondary maximum while under unstable flow conditions two maxima of almost equal strength appear, which can be explained on the basis of the appearance of a strong return flow vortex.

As was shown by Brüning for different designs of melting chamber (Figure 6.7),[3] instability is, generally speaking, more likely with smaller distances between ports and when there is a lower total momentum from the combined system of burner plus combustion air.

A comparison of the mixing behaviour of cross-fired and end-fired tank furnaces shows significant differences between them. The most intense mixing between the fuel and air flowing into the furnace and the gases resident in the combustion space occurs with end-fired furnaces (U shaped flame tanks and tanks with double crowns); the mixing in

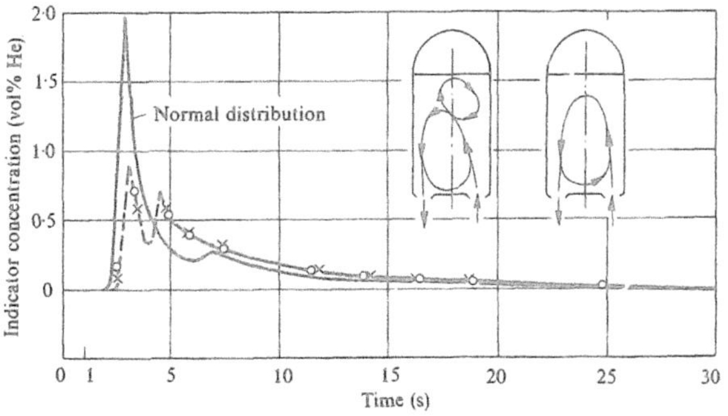

Figure 6.6. Concentrations of a tracer material in a U shaped flame tank furnace as a function of time (unstable flow through chamber)[2]

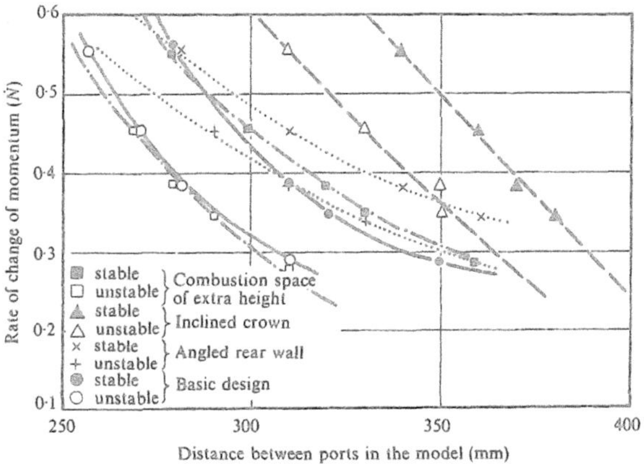

Figure 6.7. Stability diagram for different designs of combustion chamber derived from models[3]

cross-fired furnaces using oil is significantly less and that in cross-fired furnaces using gas is the smallest. Figure 6.8 gives the calculated total distributions, $E_g(t)$, against the ratio of measured to theoretical times, $t/\bar{t}_{th}$. Direct measurements of the distribution $E(t)$ are not representative of the behaviour with time of the material flowing into the furnace due to cross exchanges in the combustion space; calculated total distributions were therefore used for comparison.

Two distributions of residence times should be considered. First, the residence time distribution of the material which flows directly through the furnace. As follows from the flow balance, the integral of the distribution, $E(t)$, defined in accordance with Equation (6.1),

$$E(t) = c_{i-i}(t)\frac{\dot{V}_i}{n_{He}}, \qquad (6.3)$$

gives the proportion of material entering through the ith port which is removed through the ith port

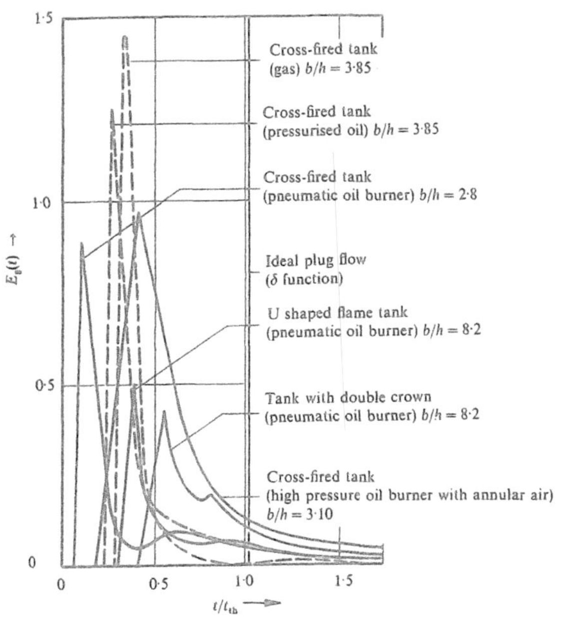

Figure 6.8. Residence time distributions for different types of furnace[2]

opposite, i.e.

$$\int_0^{\infty} E(t)\,\mathrm{d}t = k_{i-i} \leqq 1. \tag{6.4}$$

The second distribution to consider is the residence time distribution, $E_g(t)$, of the total material flow entering through one port, which takes into account the fact that parts of the material flow with different time distributions exit through neighbouring ports. Since the individual distributions taper off in parallel, this distribution appears as a superposition which takes into account the proportion of material quantities exchanged. Thus

$$E_g(t) = c_{i-(i-1)}(t)\frac{\dot{V}_{i-i}}{n_{He}} + c_{i-i}(t)\frac{\dot{V}_i}{n_{He}} + c_{i-(i+1)}(t)\frac{\dot{V}_{(i+1)}}{n_{He}} \tag{6.5}$$

applies as well as

$$\int_0^{\infty} E_g(t)\,\mathrm{d}t \equiv 1, \tag{6.6}$$

where n_{He} refers to the quantity of helium, which was used as tracer in this instance.

The average residence time is used as a characteristic quantity for describing the many different distributions; as a reference time it is strongly dependent on the shape of the curves such as are shown in Figure 6.8 and is defined as the time at which the material which entered the system at $t = 0$ can be visualised as having left the system. The calculated residence time may be obtained from the melting chamber volume above glass level and the throughput, which depends on the type of flow through the furnace; it is assumed that the total above-glass volume of the furnace, which is geometrically allocated to individual ports, is totally filled. The ratio of measured to theoretical residence times gives the utilisation of the combustion space by the flow of combustion gases.

A comparison for cross-fired tank furnaces shows that the measured residence time is considerably less than the theoretical but all end-fired furnaces show the same values for both. Depending on the port considered, cross-fired furnaces have a residence time ratio of between 0·60 and 0·85; as shown above, this means that, of the direct flow which does not take part in cross exchange, only 60–85% of the space which geometry would allocate to this port is being used.

The analysis of residence time distributions shows that the combustion space in part is traversed by a piston flow which has mixing along its sides and in part acts as a mixer on which a return flow has been superimposed. The degree of mixing is largest for end-fired tanks, which means that the mixing of fresh fuel and air with the contents of the combustion space is very intense due, in part, to the long distance available for mixing.

6.2 Flame formation in the combustion space

6.2.1 Types and arrangement of flames

Glass furnaces are hearth furnaces in which flames travel over the batch and molten glass and thus direct and indirect heat transfer from flame to molten glass are important parameters. With certain exceptions, oxidising conditions should prevail in the melting and refining zones. Temperature differences and fluctuations with time should be small. The regenerative system preferred for glass melting furnaces has the advantage that the temperature differences arising from flame firing and the changes in heat transfer due to the periodic change of flame direction, even if not completely compensated, at least occur symmetrically to the longitudinal axis of the tank furnace (Figure 6.9).

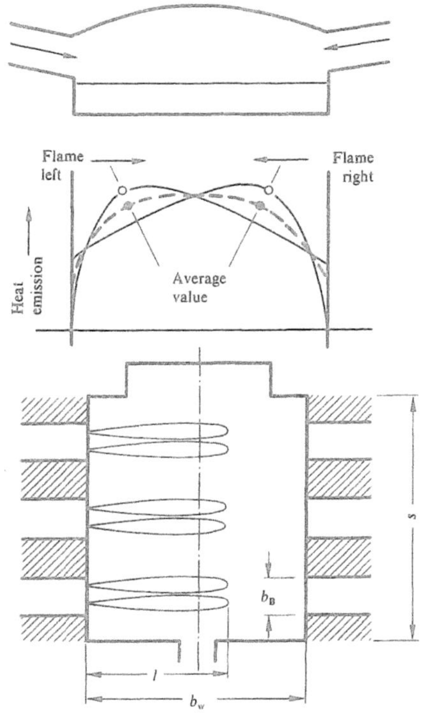

Figure 6.9. Flame position and heat emission of a cross-fired tank furnace

Important criteria are the standard flame length l/b_w and the degree of flame coverage, r, which is given by

$$r = \frac{nb_B l}{sb_w} \tag{6.7}$$

for n ports, the other quantities being given in Figure 6.9; for ports of different widths,

$$r = \frac{\sum_1^i b_{B,i} l_i}{sb_w}. \tag{6.8}$$

The value of r for cross-fired tank furnaces lies between 0·2 and 0·3.

U shaped flame tanks have only one pair of ports and the periodic temperature fluctuations are therefore greater. Jet flames, as now normally used, are seldom set to travel beyond the inversion point (Figure 6.10). Recuperative unit melters present a special case as they use premixing burners which are located across from one another (Figure 6.11), either directly opposite or offset; the latter were not successful and were abandoned.

Flame formation depends on the mixing of the combustion materials (fuel and air) and the chemical reactions between them. The mixing process is largely a technical problem involving flow, and combustion is in part a physical problem (preparation of a flow of fuel) and in part a chemical problem. The high temperatures existing in a glass furnace and the usually high temperature to which the air is preheated accelerate the chemical reactions but slow down the mixing process due to the high viscosity of the hot gases. The controlling factor for flame formation is therefore flow-dependent mixing which can be understood and described via the flow processes.

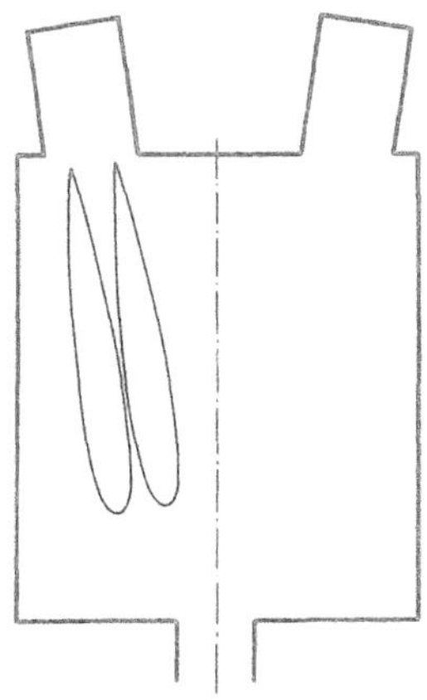

Figure 6.10. Flame position in a U shaped flame tank furnace

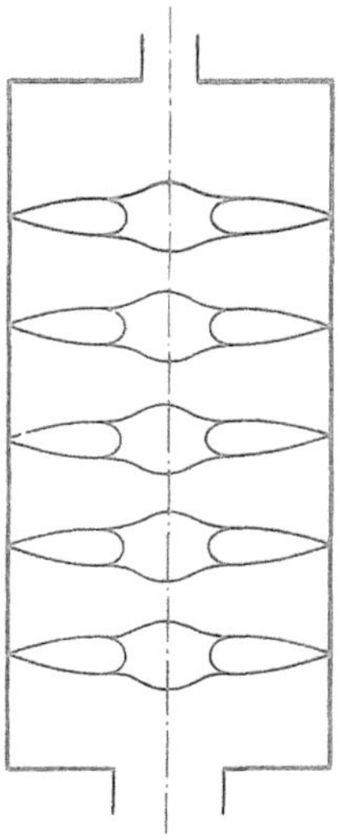

Figure 6.11. Flame positions in a unit melter

6.2.2 Turbulent jet diffusion flames[4]

6.2.2.1 Free jet

The velocity profile of a free jet flowing from a nozzle is given schematically in the top part of Figure 6.12, from which it may be seen that it is necessary to differentiate between three zones, the core, the transition zone, and the similarity zone. The core is characterised by the zone of reduced flow velocity at the free end of the nozzle and extends to about $4d_0$ (d_0 is the diameter of the nozzle). The transition zone extends to about $8d_0$ and is the zone in which the radial profiles of velocity, temperature, and concentration are formed which exist in the similarity zone. In this zone, the velocity profiles in all cross sections vertical to the main flow direction can be described by the same function as can the distributions of concentration and temperature. If a jet of fluid of

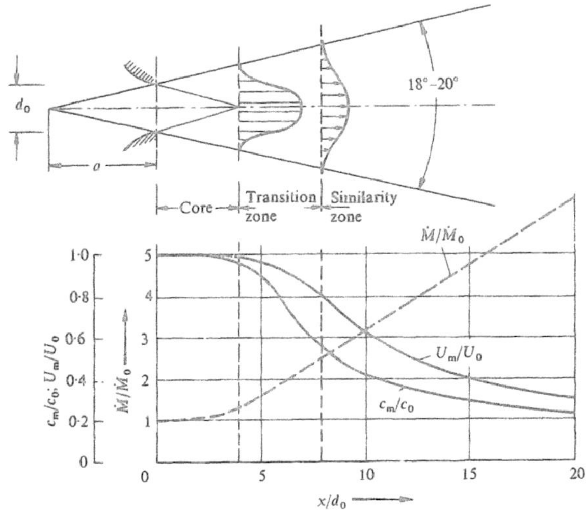

Figure 6.12. Distribution of mass, velocity, and concentration of a free jet (air in air) as a function of the normalised distance from the nozzle

density ρ_0 enters into a volume of density ρ_1, then the velocity distribution at a place (x,y) (two-dimensional analysis symmetrical in rotation) is given by

$$u = u_0 \frac{1}{2c_i x/d_0} \sqrt{\rho_0/\rho_1} \times \exp\left[-\left(\frac{1}{c_i^2} - \frac{1}{2c_i^2} \right) (y/x)^2 \right] \qquad (6.9)$$

where c_i is the momentum transfer factor which depends on the Reynolds number, and u_0 is the jet velocity at the exit from the nozzle. The distribution of the concentration is analogous:

$$c = c_0 \frac{c_i}{2c_c^2 x/d_0} \sqrt{\rho_0/\rho_1} \times \exp\left[-\left(\frac{1}{c_c^2} - \frac{1}{2c_i^2} \right) (y/x)^2 \right]. \qquad (6.10)$$

The transfer factors of the distribution functions for tubular jets of $Re \approx 10\,000$ are given in Table 6.1. Material interchange is more intense than the momentum interchange. The lower part of Figure 6.12 gives the curves for u_m/u_0 *and* c_m/c_0 (the jet axis).

Table 6.1. *Transfer factors for use with the distribution functions (6.9) and (6.10)*

Pair of gases	c_i	c_c
Town gas/air	0·076	0·088
Natural gas/air	0·075	0·087
Air/air	0·070	0·082
Co_2/air	0·066	0·077

The jet angle is 18–20°. For the velocity, it is 19° and is defined by the jet cone surface where the velocity is 1/10th of that at the axis.

With u_m as the average jet velocity, and using the principle of conservation of momentum, the mass flow distribution in the jet is

$$\frac{\dot{M}}{\dot{M}_0} = \frac{u_0}{u_m}. \qquad (6.11)$$

If the diameter of the jet at a place x is d, then, for a jet angle of 19°, $d = 0{\cdot}32\ (x + a)$. At large distances from the nozzle a is small in comparison to x, and one can write

$$\frac{\dot{M}}{\dot{M}_0} = 0{\cdot}32 \frac{x}{d_0} \sqrt{\rho_1/\rho_0} \qquad (6.12)$$

where ρ_1 is used as the density of the mixture of jet material and the materials surrounding it.

6.2.2.2 Diffusion flame

6.2.2.2.1 Free jet diffusion flame

Equation (6.10) for the concentration distribution may be used to calculate the length L of swirl-free turbulent free jet gas flames. With $x = L$ for $c = c_{st}$ (concentration of the stoichiometric mixture) at $y = 0$, and by using the applicable transfer factors and ρ_F as the density of the gases constituting the flames,

$$L = 5{\cdot}3 \frac{c_0}{c_{st}} d_0 \sqrt{\frac{\rho_0}{\rho_F}}. \qquad (6.13)$$

The flame length according to this Equation is shorter than those observed in practice due to the fact that the distribution functions are based on average values: in fact, the flame ends only where a stoichiometric concentration distribution can never occur and the real length is more like 1·3 times the calculated value of L. The factor 1·3 applies not only to the length of the flame but also to its diameter. Flame length does not depend on volume flow and for a given fuel, the diameter of the nozzle, d_0, is the deciding criterion. This lack of dependence upon volume flow is caused by the fact that with increasing velocity the rate of material exchange increases and at the same rate.

The jet momentum is given by

$$I = u_0 \dot{M}_0 = \dot{M}_0^2 \frac{4}{\pi} \frac{1}{\rho_0 d_0^2}. \qquad (6.14)$$

For a constant mass flow rate it follows that

$d_0 \sim 1/\sqrt{I}$ and, from above, $L \sim 1/\sqrt{I}$. As the momentum increases, the flame shortens.

6.2.2.2.2 Diffusion flames in a restricted space

If the nozzle is immediately adjacent to a wall, then the entrainment of surrounding air is largely prevented from one side (Figure 6.13) and in consequence its flame is longer than one burning in open surroundings. The flame length is more than doubled according to the relationship

$$L_{\mathrm{Wa}} \approx 11{\cdot}8\left(\frac{c_0}{c_{\mathrm{st}}}\right)^{2{\cdot}66} d_0\sqrt{\rho_0/\rho_F} \qquad (6.15)$$

where L_{Wa} is the length of the flame adjacent to the wall.

Jets placed close together interfere with one another in the entrainment of air (Figure 6.14) and this also causes a lengthening of the flame. In the limiting case of two flames moving closer together, $d_0 = d_0\sqrt{2}$, and the flame length becomes

$$L = 5{\cdot}3\sqrt{2}\frac{c_0}{c_{\mathrm{st}}}\sqrt{\rho_0/\rho_F}\,. \qquad (6.16)$$

For 3 or 5 adjacent jets the influence of the distance a/d_0 on the standard flame length L_m/L_e can be taken from Figure 6.15[5] in which L_m is the length of the central flame and L_e the length of each of the other flames. If the axes of the nozzles are closer than 18 nozzle diameters then the flames interfere with one another in the entrainment of air from the surroundings. In the limiting case the factors by which the flames are lengthened are $\sqrt{3}$ and $\sqrt{5}$ respectively. These results apply for coke oven gas and also approximately for natural gas.

A burning jet of flame enclosed in a cylindrical space draws in material from its restricted surroundings and creates a negative pressure which causes a ring-shaped rotating return flow to form. The return flow starts near the walls because it is there that a part of the dynamic energy is changed into pressure energy as a result of friction at the wall because the pressure gradient towards the root of the jet is at its largest here.

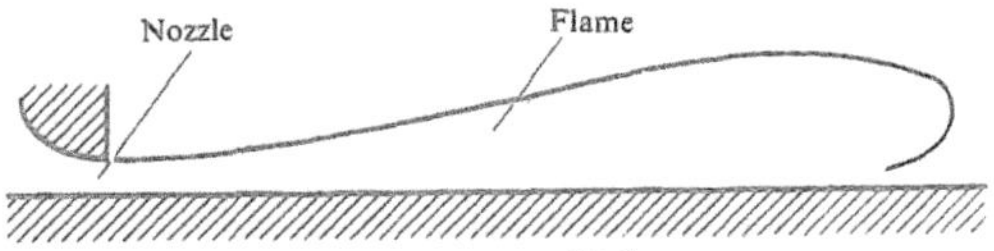

Figure 6.13. Jet formation along a wall

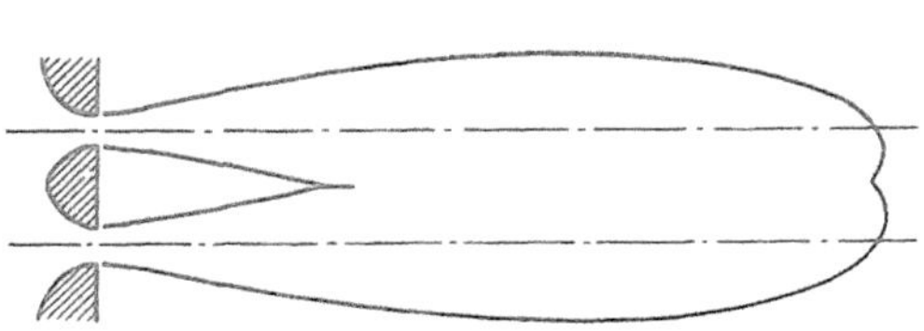
Figure 6.14. Formation of a double jet

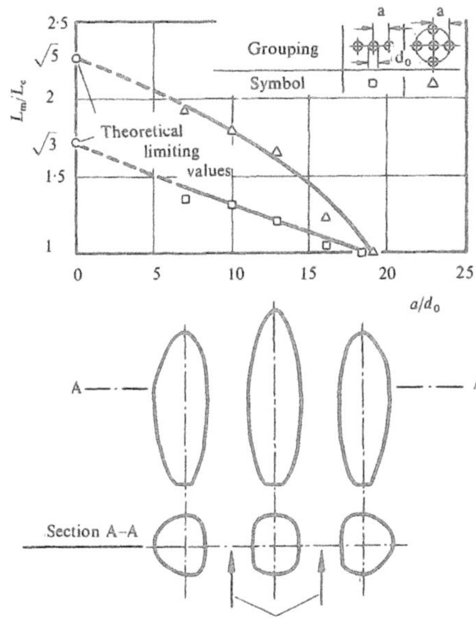

Figure 6.15. Length and lack of symmetry of neighbouring jet flames (L_e = length of an individual flame, L_m = length of central flame of a group)[5]

If the space in the plane of entry of the jet is enclosed, then the turbulent return flow reaches back to the beginning of the jet but if combustion air is introduced in the plane of entry of the jet then the return flow reaches back only as far as the zone where combustion air is entrained. Encasing a jet of flame always leads to a lengthening of the flame since combustion gases have to be mixed into the jet (Figure 6.16).

For a constant jet angle, α, the length of the zone of return flow increases as the diameter of the surrounding space increases, the relationship between the diameter and the returning mass flow being

$$\frac{\dot{M}_{R,2}}{\dot{M}_{R,1}} = \frac{D_2}{D_1}. \qquad (6.17)$$

At constant D, the jet angle will be increased by, for example, swirl as the zone of the return flow becomes smaller while $\dot{M}_R$ remains almost constant (Figure 6.16).

The length of an enclosed jet to which the combustion air is supplied parallel to the jet axis can be calculated similarly to those for free jets provided the effect of the return flow is taken into account by using the return flow parameter, $1/\Theta$ (Figure 6.17). According to Craya & Curtet[6a]

$$\Theta = \frac{\dot{M}_0 + \dot{M}_1}{\dot{M}_0} \frac{d_0}{D} \sqrt{\rho_0/\rho_F} \tag{6.18}$$

where $\dot{M}_0$ is the mass flow of the jet coming from the nozzle, $\dot{M}_1$ the mass flow of air fed to it concentrically, and D the diameter of the circular cylinder surrounding the jet. The length of the enclosed flame is

$$L_e = 5{\cdot}3 \frac{c_0}{c_{st}} d_0 f\left(\frac{1}{\Theta}\right) \tag{6.19}$$

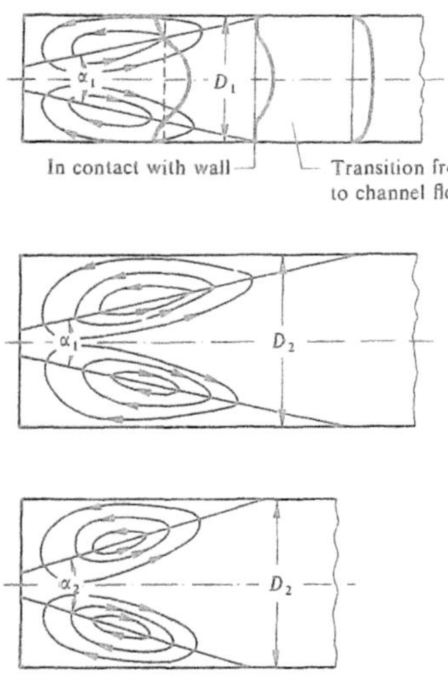

Figure 6.16. Zones of return flow for different diameters of combustion chamber and jet angles[5]

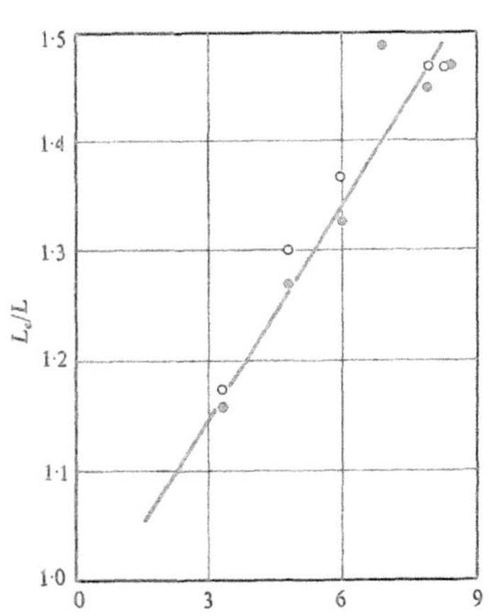

Figure 6.17. Return flow parameter and length of flame (L = length of free burning flame, L_e = length of encased flame)[7]

where $f(1/\Theta)$ can be taken from Figure 6.17 after calculating Θ.[7] The return flow lengthens the flames by a factor of 1·1–1·2, causes a reduction in the temperature differences, and improves the stability of the flame.

In glass furnaces turbulent return flows arise near the crown and on each side of the flame. Determinations of the proportions of return flow on seven cross-fired tank furnaces and their relationship to Θ are shown in Figure 6.18. The return flow parameter was about 2·2, which corresponds to a lengthening of the flame of about 1·1.[2]

Natural gas or coke oven gas is introduced symmetrically through nozzles set at an angle of 45° behind the port mouth and just above the port sill. Depending on the momentum of the gas there is some mutual interference which may reach considerable proportions (Figure 6.19). Experiments on models have shown that the sum of the momentum plus that force which acts as the base flow on the outside of the jet gives good agreement with the actual behaviour of the jet. The dispersion of the jet sideways causes a substantial increase in turbulence and thereby in mixing. Compared to the free jet, a higher fuel momentum causes a reduction in flame length of about 35%.

An increase in the entry angle from 30° through 45° to 60° results, not in a reduction, but in an increase in flame length. A large angle obviously

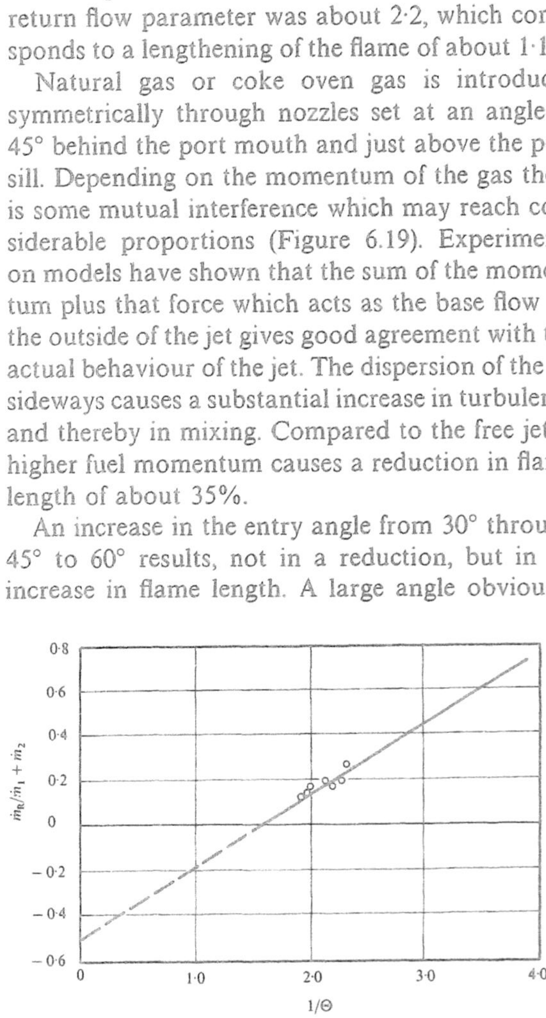

Figure 6.18. Proportion of return flow of an encased jet in glass tank furnaces with under-port firing[2]

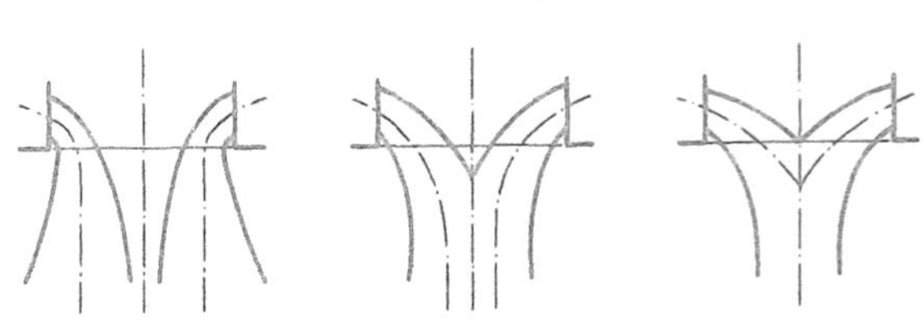

Figure 6.19. Behaviour of jets of different momentum[6]

reduces the distance travelled by the individual jets before they combine to such an extent that only a small amount of air is drawn into them. The rate of entrainment of air into the combined jet is clearly less than into two individual jets.

The use of multiple nozzles, i.e. two or three above one another, brings few advantages to the above system since the distances between the nozzles are small. Compared to a single jet of equal total momentum the reduction in flame length is about 7%.

Jets which enter a main stream with their axis at an angle α to the main stream flow are governed by the equation given by Patrick,[9]

$$\frac{y}{d_0} = \sqrt{\frac{\rho_1}{\rho_0}}\left(\frac{u_1 \sin\alpha}{u_0}\right)^{0.85}\left(\frac{x}{d_0}\right)^{0.38} \qquad (6.19a)$$

where x is the penetration in the direction of the jet, y the change in direction of the jet axis, and d_0 the nozzle diameter. For velocity u and density ρ the suffix 0 describes the jet material at its exit from the nozzle and the suffix 1 that of the main flow. This Equation describes the behaviour of jets without chemical reactions. The jet density of flames can be taken, to a first approximation, as the average density of the flame gases.

6.3 Heat emission in the combustion chamber

6.3.1 The combustion of jet flames

As has already been noted in connection with the length of flames, the combustion of jet flames depends to a major extent on external conditions. Initial information concerning the combustion process is provided by the 'half-value length', which is that length of flame where half the fuel has been burnt ($\alpha = 0.5$): for swirl-free flames the half-value length is about 0·3 and for swirling flames about 0·2. A more accurate calculation can be carried out based on the equation given by Günther[8]

$$\alpha = 1 - \exp\left[0.3 \times 10^{-3}\left(\frac{x}{d_0}\right)^{1.9}\right] \qquad (6.20)$$

where x is the length of flame and d_0 the nozzle diameter. The exponential factors are valid for natural gas but for other fuels they must be multiplied by $c_{st,natural\ gas}/c_{st}$, where these are the stoichiometric concentrations applying to natural gas and the other fuel, respectively. Figure 6.20 shows that the heat emission of a flame only begins at $x/L \approx 0.1$, so to make full use of the available melting area, it pays to retract the burner and its block away from the furnace by about this amount, as shown schematically in Figure 6.21.

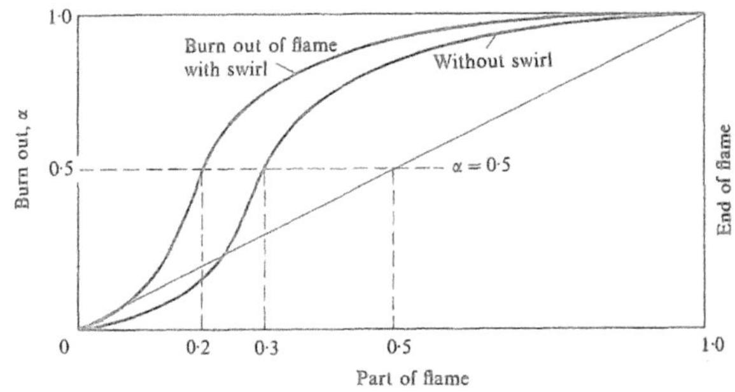

Figure 6.20. Combustion process with half-value lengths as a function of flame length[8]

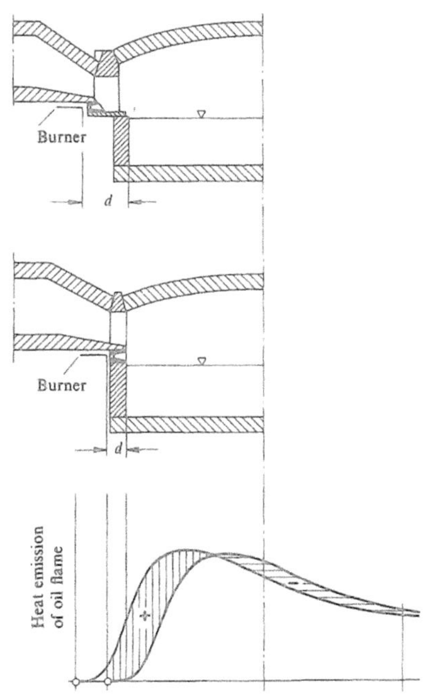

Figure 6.21. Improvement of heat transfer between flame and molten glass by setting the burner away from the furnace[5.1]

6.3.2 Heat emission and flame length

Although in an end-fired furnace the flame is set long enough to provide adequate heating of the refining zone, there is greater freedom of choice with cross-fired furnaces. Here the limits are set by the risk of overheating the refractories constituting the port mouth entry arch and the checkerwork on the waste gas side and by the effect of excessive temperature fluctuations on the firing side.

Apart from these technical points it is clearly of interest to investigate the effect of flame length on heat transfer to the molten glass in greater detail. This has been done on a mathematical model by Voss.[10]

A section of width B in a cross-fired tank furnace is examined and the flame combustion and temperature conditions in that section are shown in Figure 6.22. The furnace operates with air preheated by waste gases which left the combustion chamber at a temperature ϑ_A (efficiency of heat recovery η_k). The material to be heated, i.e. in this case batch or glass, has a constant surface temperature ϑ_0 which, in practice, is largely achieved across the width of the furnace by the periodic reversal of flames. The flame burns at right angles to the direction in which the batch and the glass flow.

Heating is divided into two zones, the flame zone (zone 1) and the waste gas zone (zone 2). The flame has a length L_F. The temperature changes in zone 1, ϑ_F, were chosen so that, with a linearly increasing combustion process, the energy liberated by the flame exactly matched the energy transferred to the surroundings; this means that ϑ_F remains constant. Before the flame enters the combustion chamber of the furnace some combustion has already taken place (α^* in Figure 6.22(b)) and this 'precombustion' causes the flame to enter the combustion chamber at ϑ_F. An equilibrium exists in zone 1 between the heat liberated by the fuel and that transferred to the material to be heated and the furnace walls. In zone 2 the decrease in temperature, ϑ_a, of the burning gases corresponds to the heat transferred to the surroundings: the heat exchange is assumed to occur only vertically to the material to be heated.

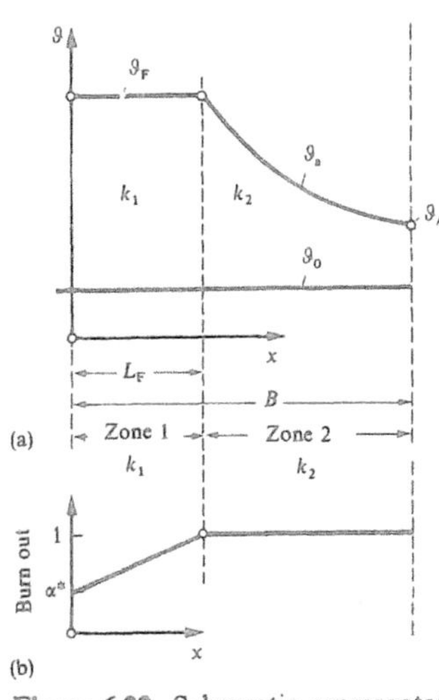

Figure 6.22. Schematic representation of temperature changes and combustion process as a function of the length of the chamber to be heated[10]
(a) temperatures
(b) combustion process

The heat exchange coefficient in zone 1, $0 \leqq x \leqq L_F$, is referred to as k_1 and that in the zone 2, $L_F < x \leqq B$, as k_2. Both represent total heat transfer quantities which result from heat exchange between flame and material, flame and crown, and crown and material.

If the quantities of heat given off by flame and waste gas are equal ($c = 1{\cdot}0$), then the heat transferred decreases with increasing flame length (L'). If the quantity of heat given off by the waste gases is only one half of that given off by the flames, then an optimum condition arises at $\approx 0{\cdot}6L'$ (Figure 6.23).

The effect of flame length on the heat transferred also depends on the heat flux density, $\dot{e}$. For a furnace of high pull ($\dot{e} = 240\ \mathrm{kW/m^2}$) the influence of flame length is significant but for one of low pull ($\dot{e} = 60\ \mathrm{kw/m^2}$) the influence is much weaker (Figure 6.23).

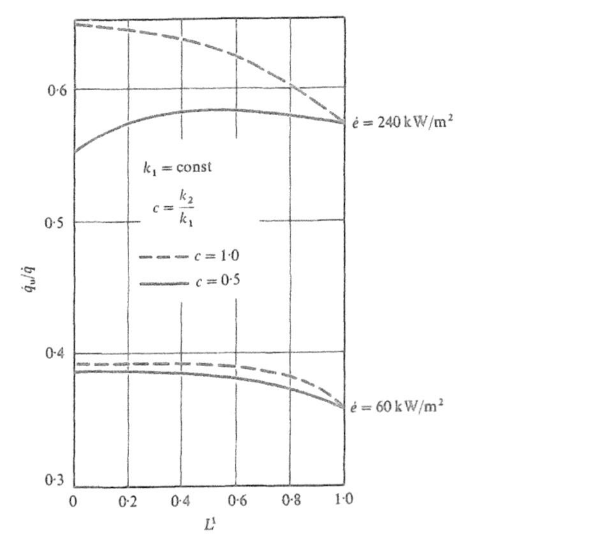

Figure 6.23. Efficiency of heat transfer as a function of normalised flame length ($k_1 = 387\ \mathrm{W/m^2K}$, $a = 23{\cdot}3\ \mathrm{kW/m^2}$ crown heat losses, $b = 15{\cdot}5\ \mathrm{W/m^2}$ heat losses through walls)

6.3.3 Heat emission of flames by radiation

6.3.3.1 Radiation of gases

Radiation from a solid is radiation from a surface; radiation from a gas, on the other hand, is radiation from a volume. Radiation from gases is not continuous over a wide part of the spectrum, but is restricted to more or less discrete wavelengths, or radiation bands. The most important gases techni-

Table 6.2. *Frequencies of radiation bands of CO_2 and H_2O (μm)*

H_2O	CO_2
1·7–2·0	2·0–3·0
2·3–3·4	4·0–4·5
4·4–8·5	14–17
12·0–30	

cally are CO_2 and H_2O and their radiation bands are given in Table 6.2.

The intensity depends on the thickness of the layer, the partial pressure, and the temperature. At low partial pressures emission and absorption increase in proportion to the partial pressure (number of molecules); at high partial pressures the rate of increase is lower. When the layer is infinitely thick, a gas flame radiates as a black body: if the radiating energy or the absorption capacity of an infinitely thick layer is based on the radiating energy of a black body, then Kirchhoff's Law applies and $\alpha(\lambda) = \varepsilon(\lambda)$, i.e. the emissive and absorptive powers are identical. According to Elgeti,[10a] $\varepsilon_\infty = 0{\cdot}23$ for CO_2, 0·90 for H_2O, and 0·98 for $CO_2 + H_2O$.

Beer's Law applies for α in the form

$$\alpha(\lambda) = 1 - \exp[-\kappa(\lambda)ps] \tag{6.21}$$

where p and s are the partial pressure and thickness of the layer respectively and $\kappa(\lambda)$ is the extinction coefficient.

In order to determine the total effect of gas radiation, the emissivity of the gas covering the total radiation over all wavelengths entering the gas is needed. This emissivity, also called 'degree of blackness', can be derived from the relationship

$$\varepsilon_{total} = \varepsilon_{CO_2} + \varepsilon_{H_2O} \tag{6.22}$$

where ε_{CO_2} and ε_{H_2O} can be taken from Figures 6.24 and 6.25 based on measurements of Hottel & Egbert.[11] If a mixture of gases contains both CO_2 and H_2O, then some bands overlap, which weakens the absorption; however, this weakening is very small and can be ignored.

The radiated heat flux for grey radiation can be determined from the data of Figures 6.24 and 6.25 where

$$\dot{q}_{rad} = \varepsilon C_s \left(\frac{T}{100}\right)^4 \tag{6.23}$$

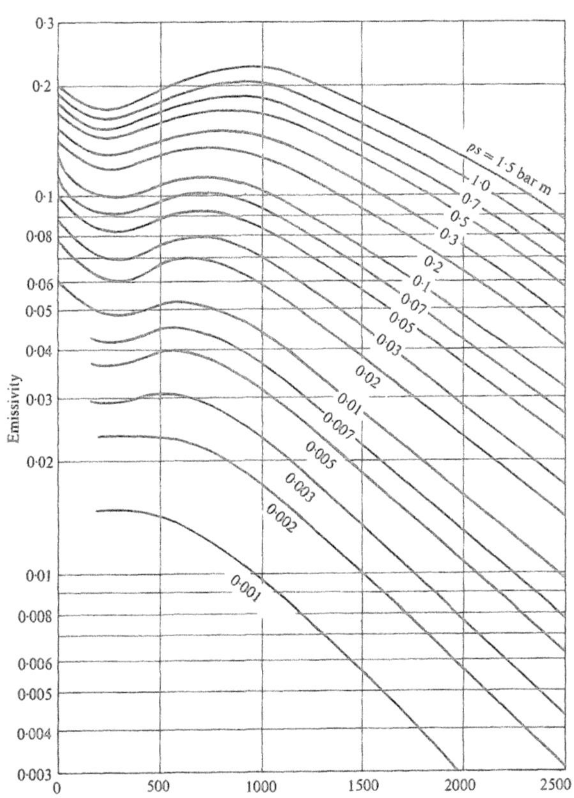

Figure 6.24. Emmisivity of CO_2 as a function of temperature[11]

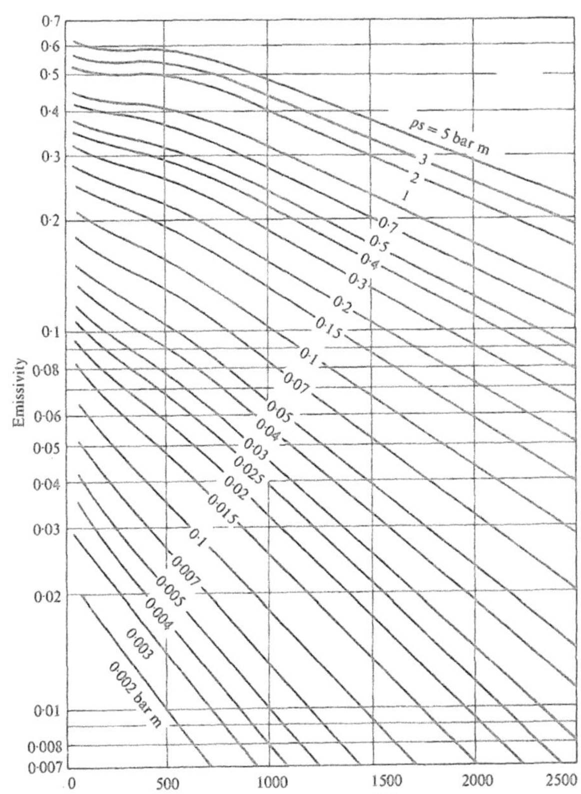

Figure 6.25. Emissivity of H_2O as a function of temperature[11]

and c_s is the radiation from a black body. Schack[12] gave the following empirical formulae for calculating the radiation flux density:

$$\dot{q}_{CO_2} = 10{\cdot}35(ps)^{0{\cdot}4}(T/100)^{3{\cdot}2}\ (\mathrm{kW/m^2}) \qquad (6.24)$$

$$\dot{q}_{H_2O} = (46{\cdot}4 - 84{\cdot}7ps)(ps)^{0{\cdot}6} \times (T/100)^{(2{\cdot}32 - 1{\cdot}37\sqrt[3]{ps})}\ (\mathrm{kW/m^2}) \qquad (6.25)$$

where s and p should be given in m and bar respectively.

In practice the determination of the layer thickness s is often difficult but according to Port[14] and Hausen,[13] it can be determined approximately from the equivalent diameter, d_{eq}. For ducts $s = 0{\cdot}9\,d_{eq}$, where $d_{eq} = 4A_q/U_q$, and for a space filled with gas $s = d_{eq}$, where $d_{eq} = 4V/A$: here A_q is the cross section of the duct, U_q the circumference of this cross section, and V the volume and A the surface of the gas space. If a layer of gas is subjected to exchange of radiation with a solid surface emitting grey radiation, then it must be remembered that Kirchhoff's Law, i.e. that emittance equals absorptance, is only valid for gases when the solid and the gas are at the same temperature. If this is not the case, which must be the situation for most cases in practice, then this fact must be taken into account when calculating the exchange of radiation. According to Hausen the exchange of radiation between a wall of emittance ε_w at a temperature T_w emitting grey radiation and a layer of gas at a temperature T_g can be calculated from

$$\dot{Q}_{rad} = AC_s\varepsilon_w[\varepsilon_g(T_g/100)^4 - \varepsilon_w(T_w/100^4], \qquad (6.26)$$

with ε_w and ε_g taken from Figures 6.24 and 6.25 for the temperatures T_g and T_w. In this case s is a thicker layer, s^*, calculated from $s^* = s/\varepsilon_w^{0{\cdot}85}$, which takes into account multiple reflections between wall and gas layer. There are also other methods for calculating $\dot{Q}_{rad}$ and the values of ε_g and ε_w, which cannot be given here for lack of space.

6.3.3.2 Radiation of luminous flames

Luminous flames are defined as flames which, as a result of their containing small solid particles, radiate over a continuous spectrum; the type, size, and number of these particles vary widely. In the case of diffusion flames they are small particles of soot, each 0.01–0.05 mm in diameter, but agglomerated into flock-like clusters of 100–1000 particles. The soot content is 10–30 g/m³ (at NTP), with a time of formation of about 20 ms, and the chemical composition is given as C_6H and C_8H;[15] the tendency to form soot increases as the C/H ratio of the fuel increases. The optical properties of clouds of soot can be calculated from the theory of opaque media. The emissivity of luminous flames is considerably increased by the luminosity of soot particles so that even for small flames, values of $\varepsilon \approx 1$ can be obtained. In practice conditions for the formation of soot exist only at the root of the flame and in consequence a clear emissivity maximum appears along the flame axis. The interrelationships covering soot formation are very complex and still defy calculation in advance: increasing the mixing of fuel and combustiôn air decreases the tendency to form soot and all adjustments which shorten the flame, such as increasing the momentum of the burner, applying swirl to the fuel/air mixture, or premixing the combustion mixture, cause a reduction in soot formation and a reduction in emissivity.

Literature data covering radiation properties and emissivities of industrial flames originate in the main from the International Flame Research Foundation at Ijmuiden in The Netherlands.[15]

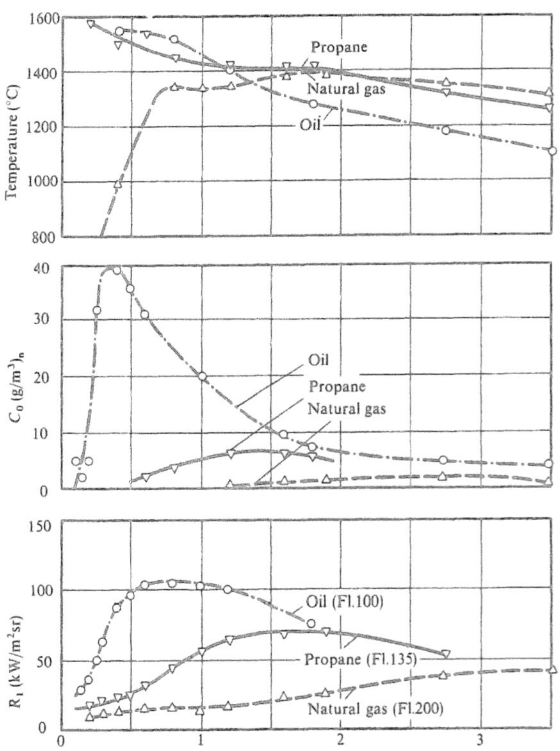

Figure 6.26. Gas temperature, solids content, and intensity of radiation of flames from different fuels and as a function of the distance from the burner[15]

Even if the results are difficult to apply to glass furnaces because of the low preheat temperature and the low temperatures of the combustion chamber, they nevertheless provide good guidance and interrelationships. A comparison of gas temperature, solid content, and intensity of radiation resulted in the data shown in Figure 6.26. The high temperature at the start of the oil flame and at the end of a gas flame are typical, the temperature distribution in the oil flame being related directly to its high soot content and the related emissivity. The oil flame attains directional radiation intensities of over 100 kW/m²sr (sr = steradian) which, when judged against the gas temperatures measured, corresponds to an emissivity of almost unity.

Combinations of different fuels, e.g. oil and natural gas, show that the emissivity is only marginally reduced from that of pure oil, even for high proportions of gas; only at gas contents exceeding 80% is a reduction in emissivity noticeable.

6.3.3.3 Interplay between flame and melting chamber

The large differences in emissivities between gas and oil flames lead one to assume that oil flames should be much more suitable for the heating of glass furnaces; in addition, the waste gas enthalpy when compared to the heating energy supplied is 3–7% higher for natural gas. Practical experience has shown, however, that the differences in specific energy consumptions and melting rates obtained are not as large as expected. Generally, the waste gas temperatures and regenerator temperatures are higher with gas than with oil, and the specific energy consumption is about 3–5% higher for natural gas; this corresponds to an increased waste gas enthalpy. The large difference in radiation properties between the two types of flame therefore has a much smaller effect than the differences in flame properties would lead one to expect. Calculations of the energy transferred to the bath of molten glass, taking into account the overall conditions existing in a glass furnace,[10] have shown that the emissivity of the flame only has an influence on heat transfer in the region $\varepsilon < 0{\cdot}6$; higher emissivities bring virtually no improvement and in fact Michelfelder's calculations[16] showed that the efficiency even declined for $\varepsilon > 0{\cdot}6$. Figure 6.27 shows the calculations by Voss: since all calculations involve major simplifications, the body of the flame was included in the calculation as an enclosed layer emitting grey radiation. It must be left undecided whether the calculations are valid in detail but it is clear that the emissivities above 0·5 are of less importance in glass melting than was previously thought. However, both sets of calculations indicate that the major factor is the loss of heat through the superstructure, particularly through the crown (see Section 7.3.3).

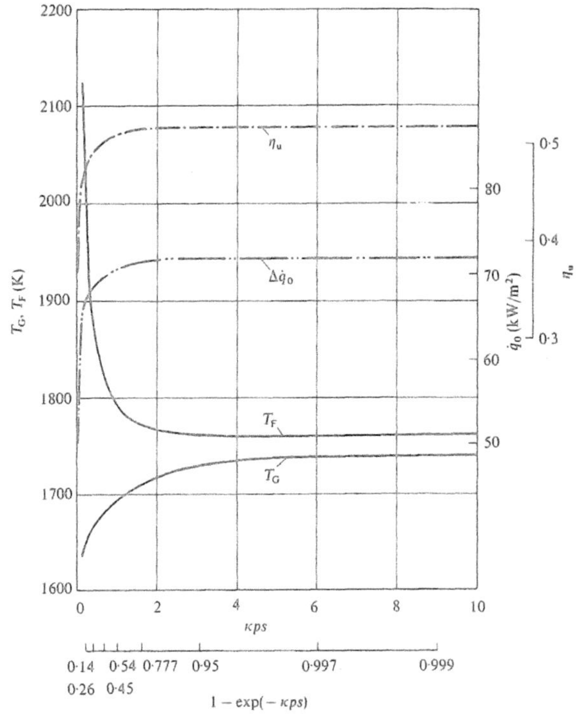

Figure 6.27. Influence of the emissivity of the combustion chamber or of the product κps on the temperature of the flame, T_F, the temperature of the crown, T_G, and the fuel utilisation, η_u (see Table 5.3); ($T_o = 1473$ K; $K_1 = 0{\cdot}43 \times 10^{-3}$ K^{-1}; $q_a = 23{\cdot}3$ kW/m²; $q_G = 7$ kW/m²; $q_B = 140$ kW/m²; $\varepsilon_o = \varepsilon_G = 0{\cdot}8$; $\eta_k = 0{\cdot}59$)[10]

What do these surprising results tell us? The explanation lies in appreciating the significance of the crown as an auxiliary heating surface, and the following are its major effects (Figure 6.28).

(a) The mirror effect

Part of the radiation reaching the crown is reflected, making the heat emission of the flames more uniform; in the extreme case, the reflection has the effect of doubling the thickness of the layer of gas. The mirror effect of the crown depends on its reflectivity; the higher the reflectivity, the greater the mirror effect and the lower the heat losses through the crown to the outside.

(b) The conversion effect

The radiation reaching the crown is partly absorbed and reradiated into the melting chamber as a

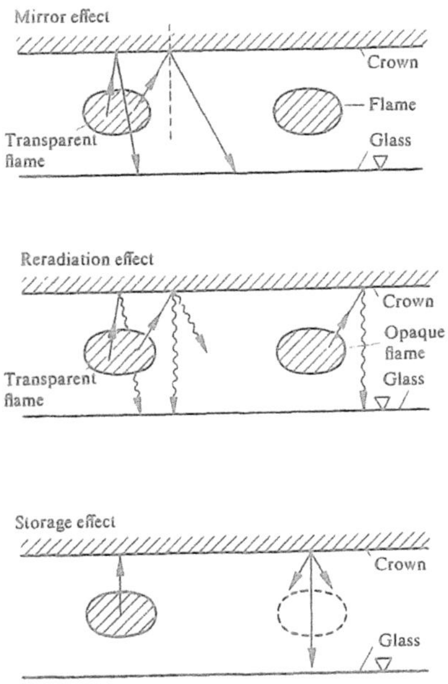

Figure 6.28. Interplay between flame, crown, and molten glass

function of the temperature of the crown. The effect is particularly effective with hot, well insulated crowns and with transparent flames.

(c) The storage effect

With intermittent flames the periodic variations in radiation are evened out, which is especially important for U shaped flame tank furnaces fired with oil. The important criteria are the temperature and heat capacity of the crown.

The more transparent the flames or, for luminous flames, the larger the distance between flames, the more noticeable is the direct and indirect crown radiation. Good thermal insulation and high reflectivity are advantages for a crown. The surprisingly large influence of heat losses through a crown have been shown in calculations by Voss;[10] a reduction of about 30% can increase the heat transferred to the molten glass by about 25% (Figure 6.29).

Superstructure heat losses should be treated in the same way as losses through the crown, and careful insulation is always to be recommended (Figure 6.29). This also applies to heat losses radiated by the superstructure into the ports and other areas such as the working end or the conditioning zone. The losses are of the same order of magnitude as the crown losses. Vertically descending ducts and hot brickwork reduce the losses due to radiation in the ports (see Figure 3.23); radiation losses in the working end can be reduced by

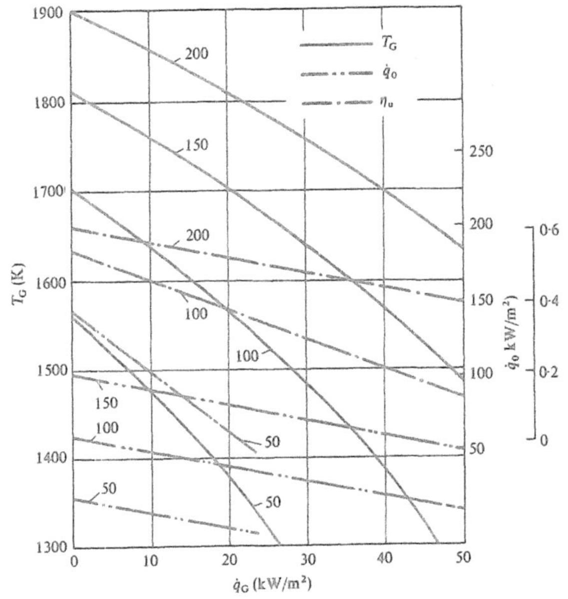

Figure 6.29. The influence of heat losses through the crown. The number on each curve is the parameter $\dot{q}_B$ in 1·163 kW/m² (other thermal data as in Figure 6.27)[10]

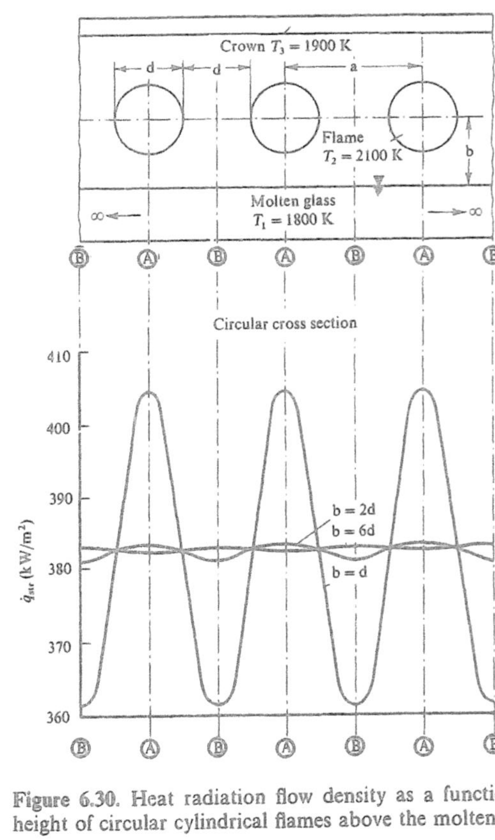

Figure 6.30. Heat radiation flow density as a function of the height of circular cylindrical flames above the molten glass[18]

operating the tank as two separate chambers; and tile curtains and constrictions should be used for conditioning zones of tank furnaces.

Increasing the distance between flame jets increases the efficiency of the secondary heating effect of the crown, especially for luminous flames; a disadvantage of this arrangement is the increasing unevenness with which the bath of molten glass will be heated. This effect has been studied by Trier & Stein[18] using calculations based on models, and the results are shown in Figure 6.30: the unevenness decreases as the distance of the flame from the glass surface increases. The calculations assumed, for the sake of simplicity, black body radiation by flame, crown, and bath of molten glass in order to accentuate the effect of geometry.

6.3.4 Heat transfer by convection between flame and bath of molten glass

The high temperatures existing in a glass furnace make it inevitable that radiation dominates the heat transfer process; convection is 6–9% if combustion reactions occur in the boundary layer but accounts for only 4–5% in their absence. The heat transfer coefficient, based on an average gas temperature covering the boundary layer without combustion, is about 35 W/m^{2}K and, with combustion, about 80–120 W/m^{2}K; these data are based on measurements by Germerdonk,[17] whose work was done using a model of a glass furnace with and without combustion in the superstructure.

The convective part of the heat transfer varies considerably from place to place due to the fact that the gaseous flames impinge on the bath of molten glass at a slight angle and to the turbulence surrounding the jet. Figure 6.31 shows the heat transfer determined for three different flow velocities in terms of the dimensionless ratio between the local Nusselt number, $Nu_x = \alpha(x)x/\lambda$, and the average Nusselt number. The highest heat transfer rate is observed at the entry zones of the flame. Measurements of hot waste gases without combustion in the superstructure volume can be represented by

$$Nu = 0{\cdot}428\,\Phi^{0\cdot123} Re^{0\cdot772 - 0\cdot669(d/L)} \qquad (6.27)$$

where Φ is the ratio between the ground area of the combustion space and the area subject to direct flame impingement (A_g/A_a), and d/L is the ratio between the hydraulic diameter of the port mouth and the length of the combustion space, which here is the width of the tank, i.e. the distance between the two port mouths. The average velocity is derived by dividing the total gas volume traversing the combustion space by its average cross section, ignoring the influence of any return flow. The reference temperature for the various values is taken as the

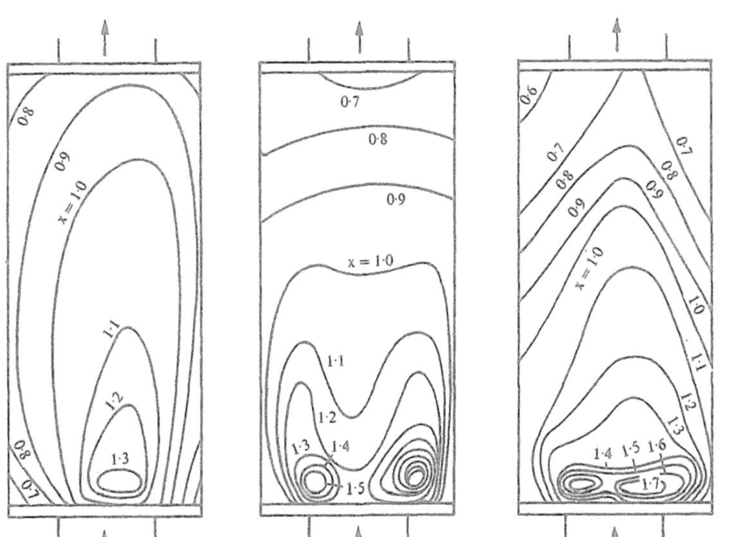

Port	A	F	H
Re_L	72706	81132	86479
Nu_L	1077	1543	1605
Δ	0·1088	0·1033	0·1036
Φ	2·94	10·5	20·7
ε	6	21	45

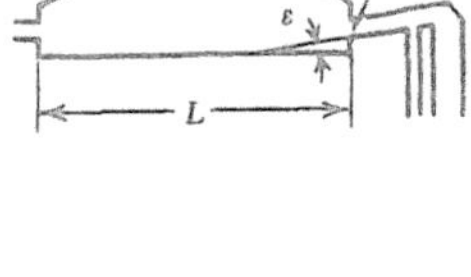

Figure 6.31. Influence of burner inclination on the local heat transfer in the base area (arrows indicate flow direction)[17]

average temperature in the combustion space. The results for different inclinations of burners are given in Figure 6.32, in which the strong increase in convective heat transfer with flow velocity is noticeable.

Experiments using firing in the combustion space instead of hot air led to the convective part of the heat transfer being increased by a factor of two to three; this is explained by combustion reactions in the surface layer.(17)

Beer worked on the firing section at the Flame Research Foundation, all his experiments involving combustion and, in some cases, oxygen enrichment; his results(19) can be summarised by the relationship

$$Nu = 0{\cdot}13\,Re^{0{\cdot}81}. \qquad (6.27a)$$

When comparing this Equation with that of Germerdonk, it should be pointed out that different characteristic lengths were used for *Nu* and *Re*. Since it is possible to assume that the thickness of the boundary layer increases as the distance from the point at which the flame impact on the glass surface increases, Beer took this distance as the characteristic length for the calculation of *Nu* and *Re*: the characteristic velocity was taken as that of the main flow. The experiments showed that the total heat emitted and absorbed by combustion could be increased by the addition of O_2 to the combustion air, and that a steeper impingement angle onto the glass surface improved heat transfer. The use of different fuels (coke oven gas, heating oil) appeared to have no influence on these results. Nor was there any correlation between the O_2 content of the combustion air and the proportion of heat

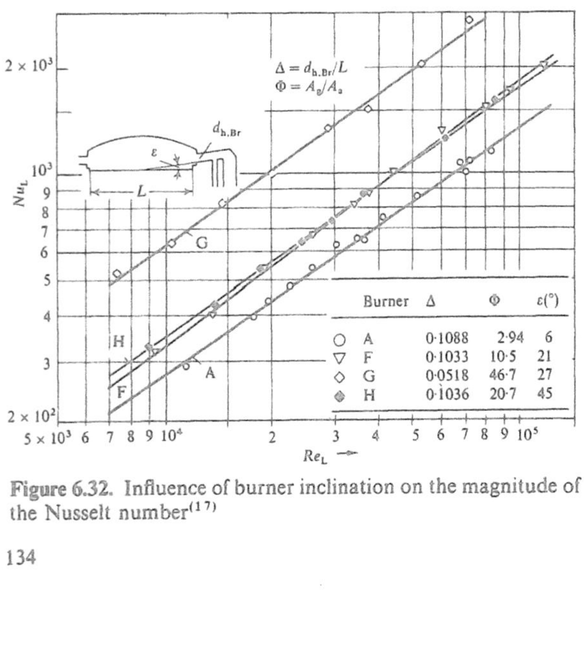

Figure 6.32. Influence of burner inclination on the magnitude of the Nusselt number(17)

transferred by convection; this can be attributed to the fact that the increase in gas temperature due to O_2 enrichment of the combustion air is offset by a reduction in the heat transfer coefficient due to the reduced quantity of waste gases.

Beer's experimental conditions and his use of oil and natural gas burners correspond more to current conditions in glass tank furnaces than Germerdonk's investigations, which were based on producer gas burners.

6.4 The charging and melting of batch

The charging of batch and cullet has a major effect on the performance of a glass melting plant and the methods for carrying it out must be carefully considered during design and construction. Many factors depend on it: attainable glass quality, maximum pull, energy requirements, and furnace life. Since charging and melting down are closely connected, the various methods of charging and their characteristics can only be understood alongside the melting down process: the melting down of the batch will therefore be considered first.

6.4.1 The melting process

Data dealing with the heat flux density required for melting batch occur in the literature only as isolated measurements.(20–25) The heat flux density between superstructure and melting batch is 45–80 kW/m^2 and that between the batch and the bath of molten glass is of the same order of magnitude. In the presence of strong currents in the glass, such as occur in electric tank furnaces, values of up to 120 kW/m^2 are possible; as actual measurements do not exist, these values have been estimated from the high melting rates (up to 5 t glass/m^{2}d) attainable with furnaces operating with a cold top.

The heat flux density between superstructure and normal powdered batch before sintering of the surface has taken place depends strongly on temperature, as has been demonstrated from radiation measurements by Costa.(26,28) The total absorbance falls markedly with increasing temperature, then rises sharply as the melting reactions commence on the surface (Figure 6.33). Costa ascribes this effect initially to absorption by water and then to the absorbance of the molten layers as they form. A different, more favourable, behaviour is shown

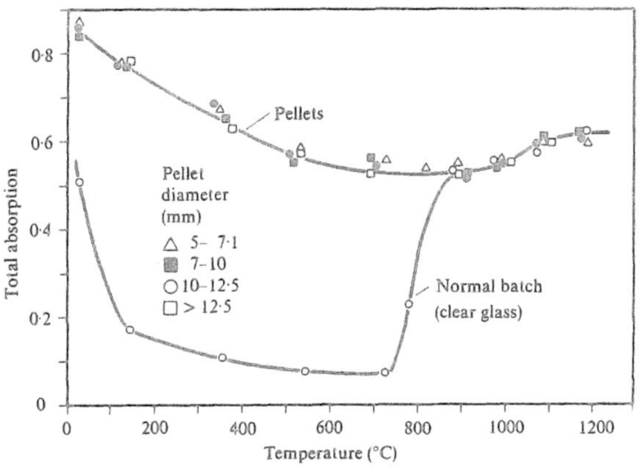

Figure 6.33. Total absorption as a function of temperature at a radiation temperature of 1000°C[28]

with batch pellets: the deeper penetration by radiation into the voids between the pellets leads to higher radiation absorption. As fusion commences this advantage is lost; however, a noticeable amount of absorption by molten layers of batch begins to take place.

A second factor influencing the melting down process is the thermal conductivity of the batch. Detailed investigations have been carried out by Kröger & Eligehausen[30] and an example of their results is given in Figure 6.34 which shows that the thermal conductivity is low and nearly constant until the first melting reactions take place; it then increases steeply, the increase being linked to heat transport by radiation. Glass itself shows a more steady increase. The data are virtually the same for different types of glass and different batch formulations.

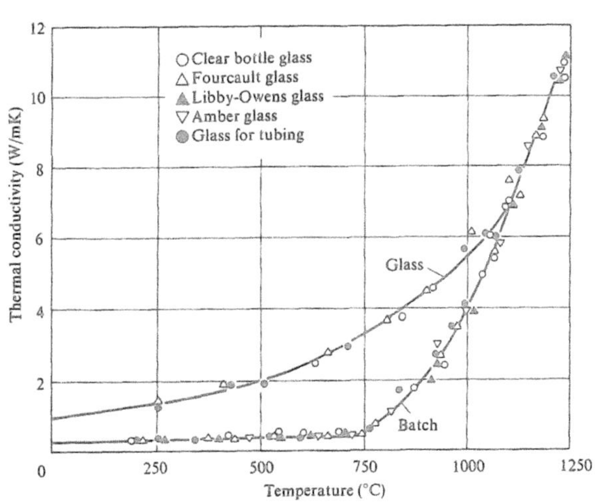

Figure 6.34. Relationship between thermal conductivity and temperature of some technical glasses[30]

The thermal conductivity of pellets was determined by Costa.[26] The relationships shown in Figure 6.35[25] were calculated assuming that heaps of pellets behaved like heaps of spherical bodies. The conductivities can be compared to those obtained by Kröger[30] (Figure 6.34).

The melting down process has been investigated theoretically by Fuhrmann[31] and experimentally by Daniels[22] and Costa and since the fundamental relationships are more easily understood from a theoretical approach, the work of Fuhrmann is considered first.

If the melting batch is considered as a homogeneous isotropic layer 10 cm thick which takes up 60 kW/m^2 of heat from each side in which fusion starts at 800°C and in which molten material is continuously removed by running off and by the action of the glass currents surrounding the layer, then the time and temperature contour lines shown in Figure 6.36 result. It is remarkable that the core

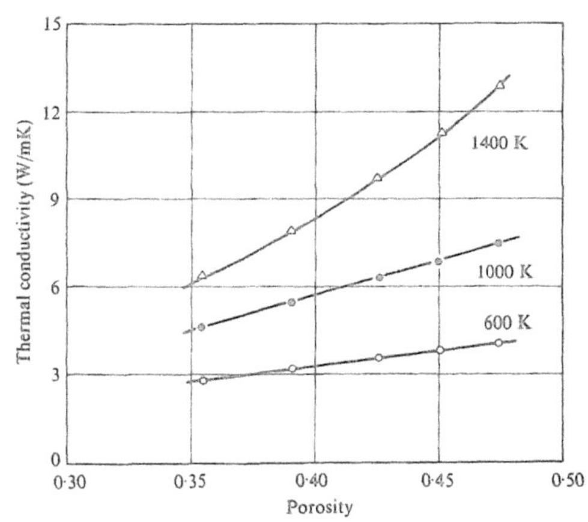

Figure 6.35. Effective thermal conductivity of a pile of pellets related to porosity[29]

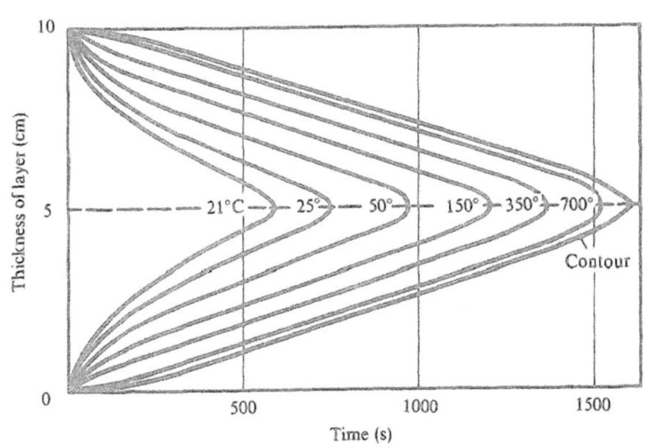

Figure 6.36. Temperature distribution in a layer of batch with no gases of reaction[31] ($\vartheta_a = 20°C$; $\vartheta_s = 800°C$; $\dot{q}_o = \dot{q}_u = 60\ kW/m^2$)

temperature even after 20 min (1200 s) has reached only 150°C. The front of the melting process, which can be considered as the geometrical location of the temperature of melting against time, indicates a process involving five stages (see also Figure 6.37).

1. During heating up the original thickness remains unchanged ($ds/dt = 0$). This is the preheating stage.

2. Once the fusion process has started, there is an initial stage during which the melting rate increases. At this stage a major part of the heat available at the surface is used to heat the zones near the edges ($ds/dt > 0$, $d^2s/dt^2 > 0$). This is the starting stage.

3. For a certain time a quasi stationary condition follows. The heat arriving at the surface is used steadily to provide the enthalpy for the transition changes and for heating the interior of the layer. The melting process can, during this condition, be described pretty accurately as a stationary invasion of a melt front into an infinitely extended layer because the heat transport processes on both sides of the layer do not yet influence one another ($ds/dt \approx$ constant 0). This is the constant stage.

4. When the phase boundaries have advanced far enough into the layer for the temperature of the batch at the centre to be substantially increased, the amount of heat flowing into the layer reduces. Since the heat available at the phase boundaries remains the same, more energy per unit time becomes available for transition changes. This therefore gives rise to an acceleration of the melting process ($ds/dt > 0$, $d^2s/dt^2 > 0$). This is the final stage.

5. The point at which the melt fronts finally meet is determined in space and time by the properties of the batch (density, specific heat, transformation enthalpy) and by the amount and distribution of heat available from the sides.

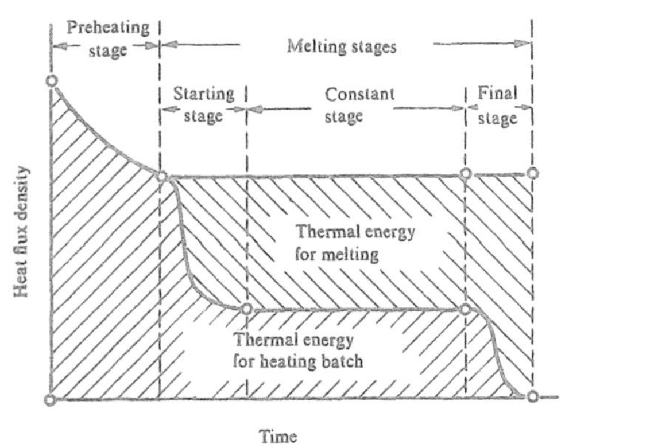

Figure 6.37. Thermal flux density as a function of time during the individual stages in the melting of a layer of batch.

A calculation encompassing these five stages in a closed analytical manner is very difficult. Stage 3, which is the most important in the process and which can be regarded as quasi stationary, can easily be treated mathematically using the stationary velocity at which a melt front advances into a body under constant heat flux density, $\dot{q}$:[3]

$$(ds/dt)_{stat} = \dot{q}/\rho[h_R + c(\vartheta_S - \vartheta_A)]. \qquad (6.28)$$

where ρ is the density of the batch, h_R the specific melting enthalpy, and c the average specific heat absorption in the temperature range ϑ_S (melt) – ϑ_A (start). It is worth noting that the thermal conductivity of the batch does not appear in the Equation as it influences the melting process only during the relatively short starting phases (1 and 2) and final phases (4 and 5).

A more thorough investigation into the influence of the thermal conductivity of batch confirmed this conclusion. Increasing the thermal conductivity to six times the values measured by Kröger only achieved a small reduction in melting time (Figure 6.38). A high thermal conductivity lengthens stages 1, 2, 4, and 5, when compared to stage 3. The melting front related to time becomes more convex. The limited influence that the thermal conductivity of the batch has on the melting process can also be seen from Figure 6.39 which includes the extreme conditions when $\lambda \to 0$ and $\lambda \to \infty$. The upper part of the figure shows surface temperature against time schematically: for a very low thermal conductivity ($\lambda \to 0$) the melting temperature is reached very quickly and remains constant for the whole of the melting process; for a very high thermal conductivity ($\lambda \to \infty$) the whole layer must be brought to the melting temperature before the melting reactions can commence, as temperature gradients

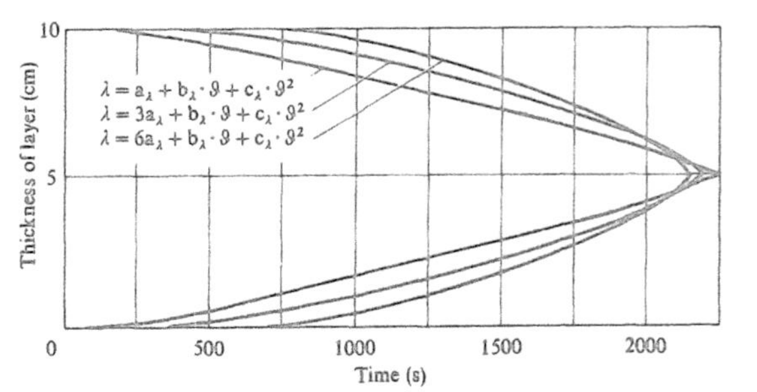

Figure 6.38. Melting fronts of a layer of batch for different thermal conductivities[31] ($\vartheta_a = 20°C$; $\vartheta_s = 800°C$; $\dot{q}_o = \alpha_o(\vartheta_u - \vartheta_o)$; $\dot{q}_u = \alpha_u(\vartheta_u - \vartheta_o)$; $\alpha_u = 75\ W/m^2K$)

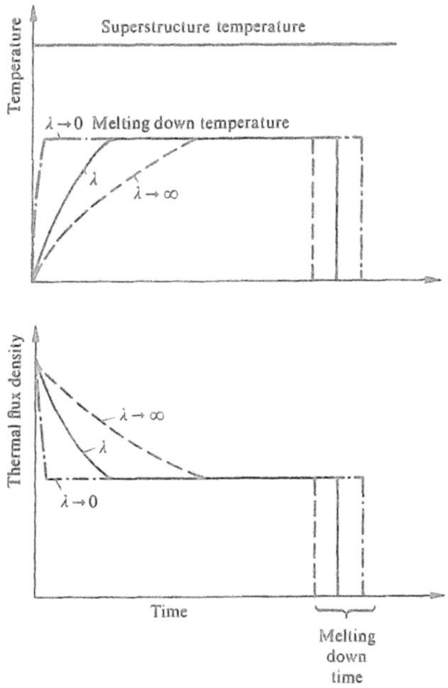

Figure 6.39. Influence of the thermal conductivity of batch on surface temperature, ϑ, heat flux density, $\dot{q}$, and time of melting down, t

cannot arise within the layer. Once ϑ_S has been reached, additional heat is used to carry out the melting reactions.

The heat flux density, obtained from the temperature difference $\vartheta - \vartheta_S$, is entered in the lower part of Figure 6.39. Since the quantity of heat necessary for melting is independent of the specific heat of the batch, the area must in all cases be the same.

The assumption by Fuhrmann[31,32] that molten material is continuously removed as it is formed[34] only applies to a limited extent. When it is realised that the heat flux density at the surface is reduced by the formation of foam and the retention of residual molten material, then calculations show that this has a considerable effect on the time required for melting (Figure 6.40). This calculation is based on the assumption that, after melting commences, there would be a linear decrease in the heat transfer coefficient between superstructure and batch surface down to some minimum value, α_{min}.

Some glasses, e.g. amber glasses, have combustible materials such as coal or oil added to the batch for technical reasons; from a thermal point of view they act as sources of heat in the melting batch and the addition of normal quantities (0·7% carbon) accelerates the melting rate by about 5%. The addition of carbon in this quantity is, in terms of energy, equivalent to about 7·5% of the enthalpy of the batch at the melting temperature.

The quantities of gases evolved during the heating and melting of the batch (0·05–0·2 m^3 gas/kg batch measured at NTP) also influence the melt front. The effect on the batch is that it melts more quickly underneath and more slowly on top since the hot gases evolved by the batch warm up the batch underneath as they flow through it, and cool it above. The sum shows both influences cancelling one another when the melting rates on top and underneath are identical.

There are significant differences between the temperature profiles calculated by Fuhrmann and those which were determined experimentally by Costa & Daniels, especially in respect of temperature changes on the inside (Figure 6.41). Measurements show that the temperature in the interior reaches 600°C after 12 min and the cause of the difference is undoubtedly the method of measurement. The experimental furnace used was equipped with thermocouples in protective sheaths which had been heated to 1400°C before commencing batch feeding; they were cooled to 300°C by the cold batch and, in consequence, the actual heating phase could not be observed. Measurements within the batch in a glass furnace give results similar to those obtained by Daniels.[22] The unavoidable preheating of the sensing equipment and the difficulties associated with its accurate positioning also raise several questions.

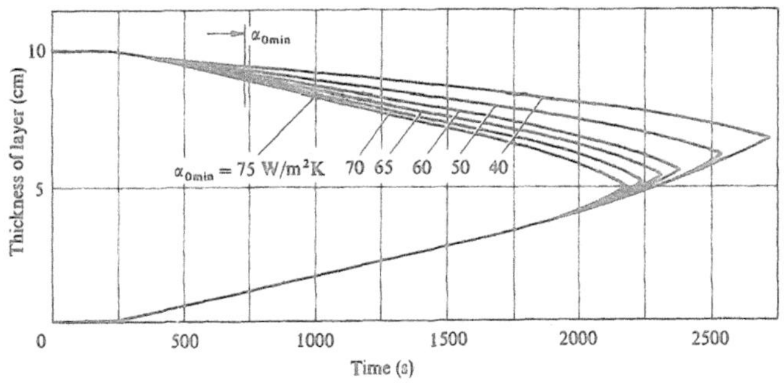

Figure 6.40. Melting process of a layer of batch for different rates of heat transfer at the surface[31] ($\vartheta_a = 20°C$; $\vartheta_s = 800°C$; $\dot{q}_o = \alpha_o\ (\vartheta_u - \vartheta_o)$; $\dot{q}_u = \alpha_u(\vartheta_u - \vartheta_o)$; $\alpha_u = 75$ W/m^2K; $\alpha_o(t = 0) = 75\ W/m^2K$)

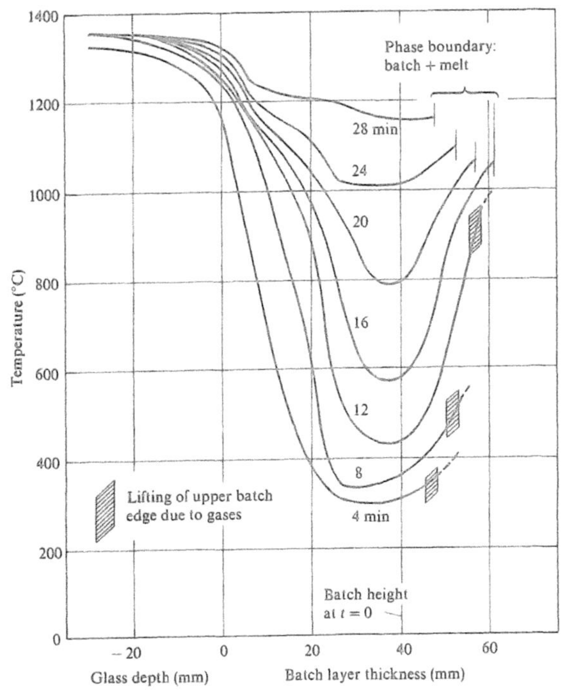

Figure 6.41. Temperature profiles of normal batch[22]

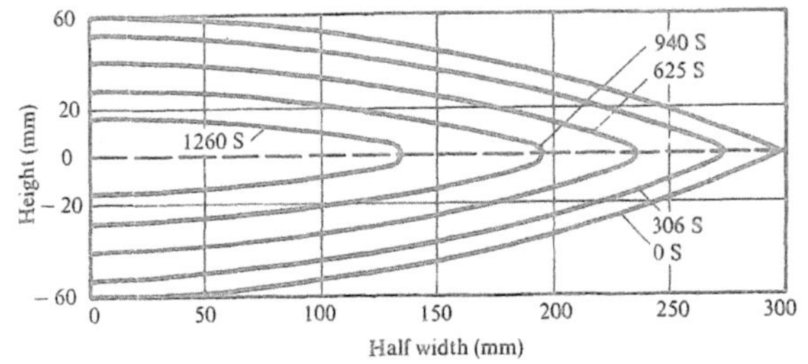

Figure 6.43. Progress of the melting front in an island of batch[31] ($\vartheta_a = 20°C$; $\vartheta_s = 800°C$; $\dot{q}_o = \dot{q}_u = 60\ kW/m^2$)

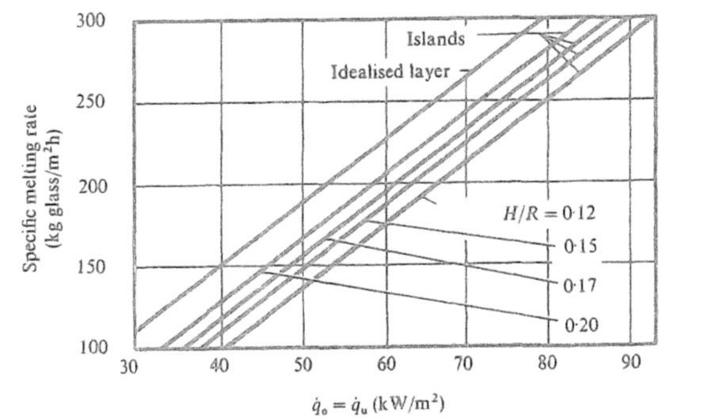

Figure 6.44. Specific melting rate as a function of the idealised layer and islands. H = half height, R = half width[31] ($\vartheta_a = 20°C$; $\vartheta_s = 800°C$; $\dot{q}_o = \dot{q}_u$)

The calculations of Fuhrmann and the measurements of Daniels & Costa can be used to provide a simplified sequence for the primary melting process of a layer of batch heated from both sides (Figure 6.42).

The melting process is of special practical interest because even a continuous blanket of batch is transformed into separate islands during this stage. Fuhrmann and others determined the movement of the melt front for an island of batch 120 mm thick and initially 600 mm wide (Figure 6.43) and found that the width decreases more quickly than the height. It is interesting to compare the melting behaviour of an island with that of an idealised plate, which may be regarded as a limiting case. If the melting rate (in kg glass/m^2h) is used as the criterion, then the result indicates that the plate achieves the higher rate and that the rate achieved by the islands is 10–20% lower. Islands with domed surfaces are somewhat better than flat ones (Figure 6.44).

Pelletised batch melts more quickly than normal batch, a fact confirmed by experience on furnaces and from the measurements of Daniels & Costa. As Costa showed, this is due to the higher absorption of a heap of pellets in the temperature range 100–800°C (Figure 6.33). The diameter of the pellets also has an influence on this effect, but only a minor one; the highest heat flux density occurs for pellet diameters of 10–12 mm, as shown by the measurements of Costa.

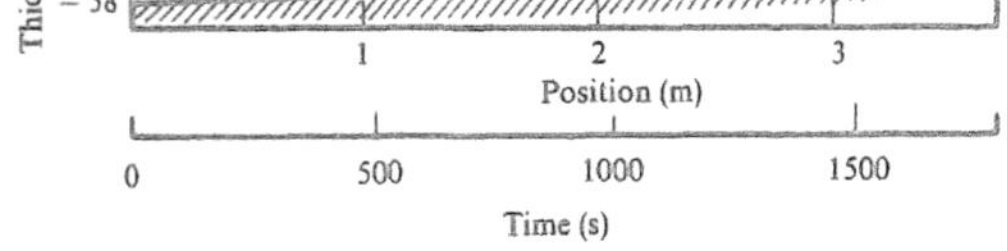

Figure 6.42. Relationship between melting down process, time, and position in furnace[31] ($\vartheta_a = 20°C$; $\vartheta_s = 800°C$; $\dot{q}_o = \dot{q}_u = 45\ kW/m^2$)

6.4.2 Dust and volatilisation

If in the superstructure there are high gas flow velocities such as those which arise from close flame contact with the batch, then normal batch tends to give rise to dust in the area between the doghouse and the place at which the first major melting reactions take place. Alkali dust can volatilise partly or completely at the high temperature in the superstructure; losses due to dusting and volatilisation occur selectively in multicomponent glasses and call for adjustment to the batch formulation.

Reactions between alkali dust and superstructure refractories are dangerous as they lead to the formation of slag which is usually enriched with Al_2O_3 and ZrO_2 and flows into the bath of molten glass, causing defects. In addition, it also leads to the formation of deposits in the checkerwork. Wetting of the batch with 3·5–4% of water or caustic soda significantly reduces the tendency to evolve dust. In exceptional cases, e.g. a batch formulated to give reducing conditions and which is not affected by heavy metals, oil can be used as a binder instead.

In any case, it is important that batch charging is carried out in a manner that allows the batch to pass into the area of the flames only after its surface has sintered: this is easier to achieve on cross-fired furnaces than on U shaped flame furnaces. Caustic soda has been observed to cause condensation and blockages in the central zones of regenerators on large flat glass tanks; this could be linked to a higher rate of alkali evaporation due to the higher vapour pressure of NaOH.

6.4.3 Methods of batch charging

In practice, methods of charging can be divided into charging in piles, charging as a ribbon, and charging as a blanket (Figure 6.45). The object of feeding the batch in the form of a blanket is to obtain the maximum utilisation of the available melting area. The batch is fed either as a blanket or as a number of ribbons which fuse into a blanket; this is now the preferred method and is used for cross-fired tank furnaces and furnaces fired from one side.

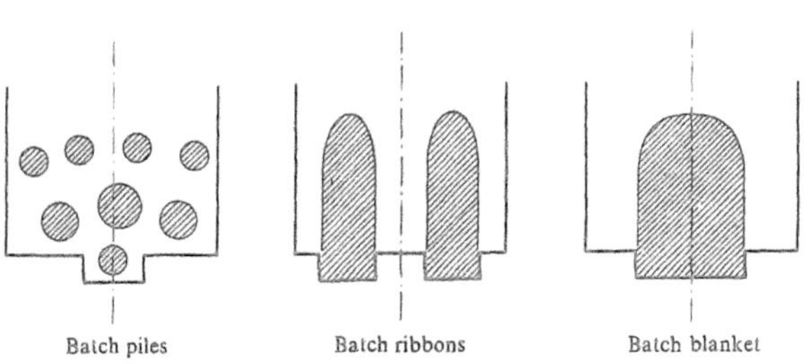

Figure 6.45. Different methods of batch charging.

6.4.3.1 Cross-fired furnaces

Batch feeders for cross-fired tank furnaces are now always of the ribbon or blanket type; the old method of feeding piles of batch, which was thermally inefficient and difficult to mechanise, has disappeared.

A relatively simple method of batch charging is based on a vibrating trough located between the outlet of the batch silo and the doghouse (Figure 6.46). Coarse adjustment of the batch supply is by means of a gate valve controlling the flow of batch from the silo, fine adjustment by altering the interval between the times when the vibrator is operated: this is controlled by the level sensor. The ribbon of batch is pushed forward by an eccentric drive acting on a water-cooled bar attached to the hot end of the trough. When correctly set and operating under high pull, the batch slides forward into the furnace without pushing due to the slight incline in glass level caused by the withdrawal of glass at the far end of the furnace. Cold batch causes a considerable temperature trough to form which, in conjunction with the hot spot, forms a strong circulating current below the melting batch. For further details see Section 6.5 which deals with glass currents.

When melting dark glasses or using small tanks the water-cooled push bar must dip well into the glass in order to create a pumping effect on the bath of molten glass and thus prevent excessive cooling of the doghouse. With clear or pale glasses only slight immersion is necessary and the pushing effect of the batch itself is often sufficient.

A different technique, but from the point of view of batch charging very similar in effect, is the use of the vibrating, or Gradenberg batch charger. This charger operates with a plate (Figure 6.47) which is set to match the angle of repose of the batch and which guides the batch onto the glass surface in a continuous stream at the same time pushing the batch forward.

The vibrating trough and the Gradenberg chargers are so similar that they can be considered

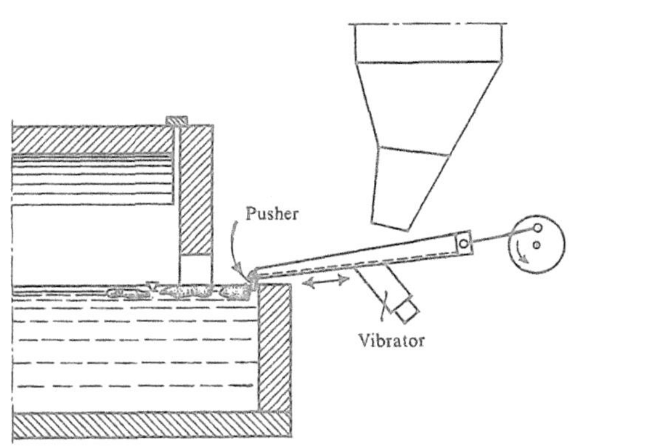

Figure 6.46. Vibrating charging trough with pushing bar and eccentric drive

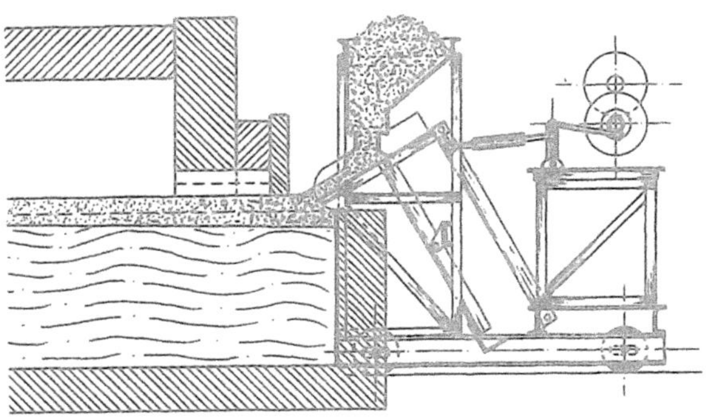

Figure 6.47. Gradenberg type of batch charger (Zippe)

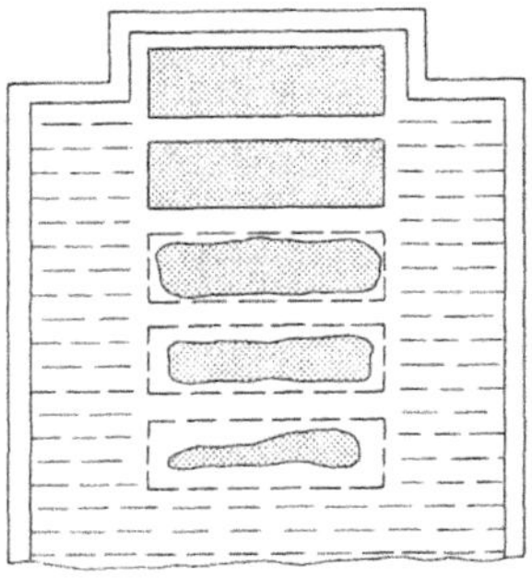

Figure 6.48. Charging method giving ribbons across the width of the tank

together. The position of the batch blanket in the melting chamber can be adjusted by altering their settings, which brings many operational advantages as it is possible to react to asymmetrical melting of the batch. Periodically switching the charger on and off helps the melting process by dividing the batch blanket into separate bands across the width of the furnace (Figure 6.48).

For glasses which have to be very homogeneous and free from seed, such as is often the case for special glasses, a batch charging technique is used which creates separate bands which diverge as they enter the furnace from the doghouse. Towards the end of the melting process residues of foam and batch are drawn back into the melting zone, which avoids their flowing into the refining zone (Figure 6.49). The return flow with the V created by the diverging bands of batch is caused by a descending glass flow which is particularly strong in this area because there is no advancing blanket of batch to obstruct it.

Blanket feeding is part of the system by which the melting chamber surface is used efficiently for melting the batch by flames and which contributes to high pulls and low energy consumptions. A significant contribution to this effect is the main current flowing back from the refining zone to the doghouse. The cooling effect of the batch blanket is an essential part of the 'heat pump' responsible for the return flow, for if charging is in the form of one or two small bands, then the strength and direction of this return current are not well defined, thereby reducing the rate at which the batch is melted from below. If the batch blanket is very wide and the furnace operates at high pull then the cooling effect of the batch can have such an effect on glass viscosity that the return current cannot form and the heat pump ceases to function; in practice, this effect is indicated by the reluctance of the batch to

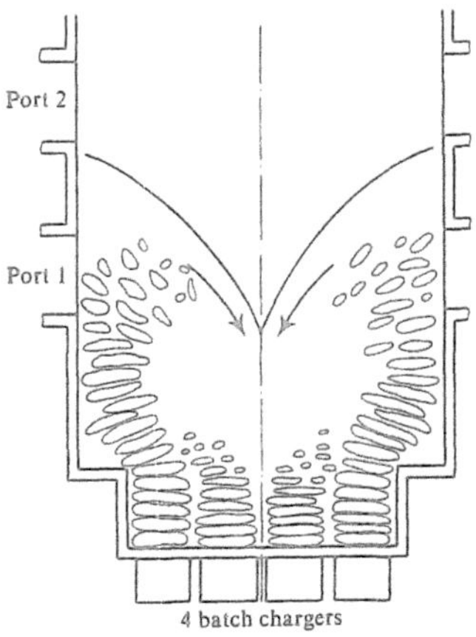

Figure 6.49. Idealised view of batch distribution using four batch chargers[35]

move forward. In this situation either the pull must be reduced or electric boosting must be installed in the batch melting zone.

Good utilisation of the melting surface can also be achieved by the use of a V shaped doghouse with two batch chargers at right angles to one another operating alternately (Figure 6.50); pusher or screw chargers can be used (Figure 6.51). The advantage of this method is that the doghouse can now be totally enclosed. Screw feeders can be arranged parallel to one another with or without a doghouse construction (Figure 6.52).

The method of charging batch into large flat glass tanks differs fundamentally from that for container glass tanks: large batch silos, containing enough material for one shift, are not used as it is believed that fresh (damp) batch should always be fed. Charging machines for flat glass tanks have their chutes at only half the normal height and these are continuously replenished with fresh batch delivered by belt conveyor or travelling batch cans. A second peculiarity, now disappearing, is the separate charging of cullet and batch. Corresponding to

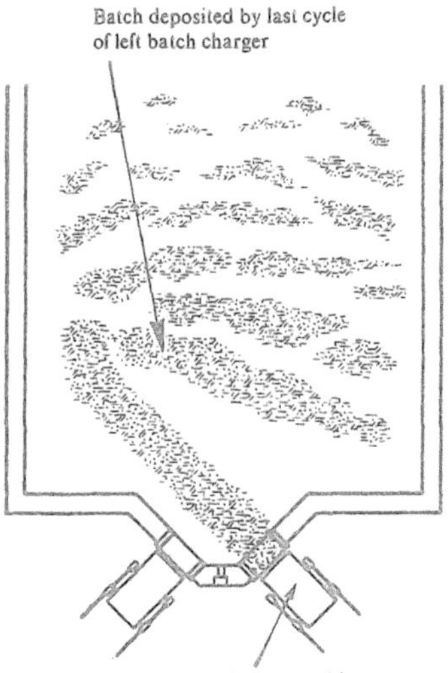

Figure 6.50. Batch distribution in separate ribbons or piles

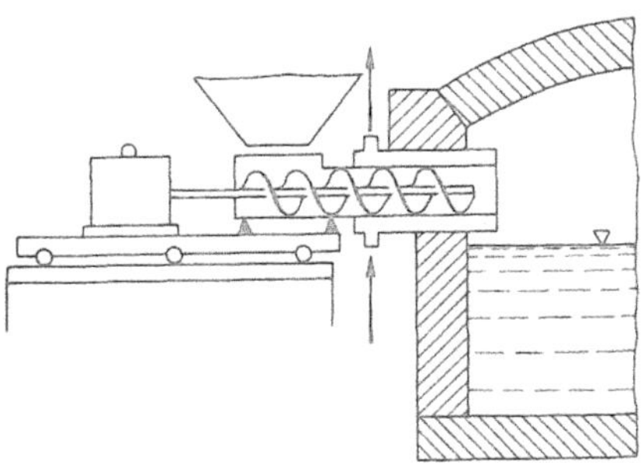
Figure 6.51. Screw feeder

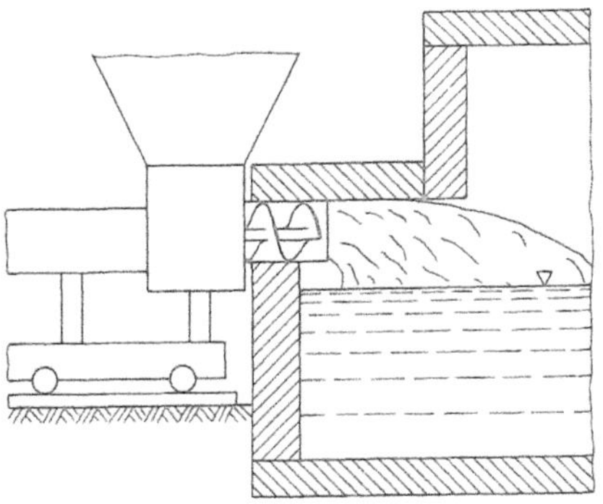
Figure 6.52. Screw feeder with doghouse

the practice of charging in piles, Univerbel has developed a compartmented charging machine for placing the batch on top of cullet.[37] The aim is to achieve a blanket cover by means of a water-cooled compartmented cylinder (Figure 6.53). Cullet is fed down an inclined chute and the compartmented cylinder places doses of batch on top of the cullet and pushes both into the melting chamber. The separate feeding of cullet and batch has now been abandoned

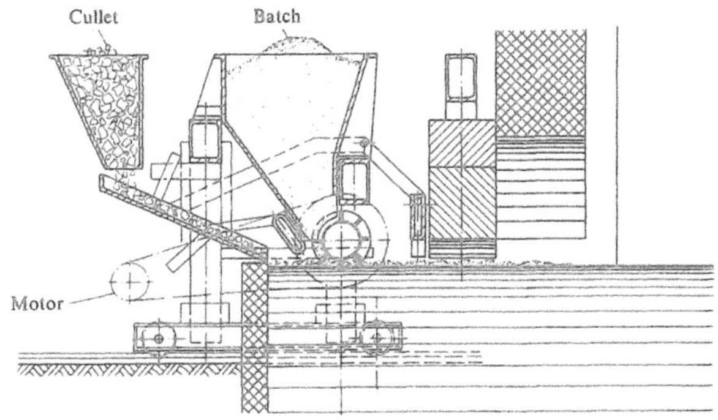

Figure 6.53. Univerbel batch charger with compartmented cylinder

on large Float glass installations since it cannot be shown to provide any advantages.

As already indicated in Section 6.4.2, the distance from the doghouse to the first port or burner is very important; if it is made too small, unacceptable levels of dust are created, if too large, the melting rate suffers. A distance of 1·2–1·8 m is believed to be correct for container glass tanks.

6.4.3.2 U shaped flame tanks

The charging of end-fired U shaped flame tanks is more difficult and less satisfactory than that of cross-fired tank furnaces. Charging must be from one or both sides and, furthermore, it can only be as a narrow band of batch so that the coverage achieved with the blanket charging of a cross-fired furnace cannot be achieved in a U shaped flame furnace. There is a further drawback with a U shaped flame furnace with a regenerator: depending on the position of the flame, the rate at which batch melts in that part of the melting area not covered by the flame decreases, allowing batch to accumulate. This has been overcome by more frequent reversals, with a period of 15 min generally recommended compared to the 20 min which is regarded as optimum for cross-fired tanks. Some improvement can be brought about by widening the doghouse and angling its entry towards the melting zone and by providing means for altering the position angle of the machine or for moving it sideways (Figure 6.54).[36,40]

A further difficulty is that the presintering zone is too short; the doghouses are usually built too small, exposing unsintered batch on the surface to high-velocity flames so that the evolution of dust with all its consequences follows. It is strongly recommended that the doghouse be of adequate size and that it incorporate a doghouse arch; furthermore, the batch should contain sufficient moisture.

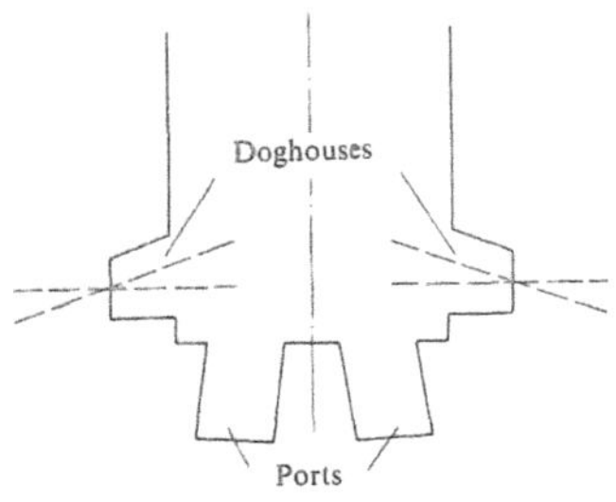

Figure 6.54. Arrangement of batch chargers for U shaped flame furnaces with facility to rotate the chargers through the angles shown

6.4.3.3 Small, continuously fired tank furnaces

The choice of a charging system for small, narrow tanks such as unit melters calls for special consideration. The most successful method of charging has been the enfolding batch feeder developed by Hartford (Figure 6.55). The folding of the batch and the strong pushing action of the rabble give good charging conditions even when there are weak convection currents in the batch charging area. A slight widening of the doghouse together with a periodic horizontal angular movement of the feeder gives better utilisation of the melting zone and increases the melting rate.

The feeding of batch for special glasses has to be considered separately for each case. For example, E glass for the production of continuous glass fibres is melted from finely powdered raw materials (50–100 μm particle size) and the problem of dust being carried over is considerable; in this case batch is almost invariably fed by a screw feeder through the wall of a totally enclosed doghouse.[122]

6.4.3.4 Electrically heated tank furnaces[38]

The charging of electric furnaces is more difficult than flame-heated furnaces since complete coverage of the melting area must be achieved without the help of glass currents (Figure 6.56). Small furnaces making easily melted batch compositions

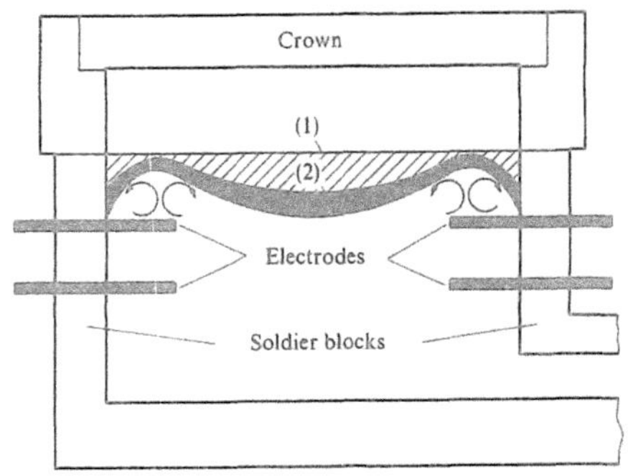

Figure 6.56. Formation of batch blanket for uniform batch distribution and delta connecting of electrodes[38]
(1) batch surface resulting from uniform distribution
(2) shape of batch blanket as a result of the melting process

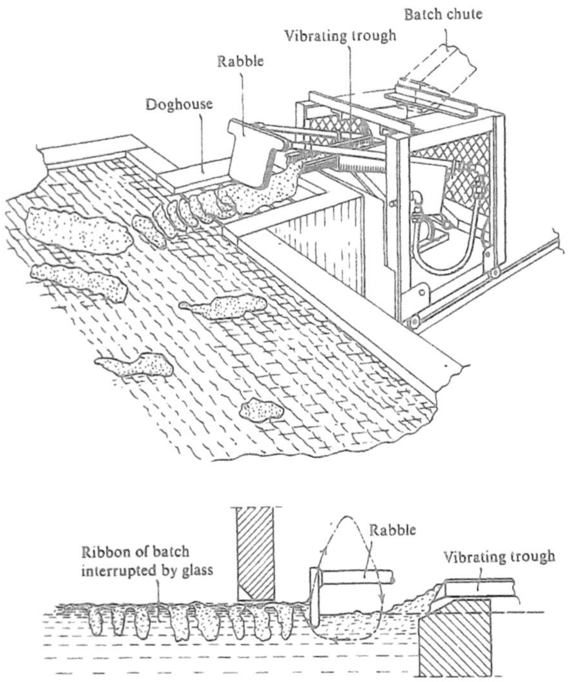

Figure 6.55. Immersion or enfolding batch feeder by Hartford

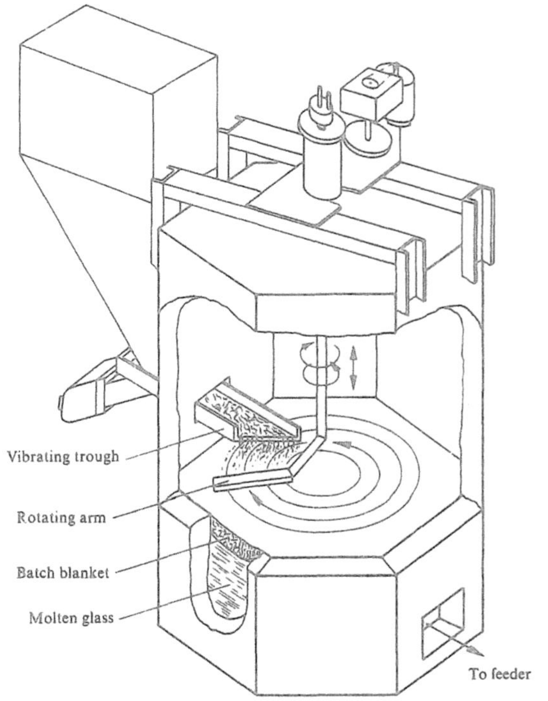

Figure 6.57. Batch distributor arm[38]

such as those for fluoride opal or soda–lime glasses and furnaces of hexagonal or nearly circular plan can use a water-cooled rotating batch distributor arm (Figure 6.57).

If the batch tends to form a crust, then it pays to replace the distributor arm by a tipping arm, rotating at a greater height above the batch blanket, which deposits the batch without making contact with it. The disadvantage of this method is that the tipping disturbs the glass level. A further improvement is a rotating cover on the furnace which is sealed to the outside. A series of reciprocating troughs cause ribbons of batch to be deposited in the form of rings (Figures 6.58 and 6.59).

A rectangular melting surface can be covered by using a charging system in which the batch is distributed by a conveyor on one side of the furnace; this conveyor is programmed to move forwards and backwards as well as sideways so that its successive discharge positions effectively cover the whole surface[39] (Figure 6.60).

The use of this type of machine presupposes that

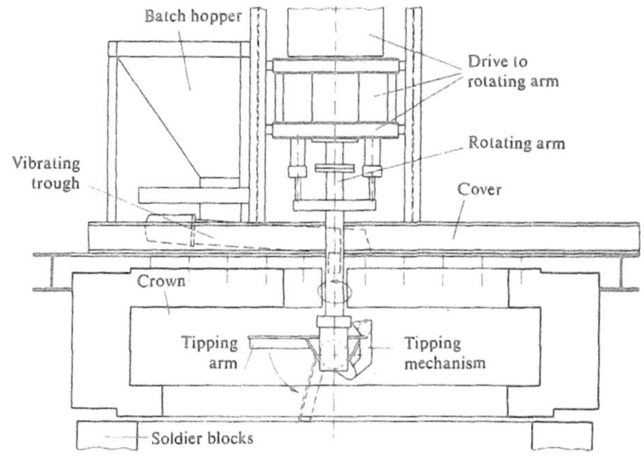

Figure 6.58. Tippable rotating batch distributor[38]

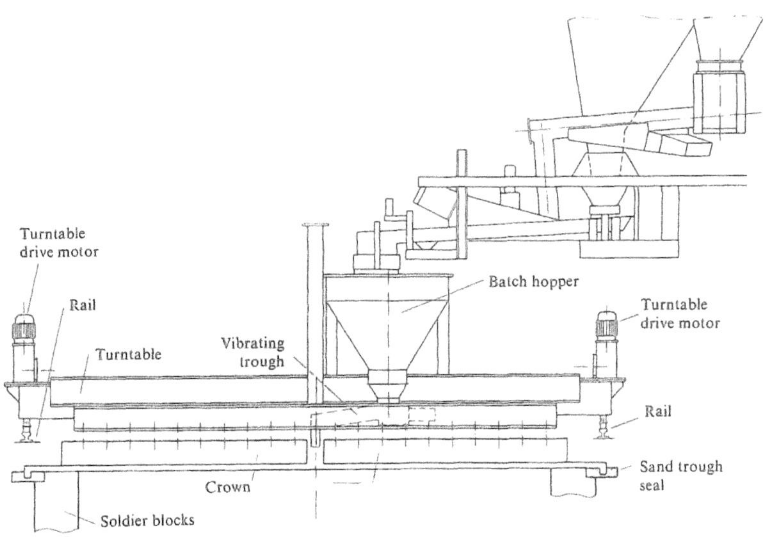

Figure 6.59. Turntable of the first version of the newly developed batch charger[38]

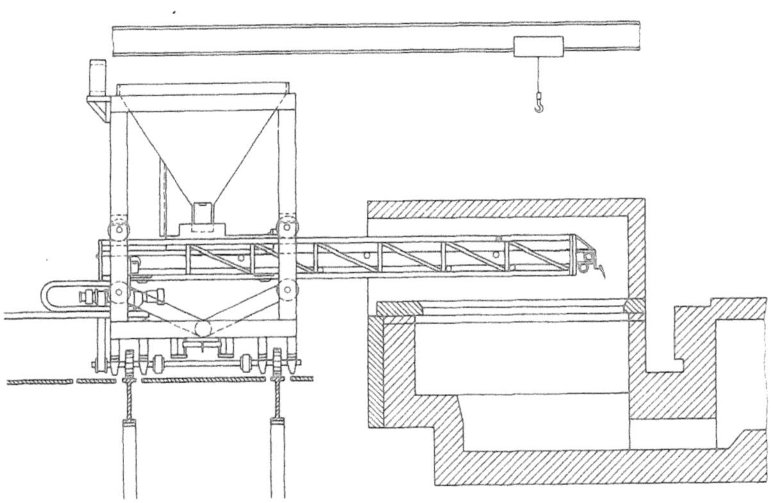

Figure 6.60. Batch charger for a rectangular electric furnace with travelling belt conveyor[39]

complete coverage of the melting area can be achieved under all normal conditions. Should large exposed surfaces form these would immediately present a danger to the batch charging equipment which would be overheated.

6.5 Glass currents

6.5.1 The origin of glass currents

The continuous operation of glass tank furnaces involves the removal of glass at one end and the charging of batch at the other; this gives rise to the withdrawal current.

The effect of heating from above, heat losses from the tank, and the processes of melting and refining cause differences in density in the bath of molten glass which give rise to convection currents. Also acting on the glass although of less importance, are external forces such as the impelling force produced by the movement of flames over the glass, and the pushing effect of the batch charger. If bubblers and electric boosting are used, these also influence glass currents.

Altogether these processes create in the furnace glass currents of great complexity which depends on the design of the furnace and its mode of operation. The proper formation of these currents is not just of major importance, it is of critical importance for the satisfactory operation of the tank furnace. The currents ensure that the individual stages of glass melting, such as the melting of the batch, the refining, and the conditioning of the glass, take place at the correct time and in the correct location in the tank. Since parts of the glass surface are shielded from flame radiation by being covered with batch and foam, the heat losses in these zones must be compensated for by glass currents. In addition, the glass currents caused by the flow gradients create stretching and stratification of local inhomogeneities. In this way the currents, in addition to promoting diffusion, contribute significantly to the homogenisation of the inhomogeneous primary melt. This triple function of the currents points to their importance for the final quality of the glass as well as for furnace pull and energy consumption.

The most important factors in the formation of glass currents are the density differences due to thermal effects and the withdrawal current. The shear forces created by the flames in regeneratively fired tanks, accentuated by the periodic reversal of flame direction, are of only minor importance. In end-fired, recuperatively heated tank furnaces these shear forces are usually a drawback since they operate in the same direction as the withdrawal current.

To obtain data giving the total distribution of temperature, and thereby of density, of the glass in a tank requires a major investment of effort. The literature therefore only provides data covering certain areas, in particular those along the longitudinal axis of the tank. The most comprehensive work in this field[42] is shown in Figures 6.61 and 6.62 which give the temperature distributions in furnaces for clear container glass and amber bottle glass.

The temperature gradient between glass surface and tank bottom depends on the type of glass. Transparent glasses such as clear container glass give a gradient of 1·3–1·5 degC/cm while glasses of low transparency, such as amber or green glass, give a gradient of 2·5–3·5 degC/cm. Whereas the gradient of coloured glass is relatively uniform, that of clear glass shows a steep increase in temperature near the surface (Figure 6.61).

A knowledge of glass currents in tanks and their

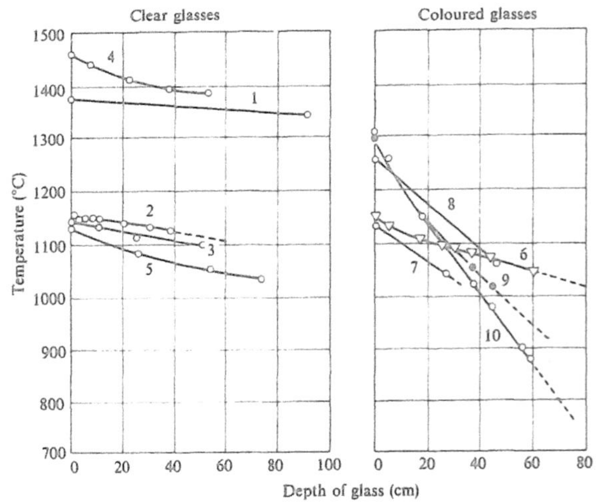

Figure 6.61. Temperature gradients in the molten glass of different tank furnaces according to measurements by Halle[41]
(1) Pyrex
(2) clear lighting glass
(3) clear Pyrex (working end)
(4) clear Pyrex (melting end)
(5) clear container glass
(6) light green bottle glass
(7) green glass
(8) smoky brown sealing glass
(9) amber glass
(10) dark green glass

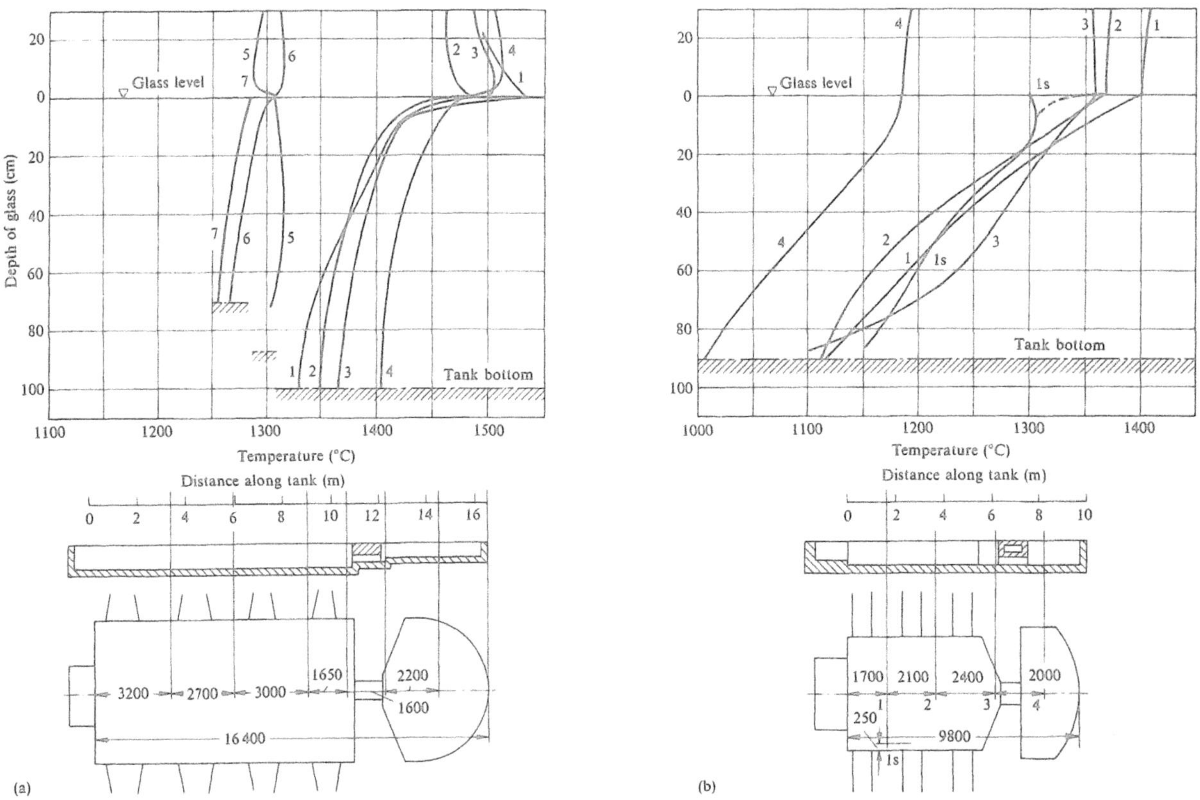

Figure 6.62. Temperature gradients in the molten glass along the tank axis[42] (dimensions in mm)
(a) clear container glass tank (b) tank melting amber bottle glass

influence and a control over them by design and operational procedures presuppose an understanding of the interrelated causes which are the origin of the glass currents. The usual statement that the cause of glass currents is the fact that glass will flow from zones of high temperature to zones of low temperature describes the process very inadequately as it cannot explain the return flow of glass from cooler to hotter zones of the tank. A better approach to an understanding of the origin of glass currents under varying conditions is to study the changes in hydrostatic pressure at different depths within the glass at different positions in the tank (Figure 6.63).

If one starts with the assumption that, initially, the temperature of the glass is the same everywhere and that no glass is fed into the tank and none flows out of it, then the hydrostatic pressure at identical depths is the same everywhere. No currents can arise within the bath of molten glass since the potential planes are exactly horizontal and parallel to the glass surface. Nor can any horizontal movement of the glass take place since there is no drop in potential (drop in pressure) in this direction. Vertical movement cannot take place without the application of external forces, as any upward movement must be balanced by an equal downward movement. The sum of the differences in potential for such a situation is therefore zero.

Figure 6.63(a) is a schematic cross section of a tank vertical to the plane of drawing and extending to infinite length. If the vertical plane B at the centre of the tank is compared to A, then, for the same glass temperature throughout, the change in hydrostatic pressure with depth can be described by the curve given on the right. The change in pressure for planes A and B are identical; the two lines are superimposed on one another.

If the glass is heated from above, more strongly in the centre than at the sides of the tank, then the glass begins to expand in proportion to the uneven temperature distribution, i.e. more in the centre than near the sides, so that the surface, initially flat, now begins to bulge upwards at the centre. If one assumes that no flow has so far taken place, then the change in pressure in planes A and B can be represented by the plot shown in Figure 6.63(b). The hydrostatic pressures at the bottom of A and B

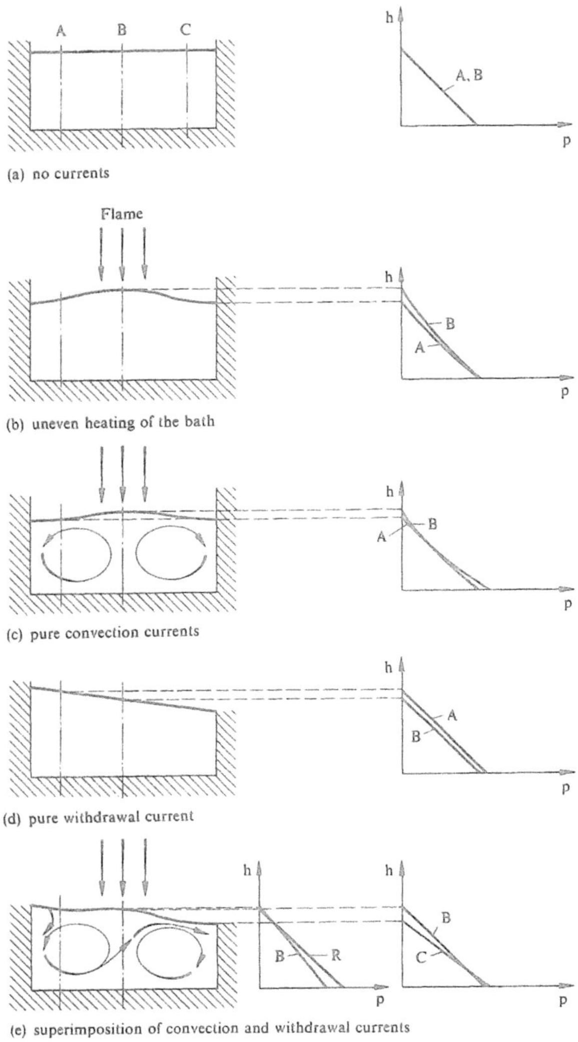

Figure 6.63. Formation of convection and withdrawal currents in a simple tank[43]

are the same and unchanged since the total weight of the column of glass above them has not changed despite the expansion that has taken place. Higher up, the differences in pressure between planes A and B in the horizontal direction are larger since the weights of the columns of glass are different as a result of the temperature differences and the bulging of the glass surface. The change in hydrostatic pressure at B deviates more markedly from a straight line since the surface zones of the glass are heated more than the bottom zones; at A this influence is smaller. The divergence of pressure lines gives rise to pressure differences between A and B, the magnitudes of which change with depth. In the centre of the bath the differences push the glass towards the sides of the tank. This flow starts at the surface since it is there that the inclination of the potential lines is greatest. Glass near the surface flows uniformly away from the surface of the bulge, reducing the weight of glass in the central column and increasing it at the sides, so that the pressures at the bottoms of A and B become different and, in consequence, the lines of pressure cross as is shown in Figure 6.63(c). Near the surface the glass is forced outwards, near the bottom inwards and this flow causes the glass at A to sink and that at B to rise. An equilibrium is set up between the initiating forces (temperature differences) and the responding forces (glass flow); as fast as heating creates differences in temperature in the bath so the flow tries to reduce them, giving rise to typical convection currents which, under completely symmetrical conditions between centre and sides, causes two opposing circular flow patterns, whose centre of rotation and shape depend on the temperature distribution in the molten glass.

Figure 6.63(d) shows a pure withdrawal current and the pressure distribution caused thereby. The charging of glass at one end and its removal at the other gives rise to a gradient on the glass surface and a flow of glass from A to B. The change in pressure with depth is, in this case, given by two parallel straight lines. If the two conditions shown in Figure 6.63 (c) and (d) are superimposed, then this leads to the simplest form of glass current such as is normally expected along the longitudinal axis of a tank. The withdrawal current causes the lines of pressure given in Figure 6.63(d) to shift from the parallel and, with sufficient current at the hot spot, a pressure distribution such as that shown in Figure 6.63(e) is formed, and an S shaped current is located between the two opposing currents. When charging batch, it should be remembered that the melting glass with its foam and residual sand remains on the surface due to its lower density and that only glass largely free of sand and gases is drawn into the descending part of the rotating convection cylinder.

The surface of the glass has the shape of a gentle S, the slope of the surface being least just before the hot spot, which leads to a reduction in the surface speed of the glass, and at its greatest beyond the hot spot, so that the surface flow of the glass there is also greatest. The effect of the upward currents at the hot spot is therefore very marked. Depending on the relative magnitudes of the withdrawal and convection currents, the overall pattern of glass currents will be shifted towards (c) or (d) of Figure 6.63.

Figure 6.63(e) demonstrates that fresh glass must pass into the hot area of the molten glass near the hot spot either directly as part of the surface current or indirectly via the rotating convection cylinder; this must contribute significantly to good refining and homogenisation. When studying this diagram one gains the impression that the satisfactory melting of glass in a tank without convection currents is open to question. Melting glass flows particularly slowly in the region of the highest tank temperature (hot spot), which must make a considerable contribution to the complete melting and refining of the glass. A pure withdrawal current would mean that the melting glass would flow equally rapidly everywhere along the surface, and would therefore be exposed to a higher temperature for a shorter time than is the case when convection currents are superimposed on the withdrawal current. The form and magnitude of the rotating convection currents, as emphasised in Figure 6.63, depend on the temperature distribution in the molten glass. If the direct and indirect effects of flames on the temperature distribution do not penetrate to a sufficient depth (e.g. for green glass), then, if electric boosting is not used, only weak rotating convection currents are formed. This can give rise to the risk that cold, inadequately refined glass will sink under the batch blanket to such an extent that it will flow underneath and past the hot spot directly through the throat into the working end.

6.5.2 Calculation of glass currents[44–58]

Any calculation of glass currents has to take into account equations covering continuity, energy, and momentum within a three dimensional space with boundary conditions and the properties of the molten glass also being taken into account. So far no solution has been possible within an acceptable time even with the assistance of computers. Leyens[50] has shown that some results can be obtained if the problem is restricted to two dimensions, use is made of an effective thermal conductivity which excludes the influence of radiation, and the boundary effects are simplified. Using Stokes's equation for flow and an average constant density for the molten glass for the equations covering energy and continuity, and ignoring inertia forces, the glass currents in a vertical plane through the longitudinal axis of the tank can be described by the energy equation

$$\rho c\left[\frac{\partial \vartheta}{\partial t}+v_x\frac{\partial \vartheta}{\partial x}+v_y\frac{\partial \vartheta}{\partial y}\right]=\lambda_{\text{eff}}\left[\frac{\partial^2 \vartheta}{\partial x^2}+\frac{\partial^2 \vartheta}{\partial y^2}\right] \tag{6.29}$$

and the flow equation

$$\mu\left[\frac{\partial^4 \Psi}{\partial x^4}+2\frac{\partial^4 \Psi}{\partial x^2 \partial y^2}+\frac{\partial^4 \Psi}{\partial y^4}\right]+g\rho_0\beta\frac{\partial}{\partial x}=0 \tag{6.30}$$

where μ is the dynamic viscosity of molten glass and β the thermal expansion coefficient, i.e.

$$\beta=\frac{1}{\rho_0}\frac{\partial \rho}{\partial \vartheta}.$$

In these equations all properties of molten glass are regarded as constant. Since they can only be solved by inserting numerical values, the temperature dependence of properties can be taken into account during the individual steps of the calculation.

From the point of view of similarity it can be shown that the field of currents can be characterised by the Rayleigh number

$$Ra = Gr\,Pr=\frac{g\cdot\beta\Delta\vartheta_{\text{max}}L^3}{\nu a} \tag{6.31}$$

in which $\Delta\vartheta_{\text{max}}$ stands for the maximum temperature difference that occurs and L the relevant length; for glass furnaces, Ra is in the region of $5–100 \times 10^3$. It is clear that convection currents become stronger as the thermal expansion of the glass, the temperature differences, and, especially, the size of the installation increase; increases in kinematic viscosity, ν, and thermal conductivity reduce convection currents.

Leyens points out in his papers that, since the dynamic viscosity changes much more with temperature than the thermal conductivity (Figure 6.64), it is the dynamic viscosity rather than the thermal conductivity which governs the field of currents. He suggests that the Grashof number should be used as the relevant parameter rather than the product of Grashof and Prandtl numbers. The Grashof number is given by

$$Gr=\frac{g\beta\Delta\vartheta_{\text{max}}L^3}{\nu^2}. \tag{6.32}$$

Compared to the Rayleigh number, the kinematic viscosity is squared and the thermal conductivity (a) is excluded, so a change in glass viscosity will give rise to a larger change in Gr than in Ra. The

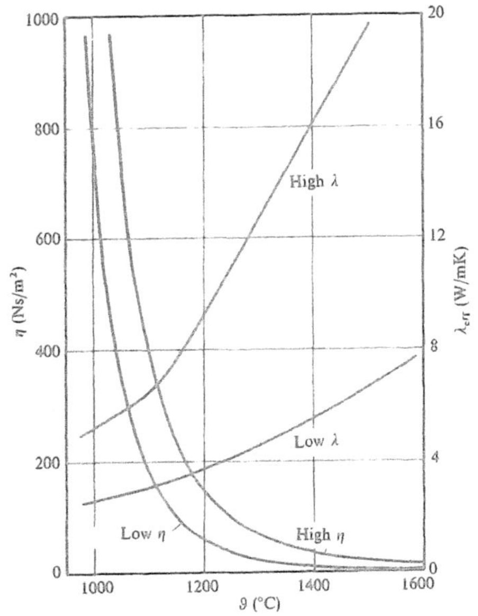

Figure 6.64. Dynamic viscosity and effective thermal conductivity of molten glasses[50]

Grashof number for glass furnaces lies between 6 and 160 × 10^3 under normal conditions. As with the Rayleigh number, the intensity of convection currents increases with increasing Grashof number. The relevant length dimension, L, can be taken from $L^3 = H^2S$, where H is the depth of glass and S the length dimension in the direction of the glass current.

Present knowledge covering glass currents is based primarily on investigations using physical models and calculations; recent results have provided new explanations. The results of these calculations will be given in the next section in conjunction with the different parameters which influence them.

Studies of glass currents on operating tanks are usually limited to experiments using floats on the glass surface[43] and measurements within the depth of the glass using radioactive isotopes and immersed pendulum systems;[51,62] the immersed pendulum provides data on currents at a precise location. The equipment required is substantial so that only point measurements are possible and this method is therefore suitable only for obtaining supporting data for calculations and investigations on models.

6.5.3 The effect of glass currents

6.5.3.1 The boundary between melting and refining zones

Among the most important functions of glass currents is that which separates the melting zone from the refining zone, so that only fully refined glass reaches the working end. Refining is taken to mean, as in the literature, the almost complete degassing and homogenisation of the primary melt. Since some refining takes place even in the melting zone, a precise separation into melting and refining zones is not possible. Having stated this reservation, this book will nevertheless retain the historical distinction between melting and refining zones.

Firing along the length of the furnace should be arranged so that a clear temperature maximum occurs in the refining zone.[59] This, in conjunction with the cooling effect produced by the charging of batch, creates an upward current (Figure 6.65) which prevents the batch blanket and foam from advancing too far into the refining zone and which forces recently melted glass upwards into the zone of high temperature: this creates what is referred to as the thermal barrier. Longitudinal currents are also indicated in Figure 6.65. The heat required to melt the batch from below flows through plane A-A as a counterclockwise rotating current which prevents batch from advancing too far towards the hot spot.

The magnitude of the cooling effect of the batch on the formation of the rotating current can be calculated for a cross-fired furnace with and without glass withdrawal (Figure 6.66).[60] During withdrawal the two opposing rotating currents are clearly deformed by the Z form of the withdrawal current (Figure 6.67) and without the withdrawal current are very weak. Temperature measurements within the molten glass and the calculated temperature profiles are in general agreement, so that there can be little doubt that the calculations are correct.

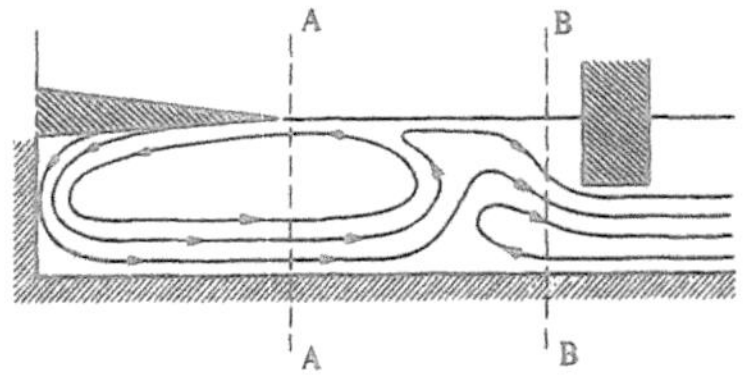

Figure 6.65. Schematic central longitudinal cross section of glass currents in a tank with throat[59]

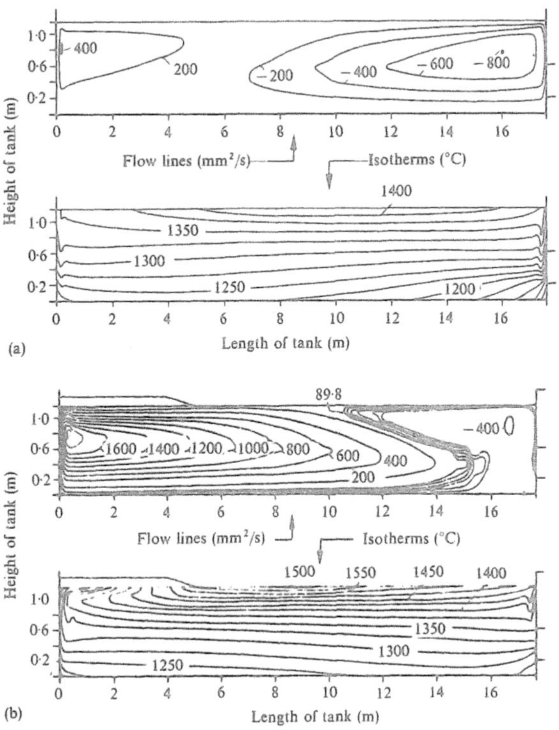

Figure 6.66. Calculated flow lines and isotherms in a cross-fired glass tank furnace[60]
(a) no glass withdrawal (b) pull of 1·33 t/m²d

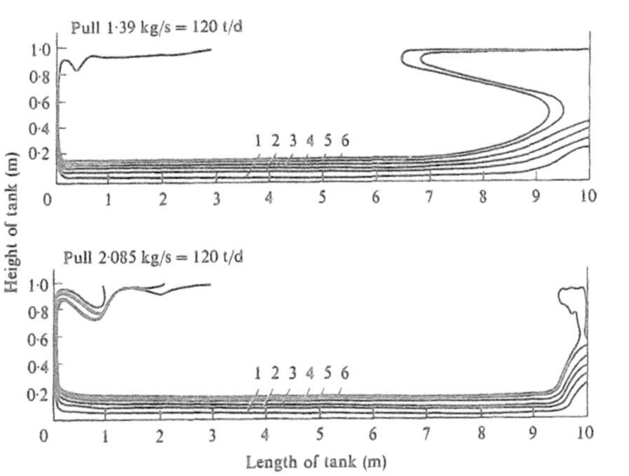

Figure 6.67. Influence of pull on glass currents in a tank furnace

6.5.3.1.1 Increasing the withdrawal current

An increase in pull by increased firing can only be achieved to a limited extent. If the pull is increased beyond this point, then the clockwise rotating current in the refining zone is weakened to an extent that makes it impossible to achieve adequate homogenisation and refining of the glass. The results of the relevant calculations can be seen in Figure 6.67.[50]

Although glass flow into the throat cannot be considered without taking the working end into account, the glass currents shown in Figure 6.67 nevertheless demonstrate that an increased pull without corresponding and adequate compensation in the form of heating must lead to a reduction in the temperature of the glass in the throat despite an increase in the above-glass temperature of the furnace. Measurements by Greschat[61] on a clear container glass tank confirm these findings; with increasing pull the glass temperature in the throat decreased despite an increase in crown temperature (Figure 6.68).

More recent work by Leyens & Smrcek[52] confirms the suspicion that a large increase in pull, albeit starting from a low level of 0·7 t glass/m²d, leads to a slight increase in the main rotating current under the batch blanket but a considerable reduction in the size of the associated current in the refining zone. This is shown in Figure 6.69, where the tendency for the glass to take a short cut to the throat may also be seen clearly. The fuel consumption had increased correspondingly. The results given in Figure 6.69 are based on two-dimensional calculations for the glass currents, but are supported by measurements of residence times.

6.5.3.1.2 Electric boosting in the refining zone

The addition of direct electric heating (Section 8.9) in the zone of upward flowing glass currents can stabilise the clockwise rotating currents and thereby enable the pull to be increased. This method is very effective for dark glasses where heating from above is less able to penetrate into the depth of the glass than is the case for clear glasses.

There are no quantitative data covering the electric heating required in relation to pull but it is

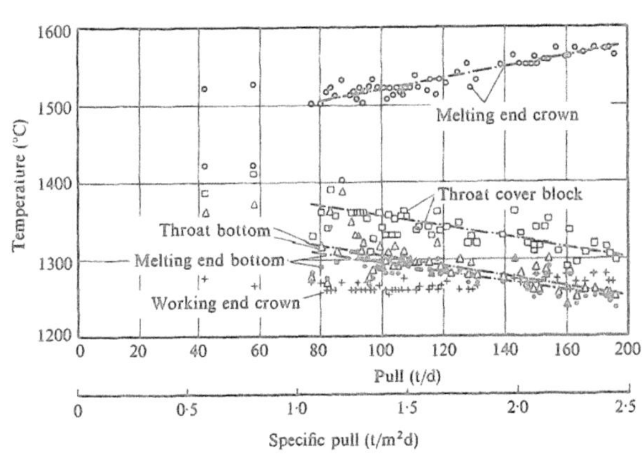

Figure 6.68. Temperatures measured at a variety of points in tank melting clear container glass[61]

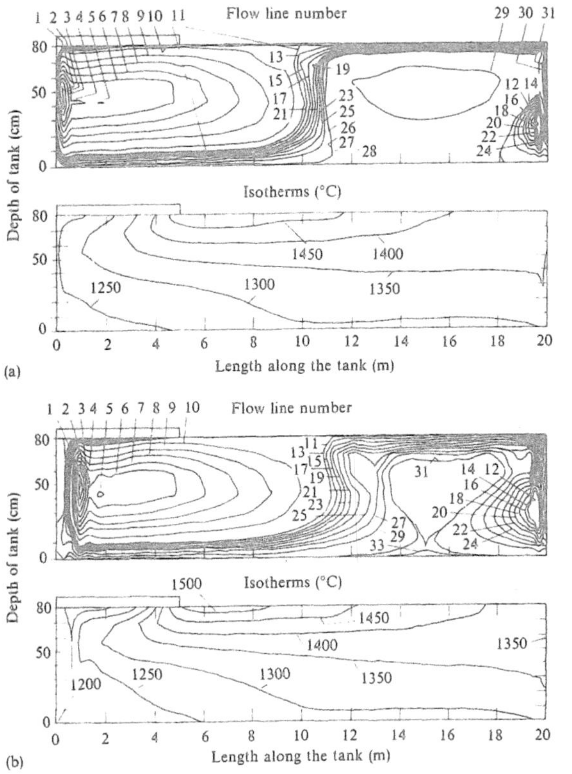

Figure 6.69. Calculated flow lines and isotherms in the vertical central longitudinal cross section of a glass tank at different specific melting rates[52]
(a) 0·7 t/m^2d (b) 1·4 t/m^2d

possible to estimate that about 5% of the energy provided by fossil fuel must be introduced as electric heating to increase the pull by about 10%.

6.5.3.1.3 Electric boosting in the melting zone

The advance of the batch blanket and foam into the refining zone as a result of increased pull, and the consequent reduction in the area of clear glass and the heating of the glass from above, can be prevented up to a point by techniques similar to those employed in the refining zone. The fundamental difference between the two is that electric boosting weakens the counterclockwise current under the batch blanket. In the limiting situation the effect of this current is no longer sufficient to prevent batch from reaching the refining zone (see Section 8.9) and the quality of glass suffers as a result of poor refining. Electric boosting is therefore generally more effective in the refining zone than in the melting zone, but a combination of both enables the largest increases in pull to be obtained. The limits are set by the cost of electricity as a source of energy.

6.5.3.1.4 Bubbling

The upward glass currents at the hot spot can also be stabilised to some extent by the controlled formation of bubbles at the tank bottom. The main benefit of bubbling is to improve homogenisation rather than increase the pull, as the gas bubbles rising from the bottom act as exterior shear forces on the surrounding glass and thereby reinforce the upward currents at the hot spot.

Bubbling differs from electric boosting in that the rising chain of bubbles has only a local effect on the surrounding glass, a rising flow cylinder being formed above each nozzle surrounded by a descending flow cylinder (Figure 6.70(a)). If several bubblers are installed in a single row or a pair of rows, then the separate cylinders combine in accordance with the distance between them (Figure 6.70(b)). When designing for the installation of rows of bubblers, care must be taken to consider carefully the distance between adjacent bubblers (l') and

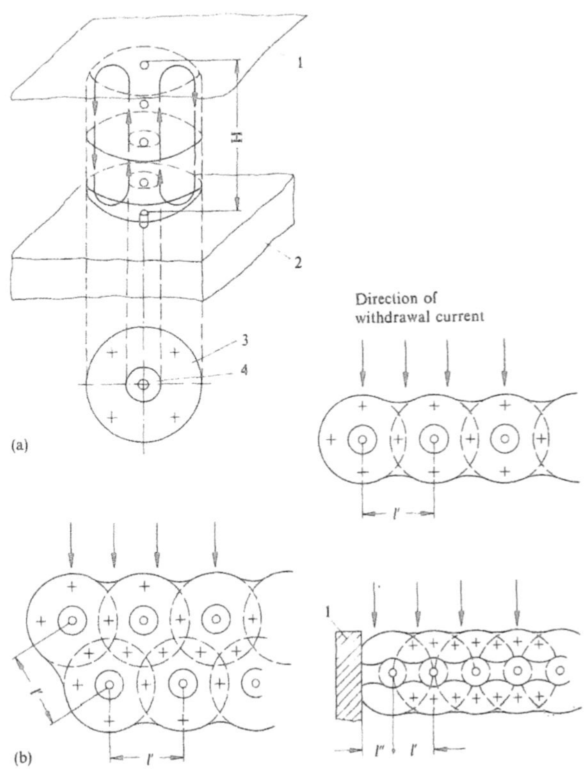

Figure 6.70. Flow cylinders with rising and descending currents (a) and the blending of individual cylinders into a unified flow front (b)[63]
(1) glass surface
(2) tank bottom
(3) descending glass currents
(4) ascending glass currents

the distances of the first and last bubblers from the sidewall (l'').

According to Pchelyakov *et al.*[63] the formation of unwanted downward currents between individual bubblers can be avoided provided that $l \leqslant (l_1 + 0{\cdot}3H)$ if l_1 is chosen as 200 mm and H is the glass depth; for l'' the recommendation is $l'' \leqslant (l_2 + 0{\cdot}2H)$ for $l_2 = 130$ mm. In practice, the distance between bubblers lies between 300 and 600 mm, with larger distances in the deeper tanks and smaller distances in the shallower tanks. The distance between the end bubbler and the sidewall is of the same order as it will accelerate corrosion of the adjacent tank block if placed too close. If two rows of bubblers are installed one behind the other, then the distance between rows should be about 300 mm.

The formation of the bubbler-induced rotating current is opposed near the sidewall by friction and, in consequence, when operating at large pulls, a breakthrough of insufficiently melted glass will take place along the sidewall.[65]

If a row of bubblers is located at the hot spot where the upward thermal currents are large compared to the flow cylinders being formed, then secondary rotating currents are formed between the thermal and the mechanically induced currents, as shown in Figure 6.71. Currents of this type have been demonstrated on physical flow models.[64] If the mechanical and thermal hot spots are only slightly separated ($< 0{\cdot}3$ of the length of the tank), then the thermal hot spot is strongly deformed and is pulled into the mechanical hot spot; if the distance is greater, the two hot spots will coexist independently.

A row of bubblers reinforces vertical currents and leads to a noticeable levelling out of temperature, layers of glass near the surface being cooled and those near the bottom heated up.

Increasing the air pressure and the diameter of the bubbler will increase the number and size of the bubbles and the rate at which they are formed. A reduction in the viscosity of the molten glass (equivalent to an increase in temperature) reduces the size of bubbles and increases their frequency. In practice, bubbling at a rate of 10–50 per minute is normal; at the higher frequencies a bubble on the surface does not burst until the succeeding bubble has formed inside it (Figure 6.72).

Figures 6.73 and 6.74 show the design and installation of a bubbler in the tank bottom. The bubbler blocks are fusion cast and project about 100 mm into the molten glass; this prevents accelerated attack on the tank bottom blocks by the

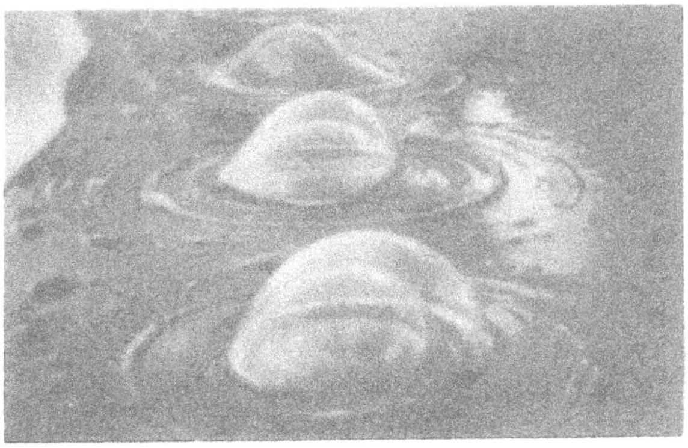

Figure 6.72. Formation of bubbles on the glass surface[67]

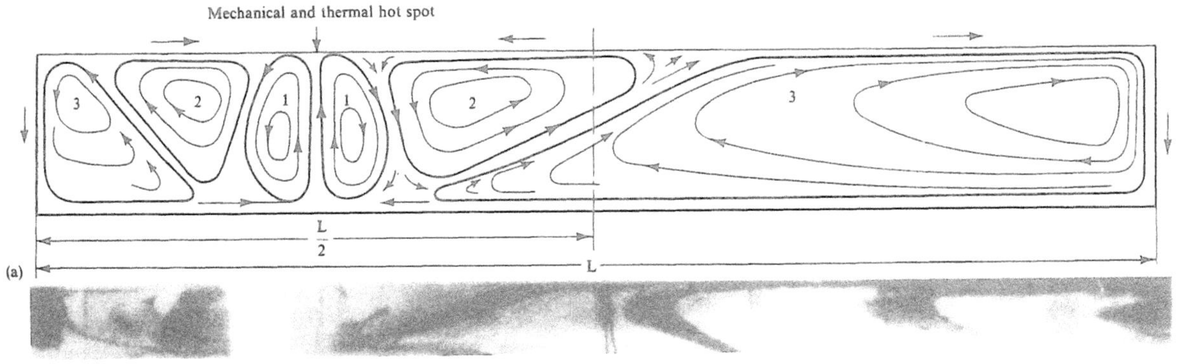

Figure 6.71. Diagram (a) and photograph (b) showing the superimposition of thermal and mechanical currents (the thermal and mechanical hot spots coincide)[64]
(1) rotating currents
(2) secondary rotating currents
(3) thermal convection currents

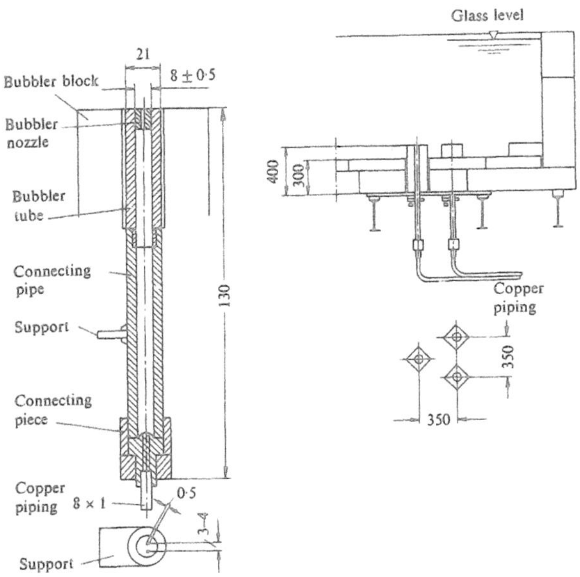

Figure 6.73. Arrangement of bubbler block and bubbler (dimensions in mm)

strong currents set up near the bubbler. The bubbler blocks should be provided with drilled holes through which heat resisting steel tubes of 20 mm outer and 8 mm internal diameter are led; the bubbler nozzle itself consists of a piece of sillimanite inserted into the steel tube with one or two drilled holes of 0·5 mm diameter. In some cases a multiplicity of holes arranged concentrically have been recommended and one particular example has 24 holes each of 0·7 mm diameter.[66]

The pressures used range from 0·5 to 2·0 bar and the amount of air passed through a bubbler is between 80 and 200 dm^3/h. The risk of forming nitrogen bubbles in the glass is sometimes avoided by using oxygen as the bubbling gas. If compressed air from the general plant supply is used care must be taken to remove all oil as otherwise nozzle blockage due to coking can occur.

A bubbling installation is shown schematically in Figure 6.74; small rotometers have been used successfully for measuring gas flows. As with thermocouples, the use of replaceable bubblers in water-cooled holders is also possible (Figure 6.75).

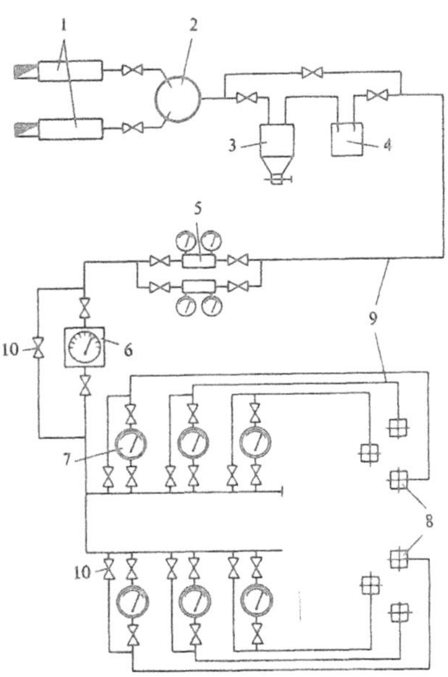

Figure 6.74. Schematic of a bubbler installation
(1) compressor 8–12 m^3/h, 5–7 bar
(2) condensor, 2·5–3 m^3 for 5–7 bar operating pressure
(3) trap for oil and water
(4) filter
(5) pressure regulating valve
(6) flow meter for total quantity of air (l/h)
(7) flow meters for individual bubblers
(8) bubbler blocks with bubblers inserted
(9) supply lines of copper piping (8 mm diameter, 1 mm wall thickness)
(10) valve, positive pressure

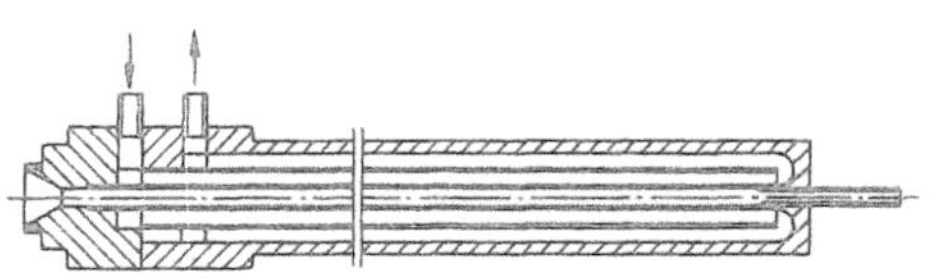

Figure 6.75. Water-cooled bubbler in its holder

Bubbling is started when melting begins in the furnace after the heat up, the rate of bubbling being decided after filling the tank with glass and fixing the operating temperature.

The glass currents shown in Figure 6.71 explain why the bubbling of gases through the tank bottom must be done with care. If bubbling is too strong, foam may be formed above the secondary rotating currents at high pulls or incompletely melted batch may become entrained in the ascending glass currents; these will lead to production problems. If the rate of bubbling is too high, new bubbles form on the inside of those on the surface before they burst (Figure 6.72) and this can lead to seed in the glass which also interferes with production. The best results are achieved with dark glasses because bubbling improves their homogeneity. Opinions differ about the increase in pull that can be achieved as a result of bubbling but in the most favourable circumstances it could be between 10 and 20%. The best performance is achieved when the bubblers are located in line with the hot spot.

6.5.3.1.5 Weirs

The construction of a weir between melting and refining zones enables both to be separated into clearly defined spaces and prevents the formation of the short cut current and therefore the ingress of contaminated glass on the floor into the throat. For many years weirs failed quickly as the refractories were unable to resist the attack by glass at the operating temperatures but the development of fusion cast AZS blocks changed this situation, and although even this material leaves much to be desired, it has made the use of weirs more widespread. However, there are not many papers in the literature reporting on their performance.[68]

In the course of their work on currents, using models, Leyens and Woelk also investigated the effect of a weir on glass currents.[69] A weir forces the molten glass upwards into an area of high temperature and refining is undoubtedly improved. The currents resulting from low and high weirs are shown in Figure 6.76 and it may be seen that a high weir, 0·8H (H = glass depth), gives better refining than a low weir. It is important that the glass at the hot spot, which must lie upstream of the weir, is adequately heated so that, even with the additional heating it receives when flowing across the weir, it does not immediately sink when entering the refining zone, but stays on the surface. Bubblers and electric boosting at the hot spot significantly improve the effect of the weir but, unfortunately, glass attack on the walls of the weir then increases as well.

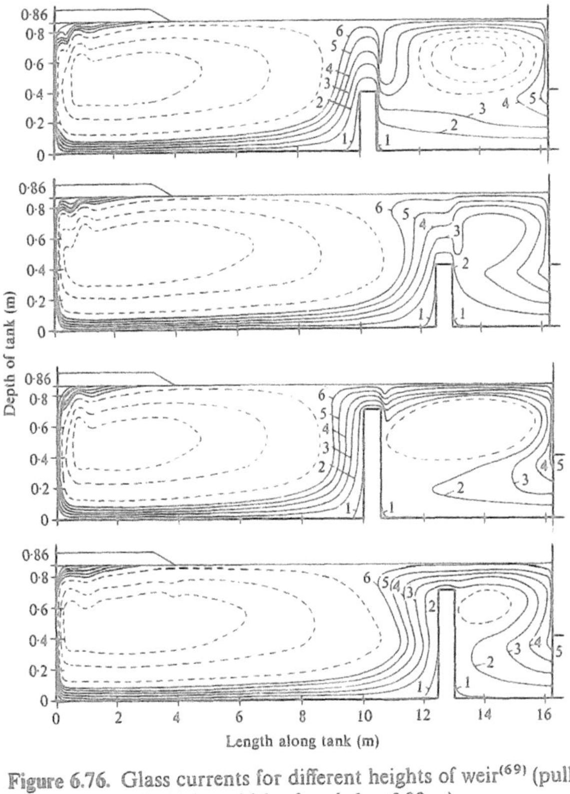

Figure 6.76. Glass currents for different heights of weir[69] (pull = 1·5 kg/s = 130 t/d for width of tank b = 5·20 m)

A combination of bubblers before the weir and a relatively low weir ($\leqslant 0{\cdot}6H$) can cause the secondary rotating currents, typical of bubblers, to form on the downstream side of the weir (Figure 6.77); this forces the withdrawal current downwards into the colder parts of the tank.

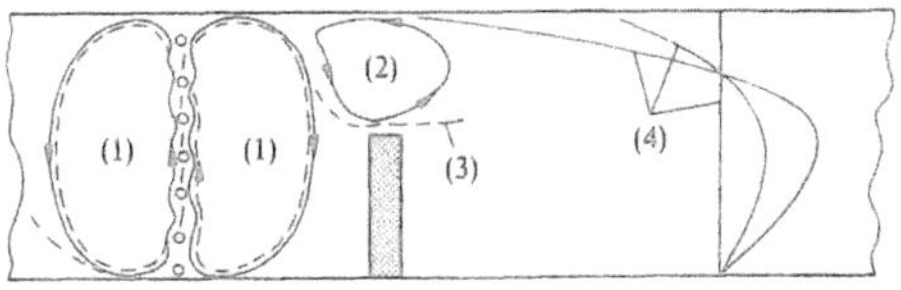

Figure 6.77. Formation of secondary rotating current above the weir[65]
(1) main rotating currents
(2) secondary rotating current
(3) withdrawal current
(4) traces of colour in glass

6.5.3.1.6 The provision of drain holes in a tank bottom

Molten glass and the reactions which take place in the combustion space attack the furnace refractories. Depending on the type and location of the corrosion, the corrosion products cannot always be fully dissolved in the molten glass, fusion cast AZS in particular causing the deposition of primary and secondary crystals and cords near the tank bottom. Glass enriched with corrosion products will be referred to as 'contaminated glass'.

Except in the case of lead glasses, the contaminated glass is denser and of higher viscosity than the bulk of the glass and the thickness of the layer of contaminanted glass has a considerable effect on its flow behaviour. The flow of layers thicker than about 5 cm depends almost entirely on their higher density and the flow and currents in glass above (which is about 1·2 m deep) exert no influence on the flow of the contaminated glass. However, thin layers of less than about 3 cm are strongly influenced by the glass currents above and very thin layers, a few mm thick, behave like thin boundary layers of the same thickness, following the currents near the bottom even up slopes and over steps.

The danger of downgrading the bulk of the glass by entraining contaminated glass from the bottom is particularly acute with layers 3 cm thick and less. The contamination appears in the glass as small stones or bands of parallel cords (cat scratches); these are highly undesirable in tanks manufacturing tableware where visually perfect glass is required. Also, cords in thin-walled articles are particularly objectionable.

The problem can be solved by the installation of a properly designed floor drain to remove the glass near the bottom which is contaminated with alumina and zircon; a special drain block with a hole lined with platinum to prevent wear by glass corrosion is appropriate. A suitable location for such a drain would be a cavity in the bottom near the throat (see Figure 2.3, Section 2.4.1), or in the dead regions immediately behind a weir. If the quality of the glass is the primary criterion, then a continuously operating drain is recommended. However, it is a drawback that the removal of contaminated glass also removes substantial quantities of good glass; the amount drained off depends on circumstances but 3–6% of the pull can be taken as normal.

If it has been decided to drain the tank only periodically, then the drain should be installed where the layer of contaminated glass is thick; this applies particularly to the regions near the tank bottom where there are glass currents. An increase in the thickness of the contaminated layer can be achieved by providing a sunken section in the tank bottom. In order to keep residues of contaminated glass out of the working end it pays to arrange for a weak return current through the throat into the refining zone.

A simple platinum nozzle is usually not suitable for draining continuously at a controlled rate as the quantity being drained off increases with time due to an increase in the temperature of the glass in the area of the drain. The flow rate can be controlled by the use of an electrically heated nozzle in conjunction with some means of cooling. A platinum tube of 18 mm internal diameter has been found suitable for a flow rate of 50 kg/h. The exit nozzle may consist of an electrically heated, heat resisting steel plate with a hole 8–10 mm in diameter; this nozzle plate should be pressed against the refractory drain block and 1·5–2·0 kW are adequate for heating it (see Figure 5.59).

The effect of heating or cooling the area surrounding the drain block greatly depends on the thickness of the layer of contaminated glass. If the bottom is cooled over a wide area near the drain, this increases the flow velocities near the bottom in the direction of the drain, i.e. the 'drain funnel' becomes wider. However, the accelerated flow of contaminated glass is opposed by the higher viscosity of the colder material. For a layer of contaminated glass about 3 cm thick these effects are balanced, for thinner layers the draining effect improves, and for thicker layers the effect worsens. If the bottom temperature is increased, the flow rate of contaminated glass into the drain is reduced for a layer thickness of about 3 cm, but it increases markedly for a layer thickness of 5 cm.

6.5.3.1.7 Depth and colour of glass, and insulation of the tank

Views on the effect of glass depth and insulation of the tank vary due to lack of relevant information. We can assume with some certainty that convection currents are stronger in deep tanks than in shallow tanks from the fact that the Grashof number (Section 6.5.2), which contains the furnace dimensions to the third power, applies. As convection currents become more intense, so the flow velocities increase and the larger volume of molten glass leads to better mixing and reduces the effect of variations in the compositions of batch and cullet: recent work by Leyens[52] and Smrcek[53] have confirmed these points. Two dimensional calculations of the flow lines and isotherms along the axis of symmetry of a glass tank gave the results shown in Figure 6.78 and they have been confirmed by measurements of residence times in an operating furnace. A reduction in glass depth from 0·8 to 0·4 m only slightly affected the currents in the refining zone but significantly weakened the convection current under the batch blanket so that the area covered by the batch blanket had to be increased to achieve the same pull. The transport of heat by convection currents under the batch blanket is substantially reduced, which causes a reduction in energy consumption, but also a certain fluidity of the batch (see Section 7.3.3.2.7). The calculated residence time of the glass is reduced but not as much as the volume of glass.

Good insulation of the tank bottom reduces vertical temperature differences in the molten glass although the horizontal temperature differences, which are governed by the batch blanket and the refining zone, remain unchanged. The retarding viscosity forces are reduced overall, leading to increased convection currents under the batch

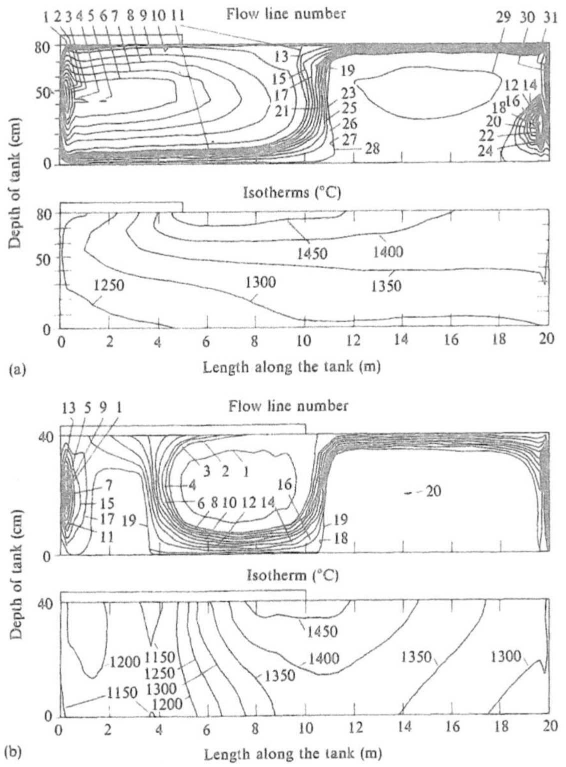

Figure 6.78. Calculated flow lines and isotherms in the vertical central longitudinal cross section of a glass tank[52]
(a) depth of glass 0·8 m
(b) depth of glass 0·4 m

blanket. The hot spot moves towards the throat. Insulation of the tank bottom therefore has an effect similar to that of increasing the depth of glass.

A change in glass colour, if coupled with an increase in iron content, acts like a change in the effective thermal conductivity of the glass. If, as can be assumed, the temperature–viscosity relationships remain unchanged, calculations show that the diagrams of currents in clear and coloured glasses are identical; however, the flow rates for coloured glasses are lower due to their higher viscosity.

6.5.3.1.8 Currents along the sidewalls of a tank

Since a three dimensional analysis has so far not been possible, calculations of glass currents have been restricted to the axis of symmetry of the tank. Scientific knowledge of currents near sidewalls is therefore limited to observations on flow models.

Heat losses through the sidewalls cause the glass along the sides to be cooler than in the centre of the tank, giving rise to convection currents across the tank which are superimposed on the longitudinal currents. Lennertz[65] observed that, when batch feeding causes intense cooling of the glass, there are currents towards the sidewalls just under the glass surface; at the same time a balancing return current flowed towards the batch blanket. If the cooling effect of the batch blanket was small (due to its small size or for similar reasons), currents were observed on and near the glass surface. The partial insulation of sidewall blocks which is now usual means that crosswise currents are only of local significance but confirmation of this view has yet to be obtained.

The erosion lines often observed on tank blocks have so far not contributed to a clarification of currents in the molten glass.

6.5.3.2 Delimitation of working and conditioning zones

After melting and refining, the glass must be reduced to a temperature suitable for working before being fed to machines; in container glass tanks this takes place in the working end or a distributing head and, in flat glass tanks, in the conditioning end.

If, as used to be the situation, the working end follows directly on from the refining zone without a throat, then a strong clockwise rotating current flows from the refining zone right into the working end. In consequence much more glass flows into the working end than is required (Figure 6.79 shows that the flow is about twice the quantity needed). The excess returns to the refining zone and its temperature has to be increased back to the refining temperature, which not only costs energy but also can have adverse effects on the refining process itself. The drawback of return currents was recognised early and the use of a throat reduces the return currents for container and special glasses; in its extreme form, i.e. the sunken throat (Section 6.5.3.3), the return currents can be prevented altogether.

The complete blocking of the return current from the working end into the refining zone can in some cases be a disadvantage in the operation of a furnace. If glass of high quality is required, especially if, as for tableware, it has to be free from cords, then

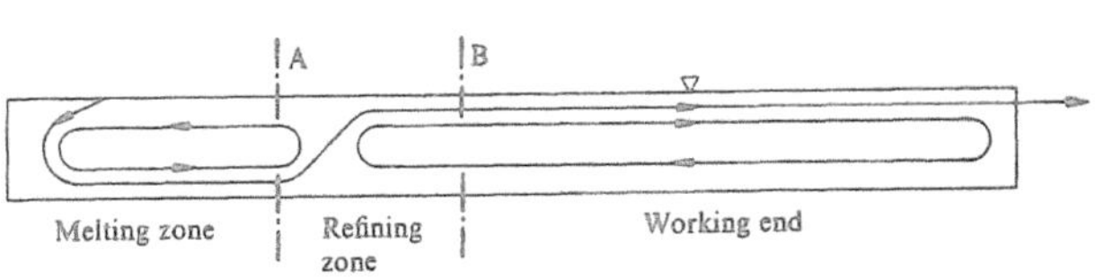

Figure 6.79. Schematic of main glass currents in an open tank

a gentle return current can prevent glass on the bottom, which contains stones and cords, from appearing in the glass to be worked. However, this return current is difficult to control and a better solution to this problem is to install a drain (see Section 6.5.3.1.6).

Flat glass furnaces are operated on the assumption that, for reasons of homogeneity, throats cannot be used; flat glass requires a parallel layer structure which must not be disturbed and forcing the glass through a throat would cause such a disturbance. This called for methods to be designed for separating the melting and the working ends but none of them has so far proved to be nearly as effective as a throat. As a result it has long been considered that the energy consumption of flat glass tanks is considerably higher than that of container glass tanks and, even today, it has not yet reached values normal for container tanks.

A separation of the conditioning zone from the melting end in terms of glass currents is achieved by the use of weirs which are similar in construction and effect to those used in container glass tanks (Section 6.5.3.1.5); designs suggested by Plumat are given in Figure 6.80.

Skimmers have an effect similar to weirs but since they can only be made of fireclay and not fusion cast materials and must float on the glass, they dissolve rapidly and contaminate the glass; their use has therefore been abandoned. Water-cooled steel tubes immersed just below the surface of the glass and used in conjunction with a narrower furnace waist act as skimmers, while their cooling effect is equivalent to a deep barrier.[71] A cross sectional view of a tank with double cooler is shown in Figure 6.81 and Figure 6.82 shows a longitudinal elevation of a tank in which coolers are combined with a sunken barrier.

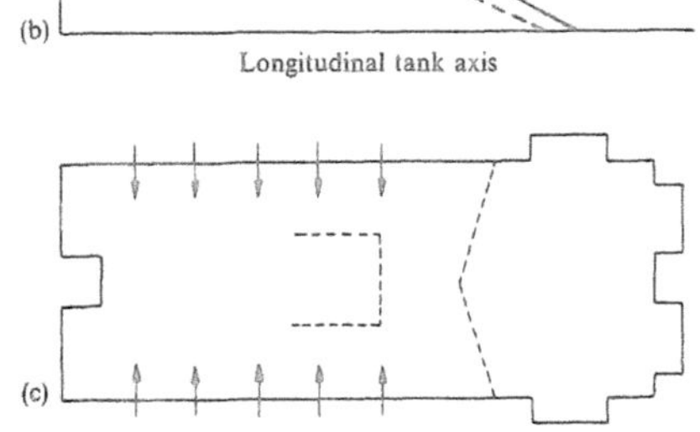

Figure 6.80. Effect of a sunken barrier on currents in a tank furnace[70]
(a) barrier in the return current
(b) temperatures on the glass surface
(c) form of barrier in plan view

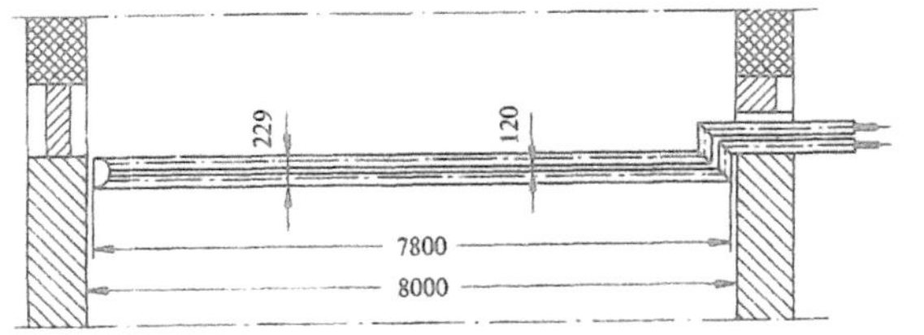

Figure 6.81. Arrangement of water cooling pipes between melting and conditioning ends of a flat glass tank[71] (dimensions in mm)

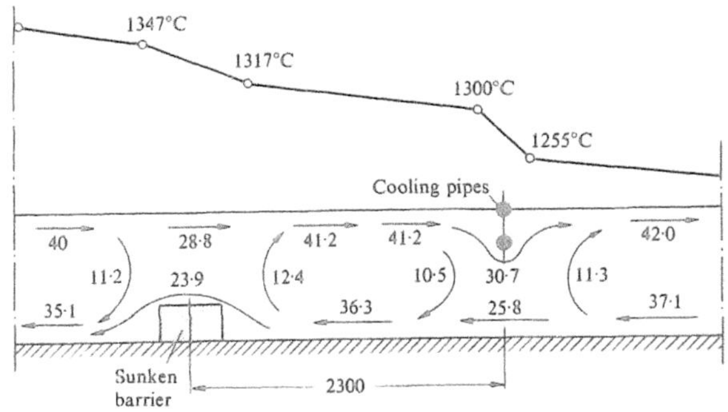

Figure 6.82. Volume flow (m^3/h) and temperatures of a flat glass tank in the zone between the melting and conditioning ends[71]

6.5.3.3 Currents in the throat of a container glass tank

The form taken by currents in the throat of container glass tanks is of major importance for their operation; the most important factor is the mechanical effect of corrosion, the second the magnitude of the return current. The available data apply only to straight throats and were obtained from mathematical considerations on flow models.

Peyches[72] calculated the currents in the throat for a two dimensional situation and gave analytical relationships; Naruse[73] improved Peyches's formula using measurements on models and calculations; the equations of Peyches and Naruse were checked by Burley, Moult & Rawson using modern methods for a two dimensional flow and were found to be in qualitative agreement.[75]

If the height of the throat cross section is h_0, the length l_0, the average temperature difference between entry and exit $\Delta\vartheta$, the viscosty η, the kinematic viscosity ν, the (volume) coefficient of expansion β, and the vertical coordinate of height at the centre of the throat y, then the flow caused by convection is subject to the following velocity distribution $v_K(y)$:

$$v_K(y) = \frac{-g\beta\Delta\vartheta}{6\nu l_0} y\left(y^2 - \frac{h_0^2}{4}\right). \tag{6.33}$$

The velocity distribution caused by the furnace output $v_D(y)$ is given by Naruse as

$$v_D(y) = \frac{-3\dot{M}}{4\rho b(h_0/2)^3}\left(y^2 - \frac{h_0^2}{4}\right) \tag{6.34}$$

where $\dot{M}$ is the average mass flow through the throat, ρ the density of the glass, and b the width of the throat. The velocity distribution for the total flow is approximately

$$v(y) = v_K + v_D. \tag{6.35}$$

The development of Equation (6.33) included the assumption that sidewall friction in the throat could be ignored but Stanek[76] has given a correction factor covering this effect. Furthermore, it assumed that the viscosity is constant within the throat. The velocity maxima often observed in the upper part of the throat on physical models are therefore only included in Equations (6.33) and (6.34) as a very small effect (Figure 6.83).

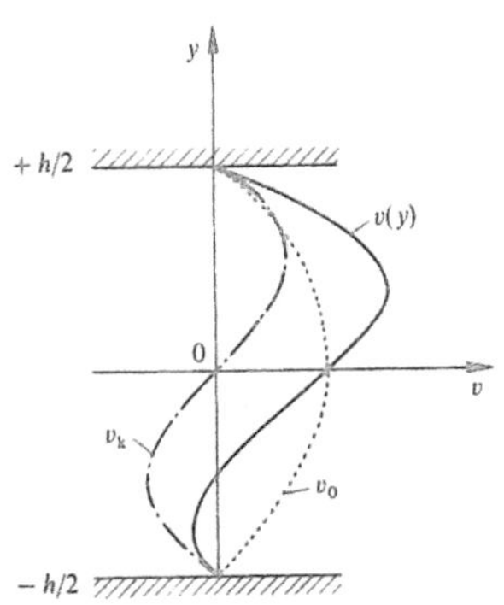

Figure 6.83. Flow profile in the throat; $v(y)$ is the velocity distribution and v_k and v_0 are the components due to convection and withdrawal respectively

6.5.3.4 Currents in the working end and conditioning zone

Little information is available concerning currents in the working end. The rise of currents of hot glass entering from the melting end is well known from observation on furnaces after shut-down (Figure 6.84) and has been observed by Schild[77] on physical flow models. The rising current produces a rotating current known as the Schild current.

The currents in the conditioning zone of a flat glass tank have been investigated in detail[78,80,81] since the quality of the glass flowing to, in front of, and in the drawing machines is strongly influenced by them. There is no homogenising effect of forehearth and feeder, as with container glass tanks, and this function must be taken over by the conditioning zone. Characteristic of all flat glass tanks are skimming pockets between the melting end and the beginning of the drawing channel; the function of these pockets is to prevent glass which has been in contact with refractories from reaching the machines by augmenting the currents across the furnace (Figure 6.85). Some of the contaminated glass enters the return current but some collects on the surface and needs to be removed by skimming. Continuous draining of surface glass is recommended. Skimming pockets may be necessary to ensure that the required level of glass quality is attained; from the point of view of energy consumption their existence is a definite drawback.

The glass currents in the distribution section of a Fourcault tank can be described as forming a fan

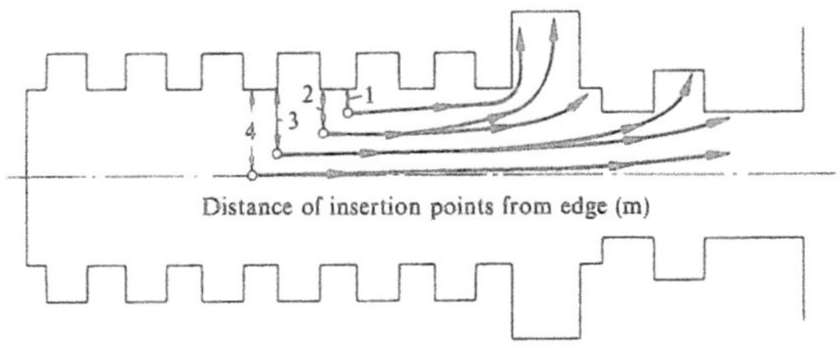

Figure 6.85. Course of surface currents near the skimmer pockets[82]

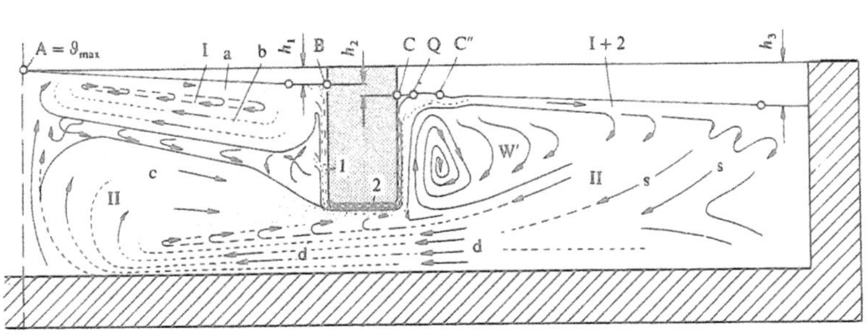

Figure 6.84. Thermal currents in the central zone of a tank on both sides of the throat[77]

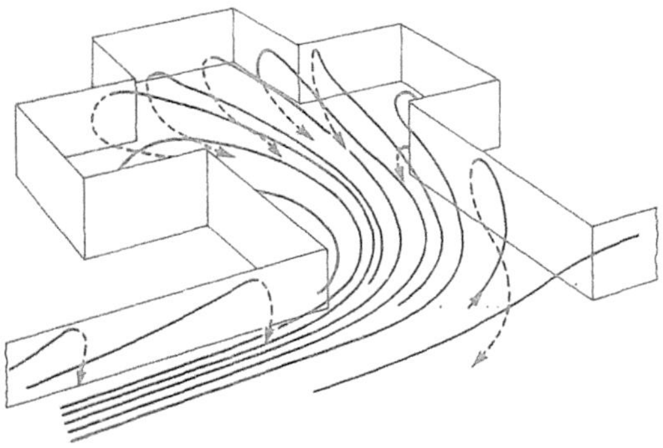

Figure 6.86. Three dimensional diagram of flow in the drawing channel of a Fourcault tank with nine machines[78]

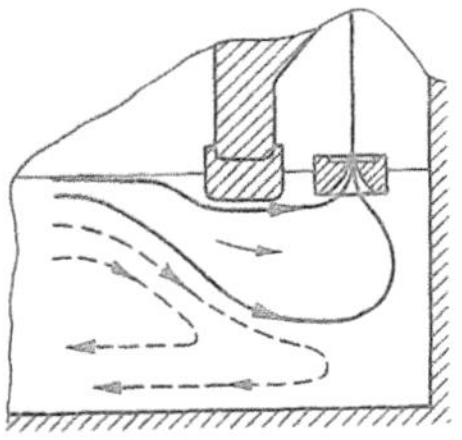

Figure 6.87. Schematic of glass flow in a Fourcault plate glass drawing machine[81]

since, for a correct setting of the furnace with respect to the quantities of glass required for the machines, the distribution flows match the glass consumptions of the individual machines (Figure 6.86). The currents flowing into a drawing machine itself are given schematically in Figure 6.87.

6.5.3.5 Glass currents in electric furnaces

The currents in cold top electric tank furnaces are fundamentally different from those of flame heated tanks. In this type of furnace density and convection currents operate in the same direction and these currents are discussed in Chapter 8.

6.6 The homogenisation and refining of molten glass

As mentioned in the previous Section, currents make a major contribution to the homogenisation of glass. However, the effect is indirect since the high viscosity of the glass and the low flow velocities are not sufficient to create turbulence and convective exchange currents, so the laws of creeping currents apply. The only process which leads to a levelling out of concentrations, and thereby a dissolution of cords, is diffusion. It is the function of glass currents to deform any cords present in the glass and thereby reduce the diffusion times needed for homogenisation so that they lie within the residence time of the glass in the furnace.

The effect of creeping currents on homogenisation requires a consideration of diffusion processes and one of the first to consider this problem was Geffcken[83] who demonstrated that a shear current would produce an effective deformation of cords (Figure 6.88). However, the deformation shown in this Figure leads to an increase in the concentration gradient in the direction in which the thickness of the cords is reduced (Figure 6.89).

During the deformation the lines of equal concentration (1–5 in Figure 6.89) move closer together as the thickness of the cord decreases:

$$\frac{x}{x_0} = \frac{\delta}{\delta_0}. \qquad (6.36)$$

This progressively accelerates the diffusion process and significantly reduces the time necessary for levelling out the concentrations. Cooper[85] showed that the deformation of cords also takes place when the flow cross section is changed, independent of any shear currents that may be operative (Figure 6.90). Rhiel[86,87] was able to show that the relationships given by Geffcken[83] and Cooper[84] were special cases of a generally applicable equation covering deformation. The solution of the differential equation valid for both types of cord

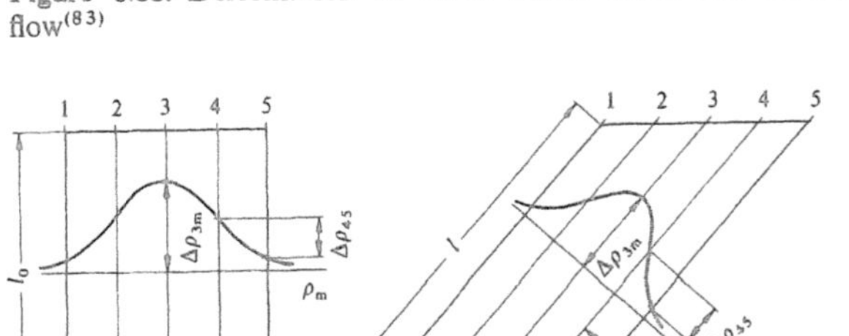

Figure 6.88. Deformation of cords under shear and linear flow[83]

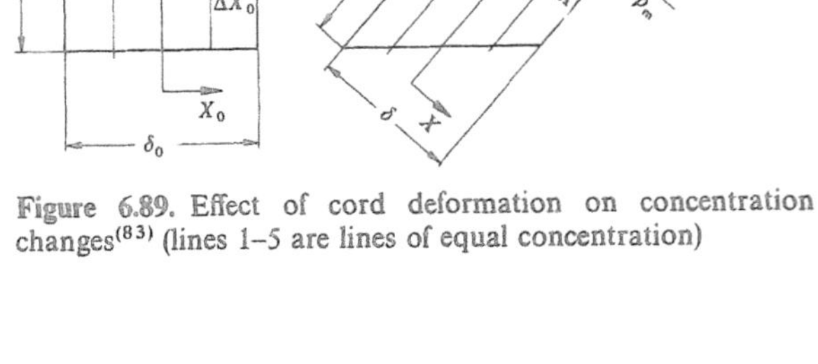

Figure 6.89. Effect of cord deformation on concentration changes[83] (lines 1–5 are lines of equal concentration)

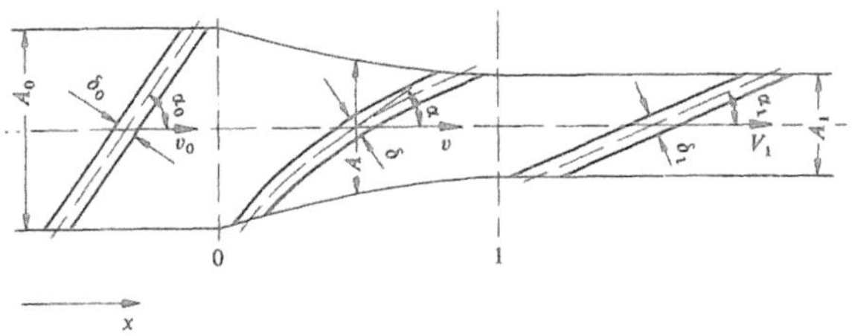

Figure 6.90. Deformation of cords as a result of constricting the flow cross section[86]

deformation is

$$\frac{\delta}{\delta_0} = \left[\left(\frac{A_0}{A}\right)^2 \frac{1+\cot^2\alpha_0}{1+[(A_0/A)^2(\cot\alpha_0 + b_0 t)]^2}\right]^{1/2}. \qquad (6.37)$$

The initial values at $t = 0$ (t = time) are identified by the subscript 0: cord thickness, δ_0, flow cross section, A_0, angle between cord and flow direction, α_0, and shear flow gradient, $b_0 = \delta v/\delta y$.

The relationships given by the general equation for deformation are represented graphically for a reduction in flow cross section in Figure 6.91: a reduction in cross section causes an increase in shear effect while an increase in cross section causes a decrease in shear effect. The deformation of a cord resulting from shear and changes in flow cross section is not the sum of the two individual effects. If, for example, the cord is initially vertical to the direction of flow, then a reduction in flow cross section first produces a thickening of the cord. If shear flow is superimposed then the cord is stretched in the direction of flow so that a reduction in flow cross section results in a thinning out of the cord. The concentration exchange within the cord vertical to its linear expansion is given by

$$\frac{\partial \rho}{\partial t} = D\frac{\partial^2 \rho}{\partial x^2} \qquad (6.38)$$

provided the diffusion coefficient, D, is taken as independent of concentration, which is admissible as a second order approximation. If Equation (6.38) is transformed by using

$$z = \frac{\int_0^t D\left(\frac{\delta_0}{\delta(t)}\right)^2 dt}{\delta_0^2} \qquad (6.39)$$

and the substitution

$$\xi = \frac{x}{\delta} = \frac{x_0}{\delta_0}, \qquad (6.40)$$

then the equation covering diffusion can be written as

$$\frac{\partial \rho}{\partial z} = \frac{\partial^2 \rho}{\partial \xi^2}. \qquad (6.41)$$

If the initial concentration is given by a rectangular distribution as shown in Figure 6.92, then the standardised excess concentration $\Delta\rho/\Delta\rho_0$ at the centre of the cord ($\xi = 0$) is given by

$$\frac{\Delta\rho}{\Delta\rho_0} = \text{erf}\left(\frac{1}{4\sqrt{z}}\right). \qquad (6.42)$$

In practice, the deformation of cords and the diffusion intensity at specific locations in a tank are of little interest. The main interest is to establish how homogenisation takes place along the path of

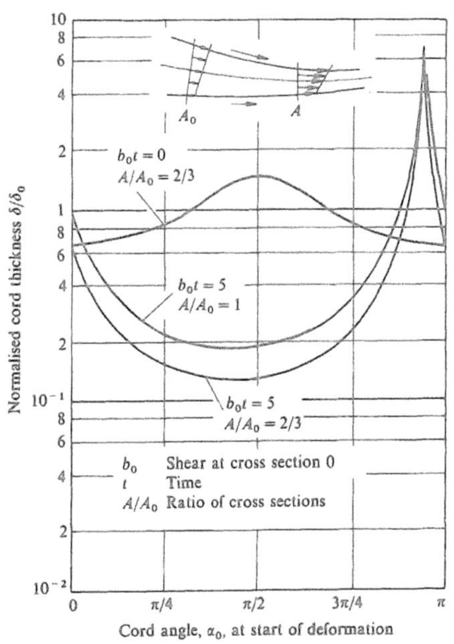

Figure 6.91. Deformation of cords under conditions of shear and constriction of the flow cross section simultaneously[86]

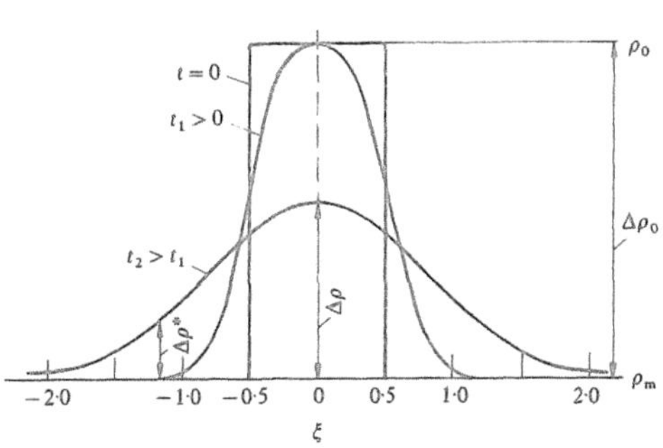

Figure 6.92. Concentration changes for a rectangular distribution of initial concentration[86]

a particle as a function of furnace design and operation. This means that the change in the value of z or, more precisely, of the standardised excess concentration $\Delta\rho/\Delta\rho_0$ must be determined along its flow path in the tank. This presupposes a knowledge of the fields of flow and temperature. According to Leyens & Rhiel[88] it is possible to derive an 'index of homogenisation' (H_k). If a definite figure is assumed for the initial thickness of a cord ($\delta_0 = 1$ mm), and if the residence time of a particle i along its flow path in the tank is t_i, then the index of homogenisation for this flow path is

$$H_{k,i} = \Delta\rho/\Delta\rho_0|_{\delta_0 = 1\,\mathrm{mm},\, t = t_i}, \qquad (6.43)$$

and with Equations (6.39) and (6.42) this gives

$$H_{k,i} = \operatorname{erf}\frac{1}{4\sqrt{\dfrac{\int_0^t D(\delta_0/\delta(t))^2\,dt}{\delta_0^2}}}. \qquad (6.44)$$

Rhiel calculated the index of homogenisation for the flow model of Leyens[50] which is valid under simplified assumptions for the central plane of a tank. The cord was assumed to be of alumina whose diffusion coefficient can be calculated to be

$$D = 10^{-6}\exp\left(-4{\cdot}228 - \frac{16830}{T(\mathrm{K})}\right) \quad (\mathrm{m}^2\,\mathrm{s}^{-1}). \qquad (6.45)$$

The flow patterns on which the calculations are based are given in Figure 6.93 and a calculation of cord deformation in the melting end along flow line 2 gave the progress shown in Figure 6.94. Attention is drawn to the strong deformation in the initial phase caused by the strong shear currents in the left descending branch of the rotating current under the batch blanket. For this reason it was assumed that the cord was formed in the region of melting batch under the blanket. A calculation of the index of homogenisation as a function of the position of the doghouse leads to the relationship shown in Figure 6.95.

Currents near the sidewalls have low velocity and high shear, which causes variations in concentration to reduce quickly and gives good homogenisation. The deciding factor is the influence of the currents below the main anticlockwise currents. As the distance from the tank bottom increases so shear and residence time are reduced and the index of homogenisation becomes larger, indicating a worsening situation.

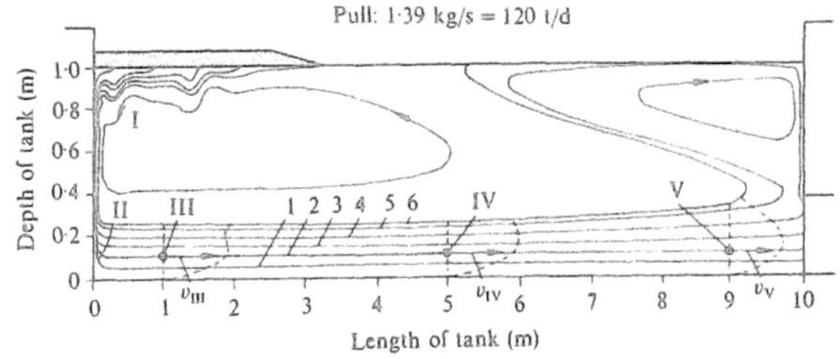

Figure 6.93. Flow lines in a plane of symmetry of a glass tank[86]

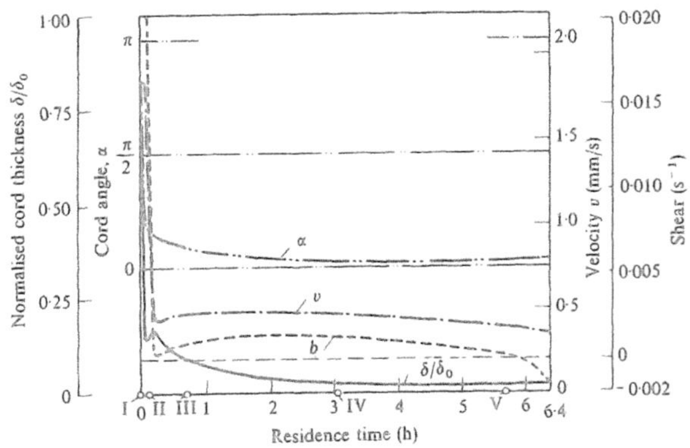

Figure 6.94. Deformation of cords in the tank furnace along flow line 2[86]

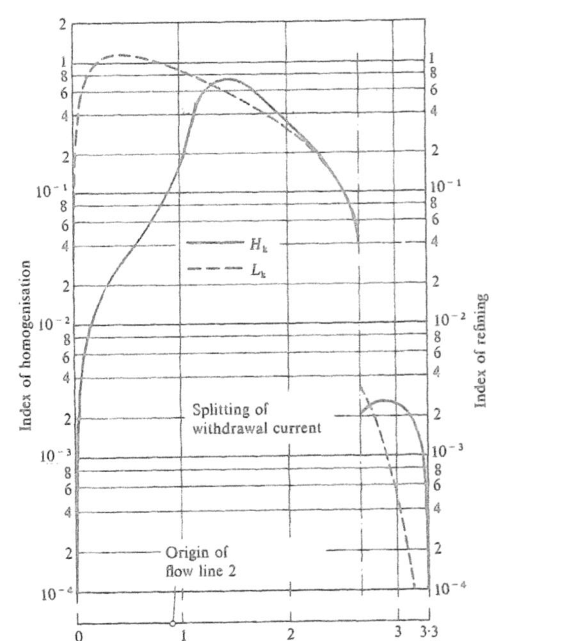

Figure 6.95. Indices of refining and homogenisation[86]

In the above flow model the lines of flow divide before the throat, and a cross section between these lines can be defined for which the following applies. All particles which melt to the left of this cross section pass directly to the throat whereas particles to the right follow the path along the open exposed glass surface. The longer path and the longer residence time that this implies as well as the higher temperature prevailing at the glass surface accelerate the diffusion process and reduce the index of homogenisation. The initial increase in concentration differences for currents forming to the right of the cross section arises from the fact that shear at the exposed glass surface tends to zero due to the absence of shear forces. The subsequent rapid decrease can be explained on the basis of the high shear to which the current following a path near the sidewall towards the throat is subjected.

6.7 Mixing processes in glass tank furnaces

6.7.1 Experimental results

The overall picture, which can only be described in outline, is one of complex mixing processes brought about by the interplay between withdrawal and convection currents and governed by the geometry of the furnace which imposes certain limitations. In operation, the mixing processes are important when a change in batch formulation, whether intended or accidental, occurs, when the proportion of cullet is changed, or when the colour of the glass is changed. Taken together, the mixing behaviour of a furnace provides information about glass currents, rotating currents, and stagnant areas where there is little or no flow.

Mixing can be investigated experimentally by adding tracers or by monitoring progress during batch changes. Basically, the mixing processes in a bath of glass are analogous to the mixing of gases in the combustion chamber (Section 6.1) and consequently the mathematical formulae are similar.

The mixing process has been investigated under operating conditions on a number of occasions. As an example, Figure 6.96 shows the residence time distribution, $C'(\tau)$, obtained from measurements using isotopes, the transition function, $C(\tau)$, and, for comparison, the mixing curve for a perfect mixer $1-\exp-\tau$; the data come from a cross-fired tank with electric boosting, a melting area of 133 m^2,

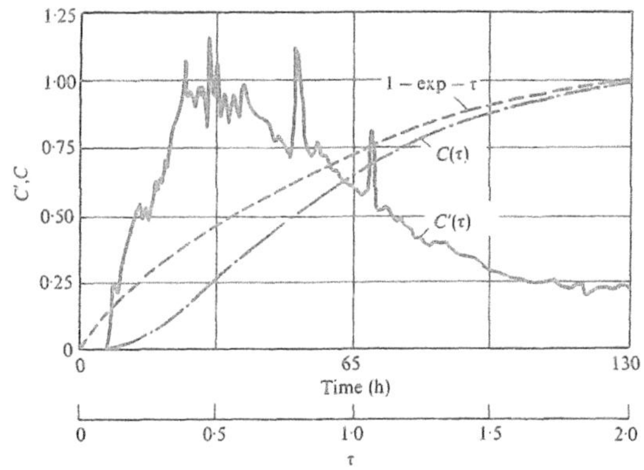

Figure 6.96. The residence time and transition function of a cross-fired tank furnace measured by means of isotopes[52,53]

and a very low pull of 0·8 t/m^2d.[52,53,62] The abscissa shows the measured time from when the tracer was added and the normalised time, $\tau = t/t'$, where t' is the average residence time, independent of the residence time distribution, and is equal to the ratio of tank volume to volume flow ($t' = V/\dot{V}$). The ordinate gives $C'(\tau)$ and $C(\tau)$ taken in a feeder. The following equation applies:

$$C'(\tau) = (c_\tau - c_0)\frac{V}{M_c} \qquad (6.46)$$

where c_0 is the concentration of the substance at a time $\tau = 0$ (before feeding), c_τ the concentration of the substance at time τ of its withdrawal, and V/M_c the ratio of the volume of the tank to the mass of the tracer. The literature often uses the terms $E(\tau)$ and $F(\tau)$ instead of $C'(\tau)$ and $C(\tau)$, especially for the mixing of gases, and in Section 6.1, where analogous formulae appear, $E(\tau)$ is used; in this section $C'(\tau)$ and $C(\tau)$ are used. The term $C'(\tau)$ gives the factor of the concentration which would result if the tracer were distributed uniformly throughout the tank immediately after feeding. The rapid increase in concentration, after some delay, is readily seen as well as the decline after passing through a maximum. The curve for $C'(\tau)$, the residence time distribution, gives information about currents in the tank; the shorter the dead time, T, the faster the flow of a part of the glass through the tank but on average $T \approx 0{\cdot}1t'$. Electric boosting, which acts as a barrier, extends the flow time through the tank while a row of bubblers can reduce it.[52]

Integration of C' gives the transition function

$$C(\tau) = \int_0^\tau C'(\tau)\,d\tau \qquad (6.47)$$

which corresponds to the integral of the residence

time distribution divided by time, and gives the total quantity of molten glass which has been withdrawn from the tank by the time τ.

If at time $t=0$ or $\tau=0$ the change in concentration is introduced, not temporarily, but on a permanent basis, then it is recommended that the dimensionless concentration, $\hat{C}(\tau)$, should be used, which corresponds to Equation (6.46),

$$\hat{C}(\tau) = \frac{c(\tau) - c_0}{c_\infty - c_0} \qquad (6.48)$$

where $c(\tau)$ is the concentration of the tracer at the time of its withdrawal, τ, c_0 the initial concentration, and c_∞ the concentration in the batch (adjusted for glass) which would exist after an infinitely long time after making the change. The literature seldom provides examples of dimensionless concentrations since long-term batch changes are usually associated with colour changes which are usually carried out by initially overcompensating (see Section 6.8).

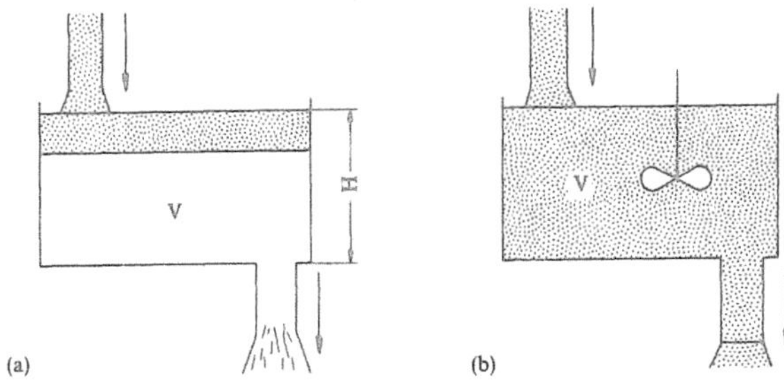

Figure 6.97. Simple mixing processes in containers
(a) no mixing (piston flow)
(b) ideal mixing

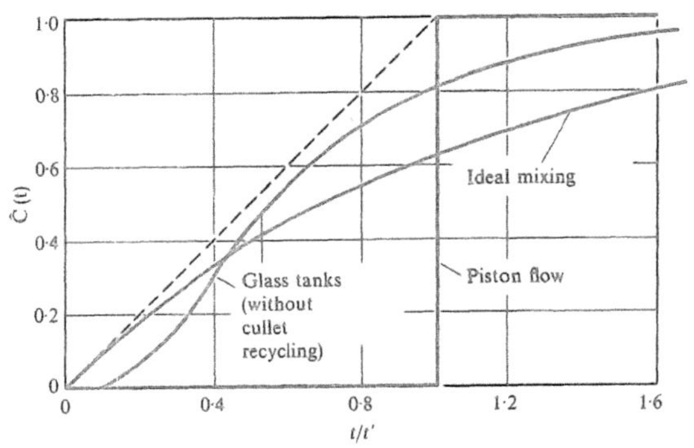

Figure 6.98. Response curves[92]

6.7.2 The performance of a perfect mixer

The experimental data collected on operating furnaces and covering changes in the concentration of a tracer with time can only be applied to a limited extent to the mixing processes and currents existing in a tank. In addition, the integration of C' provides values which, after analysing the individual processes involved, do not provide an unambiguous solution. Despite these limitations, valuable conclusions can be drawn.

The two simplest theoretical and formulated mixing processes in cells are piston flow and perfect mixing (Figure 6.97): with piston flow no mixing whatsoever occurs between the new material and that already in the cell while with perfect mixing the opposite applies, i.e. the new material mixes completely with that in the cell. The two response curves, $\hat{C}(t)$, are given in Figure 6.98. In perfect first order mixing (only one cell)

$$\hat{C}(t) = 1 - \exp(-t/t'). \qquad (6.49)$$

If the impulse is only given for a short period then the measurement of concentration with time at the point of withdrawal will give

$$\hat{C}'(t) = \frac{d\hat{C}(t)}{dt} = \frac{1}{t'}\exp(-t/t'). \qquad (6.50)$$

The mixing characteristics of a tank furnace expressed as $\hat{C}(t)$ need not always be smaller than those for the perfect mixer of one cell for all t. There are crossover points as shown in Figure 6.98 and this is easily understood from the fact that the material being fed into the furnace moves preferentially along a central zone to the point of withdrawal and that there are many stagnant zones. This behaviour can be explained by means of mixers of higher order (in which several cells are coupled one behind the other) or by using a perfect mixer of the first order with time delays and stagnant zones.

For mixers with several cells (n) coupled in series Cooper[92] gives

$$\hat{C}(\tau) = 1 - \sum_{j=1}^{n} \frac{(n^{t/t'})^{j-1}}{(j-1)!}\exp(-nt/t'). \qquad (6.51)$$

For an impulse of short duration, differentiation gives

$$\hat{C}'(\tau) = \frac{n^n (t/t')^{n-1}\exp(-nt/t')}{(n-1)!}. \qquad (6.52)$$

The number of cells can be used to define a coefficient of mixing, i.e.

$$M_{comb} = \frac{1}{\sqrt{n}}. \qquad (6.53)$$

As shown in Figure 6.99, the increase in concentration becomes steeper as the number of cells increases and, in the limiting case of an infinite number of cells, becomes piston flow.

If the problem is approached from the position of a perfect mixer and includes contributions from piston flow and stagnant zones, then

$$\hat{C}(t) = 1 - \exp\left[-\frac{V}{V_M}\left(\frac{t}{t'} - \frac{V_K}{V}\right)\right] \qquad (6.54)$$

where V_K/V is a measure of the contribution of piston flow and V_M/V is the portion allocated to perfect mixing, with reference to the total volume. Of the total volume, V, the flow in the partial volume, V_K, behaves as piston flow and the term V_K/V acts as a time delay for $\hat{C}(t)$. Since $t' = V/\dot{V}$, and $\dot{V}$ is the same for regions of piston flow and perfect mixing,

$$\frac{V_M}{V} = \frac{t_M}{t'} \text{ and } \frac{V_K}{V} = \frac{t_K}{t'}. \qquad (6.55)$$

Hoetink[93] carried out residence time measurements on a television tube tank and determined times for t_M and t_K: in the absence of cullet and with $t'_{total} = 50$ h, $t_M = 40$ h, and $t_K = 6$ h

$$\hat{C}(t) = 1 - \exp\left(-\frac{t(\text{h}) - 6}{40}\right) \qquad (6.56)$$

and when feeding with 50% recycled cullet

$$\hat{C}(t) = 1 - \exp\left(-\frac{t(\text{h}) - 10}{100}\right), \qquad (6.57)$$

for $t_K = (4 + 6)$ h with an average residence of the cullet in the storage silo of 4 h. At 100 h, t_M is significantly higher when operating with cullet and the recycling of cullet therefore significantly reduces the increase in concentration. In principle the single cell mixer with partial piston flow can be replaced by a multicell mixer as was shown by Joosen[94] but the calculations become much more complicated.

Cooper[92] has pointed out that the influence of laminar flow on mixing should not be ignored but for the purpose of calculations used in practice, the concept of a single cell mixer with time delay should be adequate.

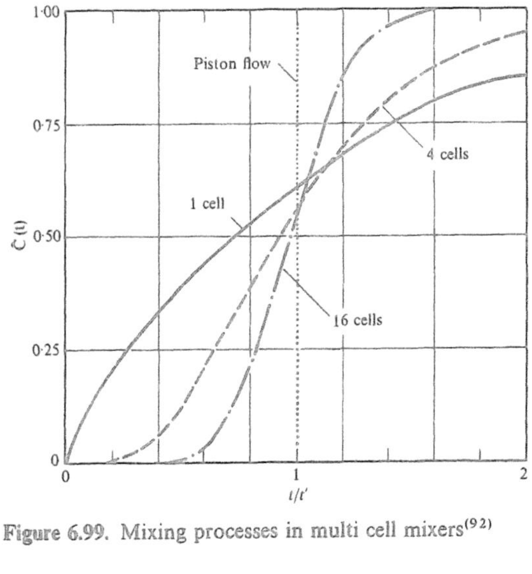

Figure 6.99. Mixing processes in multi cell mixers[92]

6.7.3 The mixing process during accidental variations in the composition of batch or cullet

It is impossible in practice to avoid occasional variations in the composition of the batch charged into a furnace and if foreign cullet is used, these variations become larger. It is important to know the effect that these variations will have on the composition of the glass withdrawn and Nölle[95] has investigated this question.

If it is assumed that each mix of batch is itself homogeneous, but that there are occasional variations between mixes, that the time of charging is t_0, and that the glass composition varies around an average value of σ_1 (Figure 6.100), then a charging function such as that shown in Figure 6.101 is applicable. The consequences of variations in batch composition are periodic fluctuations in the composition of the glass withdrawn (the withdrawal time function). The occasional variation in the composition of the glass withdrawn cannot be predicted exactly, but the standard deviation, σ_2, of these variations can be calculated provided that information on the mixing behaviour of the tank can be obtained with the help of the theory of

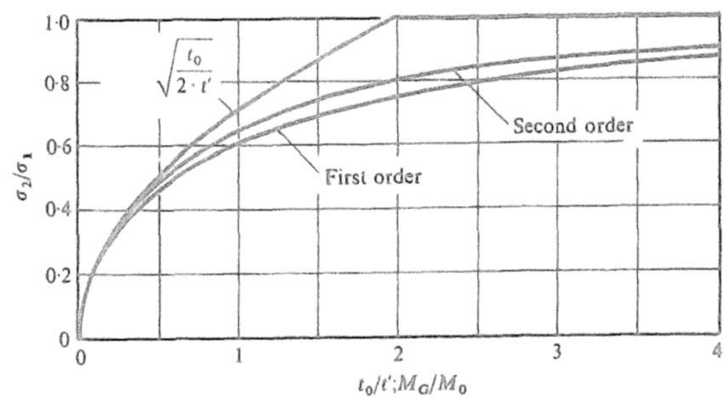

Figure 6.100. Relationship between the standard deviation of different mixers and mass ratios[95]

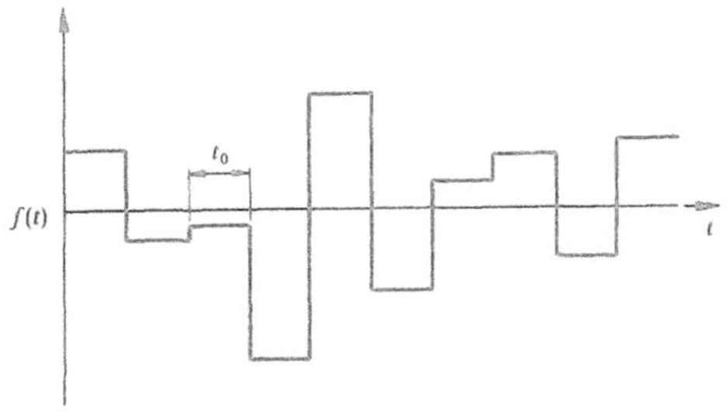

Figure 6.101. Charging functions of individual batch mixes[95] (t_0 is the time needed for charging one mix)

stochastic processes. The frequency spectrum of the feeding process can be used in conjunction with the frequency of the system (in this case the tank) to calculate the frequency spectrum of the withdrawal process. Although the mathematical skills required are considerable, the results are simple and easily appreciated.

For first order mixing

$$\left.\left(\frac{\sigma_2}{\sigma_1}\right)^2\right|_1 = 1 - \frac{t'}{t_0}[1 - \exp(-t_0/t')] \qquad (6.58)$$

where t' is the average residence time corresponding to $t' = V/\dot{V}$, or $t' = M_0/\dot{M}$ with M_0 being the mass content of the tank. For second order mixing

$$\left.\left(\frac{\sigma_2}{\sigma_1}\right)^2\right|_2 = \left[1 + \frac{1}{2}\exp\left(-2\frac{t_0}{t'}\right)\right] - \frac{3\,t'}{4\,t_0}\left[1 - \exp\left(-2\frac{t_0}{t'}\right)\right], \qquad (6.59)$$

while, for both mixing systems,

$$\lim_{(t_0/t') \to 0} \left(\frac{\sigma_2}{\sigma_1}\right)^2 = \frac{1}{2}\frac{t_0}{t'}. \qquad (6.60)$$

For higher orders and for small t_0/t' the factor $\frac{1}{2}$ is replaced by 0·56 for $n = 3$ and 0·62 for $n = 4$. As an approximation for $t_0 < 2t'$

$$\frac{\sigma_2}{\sigma_1} = \sqrt{\frac{1}{2}\frac{t_0}{t'}} \qquad (6.61)$$

or, if it is recalled that $\dot{M} = M_G/t_0 = M_0/t'$, where M_G is the mass of a batch mix, then

$$\frac{\sigma_2}{\sigma_1} = \sqrt{\frac{M_G}{2M_0}}. \qquad (6.62)$$

Figure 6.101 shows that, for small t_0/t', the mixing characteristics of the system (the tank) have a small, and at the limit a negligibly small, influence; this conclusion is important in practice since it confirms the validity of the interrelationships. The effect of local differences, e.g. cords, although of practical importance, cannot be assessed by this method.

The effect of a closed recycling circuit for a tank's own cullet (cullet proportion s) without storage time can be assessed by the above equations if t' is replaced by

$$t'_s = \frac{t'}{1 - s} \qquad (6.63)$$

from which it follows that fluctuations in batch composition are only slightly reflected in fluctuations in glass composition. If a tank contains 40 t glass, a batch mix is 200 kg, and the weight loss on melting is 23%, then the ratio σ_2/σ_1 is 0·024.

6.8 Changing glass colour in tank furnaces

Commercial or operational reasons will sometimes make it necessary to change the glass in a tank furnace from one type to another; in most cases the basic glass composition remains the same, only the colour being changed. This applies in the main to container glass but also increasingly to flat glass. In the case of special glasses most changes involve changes in the type of glass.

During the transition period from one type of glass to another, or from glass of one colour to glass of another, no saleable glass can be produced; the aim is therefore to effect the change in as short a time as possible and to obtain waste glass of a type that can be used under certain conditions. Procedures for effecting colour changes have been developed in the main on an empirical basis.

6.8.1 Draining and refilling of the tank

A changeover from one type of glass to another can be carried out relatively easily by draining the tank and refilling it with the new glass composition. Cleaning the tank with cullet can be useful for tanks of complicated geometry or if one or other of the glass compositions tends to foam. Faulty colours only arise to a minor extent. The disadvantages of this procedure are the long time required, the thermal stresses exerted on the bottom of the tank, and, in the case of large tanks, the problem of removing and storing the drained glass which becomes almost unsurmountable. Since tanks, especially for container and flat glass, are increasing in size, it is only with the smaller tanks melting special glasses and tableware that there remains a need for draining and refilling. Any such furnace installation should be fitted from the beginning with adequate and accessible draining facilities and equipment for the removal and storage of cullet. Chargers suitable for feeding batch and cullet for refilling the furnace must also be provided.

6.8.2 Colour changes during operation

Improved operational control of the melting process, the size of modern tanks, and the need to carry out a change as quickly as possible have meant that large container and flat glass tanks have to undergo changes in glass colour while production operations continue. To do this successfully it is necessary to know the interrelationship between melting and mixing characteristics of the furnace.

6.8.2.1 The mixing characteristics of glass tank furnaces[96–100]

The different fundamental mixing characteristics of a tank discussed in Section 6.7 are noted again here and Figure 6.102 gives the mixing characteristics of all the types of mixer described below.

An ideal mixer with one mixing cell will approach the new composition immediately following the charging of the new batch although, theoretically, completion is only reached after an infinitely long time. The change in concentration, c, with time of the changed constituents is given by

$$\hat{C}(t) = \frac{c - c_1}{c_2 - c_1} = 1 - \exp\left(-\frac{t}{t'}\right) \tag{6.64}$$

where c_1 is the original composition on charging and c_2 the new composition on charging, t' is the ratio of the volume of the tank to the volume rate at which glass is withdrawn, and t is time.

If piston flow occurs, the new composition pushes the old one out in front of it without any mixing taking place and the concentration c_1 jumps after a time $t = t'$ to a new concentration, c_2. Mixing may therefore be imagined as taking place in an infinite number of very small perfect mixers connected in series, as has already been indicated in Section 6.7.

Another limiting condition is presented by the perfect mixer with stagnant zones. In this type of mixer only a part of the tank is involved in the mixing process, the rest remaining largely unaffected.

Experience has shown that the mixing characteristics of glass tank furnaces are a mixture of the three types of mixer described above, which type predominates depending on the design and method of operation of the specific furnace installation. In general, stagnant zones are few and one can assume that a combination of both piston flow and perfect mixing will occur, so that it is often possible to think of the mixing as a succession of two or three perfect mixers in series.

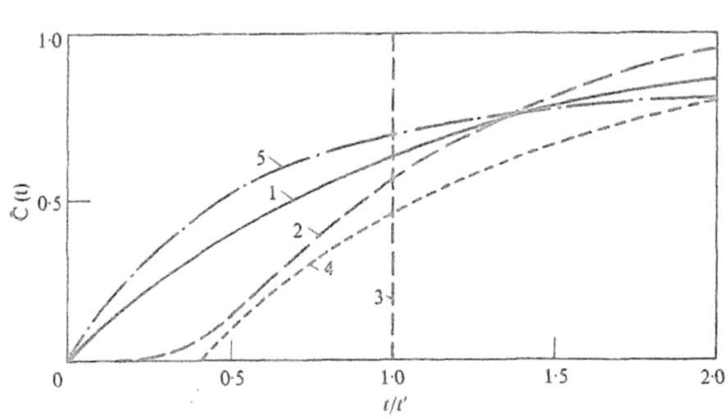

Figure 6.102. Mixing processes for different types and combinations of mixer
(1) perfect mixer, one cell
(2) perfect mixer, two cells in sequence
(3) piston flow, an infinitely large number of cells coupled one behind the other
(4) perfect mixer with initial piston flow
(5) mixer with stagnant zones

6.8.2.2 Influencing a change in colour

In practice there is the question of how to accelerate the mixing process by operational procedures. Figure 6.102 shows that piston flow is the fastest but unfortunately, no techniques exist by which this can be used although piston flow can be achieved up to a point if the density of the new glass is made higher than that of the old. However, such a procedure has been shown to produce very cordy glass, which causes major problems during working, and is now only used in special circumstances, examples of which are given by Tooley.[101]

Something approaching piston flow can be achieved by the use in series of a succession of containers with stirrers. In practice this means the use of stirrers in the conditioning zone or the working end and, as this procedure is a practical possibility, it is occasionally carried out. Similar results may be obtained by the use of bubblers or electric boosting.

A very effective procedure, and easy to implement, is the controlled overcompensation for colour for a time. If the concentration difference, $c_2 - c_1$, is enlarged by a factor B

$$\frac{c - c_1}{c_2 - c_1} = B\left[1 - \exp\left(-\frac{t}{t'}\right)\right] \tag{6.65}$$

applies: the acceleration resulting from the overcompensation is shown in Figure 6.103 in which, in terms of numbers, the reduction in time is given as a reduction from $t/t' = 2$ (see Table 6.3). To avoid excessive colour effects, overcompensation could be reduced or avoided altogether if $\hat{C} \approx 0{\cdot}9$. Overcompensation can only be used if the change can be effected by the addition of constituents which were previously absent or were present only in small quantities, e.g. a change from clear to green glass. If the change calls for the removal of a component (green glass to clear glass) then this procedure cannot be used.

On occasions it pays to overcompensate for a short time when changing glass compositions, as this enables the target composition to be reached more quickly.[93] In one instance a change to a glass containing 8·8% PbO was required after melting a lead-free glass; the changeover was started with a batch for a glass with 18·4% PbO followed by a batch for one with 10·9% PbO after 100 h. The proportion of cullet recycled was 50%, the pull was 96 t/d, and the storage time of the cullet in its silo was 4 h. With a time delay of 6 h the mixing characteristics of an ideal mixer could be reproduced in the initial phase as is shown in Figure 6.104 (see also Equations 6.56 and 6.57).

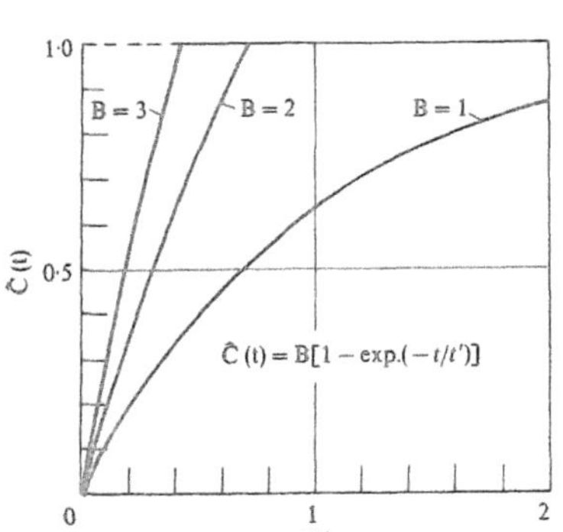

Figure 6.103. Curves for colour changes for a stirred vessel and varying overcompensation for colour[96]

Table 6.3. *Time reduction factors linked to overcompensation of colouring constituents*

B	t/t'
1	2
2	0·69
3	0·41
4	0·29

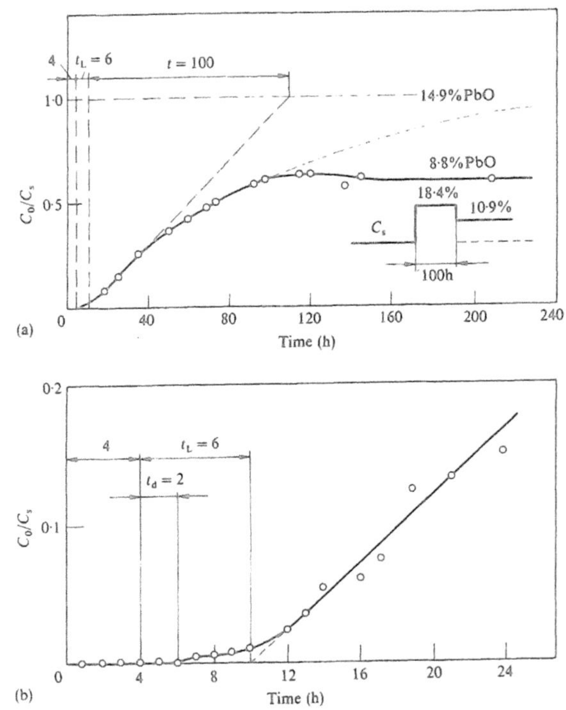

Figure 6.104. The process of mixing in a tank of 200 t capacity when changing from a lead-free to a lead glass (after Hoetink[93])

One parameter which is important in the changing of colour is the ratio t/t'; a high pull causes a reduction in t' and, for constant B, a reduction in t, and will therefore accelerate the change in colour.

6.8.2.3 Procedures for changing colour

Since draining, cleaning, and the melting of a new composition are now routine and not subject to any special difficulties, the following will only deal with colour changes carried while production continues. As already mentioned, a high pull and improved mixing by the use of bubblers or electric boosting accelerates the colour change but large changes in density are best avoided as they cause cords.

Changing colour involves not only a change in concentration of certain colouring constituents but also changes in the chemical composition and redox potential of individual glasses; consequently the changeover process can be divided into separate phases in sequence. The actual colour change is then only a part, usually the final phase, of the change. Machine operators find it easier to cope with the changing working properties of the glass if they are carried out in discrete, small steps.

In this connection the solubility of sulphur in the glass is particularly important. The work of Budd[102] and Manring[103] showed that the partial oxygen pressure of the molten glass, expressed as the $Fe^{2+}/(Fe^{2+} + Fe^{3+})$ ratio, exerts a strong influence on the solubility of sulphur and passes through a minimum in the region between oxidising and reducing situations (Figure 6.105). This means in practice that a change from an oxidising (clear or green) glass to a reducing (amber) glass or vice versa is likely to involve a considerable evolution of SO_2 together with a high level of foam.[104] Changes from reducing to oxidising formulations are likely to cause fewer problems than vice versa, since the oxidising glass, as it melts on top of the old glass, can absorb and dissolve the rising reboil bubbles due to the higher solubility of sulphur in the new glass. The oxidising glass should therefore be of lower density than the reducing glass already in the furnace. In the reverse situation the formation of a large amount of foam must be expected but by adjusting the furnace atmosphere to reducing, the foam containing SO_2 is relatively easily dissolved. If the residence time characteristics of the tank are taken into account, the transition period can be reduced to a minimum.[104] The solution mechanisms involved have been discussed by Mulfinger.[105]

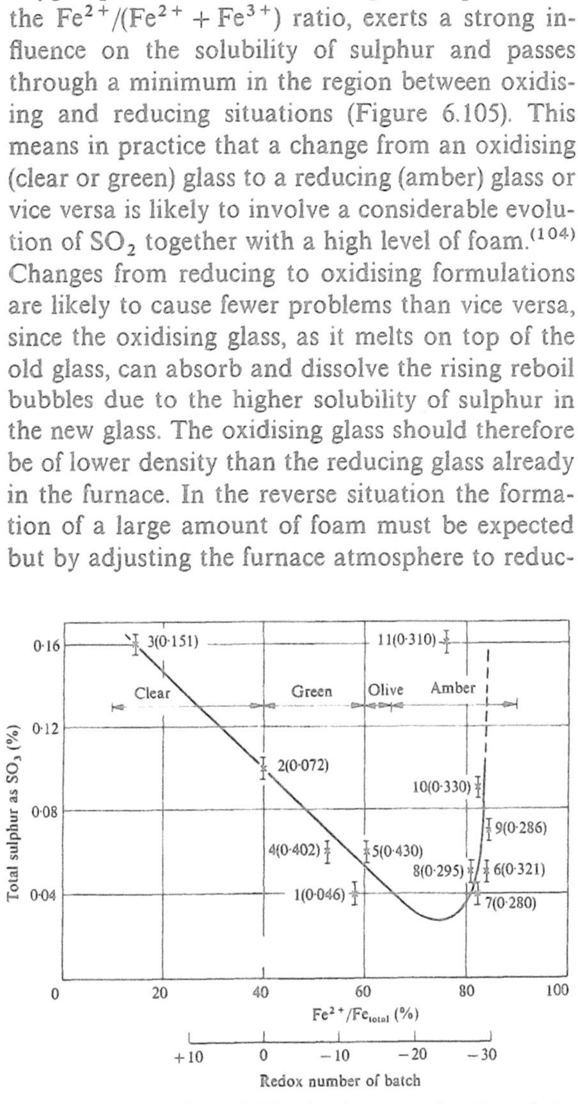

Figure 6.105. Solubility of SO_3 in glass as a function of the oxidising/reducing conditions[102,103] (the figures in parentheses are $\%Fe_2O_3$)

(1) clear (reducing)
(2) clear
(3) mid green
(4) green
(5) green (with foaming)
(6) olive
(7) light amber
(8) normal amber
(9) dark amber
(10) dark amber
(11) very dark amber

6.8.2.3.1 Examples of colour change procedures

6.8.2.3.1.1 Container glass[98,100]

Clear glass to amber (carbon–sulphur amber) glass

This transition can be carried out in five steps. The first is to increase the pull so that the total tank volume is turned over in 24 h; this means that $t/t' = 1$. All equipment which can assist mixing, e.g. bubblers, should be brought into operation.

The next three steps involve adjusting the composition of the old glass to that of the new, a reduction in the redox number by reducing the sulphate content of the batch, and removing if present all oxidising batch constituents such as sodium nitrate and arsenic.

The fifth step is the colour change itself: the redox number now sinks to values of −30 and the glass tends to foam. In order to compensate for the oxidising influence of bubblers, the carbon content must be set at 10% above normal but overcompensation using pyrites is not recommended as the risk of foaming is very much increased. This type of colour change can be carried out in 100 h, and saleable glassware can be produced after 90 h.

Amber to clear glass

The separate steps are analogous to those for the change from clear to amber, the Fe_2O_3 content being reduced step by step from 0·33 to 0·12 wt% and the sulphate increased. This colour change requires about 120 h but it must be remembered that the time required depends, apart from the behaviour of the tank itself, on an assessment of the quality of the glass melted.

Green to amber glass

Similar to those above, this change is also carried out stepwise. The redox number of the glass is reduced by reducing the sulphate content and the changeover time can be reduced by overcompensating with carbon by up to 40%. Saleable containers should be produced after 60 h.

Amber to green glass

This colour change is best carried out in one step to avoid the undesirable grey-brown colour which is caused by an increasing iron content. The process can also be accelerated by overcompensating with chrome ore up to a maximum of three times the final quantity and at the same time the sulphate content should be increased over the span of a few hours to that required for green glass. The change can be carried out in 40 h.

Clear to green glass

This change is without problems since both compositions are oxidising and the right colour can be obtained quickly by overdosing at the beginning of the transition process. Apart from increasing the iron content, the green colour is produced by chrome ore. It is worth noting that large additions of chrome ore cause the molten glass to become very viscous and the initial overdosing of chrome ore must therefore be limited to $3\frac{1}{2}$ times that normal for green glass. The changeover time for medium size tanks is 8–12 h.

When changing from clear to half white (pale) glass, overcompensation cannot be used since this would give too strong a colour and it is therefore best to reduce the decoloriser before changing to a minimum which will just produce acceptable clear glass. This is followed by adding the necessary quantity of coke, which is not used in the batch for clear glass, altering the Fe^{3+}/Fe^{2+} ratio in favour of Fe^{2+} to produce the blue tint required for half white glass. In small tank furnaces the half white colour can be established in about 15 h.

Green to clear glass

In terms of colour change this is the lengthiest process since overcompensation in a change from dark to light colours is hardly possible. The colour change requires Fe_2O_3 and Cr_2O_3 to be almost totally removed. The colour change itself is simple and is carried out by charging the batch proper for clear glass; when the Fe_2O_3 content has fallen to 0·1% and Cr_2O_3 to $<0{\cdot}005\%$, a strong dose of decoloriser is added, $1\frac{1}{2}$–$2\frac{1}{2}$ times the quantity normal for clear glass. The whole change can take 160–200 h and this very long time calls for consideration to be given as to whether it would not be preferable to carry out the change by draining, cleaning, and refilling, especially since deviations from good colour arise periodically with all tanks in which the colour has been changed by the mixing process.

A change from green to half white takes almost equally long (100 to 120 h) since only a part of the iron colour remains and Cr_2O_3 must be completely removed. The process can be accelerated by first charging with about $2\frac{1}{2}$ times the tank volume of batch for clear glass, omitting all colouring constituents; $3\frac{1}{2}$ times the tank volume gives a glass which is too light in colour. Special procedures must be followed if the green glass also contained nickel (light green).

6.8.2.3.1.2 Flat glass[99]

The frequency of colour changes in flat glass tanks is much lower than in container glass tanks: in the

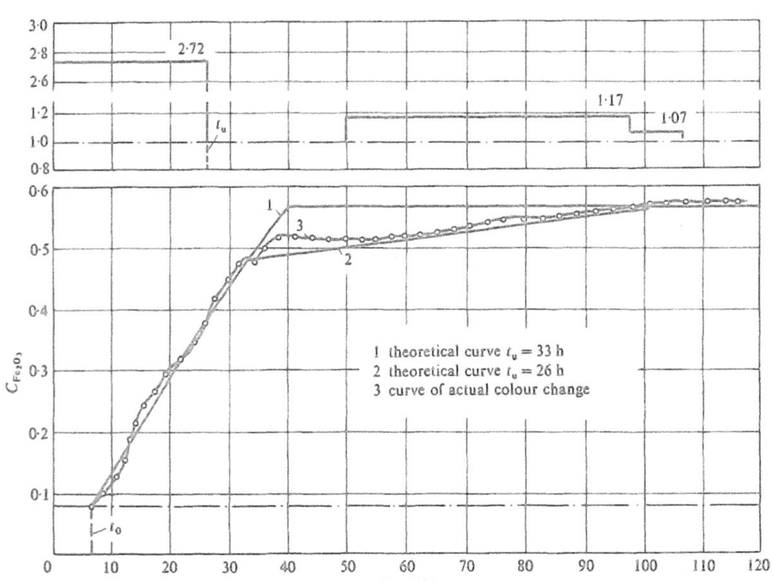

Figure 6.106. Curves of colour changes for a flat glass tank

first place the demand for coloured flat glass is much lower than for coloured container glass and in the second place flat glass is made in very large tanks, which means that a very large quantity of glass of the wrong colour will be produced in the course of a colour change.

A colour change from clear to green ($Fe_2O_3 \approx 0{\cdot}6$) is carried out relatively easily by overdosing with colouring constituents: the reverse change from green to clear glass is never carried out as the time required is too long. It is therefore recommended that coloured glass be drawn towards the end of a furnace campaign, thus avoiding the need to change back from green to clear glass; in an extreme case partial draining of the tank is recommended.

Figure 6.106 shows the theoretical curve for a colour change in a Float glass tank and the actual curve of the Fe_2O_3 content. For the purpose of calculating the mixing curves the first part is represented by a cell with piston flow and the second one with perfect mixing. As can be seen, well judged overdosing enables the required tint to be established relatively easily; a total of 100 h was needed.

6.9 The behaviour of refractories in contact with molten glass

The performance of refractories in all parts of glass furnaces is of great importance but unfortunately our knowledge is still limited to empirical observations and basic rules derived from them. Laboratory results provide only limited information on likely performance in a furnace as the effect of years of exposure in a furnace to gases and molten glass at high temperature can only be simulated to a limited extent in a short test period. The first attempts at a quantitative study were related to refractories in contact with molten glass[106] but, due to the simplifications which were necessary, the results can be used only to clarify relationships although they may be used for quantitative calculations in special circumstances.

6.9.1 The corrosion of tank blocks by diffusion without convection

The primary corrosion process is related to the solubility of the refractory or its constituents in the molten glass. Dissolution is a diffusion process which, in its simplest manifestation, is one dimensional and is not influenced by convection; it can be described by

$$\frac{\partial c}{\partial t} = D\frac{\partial^2 c}{\partial x^2} \qquad (6.66)$$

where D is the diffusion coefficient (constant), c the local concentration of the substance entering into solution, and x the coordinate of the path length. If the boundary conditions are chosen in line with Figure 6.107, the solution (ignoring the differences at phase boundaries caused by the solution process) may be described by

$$c(x, t) = c_s - c_{G,\infty}\,\Phi\left(\frac{x}{\sqrt{4Dt}}\right) \qquad (6.67)$$

where c_s is the concentration constant with respect to time (saturation concentration) at the phase boundary, and Φ is the Gaussian error integral. The concentration gradient at a location $x=0$ with respect to the molten glass is

$$\left(\frac{\partial c}{\partial x}\right)_{x=0} = -\frac{c_{G,\infty}}{\sqrt{\pi D t}}. \qquad (6.68)$$

The concentration gradient relevant to the solution process is thereby time-dependent; with increasing time $\partial c/\partial x$ becomes smaller. A boundary layer is formed which slows down the corrosive solution process.

If the concentration c_s at the phase boundary is arbitrarily set at zero, then the expression $\sqrt{\pi D t}$ can be regarded as the 'boundary layer thickness' of the diffusion. It is described as the Nernst diffusion boundary layer, δ_N, and

$$\left(\frac{\partial c}{\partial x}\right)_{x=0} = \frac{c_{G,\infty}}{\delta_N} \qquad (6.69)$$

applies.

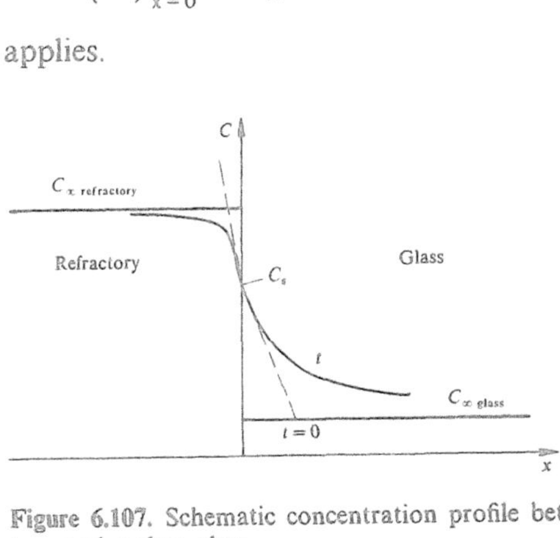

Figure 6.107. Schematic concentration profile between refractory and molten glass

6.9.2 The corrosion of tank blocks by diffusion and forced convection

In practice, conditions correspond more closely to a diffusion process upon which a laminar convective flow, parallel to the phase boundary and vertical to the direction of the diffusion, is superimposed. Of special interest (Figure 6.108) is the stationary condition which is created after a time when $\partial c/\partial t = 0$. A stationary condition implies the existence of an equilibrium and

$$D\frac{\partial^2 c}{\partial x^2} = v\frac{\partial c}{\partial y}. \tag{6.70}$$

At the wall, $v = 0$ and therefore $(\partial c/\partial x)_{x=0}$ is constant. Under these conditions a diffusion boundary layer is also formed which, in this instance, is constant with respect to time.

It is possible to define a flow boundary layer, δ_{Pr}, called the Prandtl diffusion boundary layer, analogous to the diffusion boundary layer:

$$\left(\frac{\partial v}{\partial x}\right)_{x=0} = \frac{v_\infty}{\delta_{Pr}}. \tag{6.71}$$

In flow theory, forced laminar flow is covered by

$$\delta_N = \delta_{Pr}\sqrt[3]{\frac{D}{\nu}}, \tag{6.72}$$

where ν is the kinematic viscosity of the flowing medium. Since $\nu/D \approx 1$ for gases, the two boundary layers are of nearly equal thickness; for molten glass

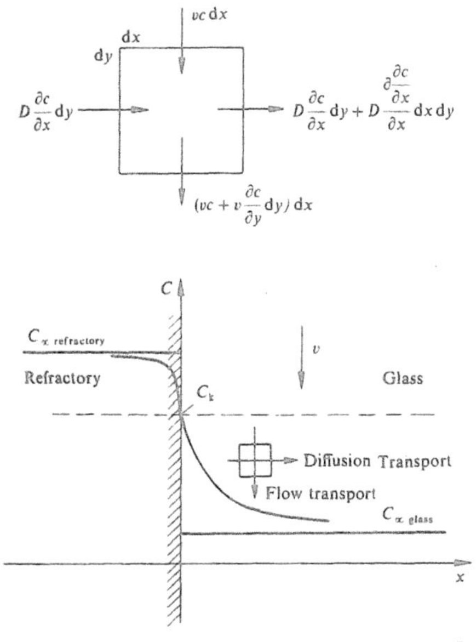

Figure 6.108. Transport processes for forced flow

at about 1400°C $\nu/D \approx 10^7$–10^9, which means that the diffusion boundary layer is only 0·002 times as thick as the flow boundary layer (Figure 6.109). This means that flow can attack the diffusion boundary layer only to a very limited extent, a fact which makes the melting of glasses in ceramic containers possible. Without this characteristic the manufacture of glass of acceptable homogeneity in furnaces having an adequate life would not be possible.

A measure of the rate of corrosion is the mass flow density of the material removed, $\dot{m}_K$: as c_s is the saturation concentration,

$$\dot{m}_K = D\left(\frac{\partial c}{\partial x}\right)_{x=0} = D\frac{c_s - c_\infty}{\delta_N} \tag{6.73}$$

and, according to Equation (6.72),

$$\dot{m}_K = D\frac{c_s - c_\infty}{\delta_{Pr}\sqrt[3]{D/\nu}}. \tag{6.74}$$

Forced convection is covered in flow theory[109,110] by

$$\delta_{Pr} = \frac{3y}{\sqrt{v_\infty y/\nu}}, \tag{6.75}$$

where y is the starting length and ν the kinematic viscosity of the fluid. The Prandtl boundary layer thickness is therefore zero at the starting point of the flow and increases as the distance from the starting point increases. When Equation (6.75) is incorporated in Equation (6.74), the mass flow density of the material removed by corrosion becomes

$$\dot{m}_K = \frac{1}{3}(c_s - c_\infty)\sqrt{\frac{v_\infty}{y}}\sqrt[3]{\frac{D^2}{\sqrt{\nu}}}. \tag{6.76}$$

The attack or the refractory corrosion at $y = 0$ is infinitely large and falls as the distance y increases

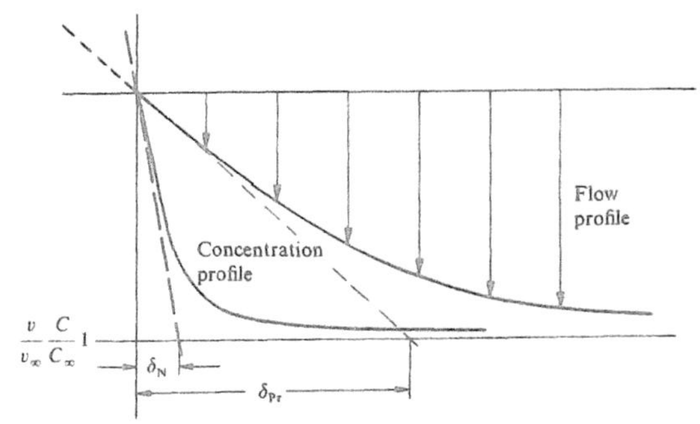

Figure 6.109. Flow and concentration profiles with flow and diffusion boundary layers

(Figure 6.110): increasing viscosity substantially reduces attack. Attack of this type takes place on the sideblocks of the throat, which are corroded away to form arcs (Figure 6.111). The glass flows parallel to the sidewall refractories. As the distance increases, flow and diffusion boundary layers of increasing thickness are formed, which act as protection against the attack on the refractories. Equation (6.76) can be used only to a limited extent for quantitive calculations but it clearly demonstrates the basic mechanism.

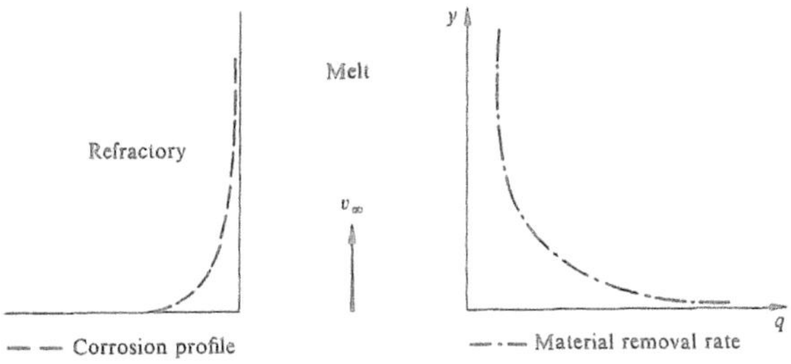

Figure 6.110. The removal of material during forced flow

Figure 6.111. Throat of a container glass tank, seen from the melting end

6.9.3 Corrosion of tank blocks by diffusion and density convection

In glass tanks the sidewall blocks are subjected to a withdrawal current, which is essentially horizontal, and, in the vertical direction, to a thermal or a density convection current resulting from the dissolution of material in the glass. For the density convection

$$\delta_N = 1{\cdot}83\, y\, Gr^{-1/4}\, Pr^{-1/4} \qquad (6.77)$$

where $Gr = y^3 g/v^2 . \Delta\rho/\rho$.

As $Pr = v/D$,

$$\dot{m}_K = \frac{c_s - c_\infty}{1{\cdot}83}\left(\frac{g\Delta\rho}{\rho}\right)^{1/4}\left(\frac{D^3}{yv}\right)^{1/4} : \qquad (6.78)$$

the rate of corrosion therefore increases with the difference in density and decreases with the flow distance.

In the case of soda–lime glasses, the density of the glass in contact with the refractories increases due to the dissolution of alumina; augmented by cooling, the density flow is directed vertically downwards as shown in Figure 6.112(a) and the attack on the block therefore decreases in the vertical direction, the decrease in temperature with depth of glass reinforcing this tendency. In the case of lead glasses, the glass in contact with refractory becomes lighter and it is therefore possible that an upward flow may be formed which overcomes all opposing effects; the corrosion profile of the block would therefore be reversed, and this is in agreement with results observed in practice (Figure 6.112(b)).

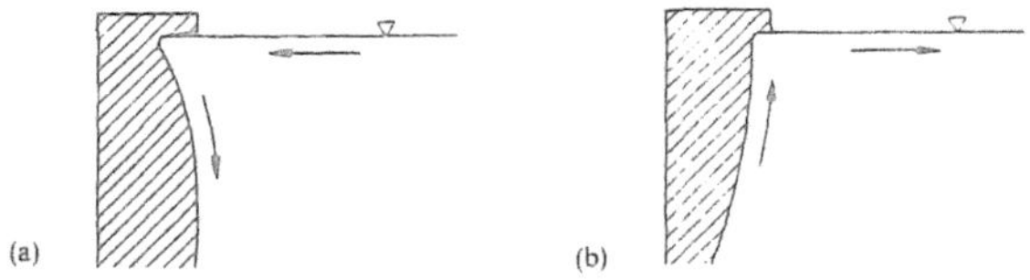

Figure 6.112. Corrosion profile in glass tanks
(a) soda–lime glass (contacting glass heavier than glass melt)
(b) lead glass (contacting glass lighter than glass melt)

6.9.4 The influence of interfacial and surface tensions

The strong corrosion observed near the flux line cannot be explained on the basis of the corrosion processes described in Sections 6.9.1 and 6.9.2, nor on the basis of forced or flow convection. It is now known that along the edges there exist currents which are responsible for these high corrosion rates; the driving forces are the surface and interfacial tensions.

The process can be explained as follows. If a fluid II is placed on top of another fluid I of different surface tension then the interaction produces a positive or negative spreading pressure (Figure 6.113) in which

$$p = \sigma_1 - (\sigma_2 + \gamma_{1,2}).$$

If $\sigma_1 > \sigma_2 + \gamma_{1,2}$, then fluid I spreads until an equilibrium condition is reached. This applies to

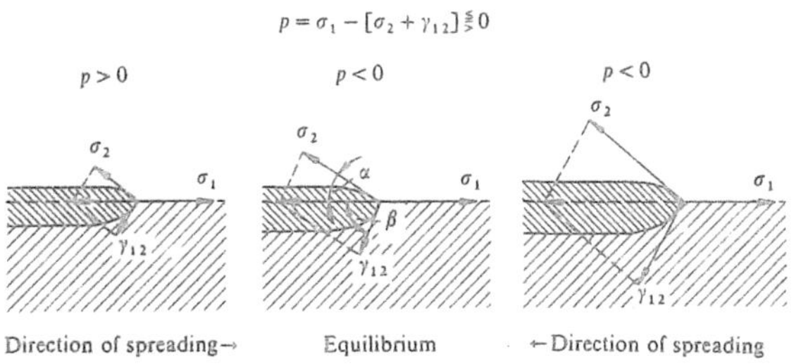

Figure 6.113. Schematic representation of spreading pressure[107]

fluids which are immiscible but if they are miscible, then σ_2 and $\gamma_{1,2}$ change with time and, as a result, the vectors turn and rotation, including eddy currents, begins.[106]

6.9.4.1 Flux line corrosion

At the flux line the bulk glass and the glass in contact with the refractory (the contacting glass) constitute two surface active substances. The surface tension and density of the contacting glass usually increase as a result of the dissolution of alumina but the glass can dissolve only a limited amount of alumina. With time the processes shown in Figure 6.114 take place.[106,112,113]

The formation of the contacting glass results in a three-phase boundary, as shown in Figure 6.114, and the dissolution of the bulk glass in the contacting glass gives rise to a rotating current; the contacting glass in the zone of the meniscus is forced downwards and fresh reactive glass is drawn towards the surface of the block. This process of continuous replacement destroys the boundary layer as it is being formed and a high degree of corroson is the result.

Convection at the contact surface also takes place when the surface tension of the glass has been reduced as a result of materials dissolved from the refractory and the direction of the convection current operating at the flux line will be completely reversed. The controlling factors governing flux line corrosion are therefore the change in the surface tension and indirectly, depending on the speed at which these processes take place, the viscosity of the glass, especially that of the contacting glass.

Occasionally, drops remain attached to the upper edge of the flux line (Figure 6.115) and either flow away on their own as saturated solutions or find their way into the contact zone convection currents at a subsequent movement of the meniscus. In view of the large amount of refractory material they contain, these drops will strengthen the convection currents at the flux line. Löffler[112] refers to the appearance of the corrosion caused by these drops as being like 'small windows'.

A mathematical description of flux line corrosion is much more difficult than one of flow corrosion governed by concentration differences. For the rate of corrosion, Hrma[114–116,120] has given an equation for the removal of material under conditions of unhindered convection at the interface but this equation is as yet unconfirmed.

The literature, especially in the field of hydrodynamics, refers to each change of dynamics involving boundary surface tensions as the 'Marangoni effect'.[117] The dimensionless Marangoni number, *Ma*, can be taken as a characteristic index for the driving forces which eventually produce interfacial convection. Changes in interfacial tension caused by changes in concentration are given by

$$Ma(c) = \Delta\sigma l/\eta D, \quad (6.79)$$

where σ is the surface or interfacial tension and l is the characteristic length.

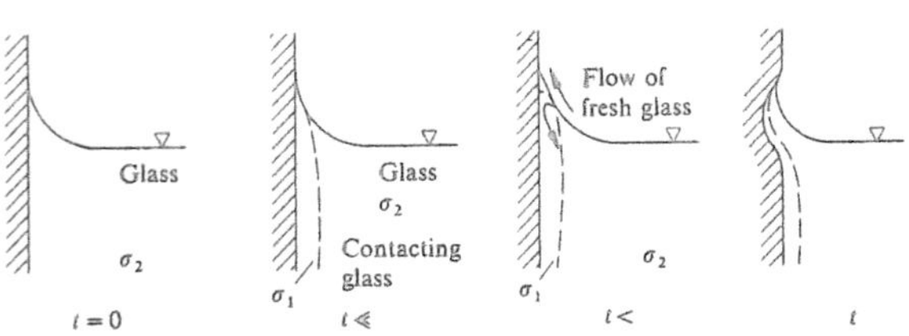

Figure 6.114. Stepwise progress of flux line erosion

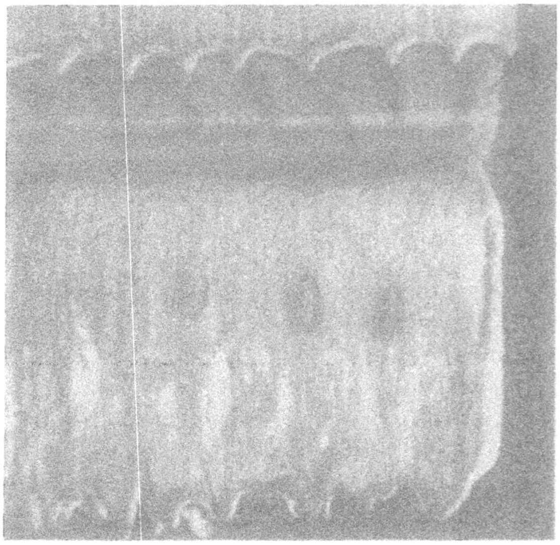

Figure 6.115. Flux line of a tank block with many 'windows'[112]

Jebsen-Marwedel[118] has introduced the concept of eddy phenomena as being responsible for interfacial convection.

6.9.4.2 Upward and downward drilling, pitting

When molten glass is in contact with the lower, horizontal surface of a refractory it will cause a type of corrosion referred to as upward drilling, or pitting (Figure 6.116). The origin of this can be ascribed to the same cause as flux line corrosion, namely interfacial corrosion (Figure 6.117(a)). If a bubble is formed at the refractory/glass interface, either as a result of the porosity of the refractory or from gases involved from the molten glass, then the bubble surface constitutes a three-phase boundary, similar to the situation at the flux line. Attack will accelerate at this point and will penetrate progressively deeper into the refractory. With large bubbles the attack proceeds more slowly in the central region than at the edges due to the formation of a diffusion boundary layer at the centre. The edge zone corrosion proceeds rapidly and leads to the formation of a central nodule which can grow in size until it projects through the bubble into the molten glass where it is dissolved under the influence of interfacial convection.

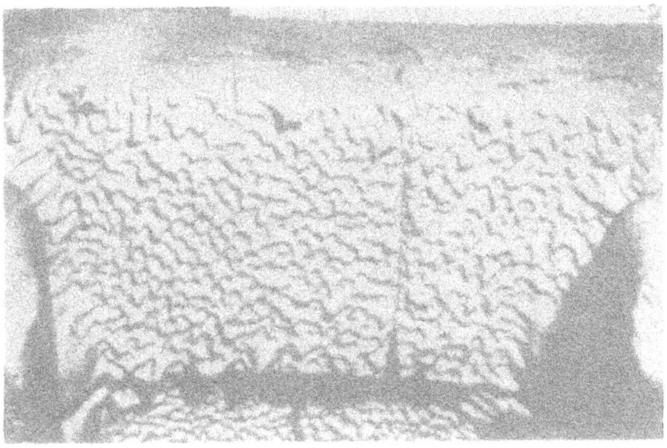

Figure 6.116. View of the lower surface of a throat cover block showing attack by upward drilling

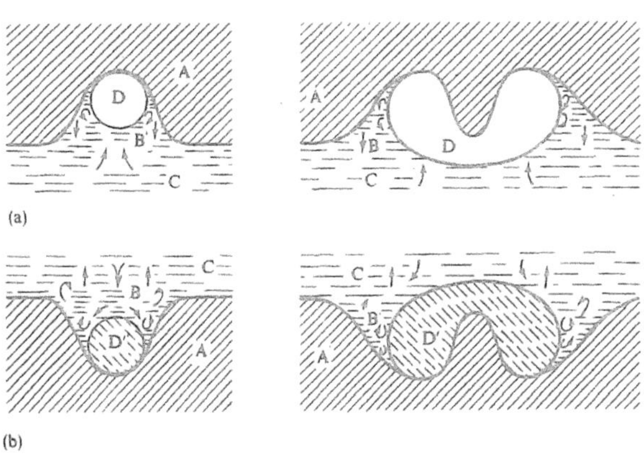

Figure 6.117. Schematic representation of accelerated vertical corrosion caused by boundary layer convection surrounding large and small gas bubbles and drops of liquid

A solid body
B solution of A in C
C liquid or melt
D gas bubble
D' drop of liquid not fully soluble in B

Substances which are heavier than glass and only slightly soluble in it (metals) and which are liquid at the operating temperatures will sink to the tank bottom where they form spherical or flat circular melts which, because of their relatively high surface tension, form a wedge between glass and refractory. A zone of interfacial convection formed in this wedge is responsible for drilling into the refractory (Figure 6.117(b)).[119]

Brückner has determined the rates of corrosion under the influence of different convection currents using a rotating disc (Table 6.4).[120] Results showed that interfacial convection at the flux line is much greater than convection caused by differences in density and is particularly strong for drops of metal drilling into a refractory. At 1400°C, corrosion rates of $7–9 \times 10^{-2}$ mm/h are attained.

Table 6.4. *Rates of corrosion (10^{-3} mm/h) of a fusion cast AZS block (ZAC 1711) by molten container glass under the influence of different types of convection flow*[15]

Type of convection flow	*Temperature (°C)* 1400	1500
Forced convection $v = 1$ m/h	20	
Density convection (average value over campaign)	0·8–0·9	11–16
Interfacial convection at the flux line	8·5	60–100

6.9.5 Possibilities of reducing corrosion

The above descriptions and occasional mathematical analyses indicate how the corroson of refractories might be reduced. The most important factors to consider are: a reduction in surface tension between molten glass and the dissolved refractory material; a reduction in the differences in density between molten glass and dissolved refractory material; an increase in viscosity of the dissolved refractory material; and a reduction in the saturation concentration, c_s, and an increase in concentration of the corresponding effective component, c_∞, of the molten glass.

6.10 The performance of refractories in the superstructure and waste gas regions of a furnace

In the superstructure and waste gas regions the attack on refractories and the consequences thereof are very complex. It is only to a very limited extent that it is possible to account for the corrosion on the basis of diffusion and solubility in a simple one dimensional form, as is possible for tank blocks. Mathematical and quantitative analyses of the major causes are not possible, only a description of the chemical and physical processes.

Depending on the different functions and conditions to be met, the superstructure of a furnace is made up of a variety of refractories and cements (see Chapter 4). In selecting and using these materials care must be taken to remember that chemically different materials can react when in contact with one another to form eutectics which flow at temperatures far below those to which the refractories are designed to be subjected during operation. This phenomenon is known as eutectic corrosion[121] and the best known example occurs in the system $SiO_2 - Al_2O_3 - Na_2O$ which has a eutectic at 800°C. The purer the refractory installed, the larger is the reduction in melting point resulting from diffusion of foreign material into its surface. For example, if a fireclay brick is installed above a silica brick, then, at furnace operating temperatures and under the influence of alkalis which take part in the reaction, a slag which progressively destroys the silica brick is rapidly formed. If the positions of the bricks are reversed, corrosion by the slag will also take place but more slowly since the slag, which in the main will flow downwards, will attack the less pure fireclay to a lesser extent. It is easy to understand that, due to their ability to absorb liquid slag, porous refractories will be attacked more severely than denser refractories. If contact between two very different refractories cannot be avoided, as is the case with silica and fusion cast AZS blocks, then the problem can be alleviated by inserting an intermediate layer which is itself more compatible with each refractory and which can absorb any slag formed (see Section 4.3.3).

Nowadays major errors hardly ever occur in the selection of refractories. However, many sins are still committed in the use of different cements and mortars, and an unsuitable mortar causes eutectic corrosion just as effectively as two incompatible refractories. Since the composition of a mortar cannot be assessed by eye, as can be done with most refractories, silica cements are now coloured green by agreement between manufacturers and users. Other cements can only be recognised by their natural colours and the description given on the packing.

A second but less serious cause of corrosion in the superstructure and regenerators is condensation. The furnace atmosphere contains alkali as well as waste gases and sulphur from fuel and batch; the resulting sulphates can condense in the cooler parts of the superstructure and in the central and lower regions of the regenerators. Condensation can also occur in the relatively cool zones of the doghouse area or in joints or cracks in the superstructure. The liquid sulphate reacts corrosively with virtually all known refractories and forms slags which, depending on the exact location, form drops which can undergo significant changes in composition with increasing temperature due to selective evaporation. Attack takes place preferentially at grain boundaries and interfaces.

In furnace zones which are above the condensation temperature of sulphates, condensation can occur if the porosity of the refractories is sufficiently large to permit sulphate to penetrate into the refractory until it reaches the zone which is at the condensation temperature. This occurs, for example, with silica bricks in crowns of tank furnaces but the penetration does seal the refractory so that the corrosion process is arrested. Open joints, however, where new sulphate is continuously being supplied, can give rise to severe corrosion, sometimes referred to as shot holing. The slag will form stalactites which fall into the tank.

A further cause of corrosion arises from changes in crystal structure which take place under the influence of temperature, time, and alkalis. The disintegration of mullite due to the formation of corundum is well known. A change in crystal structure in the reaction zone produces a change in thermal expansion and thereby causes cracking and spalling of the refractory. Similar effects can be observed with fusion cast blocks if one surface is subjected to periodic temperature changes in the range of the expansion anomaly.

There are also many other less frequently encountered causes of corrosion such as, for example, the sweating out of vitreous phase from AZS blocks if subjected to too high a temperature or if there is a manufacturing fault and the formation of slags of high iron content if parts of the furnace steelwork are overheated.

A further indirect cause of corrosion is the evaporation of individual components from a refractory. For example, if magnesite–chrome refractories are placed in very hot sections of the upper parts of regenerators of tanks heated with natural gas, chromium oxide evaporates and forms alkali chromate which then condenses in the colder parts of the regenerator chamber. The loss of chromium oxide leads in time to a crumbling of the refractory assembly.

6.11 The stirring of glass

The melting of glass from a heterogeneous mixture of raw materials causes the primary melt to be inhomogeneous but it may be homogenised by the action of shear currents which, especially at high temperatures, help to reduce the size of cords by diffusion. However, only thin cords can be dispersed by this means as diffusion is a slow process because of the high viscosity of molten glass. Only a combination of shear and diffusion can lead to the reduction of cords to a level at which they will be acceptable.

The degree of homogeneity, or, more accurately, the acceptable level of cord, is governed by the type of glassware being manufactured. The highest quality standards are set for optical glasses, glasses for lasers, and similar types of special glass. One way of reducing the cordiness of glass is by stirring and it can be readily understood why techniques for stirring developed alongside the development of optical glasses. Demands for higher glass quality in other fields have led to stirring being more widely used. Depending on the initial conditions, stirring can be used in the main to reduce chemical inhomogeneities (cords), to give thermal homogeneity, or to disperse added colouring glasses.

6.11.1 The stirring mechanism

Stirring can be treated in a manner analogous to the mixing effect of glass currents in relation to the deformation of cords. The total contribution of individual parameters is given in Equations (6.37) and (6.44) of Section 6.6. Combining both equations gives the reduction of the normalised excess concentration of a cord as

$$\frac{\Delta\rho}{\Delta\rho_0} = \operatorname{erf}\frac{1}{4\sqrt{\dfrac{\int_0^t D\left(\dfrac{A_0}{A}\right)^2 \dfrac{1+\cot^2 d_0}{1+[(A_0/A)^2\cot\alpha_0 + b_0 t]^2}\,\mathrm{d}t}{\delta_0^2}}} \qquad (6.80)$$

where the following terms correspond to those given for Equations (6.37) and (6.44): A_0/A is the change in flow cross section, D the diffusion coefficient, α_0 the original angle between cord and flow direction, b_0 the shear flow gradient, t the time during which shear is applied, and δ_0 the original thickness of the cord. If thermal homogenisation is the main objective, the diffusion coefficient can be replaced by the thermal conductivity.

The value of A_0/A can be fixed at unity since a change in the flow cross section has, in virtually all cases, no bearing on the problem. However, the original angle between cord and flow direction, α_0, is important, the cases where α_0 is 0° and 90° being of special interest.

If $A_0/A = 1$ and $\alpha_0 = 90°$, then, for $b_0 t^2 \gg 1$, Equation (6.42) gives

$$\frac{\Delta\rho}{\Delta\rho_0} = \operatorname{erf}\frac{\delta_0}{4b_0\sqrt{D\dfrac{t^3}{3}}}. \qquad (6.81)$$

A shear force gradient b_0, a long period of subjection to shear forces, and a high diffusion coefficient will therefore cause an excess concentration $\Delta\rho_0$ to disappear rapidly.

If $A_0/A = 1$ and $\alpha_0 = 0°$, then Equation (6.37) reduces to $\delta/\delta_0 = 1$, i.e. even for high shear flow gradients, the original thickness of the cord remains. Under these conditions the excess concentration can only be reduced through diffusion as shear forces make no contribution; the result is

$$\frac{\Delta\rho}{\Delta\rho_0} = \operatorname{erf}\frac{\delta_0}{4\sqrt{Dt}}.$$

The effectiveness of stirring therefore depends strongly on the angle between cord and shear flow, α_0. If they are at right angles, the effect of stirring on homogenisation is at its maximum; if in parallel, the effect of stirring is nil.

The effect of different combinations of parameters on changes in the normalised cord thickness δ/δ_0 is given in Figure 6.91.

6.11.2 Stirrers

Various types of stirrer are used depending on the specific circumstances (Figure 6.118), the oldest being a rod rotating on a vertical axis. In an improved version the stirrer moves in a planetary orbit and stirring of this type was used for many years in the manufacture of optical glass in pot furnaces.

A paddle stirrer may be used in the working end of tanks melting industrial glasses; it is inserted into the molten glass vertically from above and rotates about its vertical axis, and is mainly used for the thermal homogenisation of the molten glass. For special requirements a mixing chamber with two paddle stirrers is required; the shape of the chamber and the method of stirring must be such that the whole of the glass flow is subjected to the stirring action and that no dead flow zones are formed. Stirrers rotating in opposite directions, as shown in Figure 6.119, have given the best results in model experiments.[123]

Stirring by paddles imparts a vertical flow component to the glass which, depending on the particular circumstances, can be an advantage or a disadvantage. Rates of rotation lie between 6 and 30 rev min. The depth of immersion and distance from surrounding walls have been investigated on models.[124]

For furnaces with high pull and relatively wide flow channels, as is the case for Float glass tanks, the use of a row of stirrers placed across the width of the furnace (Figure 6.120) has been employed with success and it has been claimed that this procedure substantially reduces the cordiness of flat glass.[125] The need to provide openings in the crown is a disadvantage: material condenses on the stirrers and falls back in an uncontrolled manner

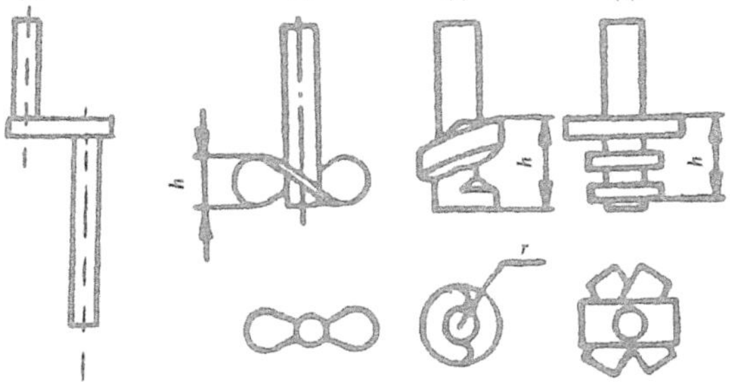

Figure 6.118. Various types of stirrer[126]
(a) rod
(b) paddle
(c) screw
(d) rod with vanes

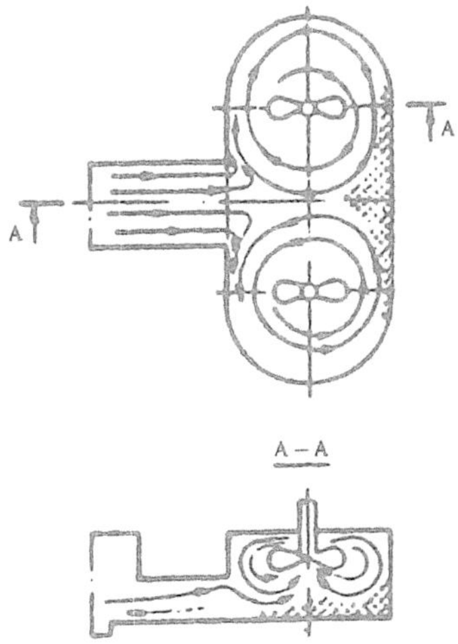

Figure 6.119. Arrangement of a pair of paddle stirrers in the working end of a tank melting hotel glassware[124]

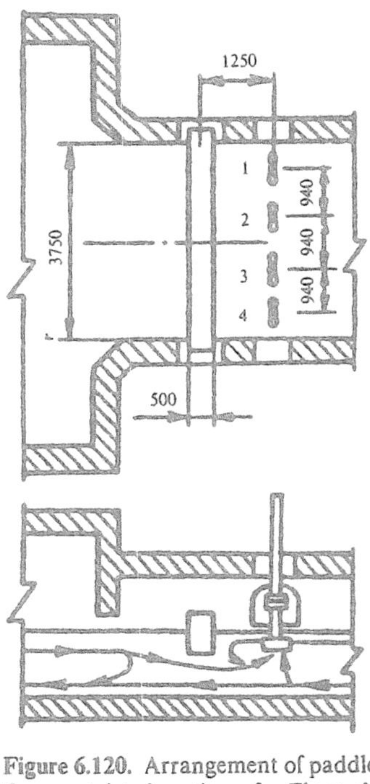

Figure 6.120. Arrangement of paddle stirrers across the width of the constricted section of a Float glass tank[125]

into the furnace and can spoil the glass. Insertion of stirrers at an angle through the sidewalls is better.

The incoming flow conditions, especially α_0 as defined in Equation (6.80), are not particularly favourable for paddle stirrers rotating on a vertical axis. A large α_0 is only obtained where the currents flow in and out while, at the sides, α_0 becomes very small and the contribution to homogenisation is correspondingly small.

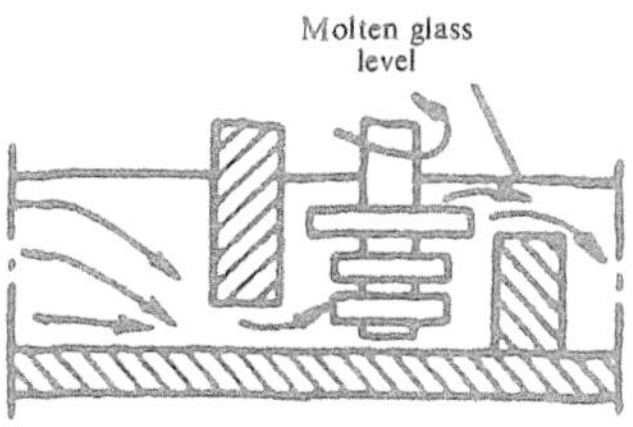

Figure 6.121. Arrangement of rod stirrer with vanes in a combined throat and weir section[126]

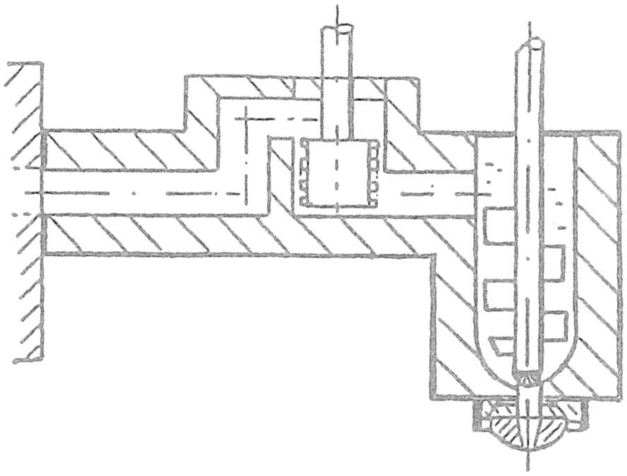

Figure 6.122. Rod stirrer with vanes in a feeder section with glass flow in the vertical direction[127]

The type of rod stirrer shown in Figure 6.121 gives rise to multiple homogenisation actions due to its multilayer construction. If, as is shown in Figure 6.121, a vertical flow component is superimposed on the stirring action, then, in all regions surrounding the stirrer, the angle α_0 is relatively large and the process of homogenisation is thereby improved.

The greatest degree of homogenisation can be achieved by a rod stirrer located in a mixing chamber through which the glass flows parallel to the axis of the stirrer (Figure 6.122); the effect can be further improved by installing stationary rods in the mixing chamber.[127]

The best mixing geometry is useless if the materials subjected to the stirring process cannot resist the chemical or physical attack of the glass. High quality fireclay and sillimanite are often used and fusion cast refractory materials have also been tried: if temperature conditions permit, water-cooled steel tubing may be used. Under adverse conditions, stirrers with platinum sheet are recommended;[127] the cladding must be done in a way that prevents glass from entering the joint between the platinum and the stirrer proper.

7 Mathematical models of glass tank furnaces

Knowledge of the behaviour of glass tank furnaces in operation, their thermal efficiency, glass quality, etc. can be obtained from specific studies and experiments on operating furnaces, from physical models, and from mathematical models. Current knowledge is based in the main on observations and experiments carried out in plants during operation. The long campaigns, the high risks surrounding such plants, and the poor accessibility of these plants for carrying out technical measurements narrow the field of application of empirical methods. Attempts are being made to obtain the necessary information by other means.

Physical models are based on the premise that an application of the theory of similarity will permit models to be constructed which resemble full size plants and which behave in a way that makes some insight into their performance possible. This method has been used for glass furnaces with some success in investigations covering glass currents. As far as the combustion chamber is concerned physical models are limited to a study of gas flow since heat transfer can only be simulated to a limited extent due to the important contribution made by radiation.

Mathematical models presuppose that the major processes of a plant can be described by mathematical equations which can be solved with respect to the parameters under investigation. The development of computers has significantly encouraged the use of mathematical models. It is extremely difficult to describe the firing system of a glass furnace and the bath of molten glass, with their complex currents, in a series of equations which have solutions. So far, success has only been achieved by radical simplifications of some parts and by the introduction of some parameters based on information obtained empirically. The models of the firing system, the glass currents, and the glass tank are of interest.

7.1 Models of the firing system

These are almost always concerned with the relationship between the liberation of chemical energy during firing, the flow processes, the distribution of the energy flow, and the heat transfer, particularly with respect to the material to be heated. The models differ in accuracy according to the investment of effort (Figure 7.1). There are three types of model.

Model I starts with the assumption that the processes taking place inside the combustion chamber can each be covered by a representative value. It is based on the concept of a cell with perfect mixing in which the gases are described in terms of

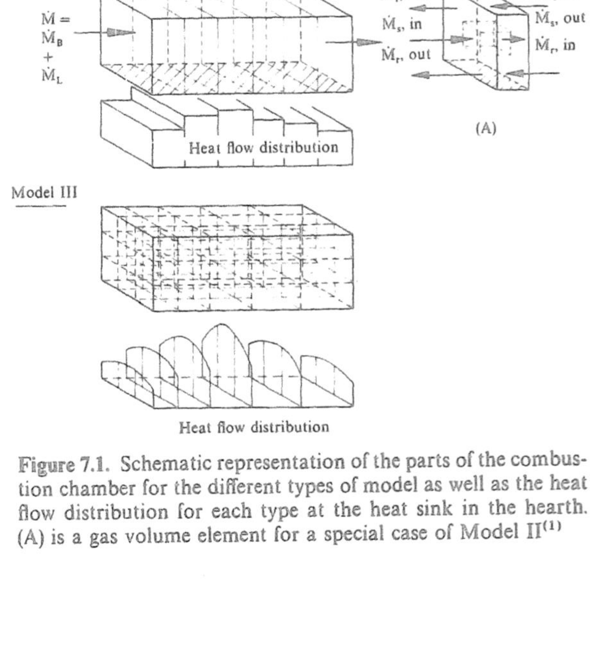

Figure 7.1. Schematic representation of the parts of the combustion chamber for the different types of model as well as the heat flow distribution for each type at the heat sink in the hearth. (A) is a gas volume element for a special case of Model II[1]

pressure, temperature, flow velocity, etc. Important work making use of this approach is given by the models of Jeschar, Pötke, and Schupe.[2,3]

Model II usually divides the combustion chamber into separate compartments longitudinally in the direction of the main flow. This approach makes it possible to determine the energy flow distribution along the length of the furnace.

A further refinement, although only obtained with considerably increased mathematical effort, is Model III. With this, an attempt is made to determine the distribution of the relevant parameters, in particular that of the heat flow.

In practice, it depends on the case in question how much mathematical effort needs to be mustered to solve it.

7.2 Models of glass currents

Up to a point, an understanding of glass currents in the melting and refining zones has been successfully achieved by mathematical means; this has already been described in Chapter 6. An exact definition of the boundary conditions, especially those of the batch melting zone, leave much to be desired. Meaningful three dimensional models have so far failed due to the scale of the mathematical effort involved. A good survey of the current position of work in the field of glass currents is given in References 46, 47, 50, and 60 of Chapter 6.

7.3 Models of glass tanks

7.3.1 General survey

A satisfactory model of a glass tank must consider the processes taking place in the combustion chamber and in the bath of molten glass. It must be a model which encompasses models of the firing system and the glass currents, and must take into account the quality of the molten glass as defined by parameters covering the refining and homogenisation processes. This demonstrates the scale of the difficulties which are faced when setting up a model of a glass tank. The models used up to now made use of major simplifications, especially for the glass currents. Refining and homogenisation and, by implication, glass quality have only been considered in one case, and this was very much simplified.

The model used by Mase & Sasagawa[4] starts from a two dimensional calculation of glass currents; the superstructure is included as a boundary region but other information, such as the effect of pull on energy consumption, is not dealt with. An improved model, which specifically includes the melting of batch, was published in 1980 but unfortunately it did not present any additional conclusions.[5]

McConnel & Goodson's model[6] starts from the firing system using Model II; glass currents are treated as two rotating cylinders of fixed magnitude but glass quality is not included. Other data covering the results which can be obtained with this model appear in only a very general form, so that no further comment is appropriate here.

Based on Model I, Trier & Voss[7] use a series of models coupled in parallel to describe the firing, with fixed values covering the thermotechnical effect of glass currents and a time factor for the refining zone for setting the glass quality at a fixed value. The answers provided by the model are presented in a form suitable for practical application and this model is therefore considered in greater depth.

7.3.2 The Trier and Voss glass tank model; its construction[7-10]

Temperature measurements of batch during melting and theoretical considerations (see Section 6.4) indicate that the surface temperature of a batch increases rapidly after entering the furnace, but then changes little over a long period. As the batch melts, the surface temperature of the molten glass rises until it reaches its maximum at the refining temperature. The change in surface temperature can therefore be assumed to approximate to that shown in Figure 7.2.

The temperature curve *g* shown in Figure 7.2 can be used as the starting point for a comprehensive system of equations to deal with the whole furnace. There are many reasons why it is not a good idea to cover the width of the tank with batch and some assumptions therefore need to be made about the temperature curve of the outer surface areas not covered with batch. One must force the issue to a certain extent and it is therefore assumed that, due to mixing and other exchange processes, the fixed point might as well be L_2, i.e. $\vartheta_{G.2}$. To keep calculations simple, the temperature at $L=0$ is taken as $\vartheta_{G.1}$; between 0 and L_1 there is a linear

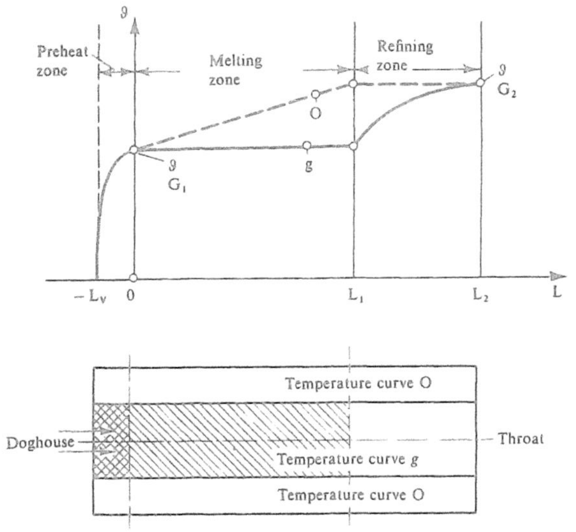

Figure 7.2. Assumed surface temperature distribution over the longitudinal axis of the melting end

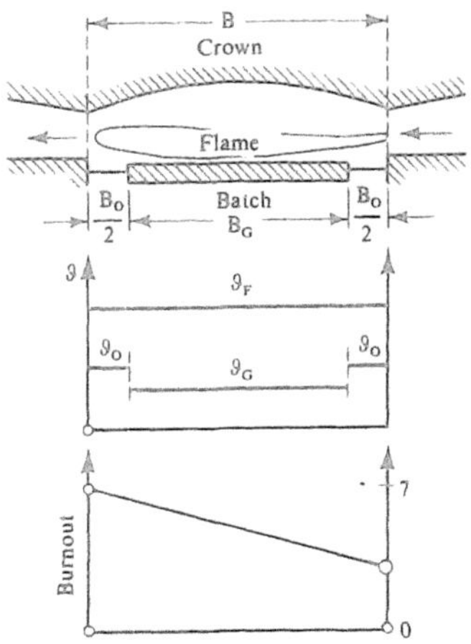

Figure 7.3. Temperature and burnout curves across the width of the furnace[7]

increase and between L_1 and L_2 the temperature is assumed to be constant.

The assumptions covering temperature do not contradict the precondition (Section 7.3.3) that temperatures must not be chosen arbitrarily and $\vartheta_{G.1}$ is a reference temperature, such as the surface temperature of molten iron. The value of $\vartheta_{G.2}$ is governed by the process technology; it signifies the temperature which in conjunction with the melting time, t_s, must be reached in order to obtain molten glass of sufficient homogeneity. More about this below. The change in temperature with time and place between the limits $\vartheta_{G.1}$ and $\vartheta_{G.2}$ is given by the laws of nonstationary heat transfer. The temperature curve marked O in Figure 7.2 is defined less precisely and is the result, using some sensible assumptions, of linking it to the temperature curve g.

As far as the temperature above the glass, that is, of the flames, is concerned, it is assumed as a first approximation that there will be a linear burnout and constant heat transfer conditions across the width of the furnace; this gives a constant flame temperature (Figure 7.3).

The flame should initially cover the surface of glass and batch and prevent a direct heat exchange between crown and glass; the radiation from the crown through the flame is included in Section 7.3.3.2. These assumptions were made to simplify the calculations but are not of a fundamental nature and can therefore be modified at any time; the same applies to the burnout of the flame.

If k is the heat transfer coefficient between flame and batch in the regions of linear temperature change ($0 < L < L_1$), then a balance can be constructed for a strip 1 m wide covering the width of the tank in the batch melting zone for each individual longitudinal unit of heat flow, $\dot{q}$:

$$\underbrace{\dot{q}_B + \dot{q}_L}_{\substack{\text{Introduced}\\ \text{as fuel and}\\ \text{combustion}\\ \text{air (kW/m)}}} = \underbrace{\dot{M}_A c_A \vartheta_F}_{\substack{\text{Via waste}\\ \text{gases into}\\ \text{regenerator}}} + \underbrace{(\vartheta_F - \vartheta_{G.1}) B_G k}_{\text{To batch}}$$

$$+ \underbrace{\left[\vartheta_F - \left(\frac{\vartheta_{G.2} + \vartheta_{G.1}}{2}\right)\right] B_0 k}_{\substack{\text{To free surface between}\\ \text{strips of batch}}}$$

$$+ \underbrace{\dot{q}_a}_{\substack{\text{Crown}\\ \text{losses}}} + \underbrace{2\dot{q}_b}_{\substack{\text{Sidewall}\\ \text{losses}}} + \underbrace{2\dot{q}_b \frac{B}{L_1}}_{\substack{\text{End wall}\\ \text{losses}}}. \qquad (7.1)$$

The total quantity of heat transferred into the batch during the time it resides in the melting zone ($0 < L < L_1$) is

$$\dot{Q}_G \varphi = G \vartheta_{G.1} \varphi = (\vartheta_F - \vartheta_{G.1}) k B_G L_1 \qquad (7.2)$$

in which the factor φ takes account of the fact that the batch melts in part from above and in part from below. Since the melting batch cannot transfer any heat downwards into the molten glass, more heat needs to be absorbed in the reagion ($L_1 < L < L_2$) than is required for increasing the glass temperature from $\vartheta_{G.1}$ to $\vartheta_{G.2}$. The excess $1 - \varphi$ corresponds to the part ($0 < L < L_1$) (Figure 7.4). This assumption is not as arbitrary as may appear at

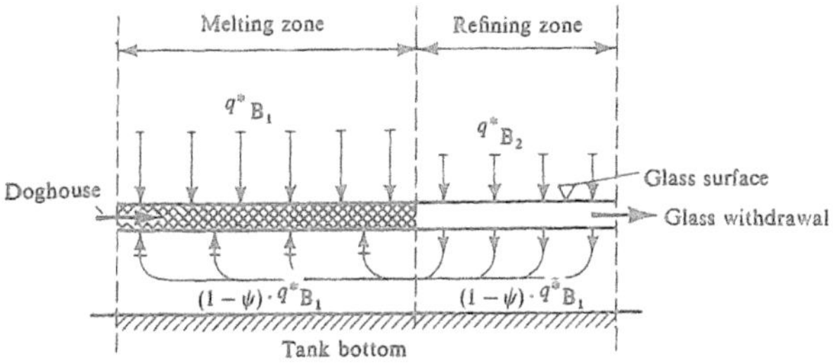

Figure 7.4. Schematic of heat inflow

first sight. In reality the heat imparted to the batch blanket from below is carried there from the refining zone by glass currents. The heat radiated from free surfaces (B_0L) is assumed to cover the heat losses of the tank basin but this assumption is not completely satisfactory since, without doubt, heat flows directly into the batch from this source. There is no difficulty in a more accurate separation of these quantities at a later stage.

Equation (7.4) for heat exchange can be constructed for the refining zone analogous to the energy balance of Equation (7.3); however, the fact that ϑ_G is a variable must be taken into account.

$$\underbrace{\dot q_B + \dot q_L}_{\substack{\text{Introduced}\\ \text{as fuel and}\\ \text{combustion}\\ \text{air (kW/m)}}} = \underbrace{\dot M_A c_A \vartheta_F}_{\substack{\text{Via waste}\\ \text{gases into}\\ \text{regenerator}}} + \underbrace{(\vartheta_F - \vartheta_G)B_G k}_{\text{To glass}} + \underbrace{(\vartheta_F - \vartheta_{G.2})B_0 k}_{\substack{\text{To free surface}\\ \text{between strips of}\\ \text{batch}}} + \underbrace{\dot q_a}_{\substack{\text{Crown}\\ \text{losses}}} + \underbrace{2\dot q_b}_{\substack{\text{Sidewall}\\ \text{losses}}} \tag{7.3}$$

$$\xi G \frac{d\vartheta_G}{dL} = \underbrace{(\vartheta_F - \vartheta_G)B_G k.}_{\substack{\text{Heat flow per unit length}\\ \text{(kW/m) to glass in}\\ \text{refining zone}}} \tag{7.4}$$

Equations (7.1) to (7.4) can be combined by substitution and elimination into

$$\left\{\frac{G}{Ak}\,\frac{\xi\left[\frac{\dot q_B}{B}\frac{K_1(1-\eta_k)}{(1-C)} + \frac{k}{(1-C)}\right]}{\left[\frac{\dot q_B}{B}K_1(1-\eta_k) + kC\right]}\cdot \ln\left(\frac{\frac{\dot q_B}{B}[1 - K_1(1-\eta_k)\vartheta_{G.1}] + kC(\vartheta_{G.2} - \vartheta_{G.1}) - \frac{\dot q_a}{B} - \frac{2\dot q_b}{B}}{\frac{\dot q_B}{B}[1 - K_1(1-\eta_k)\vartheta_{G.2}] - \frac{\dot q_a}{B} - \frac{2\dot q_b}{B}}\right) + \frac{\frac{\varphi}{(1-C)}\vartheta_{G.1}\left[\frac{\dot q_B}{B}K_1(1-\eta_k) + k\right] + \frac{2\dot q_b' Ak}{L_2 G}}{\frac{\dot q_B}{B}[1 - K_1\vartheta_{G.1}(1-\eta_k)] - \frac{kC}{2}[\vartheta_{G.1} - \vartheta_{G.2}) - \frac{\dot q_a}{B} - \frac{2\dot q_b}{B}}\right\} = 1 \tag{7.5}$$

where C is given by $B(1 - C) = B_G$, the proportion of glass surface not covered with batch (Figure 7.3): details of the calculations are given in the appendix to Reference 7.

So far no account has been taken of the fact that the melting of good quality glass requires not only a certain temperature, $\vartheta_{G.2}$, but also a specific melting time. This was included by formulating the condition that the time necessary for heating the batch from $\vartheta_{G.1}$ to $\vartheta_{G.2}$ after melting is t_S, which may be chosen as any suitable time, say 1 h. This gives rise to

$$t_S = \frac{dB(1-C)(L_2 - L_1)}{\dot V_G} \quad \text{(h)}, \tag{7.6}$$

$$\frac{t_S}{d} = \frac{\xi(\rho\cdot c)_{\text{glass}}}{k}\,\frac{\frac{\dot q_B}{B}\frac{K_1(1-\eta_k)}{(1-C)} + \frac{k}{(1-C)}}{\frac{\dot q_B}{B}K_1(1-\eta_k) + kC}(1-C) \cdot \ln\left(\frac{\frac{\dot q_B}{B}[1 - K_1(1-\eta_k)\vartheta_{G.1}] + kC(\vartheta_{G.2} - \vartheta_{G.1}) - \frac{\dot q_a}{B} - \frac{2\dot q_b}{B}}{\frac{\dot q_B}{B}[1 - K_1(1-\eta_k)\vartheta_{G.2}] - \frac{\dot q_a}{B} - \frac{2\dot q_b}{B}}\right). \tag{7.7}$$

Since t_S, the time necessary for melting, depends primarily on $\vartheta_{G.2}$, the temptation arises to combine both parameters functionally and insert them into the equations. In principle, this is possible but it was decided not to do so and to aim at clear-cut equations; in any case it is difficult to find an equation which corresponds to reality. However, this is the point at which the type of batch (grain size) and glass quality (seed count) can be fed into the equation.

A calculation of the inside crown temperature, $\vartheta_{S.i}$, calls for modifications to the equations; $\dot{q}_a$, the total superstructure heat losses, should now be divided into $\dot{q}_S$, the heat flow through the crown, and the heat losses through the sidewalls since only part of the superstructure losses are through the crown. The inside temperature of an insulated crown in the region $(0 < L < L_1)$ is

$$\vartheta_{S.i} = \vartheta_F\left(1 - \frac{1}{1 + \frac{\alpha_0 d_1}{\lambda_1} + \frac{\alpha_0 d_2}{\lambda_2} + \frac{\alpha_0}{\alpha_u}}\right), \tag{7.8}$$

where ϑ_F is determined from

$$\vartheta_F = \frac{\frac{\dot{q}_B}{B} + k(1-C)\vartheta_{G.1} + \frac{1}{2}kC(\vartheta_{G.1} + \vartheta_{G.2}) - \frac{2\dot{q}_B}{B}}{\frac{\dot{q}_b}{B}K_1(1-\eta_k) + k + \frac{1}{\frac{1}{\alpha_0} + \frac{d_1}{\lambda_1} + \frac{d_2}{\lambda_2} + \frac{1}{\alpha_u}}} \tag{7.9}$$

The highest temperature at $L = L_2$ is calculated to arise in the refining zone $(L_1 < L < L_2)$, when the glass temperature should be a uniform $\vartheta_{G.2}$ according to definition. The inside crown temperature at $L = L_2$ is also calculated from Equation (7.8), where the value for ϑ_F must be inserted at L_2, which is also derived from the modified system of equations.

This simplified equation assumes that a direct heat exchange between crown and glass bath is prevented by an opaque cover of flame. Furthermore, the spatially diffuse character of heat exchange by radiation has not been considered and only heat exchange vertical to the glass surface has been included. The crown temperature calculated with simplifications therefore corresponds more to an order of magnitude than to a real temperature. This simplification will be avoided in Section 7.3.3.2.

The ratio of zone lengths L_1/L_2 as a function of refining time, batch cover, specific melting rate (weight/time × area), and batch thickness is covered by

$$\frac{L_1}{L_2} = 1 - \frac{s}{(1-C)\,d}\,\frac{t_S}{24\rho_{\text{glass}}}\,\frac{1}{}. \tag{7.10}$$

This allows a simple system of equations to be developed using surprisingly few but well based assumptions which permit the interplay of the most important thermotechnical parameters to be observed and which represents, as a first approximation, what can be called a mathematical model of a tank furnace.

If Equations (7.5) and (7.7) are examined for their structure, then G/Ak and $\dot{q}_B/B$ stand out. The former is a dimensionless quantity involving heat exchange typical of all nonstationary processes involving the balancing of temperatures; G has the character of a specific heat value and bears some relationship to the specific melting rate, G/A, which is one of the most important parameters governing the dimensions of a tank furnace. But G/Ak is more suitable as a thermotechnical parameter because of its dimensionless nature and the inclusion of the heat transmission coefficient, k. There are, it must be admitted, some difficulties in the determination of k, but this can be done using Equation (7.5). The second important parameter, $\dot{q}_B/B$, is the quantity used for the consumption of thermal energy by the installation; it is very sensible but is, surprisingly, only used rarely in practice. Indirectly it is used as a combination of the specific energy consumption (kJ/kg glass) and the specific melting rate (kg glass/m^2d).

Surprisingly, the superstructure temperature, a parameter of considerable complexity and difficult to determine, does not appear in the final equation. The temperatures $\vartheta_{G.1}$ and $\vartheta_{G.2}$ are not superstructure temperatures but technical parameters of which $\vartheta_{G.2}$ in particular influences glass quality and the required residence time, t_S. It is possible to determine the 'flame gas' temperature via ϑ_F, a quantity which approaches the concept of superstructure temperature. In order to determine the crown temperature, which has the advantage of being precisely defined and measurable, the equations were modified by the introduction of the crown resistance.

The thickness of the glass or the batch, d, the charging speed, v, and the proportion of the glass surface not covered by batch, C, are also undoubtedly important for the calculation.

For the specific energy consumption the important quantities are the total energy used for melting down the batch and refining the glass, the heat losses from the superstructure (excluding heat losses from the tank basin), and the regenerator efficiency, η_k.

7.3.3 The Trier and Voss glass tank model: results

7.3.3.1 Fuel input, specific energy consumption, specific melting rate (related to melting end area), depth of glass

Using the equations developed above, examples were calculated and presented graphically. The following quantities were chosen from practical experience:

$$\vartheta_{G.1} = 1000°C$$
$$\vartheta_{G.2} = 1500°C$$
$$k = 0{\cdot}2\ kW/m^2K$$
$$K_1 = 0{\cdot}43 \times 10^{-3}K^{-1}$$
$$\eta_k = 0{\cdot}65$$
$$\varphi = 0{\cdot}5 \text{ or } \xi = 2$$

$t_S = 1{\cdot}7$ kJ/kg glass, including the heat of reaction and the enthalpy of the batch gases.

The term $K_1(1-\eta_k)$ repeatedly appears in Equations (7.5) and (7.7) but K_1 and η_k do not appear as separate quantities; $K_1(1-\eta_k)$ is therefore of major importance. For normal stoichiometric combustion ($K_1 = 0{\cdot}43 \times 10^{-3}K^{-1}$) with a regenerator efficiency, η_k, of 0·65 the calculated results apply equally well to combustion using air enriched to 30% O_2 ($K_1 = 0{\cdot}302 \times 10^{-3}K^{-1}$) with a regenerator efficiency of 0·5 since the value of $K_1(1-\eta_k)$ is the same in both cases. When oxygen enriched air is used, only the reduction in the quantity of waste gases for the same fuel energy density needs to be considered.

The literature often refers to the relationship between heat input and melting rate. If the quantities $\dot{q}_B/B$ and G/A are chosen (which, in this connection, is only a change in scale when compared to the relationship $Q = A + BS$), then the results obtained by using Equations (7.5) and (7.7), including variations of other parameters, are shown in Figures 7.5 and 7.6. The resultant curves bend slightly upwards, corresponding well with those derived empirically. Whereas a change in superstructure losses causes a parallel shift of the curves (due to the effect of $\dot{q}_a$ remaining constant for an increase in specific melting rate), a change in the proportion of glass surface covered with batch, C,

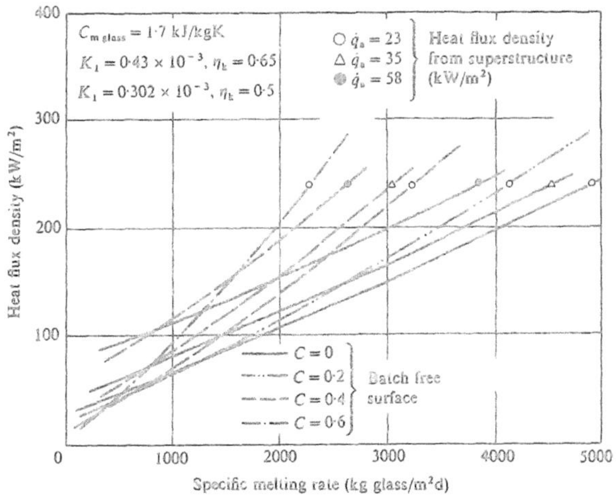

Figure 7.5. Heat flux density supplied as a function of specific melting rate for different degrees of batch cover[7]

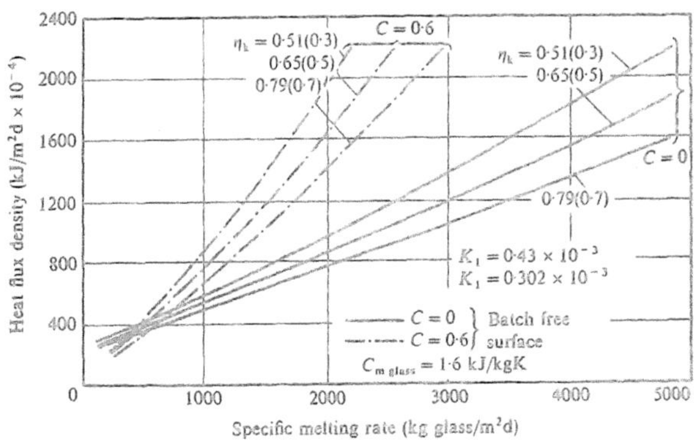

Figure 7.6. Heat flux density supplied as a function of specific melting rate for different regenerator efficiencies[7]

or in the regenerator efficiency produces a rotation (as the specific melting rate increases so the influence of C and η_k becomes stronger).

Figure 7.6 shows that an increase in regenerator efficiency has the same effect as an increase in the percentage of glass surface covered with batch. The data given in parentheses refer to results obtained with combustion air enriched to 30% O_2 ($K_1 = 0{\cdot}302 \times 10^{-3}K^{-1}$).

A pleasant surprise is Figure 7.7 which shows the relationship between specific fuel energy (specific energy consumption) and specific melting rate (referred to surface area); the well known hyperbolic relationship is the result. A comparison with corresponding data, albeit older, is given in Figure 7.8. The values derived empirically tend to be higher than the calculated values and this can be readily understood if it is remembered that the data in Figure 7.8 refer to the total furnace, whereas the

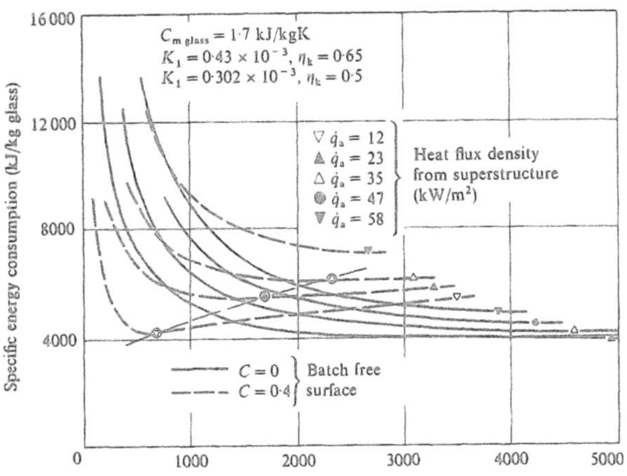

Figure 7.7. Specific energy consumption as a function of specific melting rate (calculated values)[7]

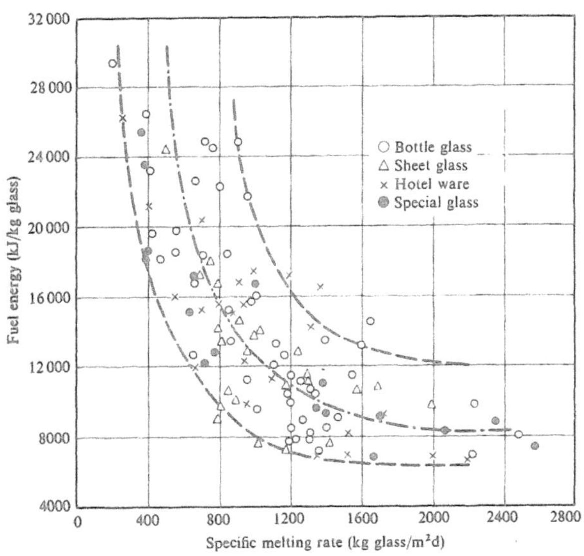

Figure 7.8. Specific fuel energy as a function of specific melting rate (measured values)[7]

calculated data refer to the melting end, except for a small preheating zone. Interesting, but as yet not theoretically based, is the minimum in the specific energy consumption, which is particularly well defined for well insulated furnaces of low pull with a low proportion of the glass surface covered with batch. A physical explanation for this can be based on a decreasing specific energy consumption with increasing pull and a localised melting rate in the batch melting zone which is higher when the surface is slightly covered than when it is ideally covered; this will be demonstrated below.

The influence of batch cover on the specific energy consumption of the furnace is particularly clear from Figure 7.9. If it is remembered that for a batch cover of $C = 0{\cdot}6$ (40% of the surface is covered with batch) and that a well designed and operated system of batch feeding in thin layers has $C = 0{\cdot}2$, then the beneficial influence of good batch cover on the specific energy consumption will be recognised, especially under conditions of high pull. On the other hand, it is surprising that for specific melting rates below 1 t glass/m²d, less well covered furnaces give better figures than those with ideal batch cover.

The importance of regenerator efficiency, particularly for furnaces operating with poor batch cover, can be seen from Figure 7.10. With ideal batch cover and a specific melting rate of, for example, 2 t glass/m²d an improvement in the regenerator effici-

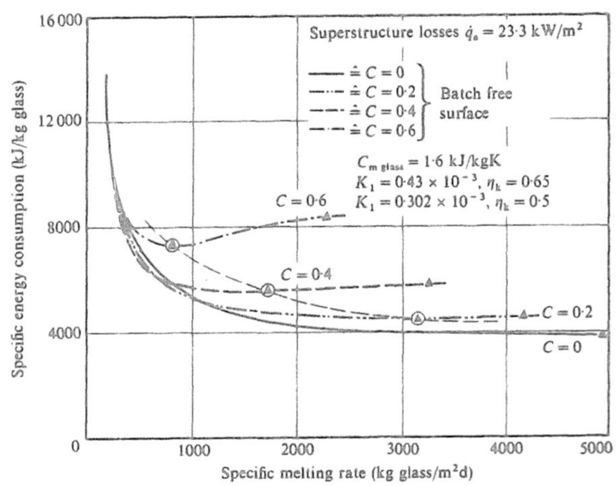

Figure 7.9. Specific energy consumption as a function of specific melting rate for constant superstructure losses[7]

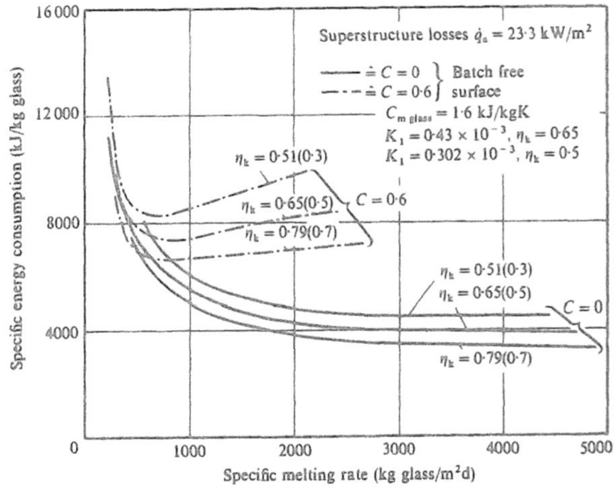

Figure 7.10. Specific energy consumption as a function of specific melting rate (effect of regenerator efficiency)[7]

ency from 0·51 to 0·79 results in a reduction of the specific energy consumption by about 1090 kJ/kg glass; when $C = 0{\cdot}6$, the reduction is 2500 kJ/kg glass.

The influence of the type of glass on the specific energy consumption and specific melting rate can be assessed from the different values of the specific heat, c_m, of the relevant glasses shown in Figure 7.11. This specific heat is characteristic of the total heat consumption (including heats of reaction and enthalpy of waste gases) during the melting of the batch followed by refining of the glass at 1500°C; for sheet glass, $C_m = 1{\cdot}9$ kJ/kgK, for tableware 1·67, and for laboratory and lead glasses 1·47. The different values of C_m becomes significant only at high specific melting rates. Lead crystal glasses appear in a very favourable light when considering the specific energy consumption due to their high density; this is not really an objective judgement since, in commercial terms, one should not be primarily concerned with the weight but with the volume of glass.

Figure 7.11 also shows to some extent the influence of an increased proportion of cullet in the batch. The studies so far have always assumed a uniform fuel energy flux density, $\dot{q}_B$, over the whole length of the tank and a constant percentage rate of the melting down of the batch blanket from above and below ($\varphi = 0{\cdot}5$).

If the energy flux density in the melting zone is chosen to be different from that of the refining zone and, in addition, the proportions of batch melting down from above and below are varied, then the multiplicity of operating conditions can be investigated and compared with one another.

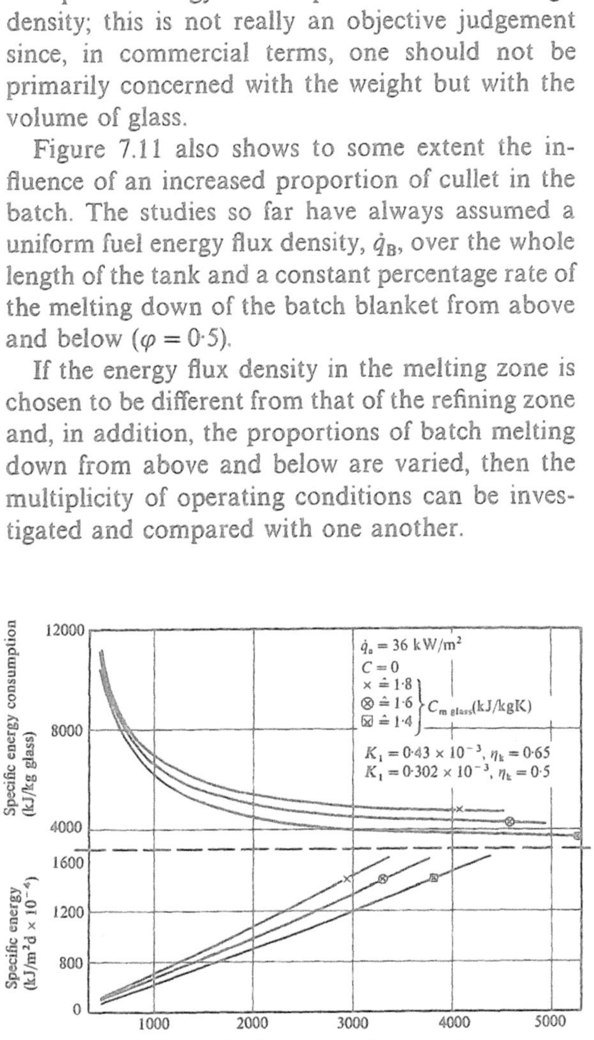

Figure 7.11. Specific energy consumption and heat input as functions of specific melting rate[7]

If the fuel energy flux densities in the refining and melting zones are $\dot{q}_{B2}$ and $\dot{q}_{B1}$ respectively, and their ratio is m, then

$$\dot{q}_{B1} = m\dot{q}_{B2}. \qquad (7.11)$$

If $m > 1$, it implies a higher energy flux density in the melting zone of the tank. If this relationship is inserted into the initial equation then the relevant quantities give the curves shown in Figure 7.12. It should be noted that as the fuel energy flux density increases in the melting zone so the specific energy consumption decreases; this almost always corresponds to practical experience. The influence decreases with increasing fuel energy flux density.

Another variable to be examined was the percentage of the batch blanket being melted down from below (by the molten glass) at values of 50% ($\varphi = 0{\cdot}5$), 30% ($\varphi = 0{\cdot}7$), and 10% ($\varphi = 0{\cdot}9$). The energy required for melting the batch from below originates in the extra energy radiated into the glass in the refining zone and carried from there by glass currents (Figure 7.4). A calculation of this influence gives the remarkable result that increased melting from below (smaller values of φ) leads to a higher specific energy consumption. With increasing heat

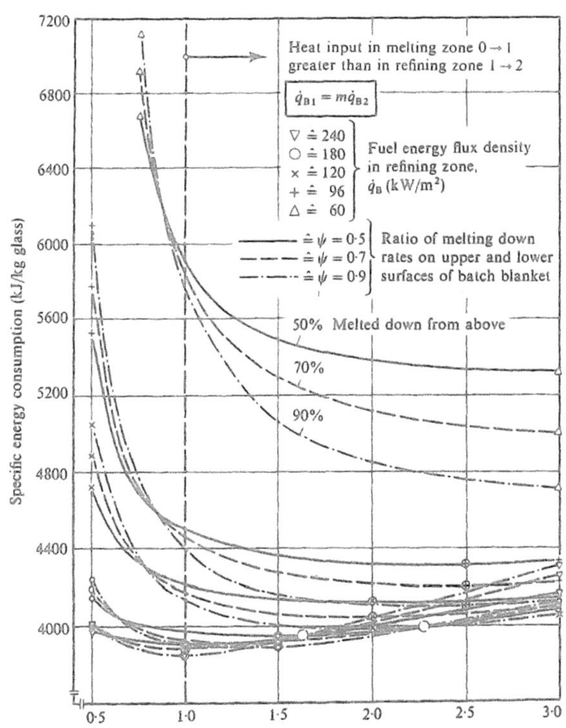

Figure 7.12. Relationship of specific energy consumption to the fuel energy distribution factor m; 80% of melting zone covered with batch ($C = 0{\cdot}2$)[8]

flux density this tendency becomes weaker. The reason for this at first sight surprising result is that the energy required for melting the batch from below must be carried by the indirect route from the refining zone ($L_2 - L_1$) which is thermally less efficient.

For technological reasons this line of thinking must not be followed to the extreme case of preventing all melting from below; the process requires that the batch be floating. At the batch/glass interface the glass is cooled and this acts as a brake on the forward movement of the batch. Only a little heat can penetrate the batch to heat the glass below, and this is especially true for a continuous batch blanket; therefore a failure in the return current from the refining zone would lead to the glass in the melting zone solidifying. For this reason heat transport from the refining zone to the underside of the batch blanket is necessary. Linked to this is the heat transfer from glass to batch and the melting down of the batch from below. The magnitude of this return heat flow depends, among other things, on the size of the tank, the pull, and the colour of the glass; it cannot yet be assessed and must be based on more accurate investigations of the heat transfer from glass to batch and the resultant glass currents.

To a certain extent the factor φ involves the depth of the glass: shallow tanks should be thermally more efficient than deep ones but this is only true under certain conditions. Further calculations show that the results given in Figure 7.12 apply only to plants operating at low to medium pulls. For plants such as container glass tanks operating at a high pull, the influence of φ on the specific energy consumption is reduced and even inverted. In this case deep tanks are thermally more efficient than shallow ones.

A change in which the refining zone is made deeper than the melting zone improves thermal efficiency, a fact confirmed in more recent work.

7.3.3.2 Radiation of the superstructure, regenerator efficiency, and crown temperature

The calculations given in Section 7.3.2 are based on a heat transfer from the melting chamber to the glass surface which is proportional to the difference in temperature and the total heat transmission coefficients k_1 and k_2. However, before considering the influence of radiation characteristics of the two bodies involved in the heat exchange process, it is necessary to involve the crown in the mathematical relationships for two reasons, first to take into consideration the heat transfer mechanism in glass tanks in general and second, and more specifically, to include more accurately the special nature of the continuous glass melting process and its consequences. This means that the model must be enlarged upwards to include the crown.

Figure 7.13 is a schematic which forms the basis of the enlarged field of calculation; it is a vertical section through the combustion chamber of a cross-fired tank at right angles to the longitudinal axis and across the width of the tank. The flames fire with periodic reversal of, say, 20 min at right angles to the longitudinal axis from left to right or from right to left across the width of the tank and therefore at right angles to the direction of material transport of batch and glass.

The glass surface temperature, T_O, the inside crown temperature, T_G, and the combustion chamber temperature, T_F, are regarded as constant across the width but not along the length of the tank; periodic reversal, in fact, almost ensures that this condition is met in practice. The burnout of the flame is set linear to its flow direction so that, under the conditions stated, it gives rise to a constant combustion chamber temperature, T_F, and the flame has completed its combustion as it leaves the superstructure. Crown, combustion chamber, and glass surface are regarded as grey emitters. The corresponding emissivities are ε_G for the crown, ε_O for the glass surface, and $\varepsilon_F = 1 - \exp(-k\kappa)$ for a combustion chamber of height h: the extinction coefficient κ is regarded as a variable.

Since the processes taking place in a glass tank are carried out at high temperatures, the convective part of the heat transfer is ignored compared to heat transfer by radiation. Only radiation in the vertical direction is being considered. Also noted

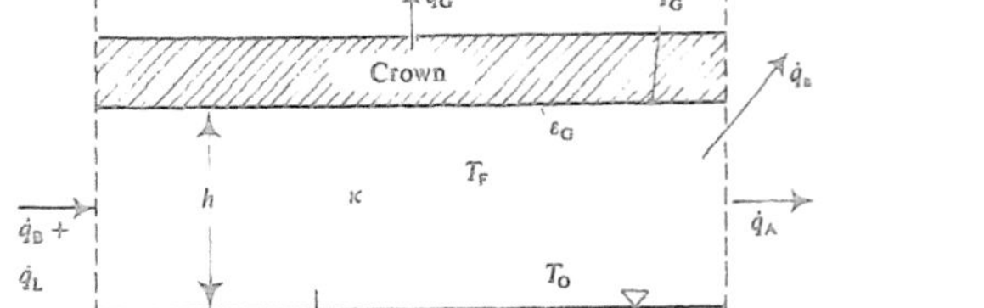

Figure 7.13. Schematic representation of heat transfer in the superstructure of a furnace[8]

are the heat losses of the crown, $\dot{q}_G$, and the heat losses of all other walls encircling the combustion chamber, $\dot{q}_a$.

The heat removed from the combustion chamber in the waste gases, $\dot{q}_A$, is partially recovered in a regenerator of efficiency, η_k, by being transferred to the incoming combustion air, $\dot{q}_L$, and returned to the system. The fuel energy flux density is $\dot{q}_B$, and the energy flux density transferred to the molten glass is $\dot{q}_O$.

Taking the absorption of flame gases and multiple reflections between glass surface and crown into account, expressions are developed which cover the heat flow from combustion chamber to molten glass, crown to molten glass, combustion chamber to crown, and molten glass to crown as well as reflections between glass surface and crown; they are used to develop equations covering the heat transferred to the molten glass, $\dot{q}_O$, the crown losses, $\dot{q}_G$, and the fuel utilisation, $\eta_u = \dot{Q}_G/\dot{E}_B = \dot{q}_O/\dot{q}_B$.

The equations themselves can be found in the original publication.[10]

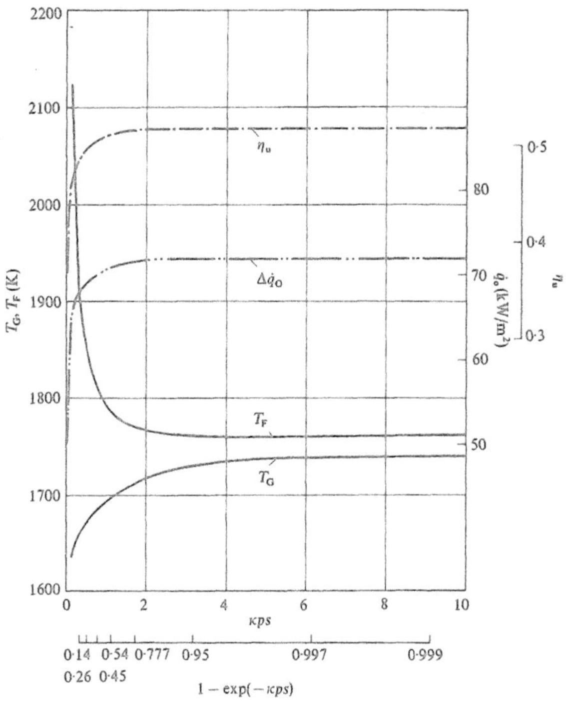

Figure 7.14. The influence of the emissivity of the combustion chamber; for thermal data, see Figure 6.27[10]

7.3.3.2.1 The influence of the emissivity of the combustion chamber, ε_F

The emissivity of the combustion chamber is defined by the expression $\varepsilon_F = 1 - \exp(-h\kappa)$, where h is the thickness of the flame and κ the extinction coefficient; as $h\kappa$ increases, so does ε_F.

It is important that both flame thickness and extinction coefficient enter into the equation for ε_F as a product so that both have an equally strong influence on the emissive power of the flame. This implies that an increase in the thickness of flames which have low extinction coefficients can effectively increase the emissivity. As far as heat transfer is concerned it is an advantage if, when flames of low soot content (small κ) enter the combustion chamber above the melt, they have a large volume (high h).

If the numerical values of the relevant parameters for the calculations are chosen in line with normal glass furnace conditions, then the relationships shown in Figure 7.14 result. In the region of low ε_F, between 0·2 and 0·5, the heat transferred to the molten glass and the fuel utilisation can be improved significantly by increasing the emissivity of the flame. Under conditions of complete flame cover (which applies to results in this diagram) an increase in ε_F between 0·5 and 0·6 brings hardly any advantage in terms of fuel utilisation (for definitions, see Table 5.3). As ε_F increases, so the crown temperature increases without increasing the heat transferred to the glass to any significant extent. At the same time, the situation as seen from the molten glass is that, as the emissivity of the flames increases, they become increasingly opaque to radiation from emissions further away and to secondary radiation from the crown in the direction of the glass.

Therefore good flame cover over the glass surface causes the increasing emissivity of flames increasingly to block the auxiliary heating effect of the crown, despite the fact that the crown temperature has increased. For this reason complete flame cover is intentionally avoided if oil firing with its high flame emissivity is used. By dividing the oil flames into separate single flames all parts of the crown have an opportunity to take part in the heat exchange process with the surface of the molten glass or batch. The distance between flame and crown must therefore be adequate to permit this to take place and must be increased as the thickness of the flames increases and as the distance between flames decreases.

7.3.3.2.2 Influence of heat losses through the crown

Optimum fuel utilisation can be achieved by reducing heat losses from the superstructure as far as the

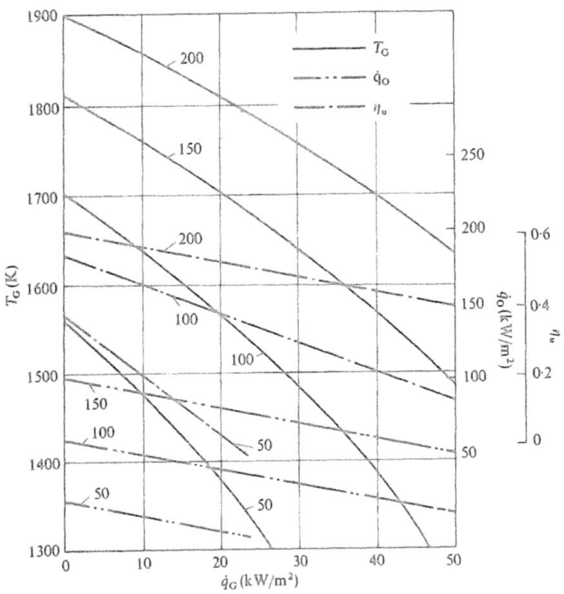

Figure 7.15. Influence of the heat losses through the crown. The number adjacent to each curve is the parameter $\dot{q}_B$ in 1·163 kW/m². (Other data as in Figure 6.27)[10]

temperature resistance or attack by molten glass on the refractories permits.

Figure 7.15 shows the influence of crown heat losses. Rising heat losses at constant fuel consumption lead to a reduction in the heat transferred to the glass and thereby to a reduction in pull and at the same time the utilisation of the fuel becomes less efficient. Furnaces with low specific fuel consumption are particularly sensitive to the effect of changes in crown heat losses on the efficiency, η_u.

The heat losses through the crown also have a major effect on the crown temperature itself. In the example cited, and for a constant rate of fuel consumption, the crown temperature decreased by an average of 40 degC as the crown heat losses increased by 5·8 kW/m². This implies, depending on the level of fuel consumption, a reduction in fuel utilisation of 5–10%. As far as fuel utilisation is concerned heat losses through the crown should be minimised as far as possible.

Figure 7.15 also demonstrates that poorly insulated tanks with high crown heat losses can achieve the same levels of heat transfer to the glass, $\dot{q}_O$, and thereby the same pull, as well insulated tanks by increasing the fuel consumption; however their fuel efficiency will always be much worse.

7.3.3.2.3 Influence of crown emissivity ε_G

A reduction in crown emissivity has three effects: the porportion of flame radiation reflected from the crown is increased; the crown heat losses at constant fuel consumption are reduced; and the possibility arises that more fuel energy could be liberated in the combustion space at constant crown temperature. As the emissivity of the crown decreases, the energy radiated towards the glass and originating in the crown temperature decreases while, at the same time, the part of flame radiation reflected at the crown increases.

The influence of crown emissivity is shown in Figure 7.16: if the heat losses are constant and only ε_G changes, then fuel efficiency, η_u, and the heat transferred to the glass, $\dot{q}_O$, remain unchanged.

It should be noted that for constant fuel consumption the crown temperature increases with increasing ε_G; this fact is technologically important since the properties of materials do not permit an unlimited increase in crown temperature. For crowns with a high emissivity the increase in temperature sets a limit to the maximum fuel consumption which is lower than that for a crown with a lower emissive power. As ε_G decreases at constant fuel consumption the crown temperature decreases. If the emissivity of the crown falls below 0·4 then, at constant fuel consumption, the crown temperature falls very rapidly. It follows that, for constant fuel consumption, the crown heat losses also fall and fuel utilisation improves.

Each curve shown is for a constant heat loss

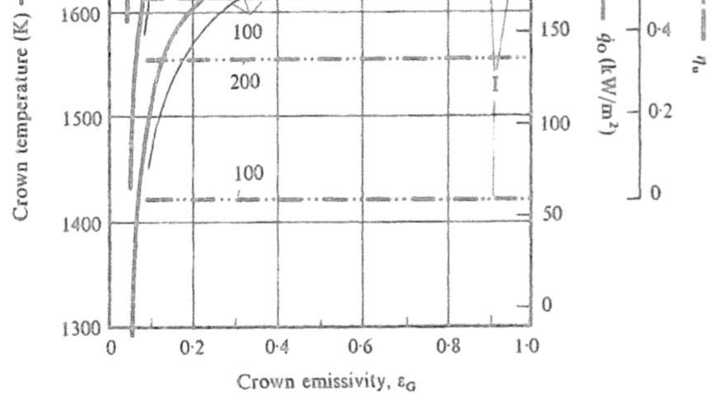

Figure 7.16. Influence of the emissivity of the crown. The numbers adjacent to the curves are the parameter $\dot{q}_B$ in 1·163 kW/m². The parameter $\dot{q}_G = 5{\cdot}8$ kW/m² for the curve I and 8·1 kW/m² for cuve II; $\varepsilon_O = \varepsilon_F = 0{\cdot}8$; $T_O = 1473$ K; $\eta_k = 0{\cdot}59$; $\dot{q}_a = 23{\cdot}3$ kW/m²; $K_1 = 0{\cdot}43 \times 10^{-3}$K^{-1}[10]

through the crown, but the heat losses change as the crown temperature changes. If the calculated heat loss of a silica crown at an inside temperature of 1500 K (1227°C) is about 5·8 kW/m², then this would rise to 7·6 kW/m² at a temperature of 1873 K (1600°C). For this reason Figure 7.16 shows the crown losses for different fuel energy flux densities at 5·8 and 8·1 kW/m². Whenever ε_G changes at constant fuel consumption, the crown temperature moves within the gap between the two curves causing only a slight change to the details discussed above.

An increase in efficiency, possibly by increasing the fuel consumption, can be obtained more easily with furnaces of low fuel consumption than with furnaces already operating at a high consumption. However, one can take advantage of an increase in fuel utilisation only if presented with a tank not subject to rapidly increasing heat losses as soon as the fuel consumption is increased.

7.3.3.2.4 The influence of the surface temperature of the molten glass ϑ_o

The influence of the surface temperature of the glass on the heat flow density directed towards the glass, the fuel efficiency, and the crown temperature is given in Figure 7.17. Changes in the temperature of

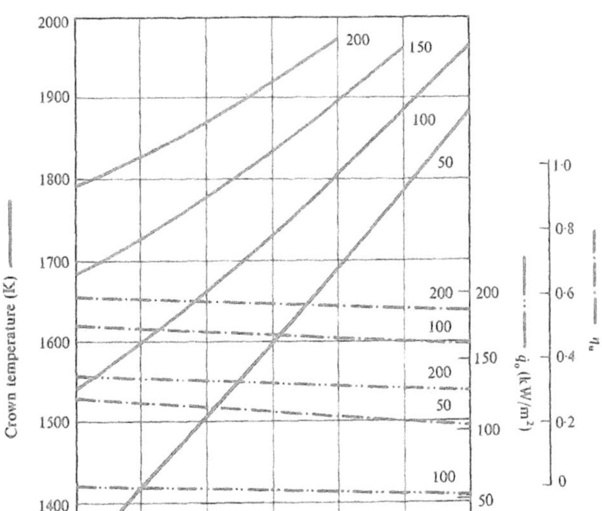

Figure 7.17. Influence of glass surface temperature. The numbers adjacent to the curves are the parameter $\dot{q}_B$ in 1·163 kW/m²: $\varepsilon_G = 0{\cdot}8$; $\dot{q}_G = 7{\cdot}0$ kW/m²; other thermal data are as in Figure 7.16[10]

the glass surface have a significant influence on the crown temperature, which is higher above zones where the glass surface temperature is higher, e.g. above regions covered in foam, and lower where the glass surface temperature is lower, e.g. above the batch melting zone. It follows that, for constant fuel input, the difference between crown and glass surface temperatures does not change much as the temperature of the glass surface increases. In parts of the furnace where the glass surface temperature is lower, more fuel energy can be liberated under a crown at about the same temperature than where the glass surface temperature is higher. In zones of low glass surface temperature this higher level of fuel energy offered leads to higher heat flow densities than in zones of high glass surface temperature.

For example, if a crown temperature of 1800 K must not be exceeded, then in the case calculated, 240 kW/m² can be introduced in the form of fuel energy where the surface temperature is 1030°C but only 60 kW/m² where it is 1520°C; at the same time the fuel utilisation effeciency decreases from 61% to 20%. Good fuel utilisation by the tank as a whole can be achieved by supplying most of the fuel to those parts where there is a low glass surface temperature (the batch melting zone).

7.3.3.2.5 The influence of the emissivity of the molten glass ε_o

An increase in the emissivity of the glass surface has, from a thermal point of view, the same effect as a reduction in the temperature of the glass surface. Figure 7.18 gives the values for crown temperature, efficiency of fuel utilisation, and heat flow density of the molten glass for difficult surface emissivities. As ε_o increases, so the bath of molten glass absorbs increasing proportions of thermal radiation from the combustion chamber and the efficiency of fuel utilisation increases. For a constant fuel energy flux density of, for example, 120 kW/m², the efficiency of fuel utilisation, η_u, increases by about 6% as ε_o increases from 0·2 to 0·8.

The glass surface emissivity has a very marked influence on the crown temperature: as ε_o increases so the proportion of flame radiation reflected by the glass surface diminishes, resulting in a much lower crown temperature in zones of high ε_o than in zones of low ε_o. This reduction is particularly marked in zones of low emissive power. For a constant glass surface temperature, more fuel may be introduced into zones of high ε_o without the risk of overheating the crown than into zones of low ε_o.

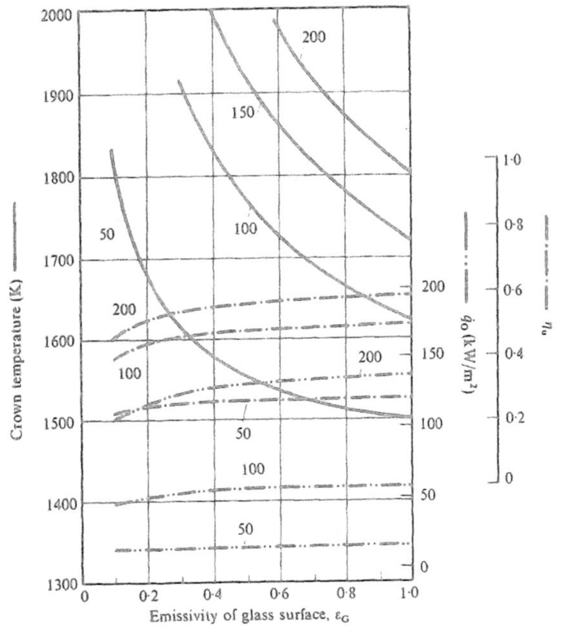

Figure 7.18. Influence of emissivity of surface of molten glass. The numbers adjacent to the curves are the parameter $\dot{q}_B$ in kW/m²; $\dot{q}_G = 5{\cdot}8$ kW/m²; other thermal data are as in Figure 7.16[10]

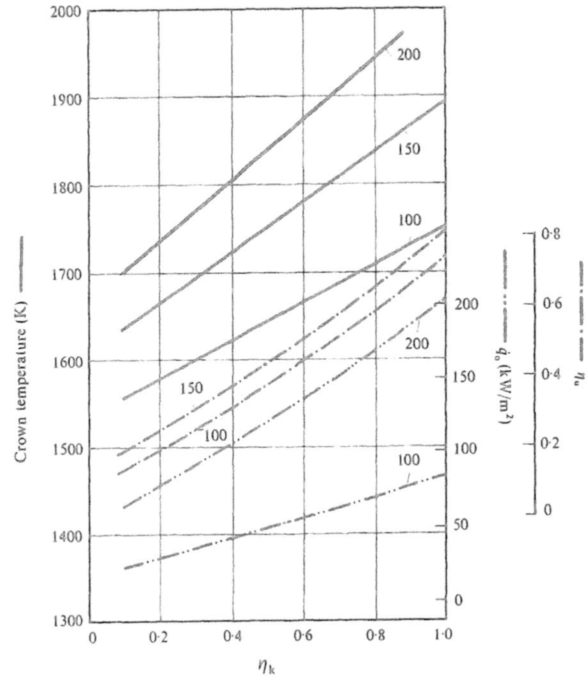

Figure 7.19. Influence of regenerator efficiency. The numbers adjacent to the curves are the parameter $\dot{q}_B$ in kW/m²; other data are as in Figure 7.16[10]

Presupposing that flame combustion is complete within the combustion chamber and that the regenerators are of sufficient size in zones of high ε_0 and low glass surface temperature (the batch melting zone), this leads to a marked increase in the heat flow into the glass or batch while, at the same time, increasing the utilisation of the fuel.

7.3.3.2.6 The influence of the regenerator efficiency η_k

The transfer of part of the waste gas heat to the combustion air not only reduces waste heat losses but at the same time increases flame temperature and thereby improves the conditions for heat transfer. The influence of regenerator efficiency on the heat transferred to the molten glass, the efficiency of fuel utilisation, and the resultant crown temperature is given in Figure 7.19. While the fuel supply stays constant, the heat transferred to the bath of glass and the efficiency of fuel utilisation increase with increasing regenerator efficiency.

It must not be overlooked that at constant fuel input but increasing regenerator efficiency the crown temperature will rise. As an example, the case calculated corresponds to a maximum permissible crown temperature of 1600°C in a furnace of high fuel consumption, namely 240 kW/m² fuel energy flux density, and a regenerator efficiency of 60%, which is normal for modern regenerators and is sometimes even exceeded. If the regenerator efficiency is increased the fuel consumption can be reduced without the crown temperature falling. At the same time the efficiency of fuel utilisation increases, which is equivalent to the plant operating more profitably.

7.3.3.2.7 The influence of the difference in melting rate of the batch blanket between upper and lower surfaces (factor φ)

In the mathematical model the factor φ established that only part of the batch is melted from above and that the remainder is melted from below by contact with molten glass; φ may have a value between 0 and 1, the proportion of batch melting from above increasing as φ increases (see Figure 7.20).

Since the layer of batch cannot pass any heat downwards into the molten glass while it is melting, it follows that, for a continuously operating furnace, the heat necessary for melting from below must be transported from batch-free zones, such as the refining zone, under the batch blanket, to the batch itself. Since φ is a measure of this heat transport, this factor can also be used to fix the

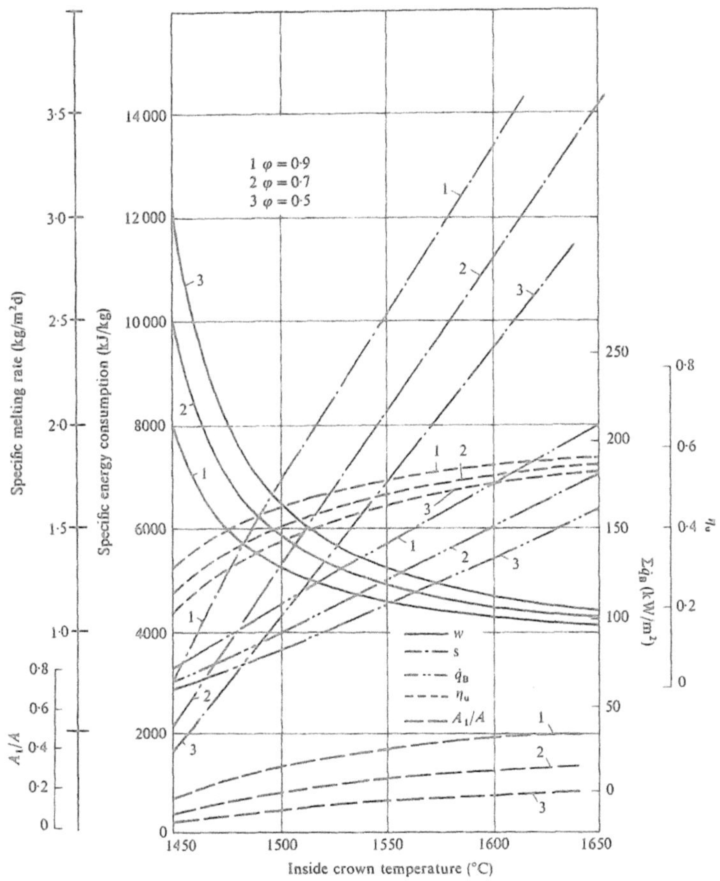

Figure 7.20. Influence of the different melting rates between the upper and lower surfaces of the batch blanket[10]

system of glass currents which exists in the melt, especially the return current from the refining zone to the melting zone.

Looking on the furnace as a whole, the influence of φ on the specific melting rate, s, the specific energy consumption, w, the fuel input, and the efficiency of fuel utilisation is shown in Figure 7.20, which uses the same parameters as Figure 7.21, for $\eta_k = 0{\cdot}65$ and different values for the crown temperature required. The shapes of the curves remain the same for different values of φ. For all crown temperatures, as φ, i.e. the proportion of batch melted from above, increases, the specific melting rate, s, increases. The area covered by batch in the melting zone, A_1, increases and the fuel energy flux density, $\dot{q}_B$, has to increase to allow the required crown temperature to be reached; this follows from the fact that, due to the relatively high emissivity of the batch surface in the melting zone, the highest fuel energy flux density can be applied there.

The higher the proportion of material melted from the top of the batch blanket and the lower the proportion of heat which has to be supplied by the thermally inefficient route via the refining zone, the higher the efficiency of fuel utilisation and the lower the specific energy consumption for melting the glass.

7.3.3.3 Results for the tank furnace as a whole

A comparison of different operating conditions was carried out by setting constant crown temperatures (one for each operating condition) along the centre line of the tank and calculating what fuel distribution would be needed to achieve whatever temperature distribution was required. At the same time the heat flux density to the batch/glass surface, $\dot{q}_O$, and the fuel utilisation efficiency were determined. The average efficiency for the whole furnace is the sum of the efficiencies of the separate zones, making due allowance in the individual efficiencies for the size of each zone. The individual fuel utilisation efficiency is largest in the batch melting zone, passes through a minimum in the foam region, and increases again at the downstream end of the tank; however, this assumes that the regenerator efficiency is the same throughout.

If the data for the whole furnace are plotted, not in relation to the centre line, but as the average values for the whole furnace, then Figure 7.21 results. As the crown temperature needs to be increased, so more fuel is required: this leads to an increase in the specific melting rate, s, and the proportion of the surface of the melting zone covered with batch, A_1/A. The specific melting rate of the tank increases linearly with increasing crown temperature for all regenerator efficiencies: in the example calculated here for the parameters selected, an increase in crown temperature above the centre line of the tank from 1500 to 1600°C results in an increase in the specific melting rate from 2 to 2·7 t/m^{2}d.

At constant regenerator efficiency the specific energy consumption for melting glass decreases with increasing crown temperature and increasing specific melting rate. As far as the reduction in the specific energy consumption for glass melting is concerned, an increase in crown temperature is more effective for a furnace operating at a low specific energy input than for one already operating at a high input; this means that the reduction in specific energy consumption with increasing melting rate is particularly marked for furnaces operating at low specific melting rates.

Regenerator efficiency has a marked effect on the specific energy consumption; at constant specific melting rate and crown temperature the specific energy consumption decreases with increasing regenerator efficiency. An increase in η_k from 0·65 to 0·8 does not produce quite as much energy saving as the step from 0·35 to 0·5, but is significant all the same. For this example and a specific melting rate of 3·5 t/m^{2}d, an improvement in regenerator efficiency from 0.35 to 0·8 would save about 40% of the fuel.

The curves drawn as thin lines in Figure 7.21 relate to the condition where, for $\eta_k = 0{\cdot}65$, the crown temperature in the melting zone is always 70 K lower than in the refining zone. The absolute values are in this case somewhat lower, but the trends of the curves remain the same.

The steady decrease in specific energy consumption shown in Figure 7.21 right into the field of very high specific melting rates only occurs in practice if the heat losses arising from firing are not too large and the regenerator chambers are sufficiently large.

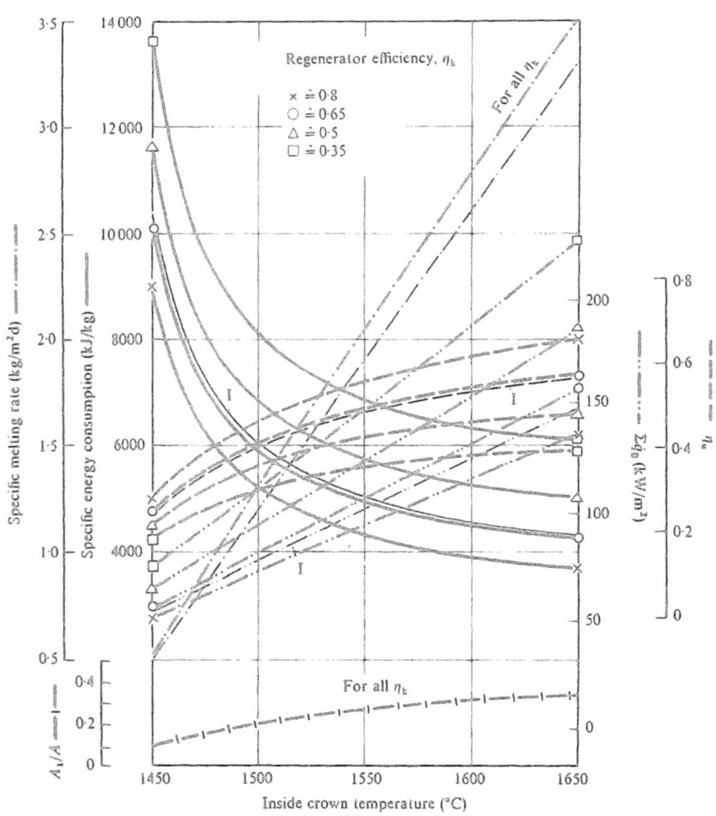

Figure 7.21. Specific melting rate and thermal data for a furnace as a whole: $\varepsilon_G = \varepsilon_F = 0{\cdot}8$; ε_O (batch) $= 0{\cdot}8$; ε_O (refining zone) = linear increase from 0·4 at L_1 to 0·8 at L_2; $T_{O1} = 1473$ K; $T_{O2} = 1773$ K according to Figure 7.7; $\dot{q}_G + \dot{q}_a = 50$ kW/m^2; $\varphi = 0{\cdot}7$; $K_1 = 0{\cdot}43 \times 10^{-3}$ K^{-1}. The thinner lines apply when, for $\eta_k = 0{\cdot}65$ the crown temperatures in the batch melting area are always 70 K lower than in the refining zone of the tank[10]

If, on an existing furnace, there is a decrease in regenerator efficiency at the same time as there are increases in specific melting rate, fuel consumption, and quantities of waste gases, then, at very high specific melting rates, the specific energy consumption will also increase.

The gradient of the curve for the specific melting rate as a function of the crown temperature in Figure 7.21 decreases if, for an increasing crown temperature, the surface temperature of the melting batch and of the glass are also included in the calculation.

7.3.3.3.1 The application of the mathematical model to process economics

The relationships explored in Sections 7.3.3.2.6 and 7.3.3.3 have demonstrated, among other things, the importance of heat recovery in the economics of a glass tank furnace. This also raises the question as to what other investment should be incurred during construction in order to increase the regenerator efficiency as required.

This question of profitability can only be assessed by comparing the financial gain arising from the saving in fuel due to increased regenerator efficiency with the cost of additional investment in larger regenerator chambers. The mathematical model gives the different specific energy consumption rates for different regenerator efficiencies. At the same time, and this applies in this instance, any determination of the size of the regenerator chamber must be based on the operating data used in the mathematical model which produces the required regenerator efficiency.

The set of parameters used for the calculations covered by Figure 7.21 are used in Figure 7.22 to show the relationship, assuming a constant inside crown temperature of 1580°C, between regenerator efficiency, η_k (which ranges from 0·4 to 0·8), and the specific energy consumption for glass melting, w, applying the fuel energy flux density required, $\dot{q}_B$, and the fuel utilisation efficiency, η_u, at each level of regenerator efficiency. The set of parameters chosen give a specific melting rate of 2·5 t/m^{2}d. If the inside width of the tank is 7 m, and if the length-to-width ratio is assumed to be 1·6, then the melting end area is 78·4 m^2 and gives a pull of 195 t/d.

Calculations covering combustion determine the waste gas flows for each level of regenerator efficiency in m^3/h and the quantities of air required for combustion which will be preheated in the regenerators. The next step is to determine the size of regenerator chamber which is capable of providing the regenerator efficiency used as a basis of these calculations for the quantities of waste gases and combustion air involved.

If the type of checkerwork, refractory, width of ducts, refractory dimensions, and entry temperatures of waste gas and air into the checkers are fixed, then the relationship between regenerator efficiency and volume of refractory, weight of refractory, and corresponding checker volume will be as shown in Figure 7.22; the three parameters increase progressively as the regenerator efficiency increases. Regenerator efficiencies of 80% can now just about be achieved with sensible levels of investment due to the low specific heat of air compared to that of the waste gases. Even with maximum preheating (in the extreme case when it is heated to the same temperature as the waste gases), the air can only absorb a limited percentage of the heat transferred from the waste gases to the regenerator.

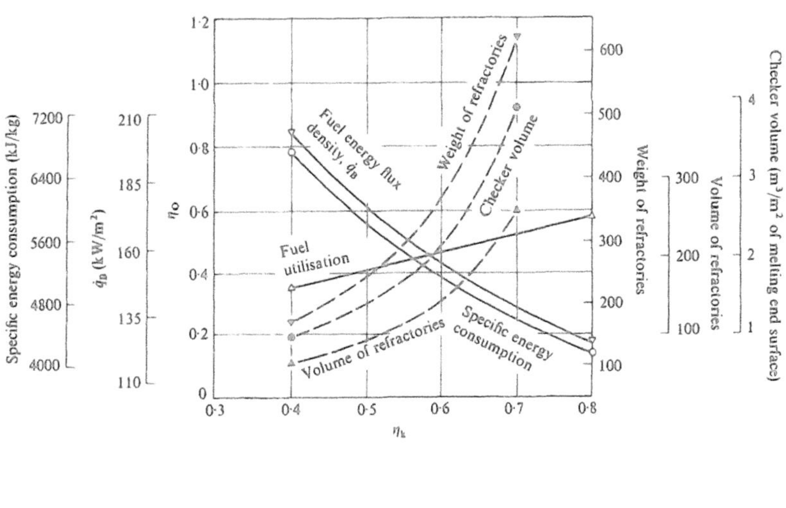

Figure 7.22. Influence of regenerator efficiency on the size of regenerator chamber and specific energy consumption. Other parameters are as in Figure 7.15. Weights and volumes of refractories cover both sides of the furnace[10]

When studying the actual economics it is necessary to compare the expected annual costs of oil fuel for a given pull at different regenerator efficiencies with the total investment in regenerators at each efficiency (materials, labour, financing, etc.). As the regenerator efficiency increases for a constant pull, so the fuel costs per year decrease; in parallel the costs of the regenerators needed to achieve these levels of efficiency progressively increase. Starting from a low efficiency, an improvement in η_k brings large benefits and a saving in fuel costs can be achieved with a relatively low investment in additional refractories. The effects are reversed if an attempt is made to improve the efficiency of a regenerator already operating at a high efficiency. This is only really profitable if the life of the checkers can be guaranteed for the length of the furnace campaign.

7.3.3.3.2 The effect of electric boosting

By making suitable adjustments, the mathematical model can also be used for studying the special thermal effects created by providing additional heating by electric boosting. This extensive and special subject cannot be treated more fully in this volume. The interested reader is referred to specialist literature.[9]

8 Electric tank furnaces

Glass furnaces can be heated by electricity, either indirectly via heating elements or directly by passing an electric current through the glass. The indirect method has only been used for very small tanks and pot furnaces or for heating part of a tank, e.g. the working end. The direct heating of glass is possible as a result of ionic conduction in molten glass, which depends strongly on the alkali and alkali earth content of the glass itself. There are all-electric glass tank furnaces, mixed melters using similar quantities of fossil fuel and electricity, and tanks operating in the main with fossil fuel but using supplementary electric heating.

In the Federal German Republic units of energy in the form of electricity are considerably more expensive than energy from fossil fuels; the use of electric melting becomes economically feasible only when particular advantages can be achieved. This applies when glasses are melted from highly volatile batch constituents or when high melting rates are needed for glasses of low infrared transmission.

The size of the furnace is also an important factor: small fuel-fired tank furnaces are much less efficient than large tanks but this is far less evident for electric furnaces, where small furnaces can operate more profitably than the equivalent fuel-fired furnace despite higher electricity costs.

Electrically heated furnaces require a lower capital investment than conventionally heated furnaces, but carry higher operating costs. In countries where electricity demand varies widely, such as, for example, in Switzerland, mixed melters are used; in autumn and winter they operate mainly on fossil fuel and in spring and summer, when the snow melts, mainly with electricity.

The possibility of directly heating glass by passing an electric current through it was recognised early, Völker taking out a patent covering this technique in 1900.[1] For many years progress was halted by the main problem of discoloration originating from the electrodes. Only with the development of molybdenum electrodes in the early 1950s did the required breakthrough occur which led to the wider application of electric melting. The increasing cost of fossil fuels relative to electricity, together with pressures aimed at the reduction of dust and harmful gas emissions, have led to an increasing use of electric melting.

8.1 Electrical properties of molten glass

8.1.1 Electrical conductivity

The electrical conductivity of glass depends, in the main, on ionic transport. The ions which have the weakest links to the glass network (network modifiers) are the alkali ions and they, therefore, become the carriers of the electric current. Fused silica and borate glasses contain little alkali and their electrical conductivity is therefore proportionally lower.

The conductivity of alkali silicate glasses is influenced not only by the quantity but also by the size of the ions involved. Compared with Na^+ ions, K^+ ions are bound to the network by much weaker forces but their size causes them to meet more resistance during migration; Li^+ ions are smaller than Na^+ ions, but because they are bound more strongly, their migration also is more difficult than that of Na^+ ions. The Na^+ ion is therefore the main contributor to electrical conductivity in molten glass. Glasses containing mixed alkalis give rise to complicated relationships in which the electrical conductivity can pass through minima.

The mobility of ions increases as temperature increases and as viscosity decreases, causing the electrical conductivity to increase significantly with increasing temperature. If the resistance of a conductor is R, its cross section A, and its length l, then the resistivity can be defined as

$$\rho_{el} = \frac{RA}{l}. \qquad (8.1)$$

Examples of the resistivities of melts of different glasses as functions of temperature are given in Figures 8.1 and 8.2; it is clear that, between the various glasses, there are large differences which can reach an order of magnitude. The chemical compositions of glasses used for Figure 8.2 are given in Table 8.1.

The determination of the resistivity or its opposite, the conductivity, of glasses from their composition has so far not been fully successful. Attempts to achieve this started with Sasek & Meissnerova[5] and are based on the relationship

$$\log|\kappa_{el}| = a + \frac{b}{T} \tag{8.2}$$

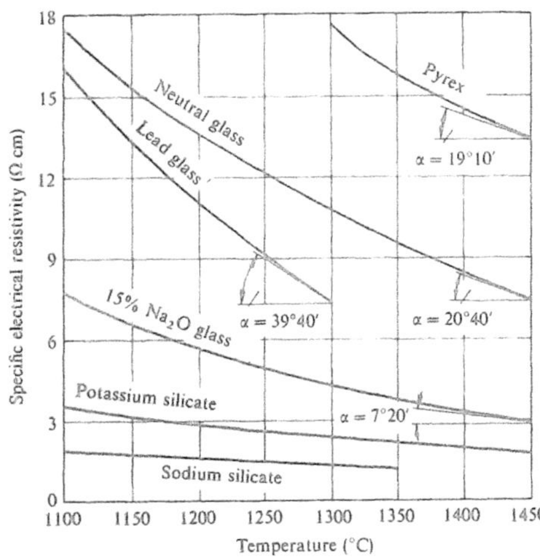

Chemical compositions (%)

Neutral glass
67·50 SiO_2
6·53 Al_2O_3
7·93 B_2O_3
3·64 K_2O
6·77 Na_2O
6·82 CaO
0·08 Fe_2O_3
0·70 ZnO
0·10 MgO

Sodium silicate
76 SiO_2
24 Na_2O

Pyrex glass
80·50 SiO_2
2·05 Al_2O_3
11·90 B_2O_3
3·91 Na_2O
1·03 K_2O
0·40 CaO
0·18 Fe_2O_3
0·04 MgO

Potassium silicate
71 SiO_2
29 K_2O

15% Na_2O glass
71·60 SiO_2
0·48 Al_2O_3
0·14 Fe_2O_3
10·70 CaO
2·02 MgO
15·00 Na_2O

Lead glass
56·50 SiO_2
0·13 Al_2O_3
5·40 Na_2O
8·60 K_2O
29·50 PbO

Figure 8.1. Specific electrical resistance of different glasses[2]

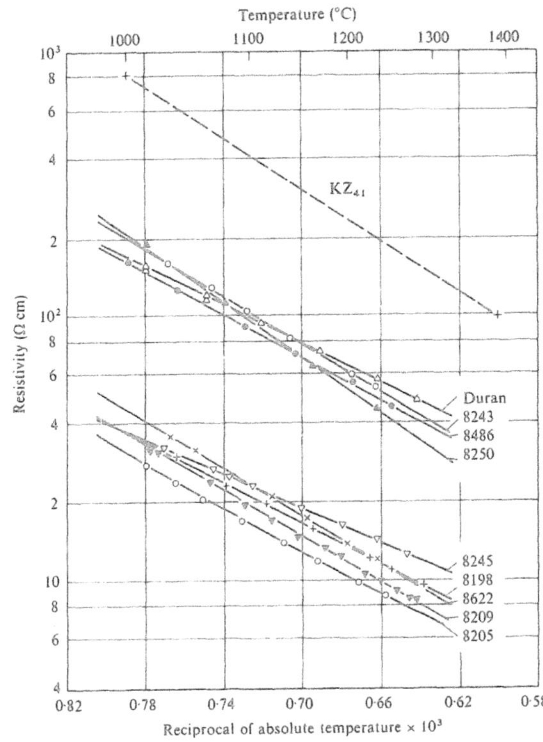

Figure 8.2. Resistivity of various glasses according to Baucke.[3] KZ_{41} is a fusion cast corundum–zirconia block with 41% ZrO_2

where a and b are constants related to the composition and T is the thermodynamic temperature. Glasses in the wt% composition range 69·45–76·0 SiO_2, 0·5–3·0 Al_2O_3, 0·05–0·5 Fe_2O_3, 7·0–8·5 CaO, 2·5–4·5 MgO, 0–0·5 K_2O, 13–16 Na_2O give the following relationships for a and b between 1273 and 1673 K:

$$\begin{aligned} a = 0{\cdot}505 &+ 0{\cdot}025(Na_2O) + 0{\cdot}08(K_2O) \\ &+ 0{\cdot}035(MgO) + 0{\cdot}056(CaO) \\ &+ 0{\cdot}004(Al_2O_3) + 0{\cdot}013(Fe_2O_3), \end{aligned} \tag{8.3a}$$

$$\begin{aligned} b = -3530 &+ 72{\cdot}72(Na_2O) - 120{\cdot}8(K_2O) \\ &- 40{\cdot}5(MgO) - 74{\cdot}27(CaO) \\ &- 11{\cdot}28(Al_2O_3) + 8{\cdot}0(Fe_2O_3). \end{aligned} \tag{8.3b}$$

The proportions of individual constituents are entered as weight percentages and κ_{el} can then be

Table 8.1. *Chemical compositions (wt%) of glasses investigated in Figure 8.2*[3]

Glass	*SiO_2*	*B_2O_3*	*Al_2O_3*	*Li_2O*	*Na_2O*	*K_2O*	*CaO*	*MgO*	*BaO*	*ZnO*	*CeO_2*	*PbO*	*TiO_2*
Duran 50	80·60	12·80	2·30	–	3·60	0·60	–	–	–	–	–	–	–
8198	66·50	–	3·80	0·25	7·20	7·20	–	–	12·40	–	–	1·70	–
8205	62·50	–	3·50	–	7·90	9·40	2·90	1·00	1·40	–	–	11·20	–
8209	62·30	–	3·30	–	8·80	6·90	2·70	1·70	13·00	–	0·20	–	0·60
8243	67·50	21·30	3·00	–	3·10	4·70	–	–	0·30	–	–	–	–
8245	69·00	17·00	5·20	0·50	7·00	0·10	–	–	–	0·80	–	–	–
8250	68·80	18·50	3·00	0·60	0·90	7·50	–	–	–	0·60	–	–	–
8486	75·60	11·00	3·80	–	4·00	1·00	1·20	–	3·00	–	–	–	–
8622	57·00	1·30	–	–	3·60	9·50	3·80	–	–	–	–	24·50	–

calculated in $(\Omega\ \mathrm{cm})^{-1}$. Values for a and b have also been given by Baucke & Frank.[3]

The electrolytic properties of molten glass are such that only alternating current can be used for heating; direct current leads to the formation of deposits and blisters on electrode surfaces, and the flow of electric current requires an activation energy. For a calculation of the electrical resistance (impedance, Z) of molten glass we have the following relationship related to Ohm's Law:

$$Z = \frac{\hat{U}}{\hat{I}} = \sqrt{R^2 + \left(\omega L - \frac{1}{\omega C}\right)^2}. \qquad (8.4)$$

where $\hat{U}$ and $\hat{I}$ are the maximum values of voltage and current, R the resistance according to Ohm's Law (the effective resistance of the conducting medium), ω the cycle frequency, L the inductance, and C the capacity of the system. When using alternating currents of 50 Hz and glass melts of normal dimensions, ω and C can be ignored so that the simple Ohm's Law relationship, used for direct currents, applies:

$$\frac{\hat{U}}{\hat{I}} = \frac{U_{\mathrm{eff}}}{I_{\mathrm{eff}}} = R. \qquad (8.5)$$

When using frequencies above 1 kHz the inductance of transformer and cables must be taken into account; this causes a phase shift to occur between U and I and the power used must take account of $\cos\phi$. The result is

$$P = U_{\mathrm{eff}} I_{\mathrm{eff}} \cos\phi. \qquad (8.6)$$

The addition of condensors can result in $\cos\phi$ being unity.[6]

8.1.2 Thermal instability[4,10,57]

Due to the decrease in resistance of molten glass with temperature small disturbances can cause a local or a general increase in power consumption and thereby lead to further temperature increases; this characteristic is called thermal instability. In the majority of cases the technical literature uses the change in electrical resistance with temperature as a measure of the tendency to thermal instability during electric melting, i.e.

$$\frac{\mathrm{d}\rho_{\mathrm{el}}}{\mathrm{d}T} = \rho'_{\mathrm{el}}. \qquad (8.7)$$

Glasses showing high values tend to be more unstable than those with small values.

Andrusieczko[7] points out with some justice that, in practice, the controlling factor is not ρ'_{el} but the ratio of the change of power to the power being absorbed; for constant current

$$\frac{\mathrm{d}P/\mathrm{d}T}{P} = \frac{\rho'_{\mathrm{el}}}{\rho_{\mathrm{el}}}. \qquad (8.8)$$

He demonstrated from examples that the ratio $\rho'_{\mathrm{el}}/\rho_{\mathrm{el}}$ tends to deviate from the characteristics of ρ'_{el}.

When using ρ'_{el} as a measure of instability, only the glass and its electrical properties, separate from the furnace itself, are being considered; on the other hand the ratio $\rho'_{\mathrm{el}}/\rho_{\mathrm{el}}$ includes the conditions arising in practice in a furnace, that is the applied electrical power, but thermal properties remain excluded in this approach.

Up to now investigations have not considered the electrical conductivity of anything except the molten glass itself, but for thermal instability, not only is the absorption of energy of importance but also its transfer. In a small unit of volume the thermal instability can be described as

$$\frac{\mathrm{d}P/\mathrm{d}T}{\mathrm{d}q/\mathrm{d}T} = D, \qquad (8.9)$$

where $\mathrm{d}P/\mathrm{d}T$ is the change in power input with change in temperature and $\mathrm{d}q/\mathrm{d}T$ the change in heat transfer with temperature in the volume element; for $D > 1$ the volume considered is thermally unstable, for $D < 1$ it is stable.

The power input to a volume element through which a current flows is governed by

$$P = \frac{I^2}{A^2}\rho_{\mathrm{el}} \quad \text{and} \quad \frac{\mathrm{d}P}{\mathrm{d}T} = \frac{I^2}{A^2}\rho'_{\mathrm{el}}. \qquad (8.10)$$

The current density is given by $i = I/A$. If the ambient temperature is designated as T_{u} and l' is the linear expansion of the volume element being considered at temperature T, then the heat, q, given off by the volume element is given by

$$q = \frac{T - T_{\mathrm{u}}}{l}\left[(\lambda_{\mathrm{L}} + \lambda_{\mathrm{S}})\frac{A}{V} = \alpha\right], \qquad (8.11)$$

where λ_{L} is the thermal conductivity of the melt, λ_{S} the infrared transmissivity, α the heat transfer coefficient which applies for convection, V the volume being considered, and A its surface. Changes in the thermal current density through this volume are then given by

$$\frac{\mathrm{d}q}{\mathrm{d}T} = \frac{(\lambda_{\mathrm{L}} + \lambda_{\mathrm{S}}) \cdot \frac{A}{V} + \alpha}{l}. \qquad (8.12)$$

The work by Genzel[8] has shown that, as a good approximation

$$\lambda_S = \frac{16}{3}\sigma n^2 T^3 F \tag{8.13}$$

where F is the mean free path length of radiation quanta, n the refractive index, and σ the radiation constant; values of F for various glasses at different temperatures are given in Table 8.2. The instability factor D, see Equation (8.9), can then be written

$$D = \frac{i^2 \rho'_{el} l}{\left(\lambda_L + \frac{16}{3}\sigma n^2 T^3 F\right)\frac{A}{V} + \alpha}. \tag{8.14}$$

In the molten glass in a tank λ_L is very much smaller than λ_S, as may be seen in Figure 8.3; when compared to radiation, the convection component, can normally be ignored. For a cube of side length l,

Table 8.2. *Mean free path of radiation quanta (cm)*

	Temperature (K) 1300	1400	1500	1600	1700	1800
Industrial glass	4·16	3·92	4·32	4·78	4·96	5·17
Window glass	5·71	5·00	3·74	3·29	3·52	3·49
Green glass (iron)	0·424	0·380	0·345	0·308	0·332	0·334
X-ray protection glass	1·43	1·103	0·691	0·508	0·594	0·537

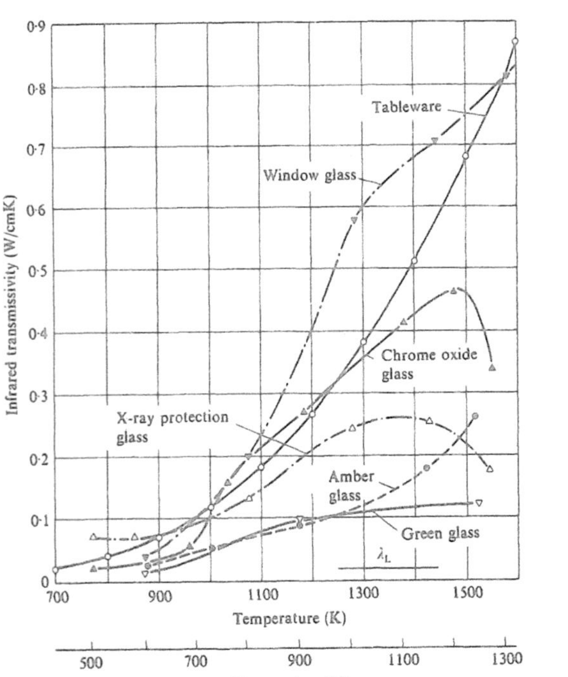

Figure 8.3. Transmission of radiation of some glasses against temperature. For comparison, λ_L is the average value of the true thermal conductivity known at low temperatures[8]

$A/V = 6/l$ and therefore, under conditions when radiation dominates, the thermal instability becomes a dimensionless expression

$$D_S = \frac{i^2 \rho'_{el} l^2}{32\sigma n^2 T^3 F} \tag{8.15}$$

where D_S has the character of a thermal property. The thermal stability $1/D_S$ increases markedly with increasing temperature and a long free radiation path, such as is found in lightly coloured glasses; it decreases with current density, the gradient of specific electrical resistance, ρ'_{el}, of the glass, and the geometric expansion of the zone considered (l^2). Hilbig[10] arrived at a similar equation.

For pure thermal conductivity such as arises with very dark glasses and, particularly, for refractories, this gives

$$D_f = \frac{i^2 \rho'_{el} l^2}{6\lambda_L}. \tag{8.16}$$

Investigations by Hammerschnidt & Hausner[9] have shown that fusion cast tank blocks, and particularly their hot surfaces into which glass has penetrated, have a considerable electrical conductivity (Figure 8.4). The temperature gradient of ρ_{el} of some fusion cast blocks is of a similar order of magnitude to that of the molten glass. On the other hand, heat transfer is limited mainly to thermal conductivity and transfer by convection is zero.

Refractories in the vicinity of electrodes where high current densities occur can give rise to local thermal instability. The danger of such thermal

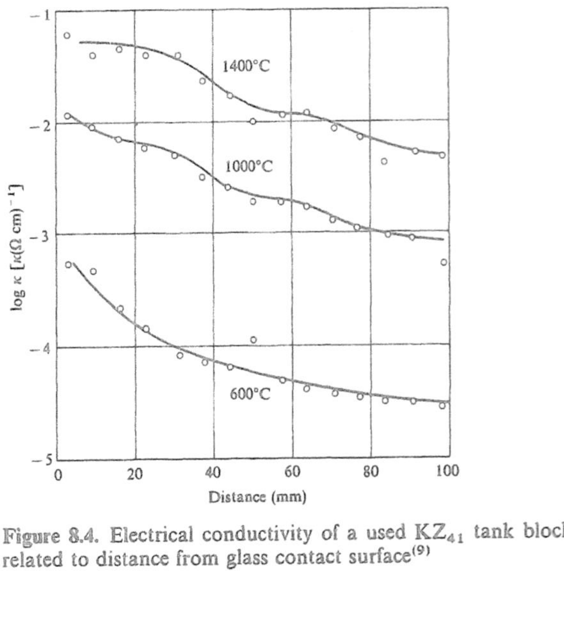

Figure 8.4. Electrical conductivity of a used KZ_{41} tank block related to distance from glass contact surface[9]

instability arising is related to the dominant influence of T^3 and often arises in regions of low temperature with relatively high current densities.[4] The dangerous aspect of such local instabilities is that, even for tanks where the current is controlled, these phenomena are not necessarily noticeable on the outside. It is very likely that thermal instabilities in fusion cast blocks located near electrodes occur when, due to external influences such as failure of an electrode, the existing equilibrium is disturbed.

Schumacher,[57] in a recent publication, has further investigated aspects of the stability of electric tank furnaces and has developed our knowledge of the problem; at the same time he has investigated the limits of stability.

The equation

$$\rho'_{el} = \frac{E_a}{RT^2}\rho^{el}$$

leads to

$$\frac{dP}{dT} = \frac{E_a}{RT^2}P,$$

if E_a/R is taken as the activation energy of the molten glass. For a system stable from the point of view of energy,

$$\frac{dP}{dT} - \frac{dq}{dT} = 0 \quad \text{and} \quad P = q$$

since, in any volume element, the thermal energy introduced must equal the energy transferred; in addition, the stability must be preserved during temperature changes. This leads to an equation in which, when compared to Equation (8.13), the infrared transmissivity, λ_S, is modified, i.e.

$$0 = \frac{A}{lv}\left\{\frac{E}{RT^2}[\lambda_L(T - T_u) + \frac{\lambda_S}{4}(T^4 - T_u^4)] - \lambda_L - \lambda_S T^3\right\}.$$

In the case of pure radiation, this gives

$$\frac{E}{RT^2}\frac{T^4 - T_u^4}{4} = T^3.$$

It is worth noting that the temperatures T and T_u in this equation depend only on the activation energy E/R of the glass which, according to Baucke & Frank,[3] lies between 6400 and 12 200 K for the glasses they investigated.

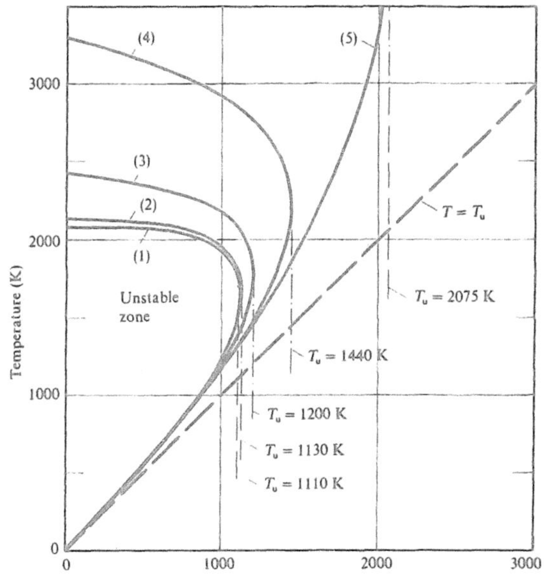

Figure 8.4A. Limit of stability of electric resistance heating[57]
(1) pure radiation
(2) to (4) radiation plus conduction
(5) pure conduction

For $E/R = 8300$ K we obtain curve 1 in Figure 8.4A, which shows an unstable zone in a region of relatively low temperature.

For pure thermal conductivity ($\lambda_S \rightarrow 0$) this leads to

$$\frac{E}{R}(T - T_u) = T^2,$$

which is shown as curve 5 in Figure 8.4A, and the region of instability is now in temperature ranges which occur in practice. Curve 2 is for white glass and curve 3 for green glass.

The results of these calculations also show that the risk of thermal instability only arises with glasses of low radiation transmission. Hot areas of refractories of relatively high electrical conductivity and high activation energy could also be dangerous.

8.2 Transfer of energy from electrodes to molten glass[11]

Knowledge of the resistance of the path to be heated is of major importance for an electrical installation. In high current situations one is generally dealing with locally concentrated load resistance in which the flow of current can be ignored; examples of this are heating coils, heating rods, and similar devices. However, in molten glass one is dealing with an extended medium. The flow of current from one electrode to another takes place in an extended space where the local current density can vary considerably. The central problem, therefore, is the calculation in advance of the local current distribution and the electrical resistance of the whole system.[11]

Analogous to conditions applying to the theory of current flow, lines of potential and currents form an orthogonal network. The current flows in the direction of the potential difference and is proportional to local changes in potential. In simple cases it is possible to estimate the potential field and this enables one to predetermine in outline the field of current densities.

In accordance with the basic laws of static and dynamic electricity, the structure of a field of currents and potentials is as follows: the lines of current begin and end vertically to the electrode surface; the distance between lines of current corresponds to the area through which a constant current flows; and the lines of potential cut the lines of current at right angles.

In a right angled container of constant cross section throughout its length, at the ends of which are plate electrodes covering the total cross section, it is easy to calculate the potential field and distribution of current density, provided the resistivity of the medium can be regarded as constant both in space and with time (Figure 8.5(a)). For a container of length, l, cross sectional area, A, and specific electrical resistance, ρ_{el}, the effective resistance is

$$R = \rho_{el}\frac{l}{A}. \tag{8.17}$$

Such a shape of container and arrangement of electrodes can be regarded as a resistance in the usual sense of applied electricity. These considerations implied, by not referring to them, that the walls of the container play no part in conducting electricity, i.e. they are ideal insulators.

If the plate electrodes are reduced in size to, for example, a quarter of the container cross section, then the lines of current become closer and lead to an increase in the current density near the electrodes. With this arrangement the backs of the electrodes make virtually no contribution to the current (Figure 8.5(b)).

Similar conditions, suitably modified, result when rod electrodes are used. The relatively small electrode surface gives a particularly high current density near the electrodes (Figure 8.5(c) and (d)).

In practice it is of fundamental importance to know the amount of heat liberated in a volume element of the molten glass by the passage of the current. The electric power per unit volume is $p = ie$, where $i = I/A$ is the current density and e the electric field strength. According to Ohm's Law, $i = e/\rho_{el}$; therefore the power given off per volume element (power density) is

$$p = \frac{I^2}{A^2}\rho_{el}. \tag{8.18}$$

When the electrodes are arranged as shown in Figure 8.5(b) and the ratio between the areas of the volume cross section and the area of the electrodes is n, then the ratio of power densities in the electrode area to that in the volume element is

$$\frac{p_E}{p_B} = n^2. \tag{8.19}$$

For an area ratio of 1/10 this gives a power density ratio of 100/1. The high energy density near electrodes becomes obvious. For simple volume geometries and electrodes configurations, $p = f(x, y, z)$ can be calculated.[12-18]

Two rod electrodes parallel to one another and separated by a distance b located in an infinitely

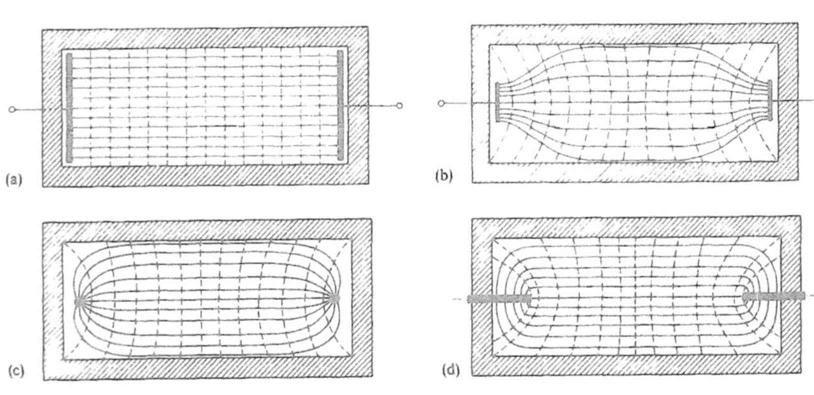

Figure 8.5. Lines of equal current and potential[11]
(a) in an ideal container for fusion
(b) with small plate electrodes
(c) with vertical rod electrodes
(d) with horizontal rod electrodes inserted at right angles to the wall

large melt of glass of constant resistivity have a power distribution lengthwise along their x axis which is easy to calculate. If the total current is I, then

$$\bar{p} = \frac{I^2 \rho_{el}}{(b/2)^2 \pi^2} \left[\frac{1}{1-(2x/b)^2} \right]^2. \tag{8.20}$$

The highest current density occurs at the inner surface of the electrodes. If the radius of the electrode is $d/2$ then, on the inner electrode surface $x_i = (b-d)/2$ and on the outer electrode surface $x_o = (b+d)/2$ (see Figure 8.6).

The total resistance of this arrangement is

$$R = \frac{\rho_{el} \ln(2b/d)}{\pi l}, \tag{8.21}$$

where l is the length of the two electrodes in parallel. Equation (8.20) and Figure 8.6 lead to the conclusion that the back of the electrodes is utilised less than the front. If the volume of the molten glass is enclosed by walls, as is the case in Figure 8.5, then the difference between front and back is even more marked. In practice it is permissible to regard the back third of a vertical rod electrode as inactive; for plate electrodes it is the whole back area. The power density distribution between two electrodes is saddle-shaped, so that thicker electrodes and a large distance between them provides a wide central area of almost constant power density distribution. But the borders of this volume (sidewalls, floor) and the temperature distribution in the molten glass do not permit the power density distribution to be taken as constant over the length of the electrodes. Short electrodes installed vertically have a current density at their top about twice that near the bottom. For electrodes installed horizontally the differences are even bigger: the ratio of total current densities could be 1:4.

Calculations of the potential distribution in advance for a bath of molten glass with walls and several electrodes is only possible in special cases and, even then, only with the help of advanced mathematics.[19,21] In practice, experiments using tanks of electrolyte or, if of lesser importance, electrically conducting paper, must suffice.

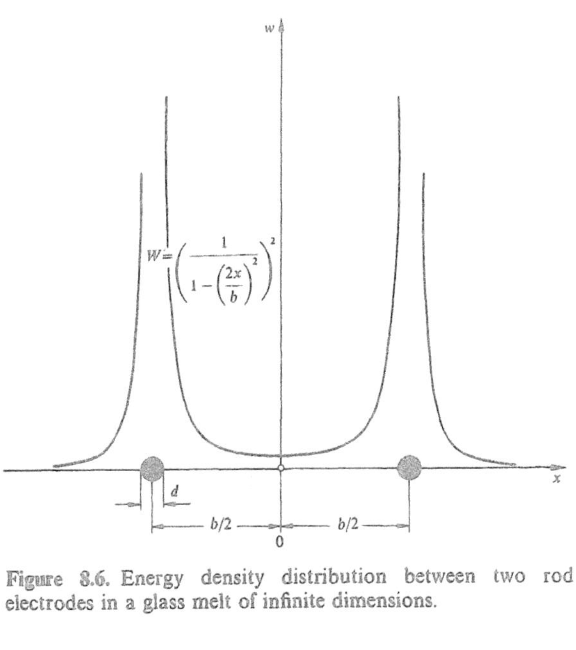

Figure 8.6. Energy density distribution between two rod electrodes in a glass melt of infinite dimensions.

8.3 Electrode circuits[23]

Electrodes can be connected in many ways, the correct circuitry depending on the geometry of the furnace, the matching of technical objectives to the expected fields of current density, the uniform loading of the circuits, and the avoidance of high working voltages.

8.3.1 Heating circuits connected in series

Heating circuits connected in series, such as those shown in Figure 8.7, can also be used for large heating distances, such as are used for electric

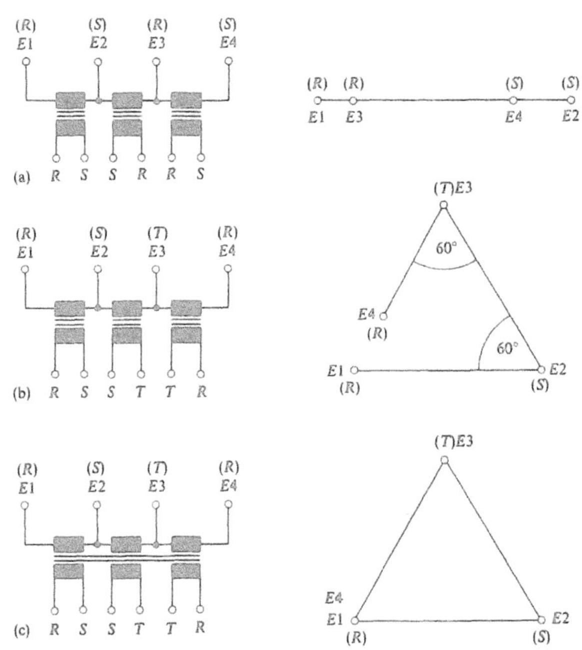

Figure 8.7. Series coupling of heating circuits[23]

boosting at the hot spot of a large tank furnace, without running into high operating voltages.[23] The circuit in Figure 8.7(a) shows three single transformers connected across one phase of the supply; this can cause phase imbalance and lead to high power consumption. It is best avoided by distributing the transformers across the three phases (Figure 8.7(b)).

Instead of three transformers, one three-phase transformer connected in delta can also be used but individual control of each heating circuit is then no longer possible (Figure 8.7(c)). The connections in accordance with Figure 8.7(c) can, in principle, also be used when electrodes are not installed in straight lines.

8.3.2 Heating circuits connected in parallel

If the objective is a uniform homogeneous temperature distribution in a large furnace, then connecting the electrodes in parallel should be considered (Figure 8.8). The most uniform temperature distribution is provided by plate electrodes but these cannot be changed during operation. Short rod electrodes, connected in parallel, almost meet the same objectives and can be changed while the furnace is in operation.

If the area to be heated is large, then a combination of connections in parallel and in series, as shown in Figure 8.9, could be an advantage.

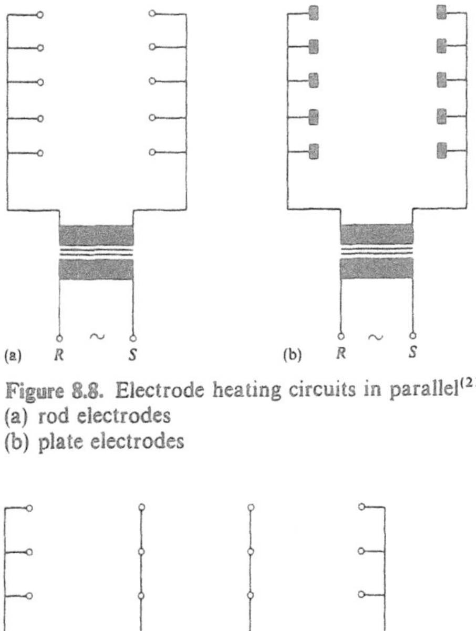

Figure 8.8. Electrode heating circuits in parallel[23]
(a) rod electrodes
(b) plate electrodes

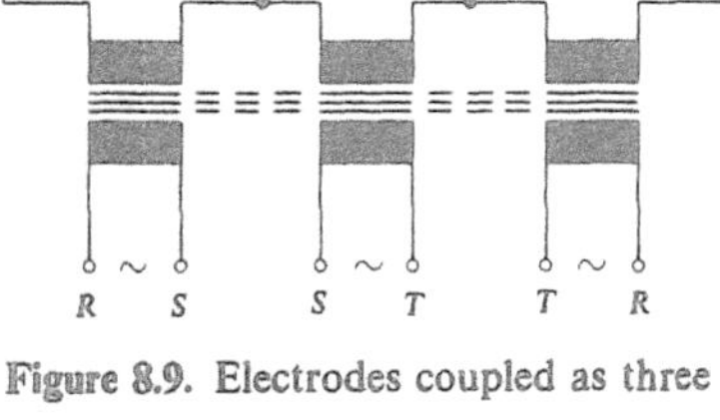

Figure 8.9. Electrodes coupled as three groups in parallel[23]

8.3.3 Delta connections used with alternating current

Connecting electrodes in delta directly to a three-phase transformer enables high electrical loads to be applied in molten glass in a restricted space. In larger furnaces a series of delta connections can be arranged in one line, which can form a very effective thermal barrier (Figure 8.10).

High energy densities and interference between individual delta connections can lead to destabilisation and consequentially to hot paths with high currents. For glasses prone to destabilisation (Section 8.12), particularly dark glasses of high energy absorption whose electrical conductivity changes markedly with temperature, this type of connection is therefore less suitable.

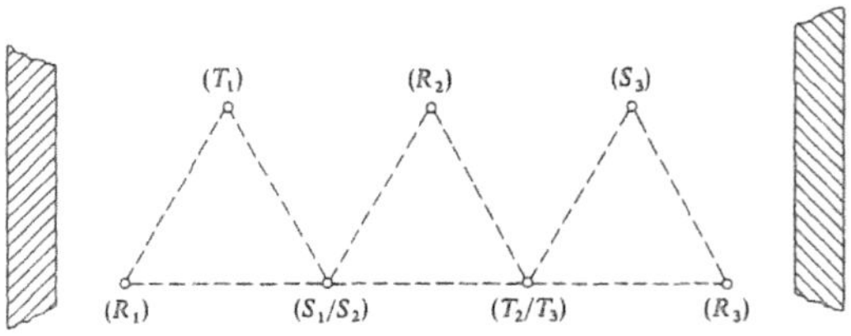

Figure 8.10. Electrodes connected in series in staggered configuration[23]

8.3.4 Star connections used with alternating current

An ideal loading using star connections can hardly ever be obtained because a direct path of current flow exists between each pair of corner electrodes; the central electrode is therefore of minor importance (Figure 8.11(a)). In practice a single-phase star connection, whose central electrode carries three times the load of the corner electrodes, is often used. Connections of this type are suitable for furnaces having many corners and a symmetrical groundplan.

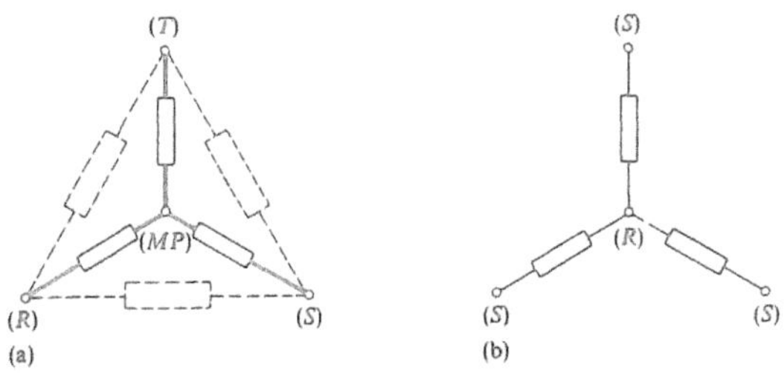

Figure 8.11. Electrodes connected in star configuration[23]
(a) connected to three-phase supply
(b) connected to single phase supply

8.3.5 Special connections

If six electrodes are to be used because the furnace has a hexagonal ground plan, or if three pairs of electrodes are to be installed one on top of the other, then changing three-phase alternating current into six phases would be an advantage. As an example of six-phase connections, Figure 8.12 shows a diagram of the SVUS furnace.[24]

For a furnace of square ground plan, or if a series of electrodes of several squares is to be constructed, the conversion of a three-phase supply into two phases by means of a Scott connection can be recommended (Figure 8.13).[25] A square furnace can be heated very uniformly if such electrodes are connected diagonally.

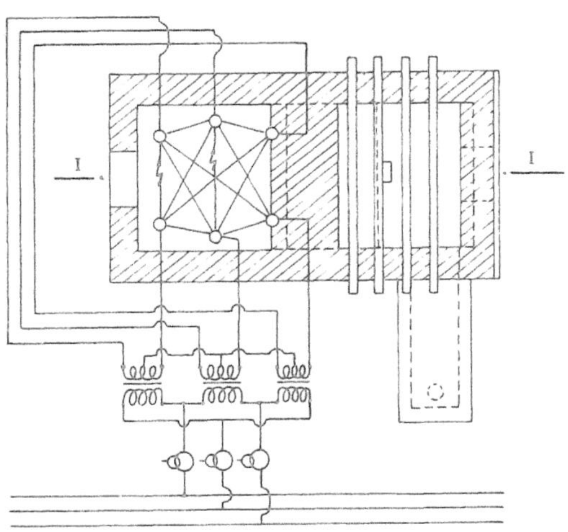

Figure 8.12. SVUS tank furnace with electrodes connected to six-phase supply

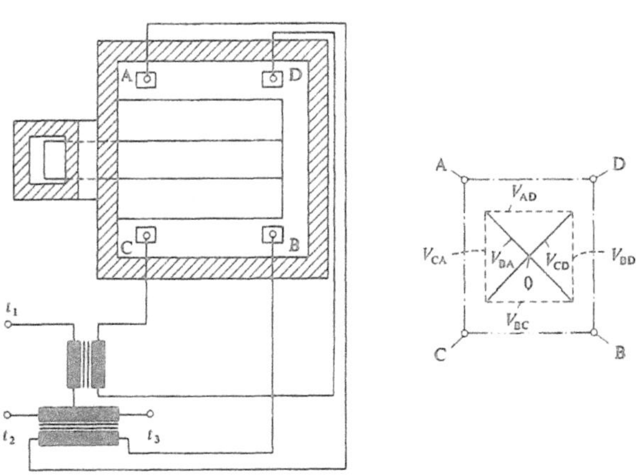

Figure 8.13. Scott circuit for converting three-phase into two-phase supply[25]

8.4 The electrical resistance of different arrangements of electrodes

A knowledge of the effective resistance between electrodes is important for the design of electric furnaces and their transformers. An exact determination in advance is not possible because of the many different electrode configurations and the influence of temperature on the conductivity of the glass. In simple borderline cases calculations can be done as has been shown by Andrusieczko[26] and Gailhbaud[19-22] but these calculations give results which can only be applied under certain conditions. Since even these are useful, selected examples will be given below.

8.4.1 Rod electrodes installed in parallel ignoring tank boundaries and using single phase power supply

8.4.1.1 Two rod electrodes in parallel

If the length of electrodes compared to their diameter is very large, then, if the resistivity of the molten glass is ρ_{el}, the effective resistance of the electrode pair is

$$R = \frac{\rho_{el}}{\pi l} \ln \frac{2b}{d} \tag{8.22}$$

where d is the diameter of the electrodes and b the distance between them (Figure 8.14).

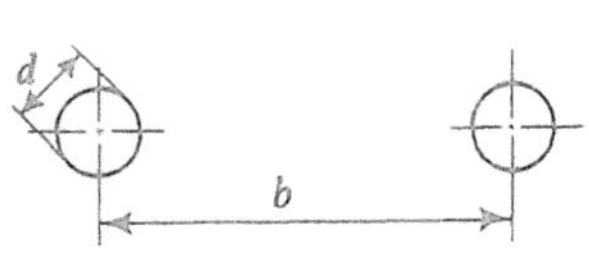

Figure 8.14. Arrangement of two electrodes[14]

8.4.1.2 Four rod electrodes in parallel

If two electrode pairs are combined as shown in Figure 8.15 and connected to a single phase supply,

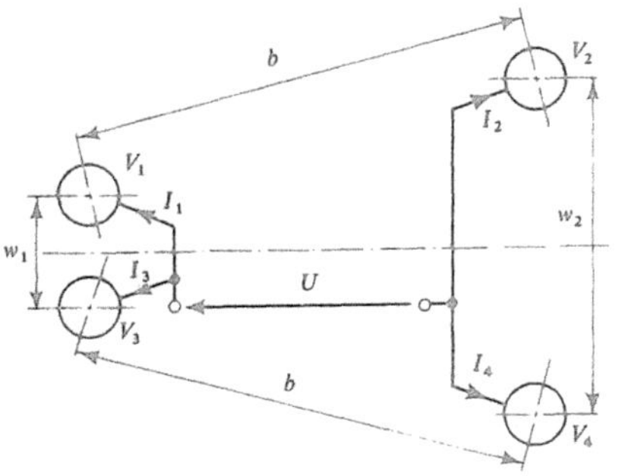

Figure 8.15. Arrangement of four electrodes[14]

then the effective resistance of this system is

$$R_{1,2} = R_{3,4} = \frac{\rho_{el}}{\pi l} \ln \frac{2b}{d} \sqrt{1 + \frac{b^2}{w_1 w_2}}. \tag{8.23}$$

8.4.1.3 Six rod electrodes in parallel

Similarly if three electrode pairs are connected as shown in Figure 8.16 using single phase alternating current supply then the effective resistances can be determined from

$$R_{1,2} = R_{5,6} = \frac{\ln(b\sqrt{b^2 + (2w)^2}/dw)\ln(2b/d) - 2[\ln(\sqrt{b^2 + w^2}/w)]^2}{\ln[2bw/(d\sqrt{b^2 + w^2})]} \tag{8.24}$$

$$R_{3,4} = R_{1,2} \frac{\ln[2bw/(d\sqrt{b^2 + w^2})]}{\ln[bw\sqrt{b^2 + (2w)^2}/d(b^2 + w^2)]}. \tag{8.25}$$

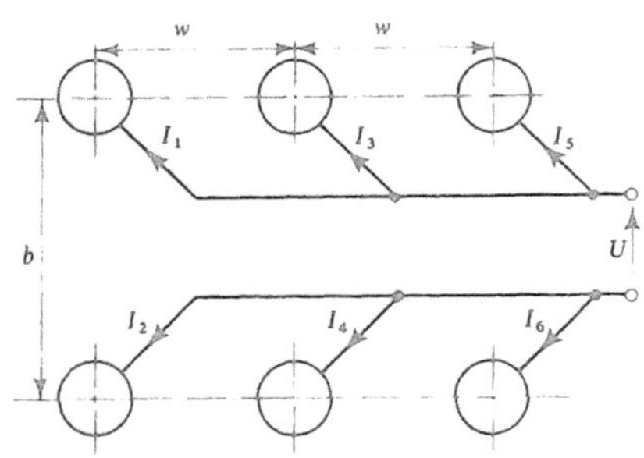

Figure 8.16. Arrangement of six electrodes[14]

8.4.2 Rod electrodes in parallel ignoring tank boundaries and using a three-phase network

8.4.2.1 Three rods in parallel connected as a delta network

The electrode arrangement given in Figure 8.17 and connected as a delta network gives

$$R = \frac{3}{2} \frac{\rho_{el}}{\pi l} \ln \frac{2b}{d} \tag{8.26}$$

for the resistances of the network diagram. In this case the electrical resistance is 1·5 times that of two electrodes using single phase alternating current.

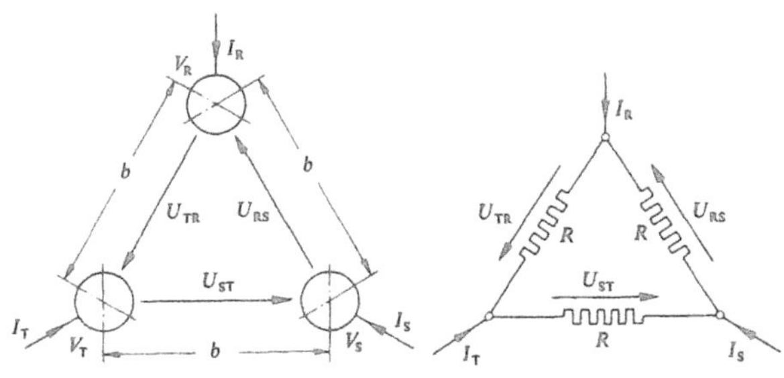

Figure 8.17. Arrangement of three electrodes for three-phase supply with network diagram[14]

8.4.2.2 Three parallel electrodes in series

If three electrodes supplied by alternating current are connected as shown in Figure 8.18, then the

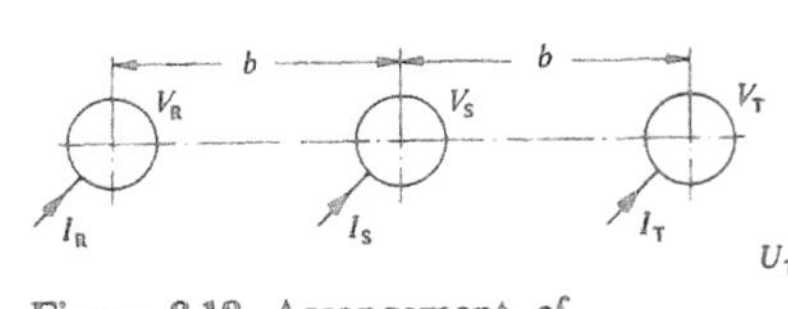

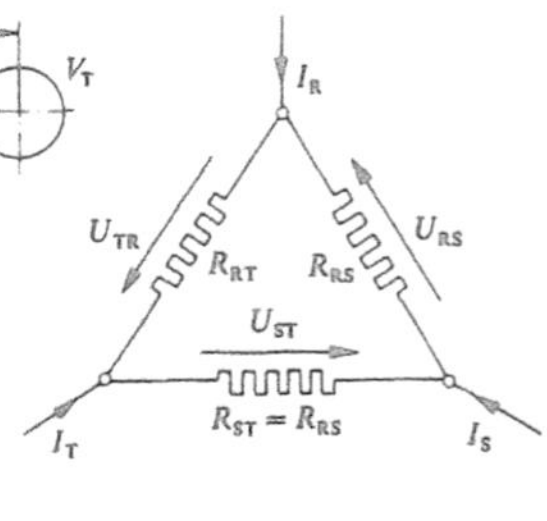

Figure 8.18. Arrangement of three electrodes in one plane for three-phase supply with network diagram[14]

resistances R_{RS} and R_{RT} can be given as

$$R_{ST} = R_{RS} = \frac{3}{2} \frac{\rho_{el}}{\pi l} \ln \frac{\sqrt[3]{4}\, b}{d} \tag{8.27a}$$

or

$$R_{RT} = R_{RS} \frac{\ln \dfrac{4b}{d}}{\ln \dfrac{b}{d}}. \tag{8.27b}$$

This arrangement leads to an uneven loading of the phases, which is undesirable in most cases, and can be avoided if four electrodes are connected in series.

8.4.2.3 Four parallel electrodes connected in series

This arrangement (Figure 8.19) can give a symmetrical loading of phases provided that the distance between the two central electrodes is kept smaller than that between the two outside electrodes: the distances required for symmetrical loading are evident from Figure 8.20. Since calculations to determine the resistances $R_{1,2}$, $R_{2,4}$, $R_{3,4}$, and $R_{1,3}$ are difficult, only the formulae for the effective resistance, $R'_{1,2}$, and the resistance, $R_{2,3}$, are given:

$$R'_{1,2} = \frac{\rho_{el}}{\pi l} \ln \frac{2b_{1,2}b_{1,3}}{d\sqrt{b_{1,4}b_{2,3}}} \tag{8.28a}$$

and

$$R_{2,3} = 2R'_{1,2} \frac{\ln \dfrac{2b_{2,3}}{d} \ln \dfrac{2b_{1,4}}{d} - \left[\ln \dfrac{b_{1,3}}{b_{1,2}}\right]^2}{\ln \dfrac{2b_{1,4}}{d} \ln \dfrac{2b_{1,2}^2 b_{2,3}^2}{d b_{2,3}^2 b_{1,4}} + \left[\ln \dfrac{b_{1,3}}{b_{1,2}}\right]^2}.$$

For a symmetrical arrangement $b_{1,4} = b_{1,2} + b_{2,3}$ applies.

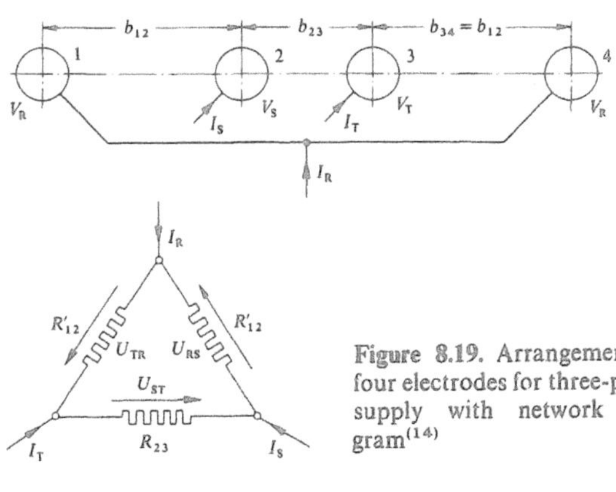

Figure 8.19. Arrangement of four electrodes for three-phase supply with network diagram[14]

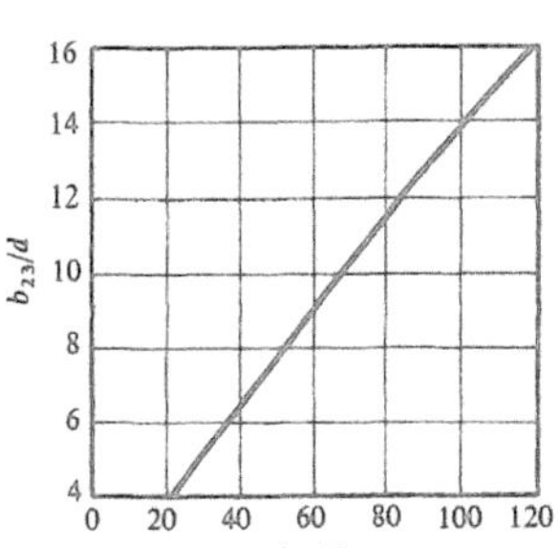

Figure 8.20. Condition for a symmetrical loading of phases[14]

8.4.3 Rod electrodes in a finite furnace volume

8.4.3.1 Two rod electrodes in parallel

If the limits of the volume of molten glass upwards and downwards are the glass level and the bottom, but the volume is otherwise of unlimited extension, then the resistance of the arrangement shown in Figure 8.21 is given by

$$R = \frac{\rho_{el}}{\pi l} \ln\left(\frac{2b}{d}\sqrt{1 + \frac{b^2}{4h_1 h_2}}\right). \tag{8.29}$$

Compared to the resistance of an infinitely extended bath of molten glass this gives an increase in resistance of

$$\ln \sqrt{1 + \frac{b^2}{4h_1 h_2}}.$$

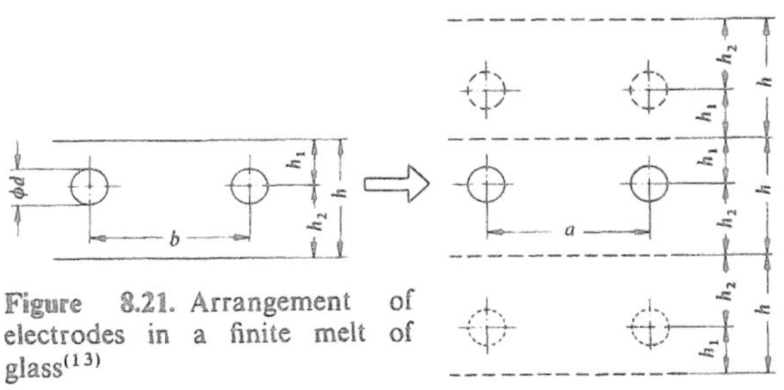

Figure 8.21. Arrangement of electrodes in a finite melt of glass[13]

8.4.3.2 Two rod electrodes installed identically but opposing one another

This example was calculated by Gailhbaud.[20] The calculation requires the use of an auxiliary function, $\gamma - \ln|kK'|$, which can be read off from Figure 8.22 for any given geometric arrangement. The following applies:

$$R = R_1 + \frac{2\rho_{el}}{\pi h}\left(\gamma - \ln|kK'| + \ln|K'/h| - \frac{0{\cdot}2h_1^2}{h^2}\right), \tag{8.30}$$

using h_1 as depth of immersion of the electrodes and h as the depth of glass. R_1 is also given by

$$R_1 = \frac{\rho_{el}}{\pi h_1} \ln\left|\frac{4h_1 + d}{(d_1 e)/2}\right|$$

where $e = 2{\cdot}7183$. The calculation is based on cylindrical electrodes but since, even after a short period, each electrode has worn to an ellipsoidal

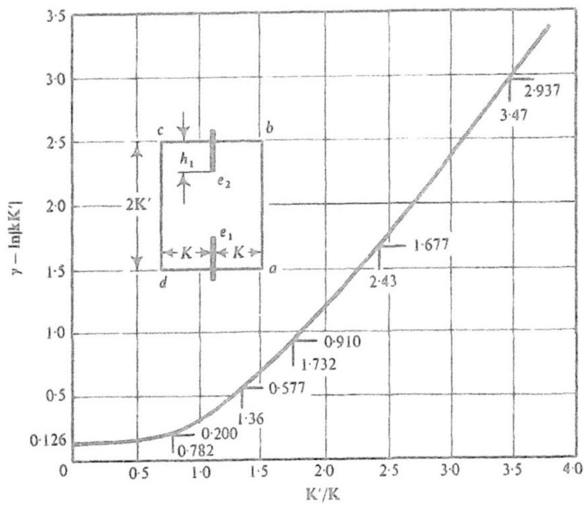

Figure 8.22. Magnitude of the auxiliary function for the determination of the resistance between two electrodes of a tank furnace[20]

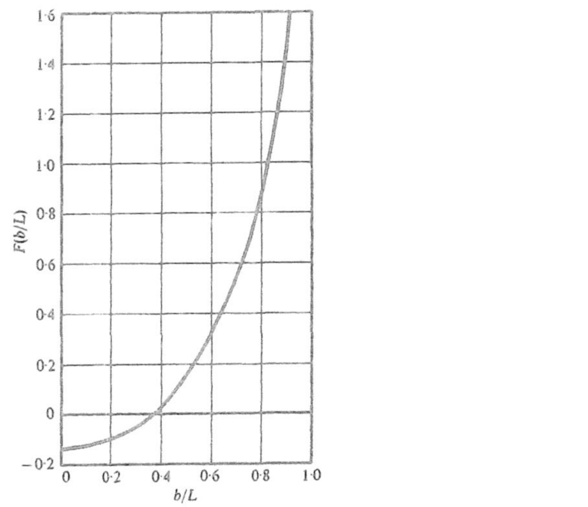

Figure 8.24. The function F in relation to the distance ratio b/L[19]

cross section, Gailhbaud recommends

$$R_1 = \frac{\rho_{el}}{\pi h_1} \ln\left(\frac{4h_1}{d}\right) \tag{8.31}$$

for used electrodes. This arrangement can give rise to very high current flux at the tips of the electrodes.

8.4.3.3 Rod electrodes forming a rectangle and inserted through the floor using a two-phase supply

A rectangular arrangement of electrodes (Figure 8.23) is suitable for a power supply via a Scott connection (see Section 8.3.5) and gives very even heating of the volume of molten glass. Gailhbaud gives the resistance of this arrangement as

$$R = R_1 + \frac{\rho_{el}}{\pi h} \ln\left|\frac{b}{h}\right| + \frac{\rho_{el}}{h} F\left(\frac{b}{L}\right). \tag{8.32}$$

where R_1 is given by the relationship described in Section 8.4.3.2; the function F can be taken from the distance ratio b/L shown in Figure 8.24.

8.4.3.4 Rod electrodes inserted through the floor arranged in a triangle and using a three-phase supply

Electrodes for a tank of three or six corners are best placed in the form of an equilateral triangle using a three phase power supply (Figure 8.25). The re-

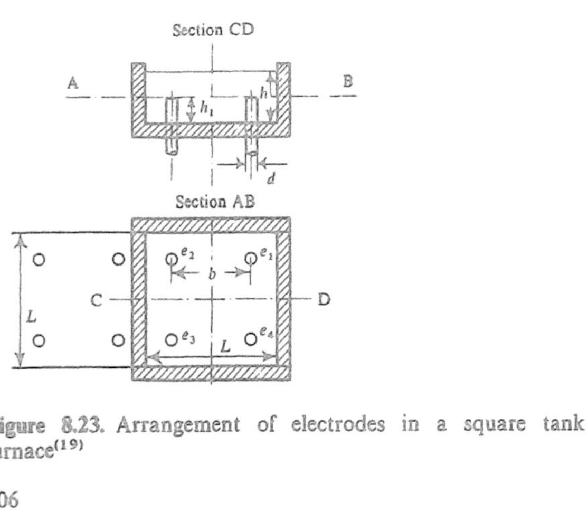

Figure 8.23. Arrangement of electrodes in a square tank furnace[19]

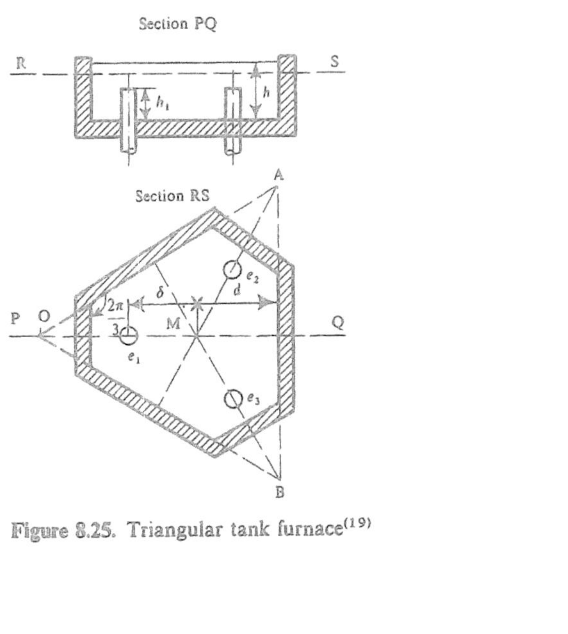

Figure 8.25. Triangular tank furnace[19]

sistance is given by Gailhbaud[19] as

$$R = \frac{3}{2}R_1 + \frac{3\rho_{el}}{2\pi h}\left(\ln\left|\frac{d}{h}\right| + 1{\cdot}441 - 5\ln\left|2 - \frac{\delta}{d}\right|\right). \quad (8.33)$$

The symbols are described in Figure 8.25 and R_1 is defined by the relationship already given in Section 8.4.3.2.

8.5 Electrode materials

The development of electric melting is largely linked to the discovery and development of electrode materials. The properties and criteria that an electrode material must meet are:

1. resistance to corrosion by molten glass and refractories, irrespective of oxygen partial pressure and current density up to 1700°C
2. resistance to oxidation in air at least up to 800°C
3. electrical conductivity similar to that of metals
4. adequate mechanical strength up to 1700°C
5. adequate thermal shock resistance
6. properties for manufacture should be similar to those of metals
7. an economically viable price.

The ideal electrode material has not yet been found and probably never will be, so compromises must be made, which means that, depending on the application and the technical and economic conditions, different electrode materials will be selected.

During the early development of electric melting, electrodes were made of pure metals such as iron and nickel or alloys of these, and they had to be intensively cooled to provide adequate mechanical properties. With a few exceptions, such materials are of no importance as electrode materials today and the field is led by molybdenum, platinum, tin oxide, and, up to a point, graphite.

8.5.1 Molybdenum

Molybdenum, a metal of high strength at high temperature and a melting point of 2610°C, meets requirements (1) and (2) up to a point and (3) to (7) adequately, and is now the most commonly used electrode material. In many ways the development of this material until it was suitable for use in molten glass has been largely responsible for the wide distribution of electric glass melting.

The physical properties of molybdenum as used for electrodes are given in Table 8.3; its electrical resistivity at room temperature is three times that of copper but by 1500°C it is thirty times that of copper.

Molybdenum electrodes are particularly suitable for glasses melted under reducing conditions but glass constituents such as lead, arsenic, antimony, and, to a certain extent, SO^{2-} have an adverse effect. They are in the electrochemical series above molybdenum and their oxides would therefore be reduced to the parent metal even if no current were flowing; even 1% PbO in a glass produces noticeable corrosion of molybdenum but it is claimed that special electrical networks (Figure 8.26) can lead to a reduction in the attack of lead on molybdenum.[24,25] In practice, the use of molybdenum as an electrode material has been found to be satisfactory provided that, for each 1000 kg of sand in the batch, the

Table 8.3. *Physical properties of molybdenum*

Melting point	(°C)	2630 ± 40
Vapour pressure	(mbar)	$8{\cdot}5 \times 10^{-9}$ at 1500°C $10{\cdot}7 \times 10^{-7}$ at 1800°C $5{\cdot}5 \times 10^{-5}$ at 2000°C
Specific heat	(J/g K)	0·27 at 20°C 0·31 at 1000°C 0·33 at 1400°C
Thermal conductivity	(W/m K)	152 at 20°C 103 at 1000°C 66 at 1600°C
Thermal expansion	(10^{-6}/K)	5·3 at 20°C
Specific electrical conductivity	(μΩ cm)	5 at 20°C 27 at 1000°C 43 at 1500°C 60 at 2000°C

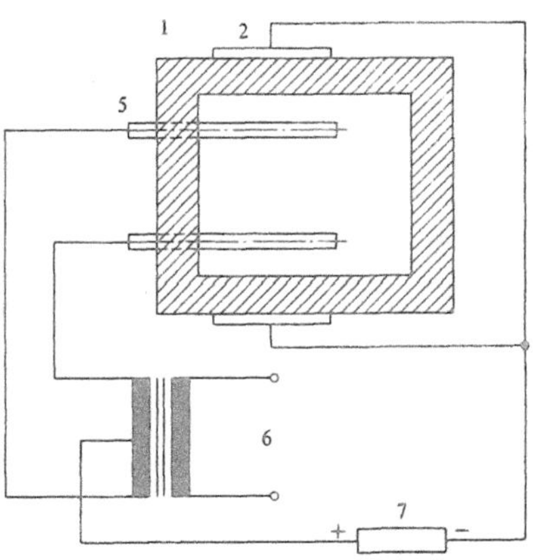

Figure 8.26. Protection of electrodes by a direct current voltage

following limits apply to the batch composition: $As_2O_3 + Sb_2O_3 < 0{\cdot}25$ kg, $Na_2SO_4 < 5$ kg, and $NaNO_3 < 3$ kg.

The reduction of these oxides causes molybdenum oxide to form on the electrode surface; under reducing conditions and when melting glasses containing SO_3 ions, MoS_2 is formed as a corrosion product; soda–potash glasses containing antimony give rise to antimony droplets on the electrode surface as well as calcium molybdate crystals; antimony reacts with molybdenum to form Mo_3Sb_7, which is removed in the molten glass. According to Ladr *et al.*[27] a protective layer of SiO_2 forms on the surface of molybdenum electrodes if, in addition to the normal power supply, the electrodes are connected to a positive direct current supply and the negative connection is made to the walls of the furnace. The electrical resistance of the heating electrodes is thereby changed according to the thickness of the protective layer created. The example in Figure 8.26 shows the electrodes, 5, and the tanks walls of fusion cast refractories, 1, to which the negative direct current supply is connected via nickel plates located opposite one another. The direct current voltage is taken from the secondary of a transformer supplying the two electrodes, 6. The relationship between the direct current supplied to the electrodes and their rate of corrosion was measured for a glass melt temperature of 1350°C, and the lowest rate was found for a current density of 1·3 mA/cm² of electrode surface.

Molybdenum is only slightly soluble in molten glass as the oxide. In soda–potash glasses it gives rise to a faint yellow colour generally so weak as to be unobjectionable even in white glasses. However, impurities in the electrode materials can produce very strong colouring during the solution of molybdenum and it is for this basic reason that a minimum molybdenum content of 99·93% is specified for electrodes for glass melting. Of particular importance is the carbon content; an excess of carbon not only impairs the mechanical properties but also causes blisters which are especially undesirable when electrodes are used in forehearths or feeders. The O_2 and N_2 contents have similar effects. The upper limits for impurities in molybdenum for electrodes are given in Table 8.4; it is assumed that the material described in the Table is manufactured by powder metallurgical processes. Vacuum-melted molybdenum is somewhat purer but is also more expensive. More extensive data

Table 8.4. *Upper limit of permitted impurities (ppm) in molybdenum electrodes for glass*

Element	*Impurity maximum*
C	50
Si	80
Al	30
Ca	10
Mg	10
W	200
O_2	40
H_2	10/20
N_2	20

covering experience with this material in the glass industry have so far not appeared. Slightly reducing conditions during glass melting cause the attack on molybdenum to be significantly less than slightly oxidising conditions.

The maximum permitted current density of molybdenum electrodes is 2–3 A/cm², but to minimise molybdenum losses, 2 A/cm² should not be exceeded. For low alkali glasses the current density should be significantly lower.

The use of molybdenum electrodes is hindered by the fact that molybdenum forms volatile oxides above 400°C; as the temperature increases the losses in air rise very steeply, leading to total destruction (Figure 8.27). It is therefore necessary to keep the temperature of electrode parts outside the molten glass below 400°C by cooling or to prevent the access of air. The design and operation of electrode holders must take account of these factors (see Section 8.7).

Today molybdenum electrodes are used mainly in the form of rods or, less frequently, plates. The

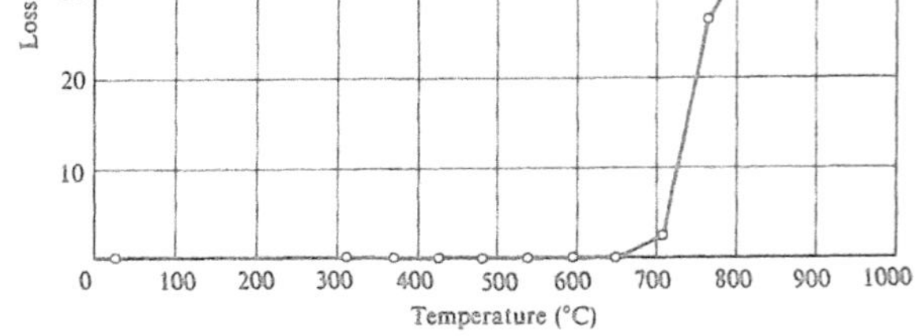

Figure 8.27. Loss in weight of molybdenum as a function of temperature (from *Molybdän-Dienst* No. 18, November 1962)

diameter normally lies between 32 and 76 mm, with a trend towards thicker electrodes. The rod lengths available are between 1 and 2 m and they can be supplied plain or threaded; the advantage of threaded rods is that compensation can be made for parts worn away. The resistance of the threaded joints must not be too high and to achieve this, the threads must be cut with care and free from impurities. The threading of one onto another can be made easier by the use of small amounts of lubricants but these must be free of lead.

Plate electrodes are constructed of sheets 2–20 mm thick and up to 1 m^2 in area joined in a suitable manner to molybdenum rods which conduct the electric current to the plates.

8.5.2 Graphite

Graphite was the preferred electrode material prior to the development of molybdenum. Its advantage is a low density of $1{\cdot}6 \times 10^3$ kg/m^3; the difference in density between graphite and molten glass is therefore much smaller than between molybdenum and glass, making the use of large horizontally mounted electrodes possible. Any pieces of graphite which become broken off will float and can be removed. The disadvantages are that graphite must be used with glasses melted under reducing conditions (amber glass), that there is a tendency to discolour the glass, and the low value of the permitted current density, i.e. 0·2–1·0 A/cm^2. Today, graphite electrodes are used only in exceptional cases for the electric melting of glass.

8.5.3 Tin oxide

Tin oxide differs from molybdenum in that it is a ceramic oxide resistant to oxidising influences and to some extent its properties lie in those areas where molybdenum is weaker; tin oxide electrodes are suitable for the melting of lead glasses but a disadvantage is that their resistance is strongly temperature-dependent and decreases rapidly with increasing temperature, while that of molybdenum increases. At low temperatures the resistance is very high, making it difficult to form proper contacts in the connections and creating more serious risks if the electrodes are heated too quickly. The breakthrough for practical application was achieved when it was discovered that its specific electrical resistance could be reduced by additions of oxides of As, Ta, Si, and other elements. By subjecting the dosed material to heating and cooling in an inert atmosphere the specific electrical resistance can be reduced significantly (Figure 8.28).

The manufacture of tin oxide electrodes is difficult but they are currently made as blocks or rods about 12 cm in diameter. Their disadvantage in use is that worn electrodes cannot be replaced by pushing in new ones as can be done with other rod electrodes. The positioning of electrode blocks and their installation must be carried out in such a way that, in the event of one electrode failing, the operation can continue using the remainder. The design and installation of electrodes should be aimed at minimising changes in their surface area as they become worn.

The density of tin oxide electrodes is about $6{\cdot}2 \times 10^3$ kg/m^3, their sensitivity to thermal shock is high, and their tensile strength is low. Their physical properties are summarised in Table 8.5.

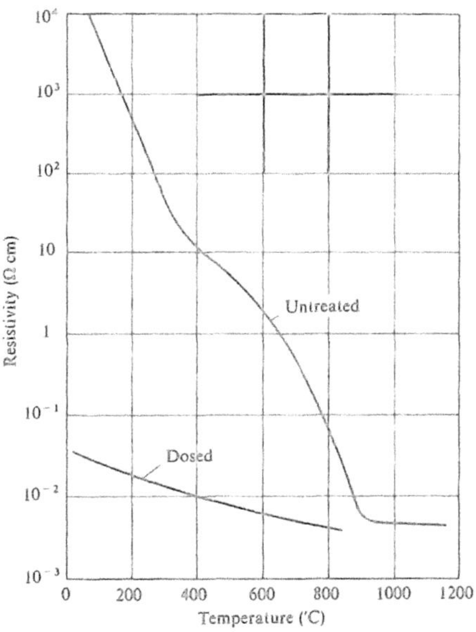

Figure 8.28. Electrical resistivity of untreated and dosed SnO_2 electrodes[23,28]

Table 8.5. *Physical properties of isostatically pressed tin oxide electrodes*

SnO_2	(%)	98
Bulk density	(g/cm^3)	6·45
Apparent porosity	(%)	7
True density	(g/cm^3)	6·94
Flexural strength at 20°C	(MN/m^2)	63
1000°C		10
1100°C		10
1200°C		9
1300°C		9
1400°C		13
Coefficient of thermal expansion 0–1000°C	(10^{-6} K^{-1})	5·3

Temperatures of 1350°C should not be exceeded if high corrosion rates are to be avoided and the current density at the surface of the active electrode must not exceed 0·5–0·7 A/cm² as the rate of corrosion increases rapidly above this. If a power failure exceeds 30 minutes, electricity to the electrodes must be restored only gradually in order to avoid too rapid an increase in temperature, which can lead to cracking.

Tin oxide electrodes can be manufactured by slip casting or isostatic pressing, the latter giving a denser ceramic. Both processes must be followed by sintering at high temperature and as tin oxide possesses poor sintering properties, materials to assist sintering must be added.

It is mainly the technical and operational limitations of tin oxide electrodes which have prevented this material from being as widely used as molybdenum; tin oxide electrodes cost more than those of molybdenum. Current objectives are to develop molybdenum electrodes and the associated circuitry so that they can also be used for oxidising glasses.

8.5.4 Platinum

Platinum or platinum–rhodium alloys are used where the demands for glass quality is especially high, e.g. for optical glasses. At 50 Hz they give off an objectionable colloidal platinum dust into the molten glass and frequencies of about 10 kHz are therefore used to avoid this problem: this attack is caused by metals in the molten glass acting as reducing agents as platinum electrodes are very sensitive to all reducing conditions. Their thermal performance is relatively high and current densities up to 2 A/cm² at 10 kHz can be used. With correct use, the rate of wear of platinum electrodes is so low that, despite their cost, their use is economically viable.

8.6 Electrode positioning

To meet the requirements laid down for the process being carried out, much depends on the proper positioning of the electrodes. The current density distribution, heat transfer, and convection currents achieved by good positioning all influence glass quality and the rate of wear and life of electrodes. The properties of the refractories must also be taken into account as well as the conditions covering construction and control.

8.6.1 Horizontal installation of rod electrodes

Horizontal electrodes are used successfully whenever an increase in temperature is wanted in the horizontal plane of a bath of molten glass. Opposing electrodes, or pairs of electrodes, create a planar distribution of current density and this gives rise to a more even heat distribution in the glass; in very wide tank furnaces this is often not sufficient. The distance between electrodes must be large enough to prevent their front ends from being subjected to current densities which are too high (Figure 8.5(d)); more than two full electrode lengths is required.

Electrodes installed horizontally are subject to bending forces. The heavy molybdenum electrodes in particular can bend and, in extreme cases, break off under high temperature conditions; to minimise this risk the length to diameter ratio should be less than 20 and, in addition, periodically turning the electrodes through 180° is advisable. In some special cases tubular electrodes have been successful.[30]

Strong glass currents are formed near the ends of an electrode but near the mounting and below the electrode the glass currents are weak (Figure 8.29). The glass currents create a trough-like corrosion pattern on the walls of tank blocks above the electrodes, especially in conjunction with upward drilling by gases (see Figure 8.31 below).

The installation and control of horizontal electrodes is simple for vertical flow furnaces and U shaped flame furnaces with electric boosting but burner construction and regenerators makes this difficult for cross-fired furnaces so that horizontal electrodes are only rarely used for these.

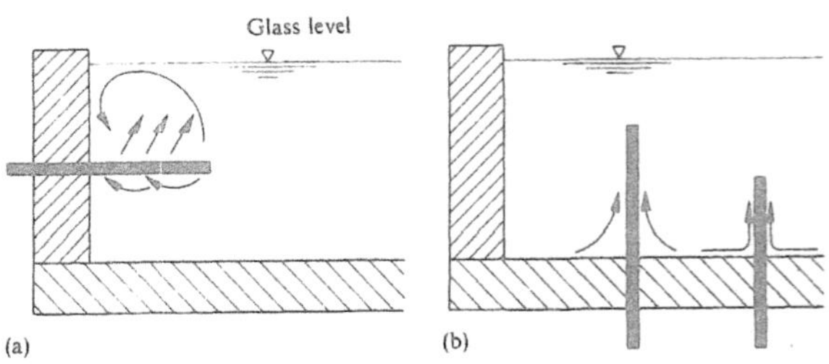

Figure 8.29. Glass currents caused by horizontal and vertical electrodes
(a) strong convection currents near the end, low near the tank block, low under the electrode
(b) long electrode, reasonable current along the bottom; short electrode, strong current along bottom

8.6.2 Vertical installation of rod electrodes

A group of electrodes set vertically can create high temperatures locally and such an installation therefore readily lends itself to creating a thermal barrier in furnaces with electric boosting. The electrodes give rise to strong convection currents right down to floor level. Short electrodes present a risk of excessive corrosion of the refractories while long electrodes are likely to bend and break off, but this risk is less than the risks associated with horizontal electrodes. The distances between electrodes must be such that, even if one does break, short circuiting cannot arise (Figure 8.29(b)).

The correct installation of electrodes through the floor of the furnace calls for more experience than through the sidewall, and the risk of glass breakout is greater. Electrodes inserted through the bottom are not as common as electrodes through the sidewalls; for boosting, however, they are widely used since access to the floor is easy and the usual objective is to create heat in the depth of the glass.

8.6.3 Plate electrodes

Plate electrodes are always installed vertically on the sidewalls and give a very even distribution of current density and therefore of heat (Figure 8.5(a)). Their rate of wear is low, particularly molybdenum electrodes used with borosilicate glasses. A serious disadvantage is that a replacement electrode cannot be fitted on the run and the furnace must be shut down and drained hot. The wear on tank blocks is low. During heat up and shut down, special precautions must be taken so that the electrodes do not make contact with the oxygen in the air.

8.7 Electrode holders

8.7.1 Rod electrodes

The proper design and construction of electrode holders are very important; apart from mechanically securing the electrode, the construction must ensure electrical insulation from the refractories and the steelwork of the furnace. Metal electrodes also require cooling and protection from air at higher temperatures.

An electrode holder for electrodes installed horizontally is shown in Figure 8.30. A tube of large

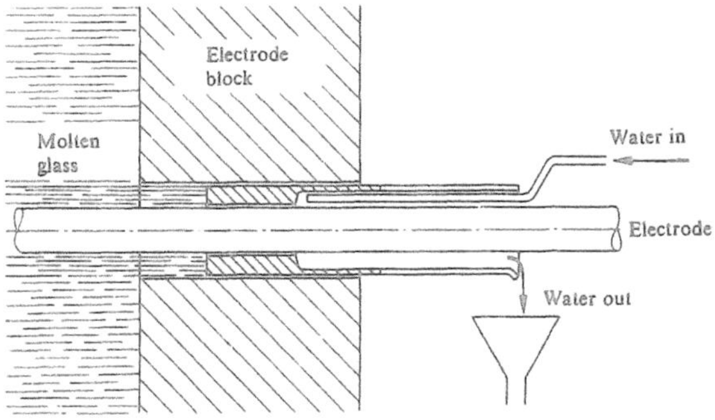

Figure 8.30. Holder for horizontal electrodes

wall thickness with a deep collar is welded onto a thinner walled tube: it has been found to be better to locate the welds outside the refractory wall because, when located well inside the wall, they may crack.[32] The thick walled tube is machined slightly conical at the end facing the glass to assist the glass to form a seal against the electrode. Cooling water enters from above and flows via an observation position into a collecting funnel, scaling being avoided by using soft water. On average, the temperature of the water will increase by about 10 degC at a flow rate of about 7 l/min. Electrode holders must be constructed of chrome–nickel alloys as normal steels are severely attacked by molybdenum oxides, the formation of which cannot be avoided. Suitable alloys are Inconel or Nimonic 90 containing 57% Ni, 20·5% Cr, 17% Co, 2·6% Ti, and 1·65% Al.

Electrodes may also be cooled indirectly, using a design which incorporates water channels drilled into the head of the electrode holder. The advantage claimed is that the water is conducted directly to where it is needed but experience has shown that, in the majority of cases, direct cooling is more intensive and certainly simpler. The indirect method can be used in special circumstances, particularly when an open water return is not possible. Electrode holders with electrodes screwed in have also been used on small furnaces.[31]

The cooling of an electrode holder must be designed so that it ensures removal of the heat from the electrode and can safely remove any electric current arising from the resistance of the electrode. The power required for cooling is in the region of 2–6 kW/h, with holders located well forward calling for more power than those located further back: these relationships have been investigated by Endress & Mommertz.[33] The temperature of the electrode near the collar can be monitored by inserting a thermocouple; the value of this is disputed.[32]

The electrode and its holder are located in the refractory in a straight drilled or stepped hole, the latter being preferred. To avoid the necessity of protecting the electrode while the furnace is being heated up, the holes are not completely drilled through, but a thin area of refractory is left at the hot face; this is then broken through when the electrode is introduced after heat up. The diameter of the hole should be 4–10 mm larger than that of the electrode in order to ensure a good seal: with the cooling water turned off, the electrode is pushed forward further than intended and then retracted so that some glass is pulled right back into the electrode holder, the cooling water is then turned on, the glass sets, and the electrode is safe from attack by air. If an air space is allowed to remain between holder and electrode, then attack here will weaken the electrode which will break off at the first opportunity. The distance from the hot face of the refractory block to the hot face of the electrode holder must be 5–10 cm and as the refractory wears, the holder must be retracted. The gap between hole and electrode often acts as a starting point for rapid refractory corrosion by upward drilling (Figure 8.31) and it is this which favours the use of stepped holes which give the smallest possible annular ring. Blocks carrying electrodes are often made thicker than the remainder in order to compensate for the more rapid corrosion.

If the cooling water supply fails, the reaction of the thermocouple is often too slow. In order to prevent critical conditions from arising, automatic warning devices have been developed and a simple, robust, and proven design is shown in Figure 8.32:

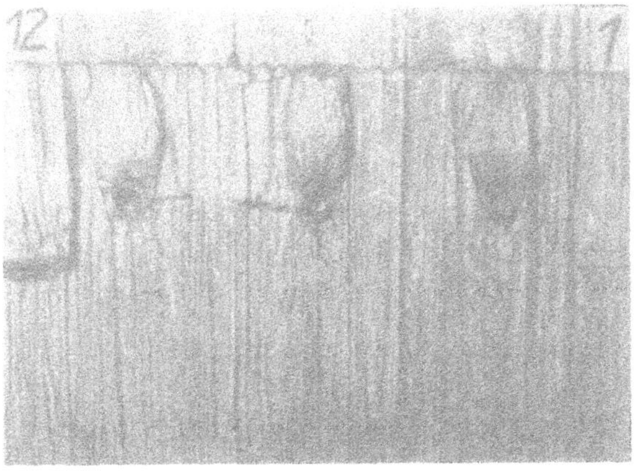

Figure 8.31. Corrosion of electrode blocks at the upper level after 20 months[41]

if the collection trough, which is normally filled with water, empties, it rotates and thereby operates a signal.

Tank blocks, including blocks for carrying electrodes, are nearly always made from fusion cast refractories which are sensitive to thermal shock. The cooling effect on the refractory by the electrode holder can be reduced by surrounding the guide tube with mat of high temperature insulating fibre; if this is not done, radial cracks develop in the block after a short time. The mat also serves as electrical insulation. When cooling has to be stopped in order to insert a new electrode, it is good practice to cool the holder with air for several minutes after positioning the new electrode, and then start water cooling. This reduces thermal stress in the electrode block.

Holders for vertical molybdenum electrodes are in principle the same as for horizontal electrodes. When open water cooling circuits are used it is important to ensure that any steam formed is free to escape (Figure 8.33).

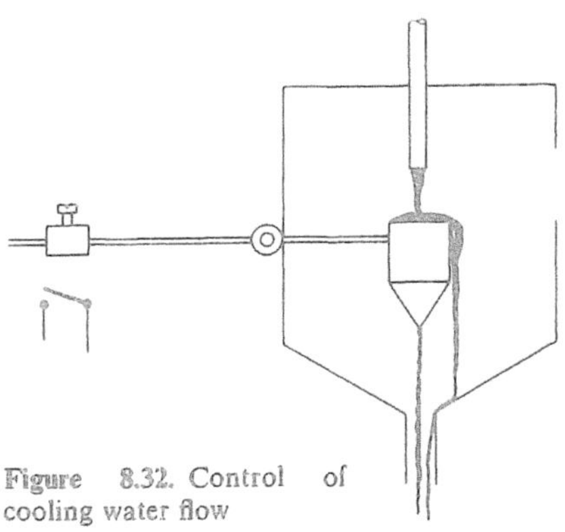

Figure 8.32. Control of cooling water flow

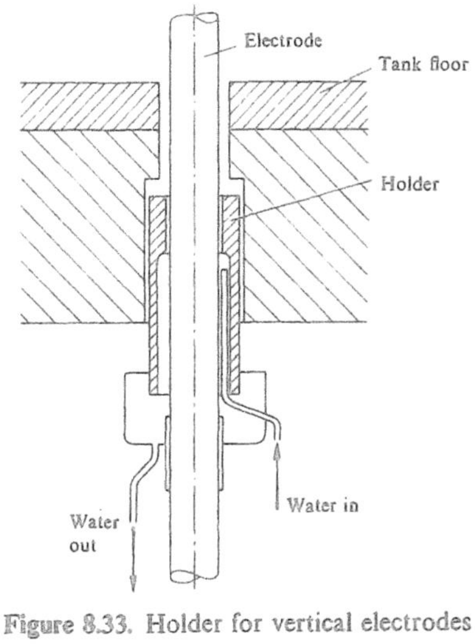

Figure 8.33. Holder for vertical electrodes

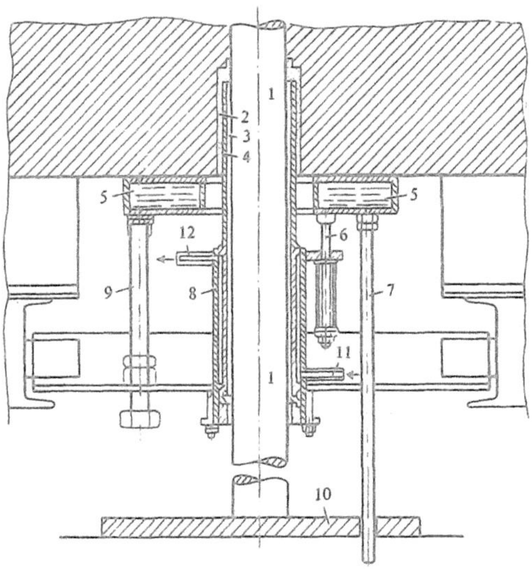

Figure 8.34. Holder for adjustable molybdenum electrode with exterior cooling system
(1) electrode
(2) exterior cavity
(3) interior cavity
(4) electrode holder
(5) electrode block cooling
(6) electrode securing bolt
(7) water inlet and outlet
(8) electrode cooling tube
(9) pressure bolt
(10) end plate
(11) water inlet
(12) water return

Electrode holders with a double cooling system are also in use and a system with cooling for both the electrode and the electrode block is shown in Figure 8.34.

8.7.2 Plate electrodes

Holders for plate electrodes are fundamentally different from those for rods due to the differences in design and an example of one of the many types of plate electrode holder is given in Figure 8.35. The electric current passes through the outermost of the two concentric tubes forming the water-cooled holder. These electrodes must be installed in the furnace when cold and must therefore be protected from the air during heat up.

8.7.3 Tin oxide electrodes

The holder for tin oxide electrodes is similar to that for molybdenum electrodes, since they are both fitted directly into the tank block; there must be no cooling since there is good electrical conductivity only at higher temperatures but thermal stresses should be kept to a minimum. Contact with the air presents no risk.

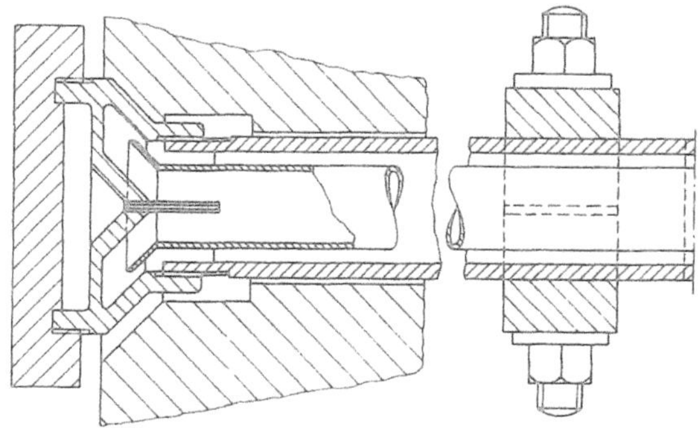

Figure 8.35. Plate electrode with water-cooled holder screwed on

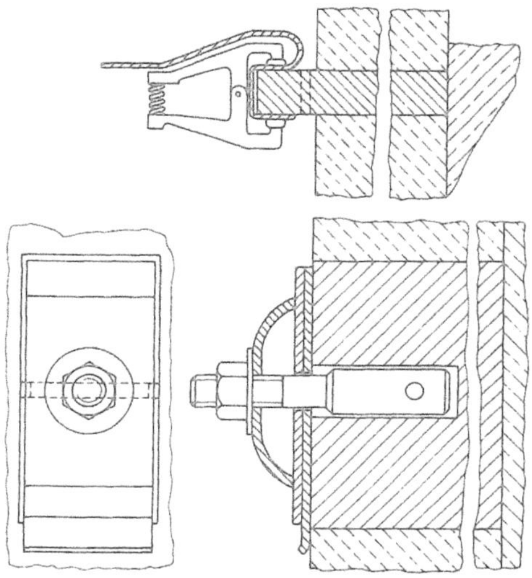

Figure 8.36. Connections to SnO_2 electrodes (Corning)[33]

Special methods need to be adopted to achieve good contact between electrode block and electric power supply and an example of such an electrical connection is shown in Figure 8.36. The protruding section of the electrode is covered in silver sheet and gripped with spring calipers[34] or, alternatively, a silver rod which melts at the operating temperature may be used but with this method a problem may arise later when the electrode has worn away in use and the silver becomes exposed to the molten glass. The use of silver sheet clamped either between two electrode blocks or between the electrode and a close fitting conical plug of tin oxide is also recommended. During heat up the tin oxide electrode must be protected from reducing conditions and this is best achieved with a coating of a slurry of ground glass which forms a protecting glaze at higher temperatures.

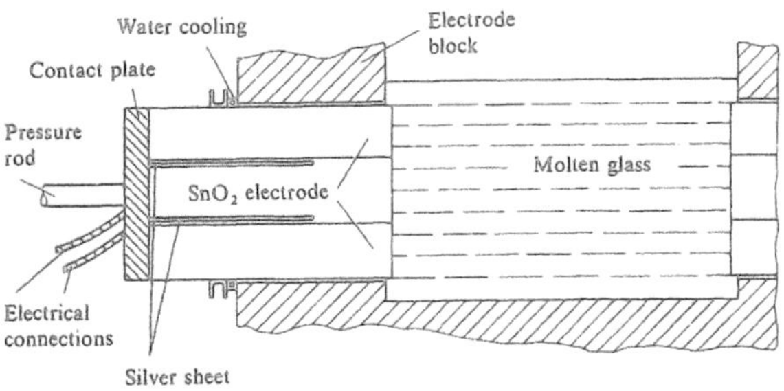

Figure 8.37. Typical cross section through a furnace with SnO_2 electrodes[52]

A further example of an installation for tin oxide electrodes is shown in Figure 8.37: the electrode is drilled and provided with a close fitting plug of tin oxide, and thin silver sheet is inserted between the main electrode and the plug. Contact is made with the silver sheet via large endplates, which may also be used to advance the electrode after wear.[32]

8.8 Design and operation of electric tank furnaces

8.8.1 Design

In the early days of electric melting the furnaces used to resemble those fired by flames. The furnace had a high melting area, of which only a small proportion was covered with batch,[2] and heat losses from the glass and power consumption were high despite a lower furnace temperature. The disadvantages of this 'hot top' furnace were removed by the development of the 'cold top' furnace, a design in which the surface area of the molten glass was reduced compared to its depth and almost totally covered with batch. This led to a significant reduction in heat losses.

In fuel fired and hot top electric furnaces the molten glass flows from batch to refined glass largely horizontally; in the cold top furnace, on the other hand, the flow is vertical. Vertical melting fundamentally changes the interplay of currents due to convection and pull. In fuel fired furnaces the convection currents assisted by the currents created at the hot spot greatly assist the refining of the glass; in a vertical, shaft type furnace this is not the case. Convection currents in shaft type furnaces are very strong and give a good homogenising effect, but with them there is the risk that insufficiently refined glass will reach the working end.[36] The correct positioning and electrical loading of electrodes in this type of furnace is therefore important and it has been claimed that sound construction and arrangement of electrodes can create piston flow in electric furnaces.[37]

One of the first furnaces with a cold top was that by Gell[38,39] but it was still shallow and its construction resembled that of flame fired furnaces (Figure 8.38). Molybdenum plate electrodes were used and as they were attacked only slightly by borosilicate glasses, their application was successful. The phase coupling devised by Scott, already described in Section 8.3.5, was used. Melting rates are given as 10·8 t/d, or 1·6 t/m^2d if related to surface area, or 2·2 t/m^3d if related to melter volume. In the case of shaft type furnaces it is more meaningful to consider the melting rate in relation to volume than to surface area.

A typical representative of the shaft furnace principle is the electric furnace by Sorg,[40] which is of hexagonal cross section with molybdenum electrodes located in two horizontal planes; Figure 8.39 shows the elevation and Figure 8.40 the plan of such an installation; the melting area is given as

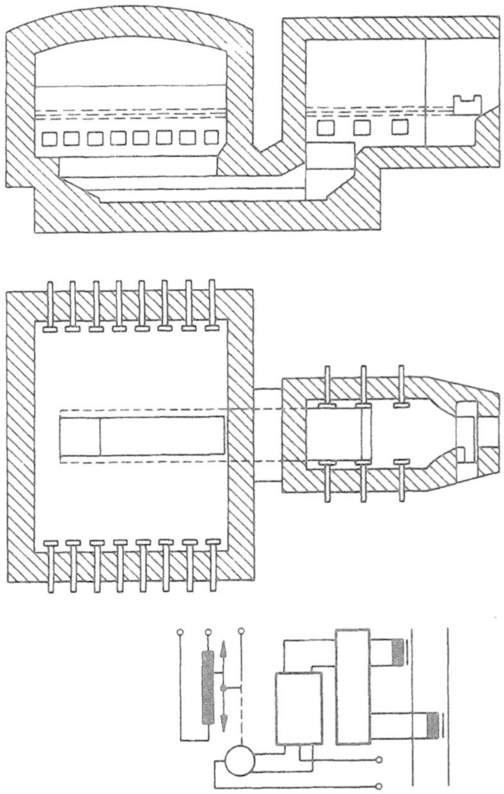

Figure 8.38. Gell furnace with plate electrodes[38]

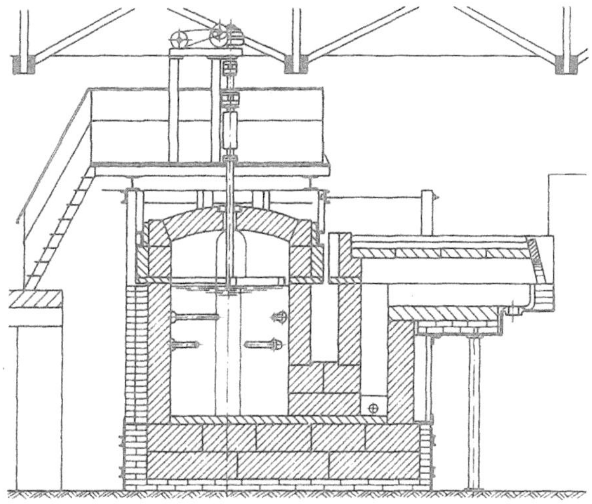

Figure 8.39. Electric furnace (Sorg)[40]

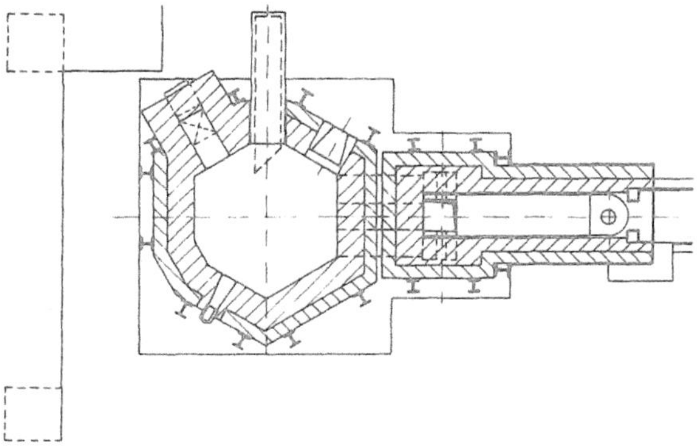

Figure 8.40. Ground plan of installation

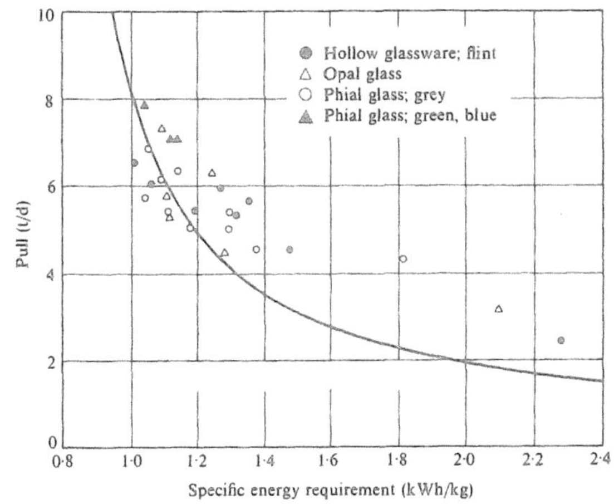

Figure 8.41. Relationship between energy consumption and pull

1·4 m^2 and the glass depth as 1·42 m. Best results are obtained when the electrical energy is divided between the upper and lower layers in the ratio 1:2. The hexagonal chamber in conjunction with a three-phase power supply creates a very even current density over the whole cross section of the shaft.

The batch is fed into the furnace through an oscillating trough and distributed over the surface of the glass by a rotating arm. The glass level in the feeder depends largely on the melting characteristics of the batch; a slowly melting batch coupled to a high pull results in an increase in batch blanket thickness and a decrease in glass level in the feeder. The difference can be controlled by matching the height of the rotating arm to the pull. The power consumption of this furnace is 375 kW, the maximum pull is 8 t/d, and the specific pull related to volume is 4 t/m^3d. The specific energy consumption is strongly dependent on the pull and decreases with increasing pull (Figure 8.41).

A development of this type of furnace for the melting of borosilicate glass for making marbles which are subsequently drawn into glass fibres in special tank furnaces, is shown in Figure 8.42.[41] The depth of the melter section is increased to 2·3 m and the pull is 55 t/d which, when related to surface area, is 2·75 t/m^2d; the crown is suspended and the height of the superstructure is 0·4 m. The blocks in the melting section are of fusion cast chrome–corundum at their maximum length of 1·2 m so that the dangerous horizontal joint is located as low as possible. The section at the top to a depth of 0·2 m is constructed of a material with good thermal shock resistance. In electric furnaces of this design the attack at the top of the refractories is very slight.

Strong corrosion advancing into the blocks appears above the holes carrying the electrodes (Figure 8.31) and it is believed that corrosion by upward drilling takes place in the annulus between electrode and tank block (see Section 6.9.4.1). The upward drilling is possibly increased by the heating effects which originate in electric melting (see Section 8.1.2). In the case quoted the leading edge of the electrode holder was 40 mm back from the hot face of the tank block.

The idea of using horizontal electrodes in several planes in a vertical melting furnace in order to achieve a more even distribution of melting energy is dismissed by some authors. They recommend the use of vertical electrodes in a melter of reasonable glass depth as this is claimed to give better control over glass currents as well as a more uniform distribution of energy.[37]

A relatively shallow tank for vertical melting has been described by Stanek;[42] it has only one plane

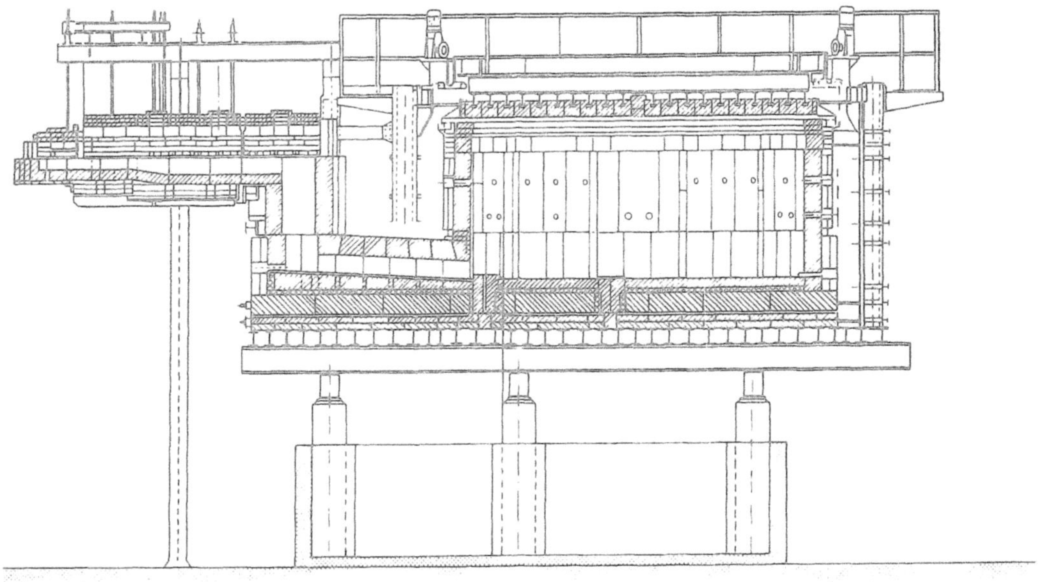

Figure 8.42. Electric tank furnace for the melting of borosilicate glasses[41]

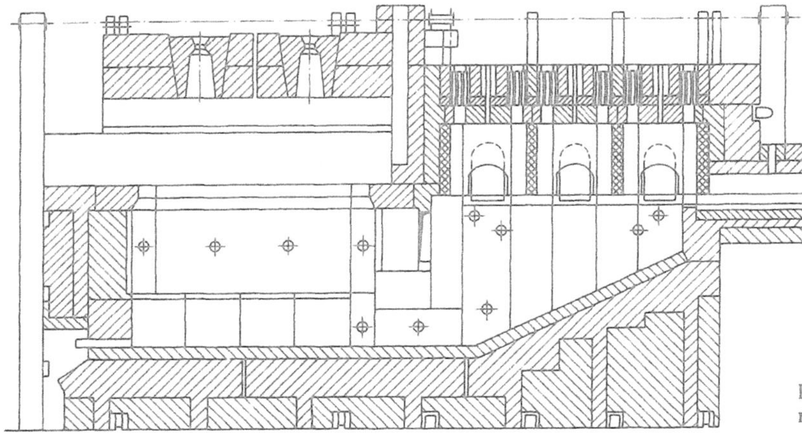

Figure 8.43. Elevation of electric tank furnace after Stanek[42]

of electrodes in the melting end and further electrodes in the region of the throat (Figure 8.43). The melting rate of borosilicate glasses is given as 6 t/d and the specific energy consumption as 1·7 kWh/kg glass. The furnace is constructed of fused quartz blocks. The electrodes in the working end are connected to earth in order to prevent any danger while working the glass at the gathering holes. The three heating circuits, melter, throat, and working end, are electrically independent and are controlled separately by three-phase induction transformers.

Another furnace for vertical melting, which combines the features of a relatively shallow cold top electric melter with a total glass depth similar to the shaft furnace described above, is the KTG Penelectro furnace. The large increase in glass depth is obtained by having a deeper unpowered 'refining zone' of small area below the powered cold top melter.

One such furnace design[58] to produce 72 t/d of borosilicate glass is based on a 12-sided melting chamber over a smaller 12-sided refining chamber as shown in Figure 8.43A. This design incorporates an intermediate shelf in the furnace through which molybdenum electrodes protrude into the melting chamber in six sets of four to heat the top melting chamber only. Batch is fed on to the top surface of the glass by a horizontal belt conveyor batch charger located through one sidewall of the super-

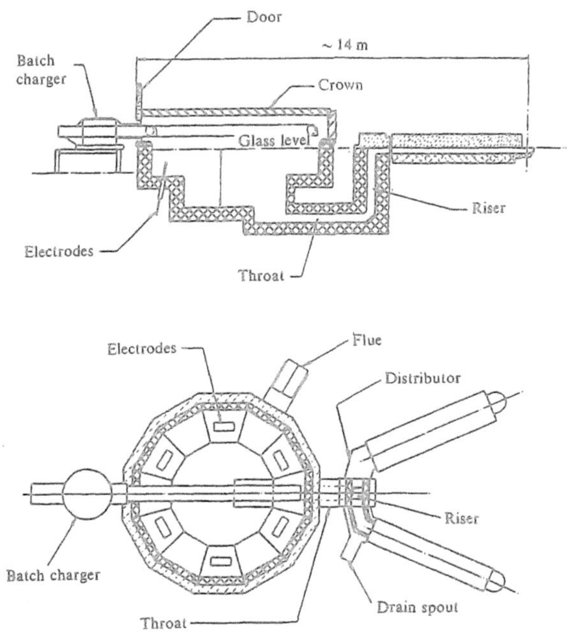

Figure 8.43A. The KTG Penelectro electric furnace for melting a wide range of compositions including borosilicate and lead crystal glasses. The illustration is of a 72 t/d borosilicate furnace with electrically heated distributor and forehearths[58]

structure programmed to spread the batch evenly over the entire surface at a rate determined by a glass level controller. The furnace superstructure normally operates at a temperature of about 100°C.

The overall depth of the furnace is 2·4 m and the furnace operates with a melting rate of 2 t/m²d. After melting, the glass passes into the unpowered refining zone and then, via a submerged throat and a vertical riser, to a distributor which feeds two forehearths. The throat and riser are also fitted with electrode heating systems which are capable of maintaining temperature under no pull conditions. The distributor and the forehearths are also heated electrically and have submerged cover tiles to eliminate the glass/air interface and losses of B_2O_3.

The glass contact refractories in the furnace and forehearth system are all void-free ZAC, insulated with sillimanite and insulation bricks over most of their outside surfaces. The furnace is powered by three 1000 kVA transformers connected in open delta to the three-phase supply, each transformer being fitted with phase angle control by silicon controlled rectifiers (SCR). The three phases of this supply are connected to the six banks of four electrodes in the furnace system. The furnace is controlled by regulating the total power input, power to the throat and the riser by current control of the various heating circuits and power to the distributor and forehearths by either current or temperature control using phase angle SCR controllers.

When operating at 92% of its designed load the power consumption over a two year period was 850 kWh/t of glass. The glass quality is significantly superior to that obtained from fossil fuel fired furnaces due at least in part to minimising the emission of B_2O_3.

This basic design has been used for many furnaces covering a wide range of pulls (2 to 120 t/d) and glasses. In particular this shelf type of furnace has been used for a large number of lead crystal glass furnaces both for hand working and for machine production. For the lead glass furnaces[59] multiple tin oxide electrodes are used and outputs ranging from 2 to 30 t/d can be achieved with power for melting ranging from 750 kWh/t for the larger furnaces to 1200 kWh/t for the smaller furnaces. As with the furnaces melting borosilicate glasses, atmospheric pollution, in this case by lead oxide, can be minimised. A typical furnace for lead glass with a designed output of 22 t/d has a depth of 1·5 m and melts at between 2·4 and 3·3 t/m²d depending on the amount of cullet used. With an average cullet usage of, say, 20%, the power consumption would be 780 kWh/t.

A further development in all-electric shaft furnaces is the molybdenum-lined decagonal furnace developed by Corning Glass works in the USA.[60] Figure 8.43B shows the principles of this furnace, whose design is aimed at producing high quality borosilicate glasses at high specific melting rates with minimum refractory corrosion and glass contamination: it is typically 1·3 m in diameter and 0·9 m deep. A molybdenum lining covers all sidewalls and bottom with the exception of the sidewalls surrounding the batch blanket. The melted glass is withdrawn through a molybdenum pipe positioned on the floor of the melter: this pipe is extended into a cooled section outside the body of the furnace before it enters upwards into the forehearth. Operating temperatures are in the region of 1700°C and performance characteristics for a selection of borosilicate glasses are given in Table 8.5A.

Melters using mixed melting have a special place in that they can be heated partly by fossil fuel and partly by electricity, with the ratio being varied according to requirements; furnaces of this type were developed in Switzerland during the 1940s. Graphite electrodes were used as molybdenum

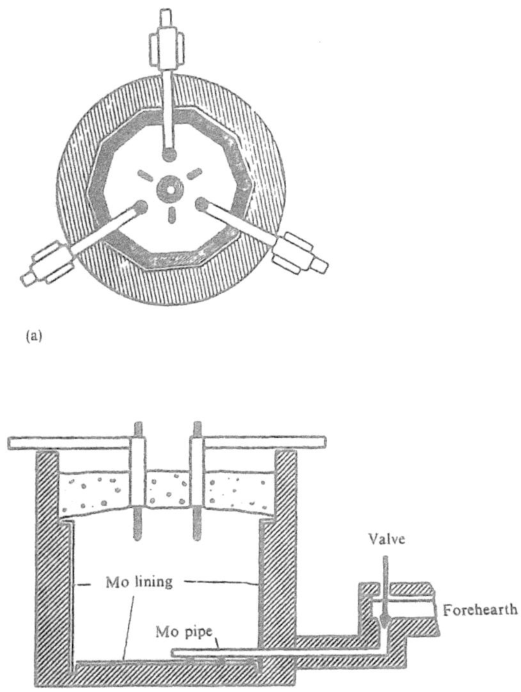

Figure 8.43B. Corning shaft furnace for melting borosilicate glasses
(a) electrode positions (b) elevation

Table 8.5A. *Some performance characteristics of Corning all-electric shaft furnaces*

Glass	*Pull (t/d)*	*Melting rate (t/m^2d)*	*Energy consumption (kWh/t)*
Borosilicate $\alpha = 55 \times 10^{-7}$	9·1	13·7	840
Borosilicate $\alpha = 32 \times 10^{-7}$	7·95	15·6	904
Glass wool	10·2	11·2	827–876
Aluminosilicate CGW Code 1720	7·95	15·6	865

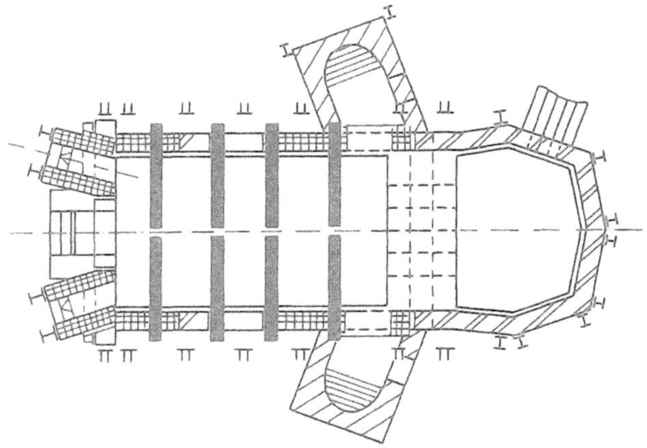

Figure 8.44. Tank with horizontal graphite electrodes[45]

electrodes were not then available. Figure 8.44 shows a furnace with two burners and four pairs of electrodes, the advantage of this firing system being the built-in flexibility which allowed the user to adjust his choice of energy in line with varying prices.[45]

A design using molybdenum electrodes inserted through the bottom is shown in Figure 8.45; this furnace has a divided superstructure so that the melting section may be heated entirely by electricity on the cold top principle while the refining section has mixed flame and electric heating.[46]

8.8.2 Operation

8.8.2.1 The melting process

Electric furnaces differ from flame fired furnaces in their mode of operation; in the latter the limit to the rate at which glass can be melted is governed by the refining rate; while in the former it is governed by

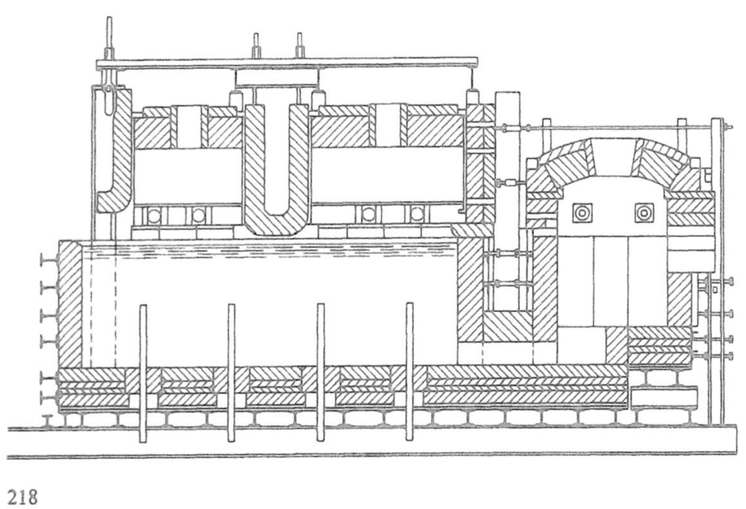

Figure 8.45. Mixed melter with two separate combustion chambers[46]

the batch melting rate. The limiting factor in the electric furnace is the heat transfer between the molten glass in the melter and the batch blanket covering it.[37] This is only adjustable within limits and it therefore follows that electric furnaces cannot be operated with pull rates which vary as widely as those of flame fired furnaces. Low pulls frequently cause problems as the batch blanket will break up, making the furnace run too hot and causing high energy losses upwards: the superstructure and batch feeder may then be damaged.

A range of pull of 1:2 has been mentioned for (container) glass of reasonable quality but for higher quality the range is about 1:1·2.[37]

The number and positioning of electrodes leads to a certain unevenness in the current density distribution, causing the batch blanket to melt unevenly and break in certain places. The batch distributor (Section 6.4.3.4) of electric furnaces must therefore be designed with particular regard for this problem.

The operation of the furnace and, at the end of the process, the seed content of the glass depend on the removal of gases from the batch as it melts. The cold superstructure of an electric furnace can cause a crust to form which seals the batch blanket as it melts so that the gases being formed have no means of escape and a zone of increased partial pressure of CO_2, H_2O, etc. is created between the molten glass and the batch blanket. In consequence, the removal of gases during refining is obstructed and the resultant glass contains blisters and seeds. This crust will be formed when the proportion of cullet in the batch exceeds 30% or when the first fusion of batch takes place before the decomposition of the carbonates. A layer of insulating foam forms between the glass and the batch and causes the rate of melting to decrease slowly. This situation can be alleviated by reducing the viscosity of the melt when it is first formed (by the addition of fluorine) or by replacing carbonates with oxides or hydrates (e.g. using caustic soda instead of soda ash).

Batch crusts are formed by batches with a high water content or those containing a high proportion of alumina. The problem can be eased by deliberately keeping part of the surface uncovered by batch so that the gases given off can escape. The only solution for high alumina batches appears to be to depart from all-electric melting and to use burners in the upper part of the furnace. Conditions for the melting of batches with lower alumina contents may be improved by increasing the proportion of hydrated alumina. New furnace designs and new batch formulations must be tested on a production scale if unpleasant surprises are to be avoided later. A change in batch materials generally gives the best results.

The type of batch feeding system used is also of importance (see Section 6.4.3).

8.8.2.2 Instrumentation for measurement and control

The factors which need to be controlled on electric furnaces are temperature, current, voltage, pull, and glass level.

The measurement of temperature is not as important in electric furnaces as it is in fuel fired furnaces. A temperature measurement in the superstructure only provides information concerning the state of the batch blanket. Temperature measurements in the glass are not satisfactory in the long run as thermocouple sheaths are damaged by the strong convection currents. The best results have been obtained by using thermocouples set in recessed holes in the tank blocks; however, even these can give only comparative readings. In special cases average temperatures can be measured indirectly by comparing the resistance, using immersion measurement techniques.

Voltage, current, and power consumption are the most important parameters to monitor and are measured by standard instruments. The early control transformers and transducers have now been replaced by silicon controlled rectifiers which have no moving parts and are small; they have two thyristors connected in opposite directions as shown in Figure 8.46. The SCR can be used in burst firing or phase angle modes[47] and the heating of glass with electrodes is best controlled by phase angle modu-

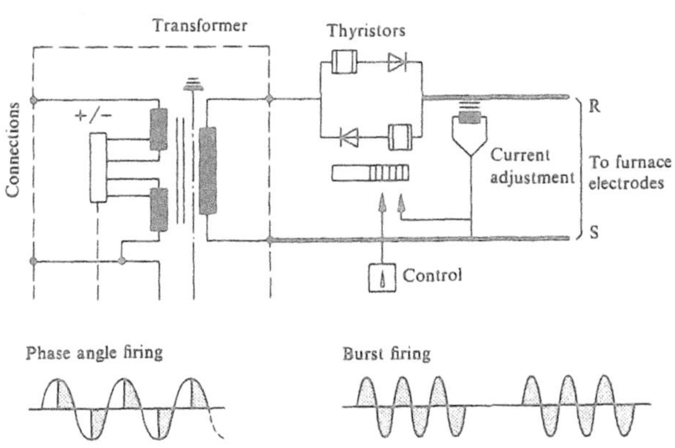

Figure 8.46. Circuit diagram of a silicon controlled rectifier[47]

lation in which control is exercised by chopping each half wave. This form of control can be linked to current limiting devices, which is important for heating paths between electrodes. Attention must also be paid to 'circuit contamination'. As already discussed in Section 8.5.1, the wear of electrodes can be reduced by the application of a small direct current voltage giving a current of 1·3 mA/cm^2 of electrode surface. The positive terminal should be connected to the electrode, the negative to the molten glass, and the maintenance of this polarity is important because if it is reversed, then not only do the electrodes corrode more rapidly but faults such as blisters appear in the glass.[44]

In practice, current, power, and, to a lesser extent, voltage can change in the heating paths between electrodes due to changes in resistance. Changes in pull cause temperature changes and thereby changes in resistance; these changes must not exceed the normal operational parameters. Errors in batch composition (e.g. a change in alkali content) or a short circuit via the refractories will lead to rapid changes in resistance. A steady increase in resistance over a period of weeks suggests corrosion and wear of electrodes or a change in their shape. When electrodes are installed one above the other, bending of the upper electrode reduces electrical resistance, whereas bending of the lower one increases it; turning the electrodes can help to establish whether bending has taken place. Extensive corrosion of refractories enlarges the volume of molten glass and results in an appreciable reduction in resistance.

A sudden change in resistance indicates fracture of an electrode.

8.8.2.2.1 Measurement of temperature

Control of melting and refining temperatures in electrically heated furnaces differs markedly from those in flame fired furnaces. Observation of batch conditions in the upper part of the furnace is, from the point of view of the process, of limited use. Temperature measurements in the molten glass require special procedures in view of the high thermal currents.

Ceramic thermocouple sheaths have a small electrical conductivity at operating temperatures. Due to the large tank surface which is earthed and despite the galvanic separation provided by the transformers to the electrodes, a potential to earth usually exists. This can be measured on the thermocouple against earth and can, depending on the circuitry of the recorder, lead to faulty measurements. Earthing of the thermocouple head is usually not successful and more success is achieved by the insertion of a low pass filter into the circuit; these are supplied by the recorder manufacturer to suit the recorder (Figure 8.47).

Another source of error arises from interference created in signal cables running parallel to high power cables, especially near connections to electrodes where a magnetic field cannot be avoided. This becomes critical when high frequencies and high currents are used. The interference can be eliminated if connections to the electrodes are made with induction heater cables which have low power losses and therefore create only a weak ambient magnetic field. Coaxial cables are used for connections to high tension or single phase transformers, the inner cable being used to transmit power to the electrodes and the outer, both ends of which are connected to earth, for screening (Figure 8.48). The current flowing in the outer conductor permits the whole cable assembly to appear magnetically neu-

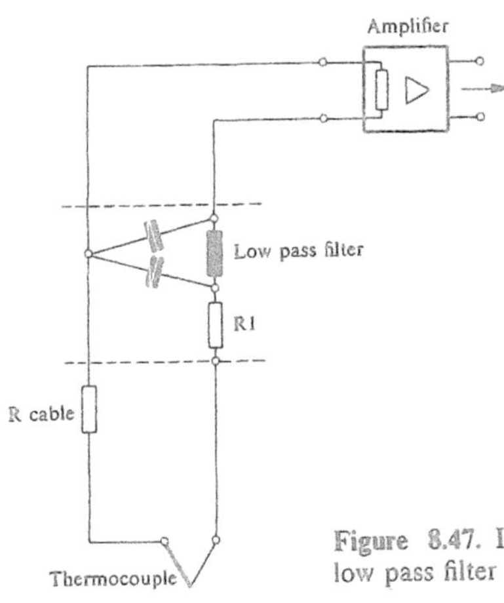

Figure 8.47. Installation of a low pass filter

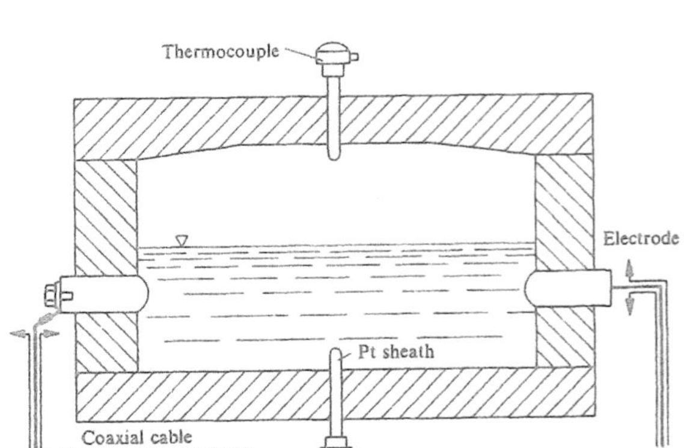

Figure 8.48. Screened electrode cable connections

tral. This type of cable also provides additional safety in that, in case of damage or accident, the earthed and not the live conductor makes contact. Additional improvement is also given by the use of a low pass filter.

The use of signal amplifiers located as near as possible to the place of measurement is also recommended. When installing instruments and control equipment it is important to separate them from high power installations; cabling to and from these must also be separate. The normal method of placing high power cables in ducts and instrument cables in conduits provides some shielding against stray influences. If desired, it is also possible to screen the instrument cabling, and screening may be added later in the form of a zip fastened sleeving. When platinum sheaths are used for the better protection of thermocouple sheaths it is important to ensure that they do not act as electrodes and cause faulty measurements during heat up; a short circuit to earth during heat up must be avoided.

The use of thermocouples can be avoided altogether by using optical pyrometers focused onto the hot end of a closed molybdenum tube inserted into the tank block. Oxidation of the metal must be prevented by flushing the tube with protective gases. Special precautions must also be taken to prevent the molybdenum tube from becoming part of the electrode heating circuit. Temperature measurements of this type are difficult and are recommended only for use in exceptional circumstances.

The electrical conductivity of molten glass is strongly dependent upon temperature and it is therefore possible to use the voltage existing between two electrodes as a measurement of glass temperature. However, it must be assumed that the active electrode surface area remains constant. A basic circuit diagram for using voltage readings as a measurement of temperature is shown in Figure 8.49 in which R_e is the resistance of the electrode pair, R_n a known resistance for comparison, and G is an applied voltage. If the loss of voltage between

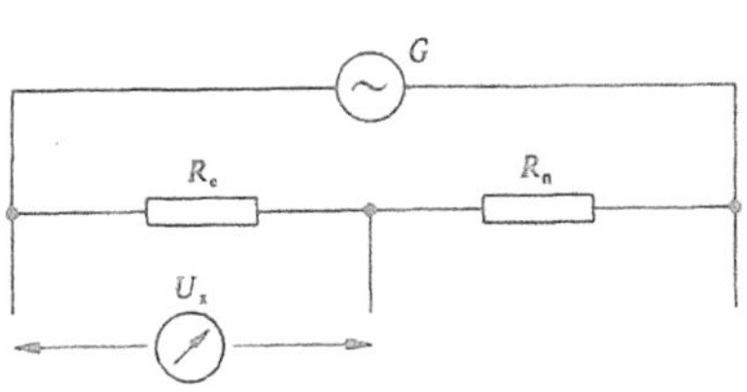

Figure 8.49. Scheme for measuring temperature via the voltage drop between a pair of electrodes

the electrodes is U, then the following simple relationship applies

$$R_e = R_n \frac{U_x}{U_n}. \tag{8.34}$$

The temperature of the glass can be determined from prior knowledge of the relationship between temperature and electrical conductivity of glass together with the geometrical relationships; alternatively, at least, a measurement is obtained which can be used as a parameter for controlling the furnace. Stanek[42] gives this method an accuracy of 0·3–1·5%. The same measurement can also be used to monitor changes in glass composition and thereby the accuracy of batch mixing; Ross & Stewart[48] defined a conductivity factor which is very sensitive to changes in alkali content.

Ampère and related voltage readings of electrodes can also provide data covering the conductivity of the glass and the average glass temperature. A further method of temperature measurement is the measurement of local voltage as already mentioned in Section 8.8.2.2.

8.8.2.2.2 Glass level

Close control of glass level is much more difficult in an electric furnace than in a flame fired furnace. The batch does not flow freely on the glass surface but tends to cake together and stick to the sidewalls and furthermore the batch blanket is thicker and has varying thicknesses depending on the extent to which gas is removed. If continuous batch feeding is not possible, then intermittent feeding will produce periodic variations in glass level. Much attention to detail is required to minimise these undesirable effects and achieve acceptable control of glass level.

For electric melting, special attention must be paid to the method of batch feeding. The surface of molten glass should, as far as possible, always be covered with a uniform layer of batch, with a narrow clear space left along the edges.

As with fossil fuel fired furnaces, the energy requirements of electric furnaces depend on the size of furnace, pull, type of glass, and thermal insulation. Finger[49] claims that the relationships shown in Figure 8.50 apply but the current view is that the values given are a little too low. Heat losses include 8–10% used in cooling the electrodes if they are of molybdenum. More details of the thermal balance of electric furnaces are given by Illig.[50]

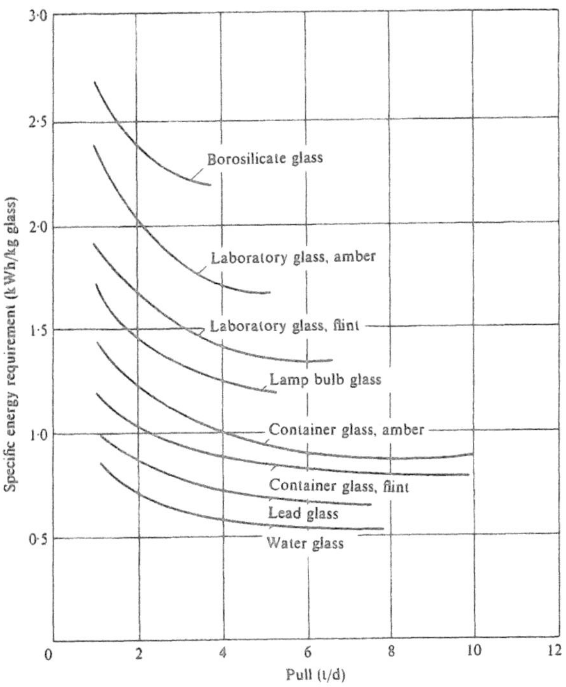

Figure 8.50. Specific energy consumption of electric tank furnaces as a function of pull[49]

8.9 Glass tank furnaces with electric boosting

Electric boosting increases the pull and/or the quality of glass from fossil fuel fired tank furnaces. If the emission of batch dust and other pollutants must be reduced then the pull can be maintained while reducing the proportion of fuel firing in favour of electric boosting. A further application is for tanks melting batch with a high proportion of cullet.

If the pull of a fuel fired furnace is gradually increased, a point is reached at which seeds and inhomogeneities appear in the glass; this constitutes the limiting pull (Figure 8.51) and electric boosting provides the means of moving this limit to higher figures. It can also be used to overcome quality problems due to the age of a furnace.

Electric boosting is very flexible. Its advantage is that heat can be applied in the molten glass precisely where it is wanted, i.e. in the refining zone, in the melting zone, or in both together. It is particularly useful when used in the melting zone with glasses of low infrared transmission. Disadvantages are the not inconsiderable investment costs and the higher cost of electricity compared to fossil fuels. The economic viability must be checked in all cases.

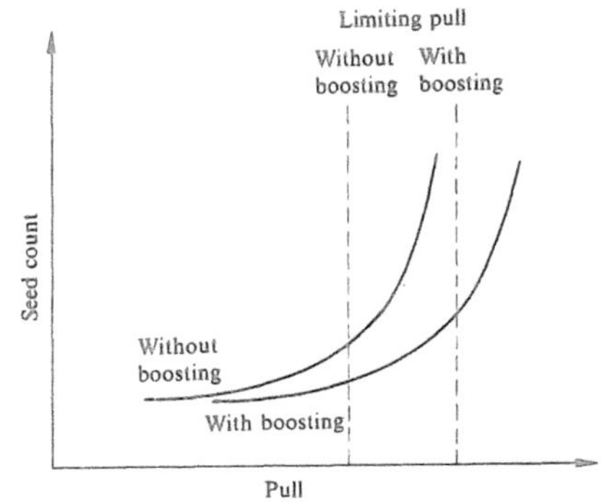

Figure 8.51. Influence of boosting on the limiting pull of a fuel fired glass tank furnace

A tank furnace with electric boosting is not an electric furnace operating like a flame fired furnace; depending on the layout, some part of the fossil fuel required for melting and refining is replaced by electric power. The heat losses incurred with boosting, i.e. the cooling of the electrodes, are low when compared to flame firing and are largely independent of temperature. Heating the glass from 1400 to 1500°C requires practically the same energy as heating it from 800 to 900°C and this characteristic makes boosting more useful for refining than for melting; the relationships are shown schematically in Figure 8.52. The efficiency of flame firing is strongly dependent on the temperature above the glass level and especially on the difference in temperature between flame and molten glass or batch. Since flame heating in the refining zone is subject to particularly high energy losses, this is the area where electric boosting is especially advantageous.

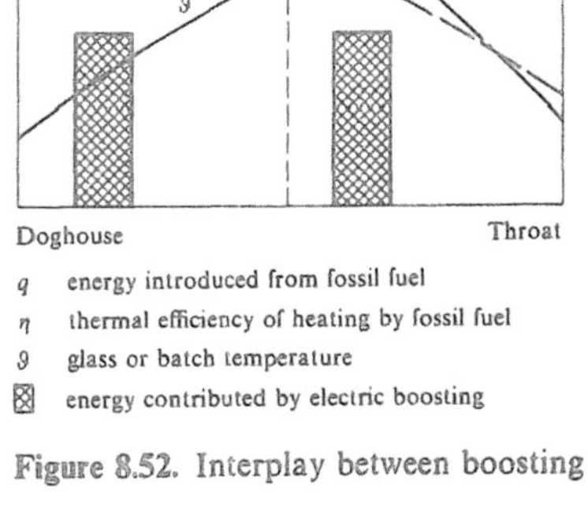

Figure 8.52. Interplay between boosting and flame heating

The location of electrodes for boosting in the plane B–B in Figure 8.53 provides heat to improve both refining and the removal of gases from the molten glass. Electrodes located under the batch in the melting zone (plane A–A) improve the rate of melting but, if the energy input from boosting is too large, weaken the anticlockwise currents in this zone. The current in the plane A–A under normal conditions ensures that the batch blanket remains in a stable position; when boosting is applied this current weakens due to the smaller difference in temperature between hot spot and batch melting area, the anticlockwise current is weakened, and the batch blanket is not pushed back as strongly as before. In addition, if the pull is increased, the difference in glass level between doghouse and throat is increased. These factors all combine to give a tendency for the batch blanket to push into the refining zone if the boosting in the melting zone is too strong (Figure 8.53(b)); the technologist is likely to be familiar with this phenomenon. The most useful arrangement is to have boosting electrodes with separate controls in both the melting and the refining zones. This provides optimum conditions for high pull and good glass quality.

Figure 8.53 shows the thermal currents schematically in stable and unstable conditions. When the batch position is stable the melting batch slides slowly on the glass surface, which is slightly inclined due to the pull, while the raising of glass at the hot spot coupled with the anticlockwise current holds the batch in a surface trough. The forces involved should be balanced in such a way that the batch does not move forward faster than the rate at which it melts. The situation shown in Figure 8.53(b) indicates the risk that the batch will slide past the trough and hot spot causing incompletely melted batch to reach the refining zone.

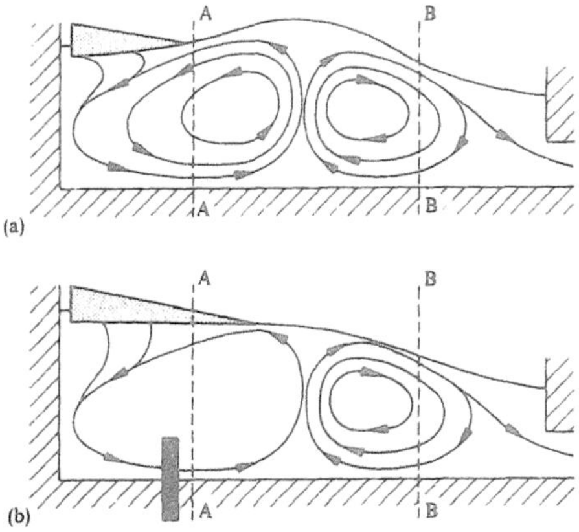

Figure 8.53. Schematic of convection currents
(a) without boosting
(b) with strong boosting in the melting zone

As boosting is increased so the efficiency with which additional pull is obtained increases, as Gneist, Hohmann & Walter[51] have pointed out. Their theoretical studies, when compared to practical experience, assume *a priori* that, compared to pure flame firing, the use of boosting increases the temperature of the glass by a constant amount ($\Delta T \approx 40$ degC) independent of the boosting power applied. This assumption is unlikely to apply in all cases. If a proportional relationship between ΔT and boosting power is assumed (which is more likely to be true), then the picture changes and the increase in efficiency with increased boosting is markedly less. Ross & Stewart[48] even suggest that an increase in energy is required as the power used for boosting increases (Table 8.6).

These energy relationships can be treated mathematically and show that, depending on the positioning of electrodes, the power used, and especially the effect on the currents in glass, the energy needed for the additional tonnage melted as a result of boosting can vary widely. Details are given in Section 7.3.3.3.2.

The electrode layouts and circuits used in practice vary widely. There are differences between the boosting of the refining zone, the melting zone, and combinations of the two. The switching arrangements should be flexible in order to control changing operating conditions. Figure 8.54 shows an example of boosting in the melting and refining zones controlled separately from independent transformers. An arrangement of this type has been used successfully for a special glass at a pull of 130 t/d with a total boosting power of 1500 kW. Suitable floor electrodes would be rods 50 mm thick inserted to up to 20 cm below the glass surface. The layout shown in Figure 8.55 for boosting the combined melting and refining zones of a tank for white hollow glassware with a specific melting rate of 3 t/m^{2}d was

Table 8.6. *Boosting energy compared to the energy required for the daily additional tonnage of glass*

Type of glass	*Boosting energy (kW)*	*Energy required (kW/t glass)*
White	0–1000	25
	1000–2000	30
	2000–2800	35
Amber	0–500	20
	500–1000	25
	1000–2000	30

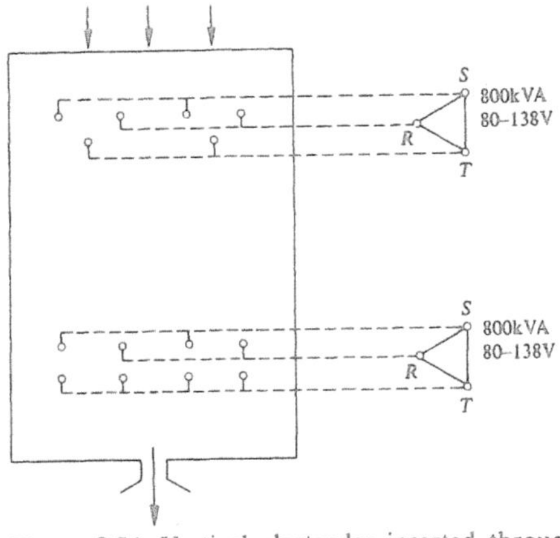

Figure 8.54. Vertical electrodes inserted through the floor in melting and refining zones

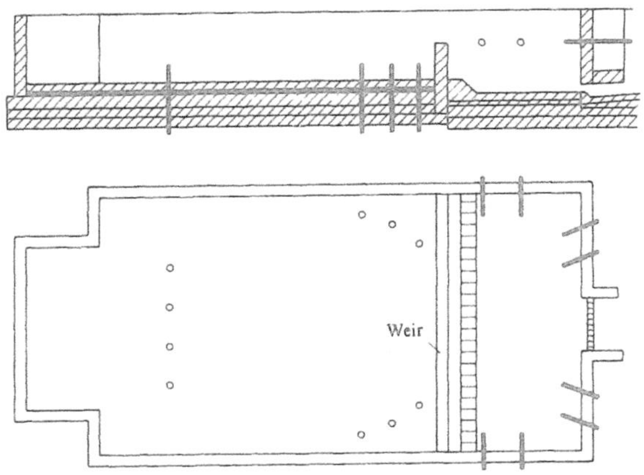

Figure 8.55. Combined arrangement of electrodes through floor and sidewalls in a tank melting flint glass[56]

judged a success; the electrodes in this tank can be coupled in different ways and Figure 8.56 gives examples.

The establishment of optimum operating conditions and coupling can be helped by installing a thermocouple through the floor of the furnace; its minimum distance from the throat should be 1 m. It is even better to install two or three such thermocouples along the centre line of the furnace (Figure 8.57).

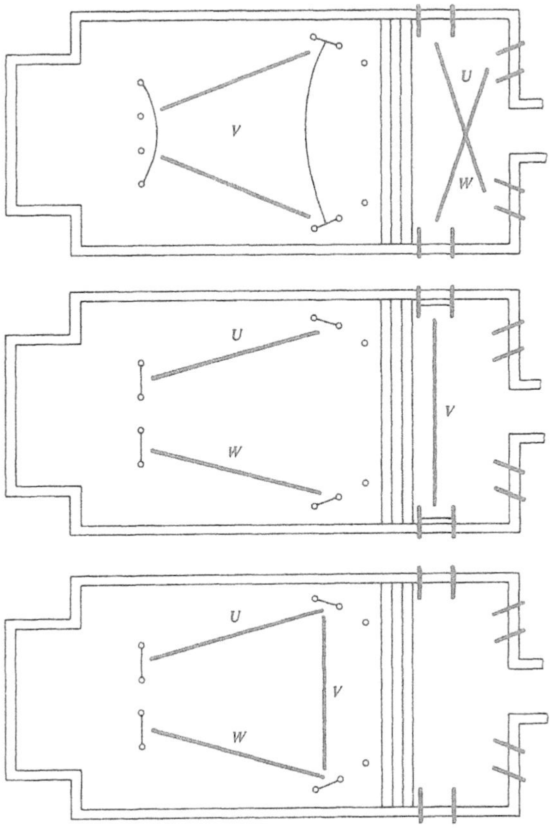

Figure 8.56. Examples of various connections[56]

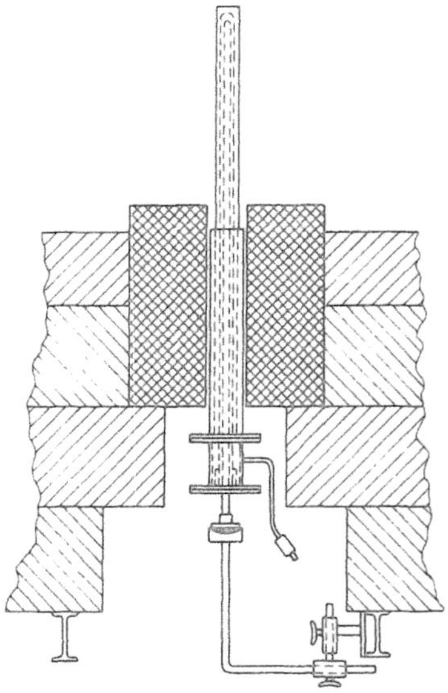

Figure 8.57. Thermocouple inserted through the floor

8.10 Safety features

8.10.1 Existing regulations

Glass furnaces require licences under (German Federal) laws covering protection from emissions. Electrically heated furnaces and furnaces with electric boosting need to comply with the relevant regulations covering high power installations of nominal voltages up to 1000 V, as well as those covering industrial heating installations.

As water cooling must be applied to electrodes, the area around an electric furnace or a furnace with electric boosting falls under the classification of a wet area in the safety regulations covering materials subject to dampness or condensation of water.

In addition, the high ambient temperature of about 35°C around glass furnaces causes humans to perspire and reduces the electrical resistance of the human body. In terms of safety regulations this means that the furnace must also be considered as a hot area.

8.10.2 Regulations covering the prevention of accidents

Cables must be laid separately to each electrode. The insulation of cables and electrode holders from the steelwork of the tank must be perfect and Gell[53] recommends double insulation of the steelwork, first with respect to the furnace proper, second with respect to earth. Dangers arising from accidental touching of live components must be anticipated by the use of protective mats of glass or refractory fibre fabrics, together with adequate heat removal or cooling. Areas where there are electrode connections, which are the tank floor or sidewalls, depending on furnace design, must be fitted out as closed electrical rooms. This means that the rooms must be hermetically sealed and the doors kept locked, with access restricted to suitably qualified personnel. These areas must also carry the necessary clear warning signs.

Since it is impossible to guard against contact with electrical conductors on electrically heated furnaces, other protective measures must be taken: an insulated floor on which to stand, insulated footwear, and insulated tools. Any gathering holes or openings for the removal of glass for working should be as far as possible from the nearest electrode; this gives a lower voltage to earth. This also applies to any opening above the molten glass where samples are taken or gates operated. Even a residual voltage of 30–40 V requires working platforms constructed of planks and grating of dry wood. Personnel must wear dry gloves and clogs. If water hoses are used, they must be of carbon-free rubber. Cooling water piping must be insulated from the place where the cooling is applied: a resistance of 30 000 Ω is regarded as adequate and this calls for an insulated length of about 2 m for a pipe of 25 mm diameter.

Maintenance work such as the tightening or loosening of screws, adjusting of electrodes, etc., on parts which, when in use, are live, must only be carried out with all power switched off. The main breaker must be opened and secured against closure. Even this work must be done on insulated floors and with personal body protection and must be done only by or under the supervision of qualified personnel.

The maintenance and upkeep of safety features must be subject to continuous verification. For example, checking the insulation of supports and foundations by measurement of total resistance to earth with an earth leakage potentiometer, checking the freedom from potential of the current entering the electrodes, checking the correct functioning of the emergency breakers, and checking the voltmeter.

9 Day Tanks

A day tank is a type of furnace which, in terms of its design and operation, stands between the continuously operating glass tank and the periodically operating pot furnace. Day tanks were developed in order to melt smaller quantities of glass without the need to arrange for shift working and without the disadvantages of a pot furnace such as pot changing and narrow limits for operating temperatures. These requirements led to the development of a tank which is a separate type of furnace with its own operating characteristics.

Although the day tank operates on a periodic cycle, the construction resembles that of a normal tank, being constructed of tank blocks, cool on the outside and hot on the inside. It may be operated at melting and refining temperatures which are not possible in pot furnaces where pots are exposed to the flames, which means that glasses which are difficult to melt can be made in day tanks. As will be shown later, the main areas of use for day tanks lie in the manufacture of special glasses which are required in only small quantities and which are usually worked by hand.[1]

9.1 Temperature schedule and the melting process

The melting process corresponds to that of a pot furnace and almost invariably the operation of a day tank is on the basis of a 24 hour cycle. When the working of the glass is completed, the furnace is heated up and this is followed by batch feeding, melting, refining, bubbling, and cooling. The daily temperature schedules of a number of different day tanks are given in Figure 9.1. The effect of the individual batch feeding events on temperature are not as noticeable as with pot furnaces, since the steps are more frequent and each feed covers only a small proportion of the melting area.

More recently, day tanks have come into operation in which the working end is separated from

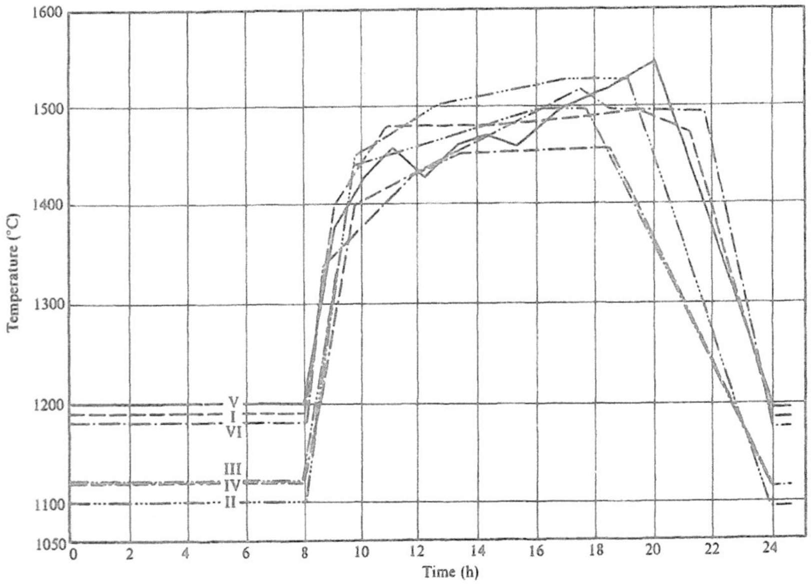

Figure 9.1. Temperature curves for different day tanks;[2] the numbers I to VI correspond to the descriptions given in Figure 9.2.

the melting and refining zones by means of a weir. This makes it possible to melt and refine on a continuous basis, but work on a periodic basis, and gives certain advantages in terms of melting and operation.

9.2 Firing system

There are many types of day tank (Figure 9.2), each depending on size and melting conditions; they are also equipped with regenerators or recuperators, the latter being preferred.

Tank I is fired longitudinally and has a trapezoidal plan view. It uses light oil fed through atomising burners from the sides into the ports and has two recuperators underneath. The gathering holes, which also serve for batch feeding, are along each side.

Tanks II, III, and IV are double crown tanks with vertical U shaped flames and the advantage of this design is that gathering holes can be located in the end wall. For very small tanks, the inner crown can be omitted, making construction easier and avoiding the risk of certain glass quality faults arising; an example of this design is tank V, Figure 9.2(c). In Tank VI the flues are located in the sidewalls and the advantage of this layout is better access to burners and ports; the drawback is loss of wall length in which to place gathering holes.

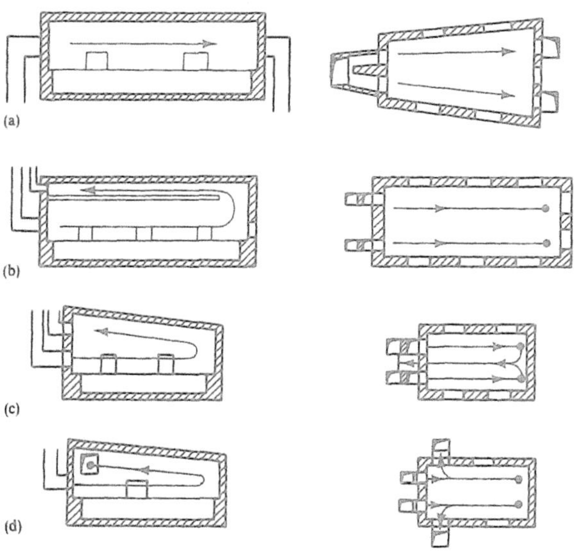

Figure 9.2. Different types of day tank[2]
(a) I, straight longitudinal flame
(b) II to IV, double crown
(c) V, port in the end wall, U shaped flame, exhaust flue in the end wall
(d) VI, as V except that there are two exhaust flues, one in each sidewall

It is more difficult to decide the dimensions of the combustion chamber and the firing positions for a day tank than for pot furnaces or continuously operating tanks. The difficulties arise from the large changes in glass level and the periodic changes in fuel consumption rates. The highest rates of fuel consumption occur when the combustion chamber has been made smaller by glass and batch. On the other hand, the largest volume of combustion space is available when the fuel consumption rate is low during glass working. Tanks built with double crowns accentuate this problem since fluctuations in their glass level are proportionally the largest; if possible, double crowns should be avoided. In practice, a firing chamber 0·7–0·8 m high when the tank is full has been found satisfactory. The distance between flame and batch should then be 0·3–0·35 m.

Burners must have a large operating range; alternatively, this requirement can be covered by the use of two burners, one of which is shut off during glass working.

9.3 Fuel consumption

The fuel consumption of day tanks is relatively high, as is the case for all periodically operating furnaces. After glass working, the residual glass, usually 40–50% of the tank content, must be reheated every time. Data from the literature relate either to the quantity of glass worked or to the

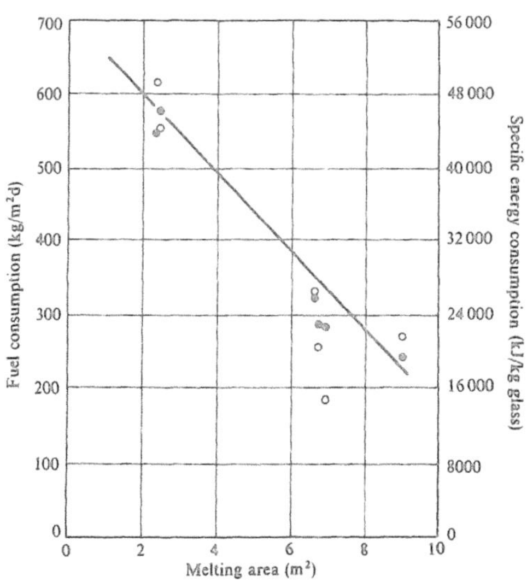

Figure 9.3. Relationship between fuel consumption per day per unit melting area, specific energy consumption, and melting area

melting area and since the proportion of glass worked out varies from plant to plant, it is best to use the melting end area as a basis for comparison (Figure 9.3). According to Starzinsky[2] the daily consumption of heating oil (kg/m²d) is given by

$$\dot{M}_{B,d} = \dot{m}_{B,0}A - cA^2. \qquad (9.1)$$

where $\dot{m}_{B,0} = 700$ kg/m²d, A is the surface area of the melting end, and $c = 58$ kg/m⁴d. With suitable adjustments this equation can also be used for natural gas.

To an increasing extent the preheating of combustion air is done in standard metal recuperators which are cheaper and more reliable in use than ceramic recuperators. Preheating the air to 450–500°C and waste gas temperatures of 900°C are regarded as acceptable.

9.4 Melting rate

Melting rates achieved in day tanks lie between 0·4 and 0·8 t/m²d, very similar to those for pot furnaces. However, it must be remembered that the melting area in pot furnaces is much smaller than that of day tanks. The higher melting temperature of day tanks probably compensates for the additional heat input in pot furnaces through the walls of the pots. The specific energy consumption is also of the same order of magnitude as that of pot furnaces.

9.5 Glass quality

The homogeneity of glass obtained from day tanks has been the subject of several investigations. Comparisons have been made with the homogeneity of glasses melted in pot furnaces and when comparable glasses are studied (potash, lead, and soda–lime) the result is not in favour of day tanks. The causes are ascribed to the flow of slag from the superstructure and the difficulty of working from rings in day tanks. The periodic operation of day tanks causes condensation in parts of the crown and sidewalls which cannot be avoided and which contributes to slag formation; this slag does not flow into the glass in pot furnaces but does so in day tanks. The periodic operation also exposes the sidewalls to very severe conditions; their surface becomes crazed (Figure 9.4) and the primary melt penetrates into the cracks to flow out during glass working and bring refractory material with it. Rings such as are used in pot furnaces cannot be used in day tanks as they sinter under the high melting and refining temperatures, become gradually denser, and then cease to float.

Figure 9.4. Cracked fusion cast AZS tanks blocks[2]

The conditions described lead to the conclusion that the glass quality attainable from a day tank is not good enough for thin walled high quality tableware. Tanks with double crowns are especially subject to slag from the superstructure. With pot furnaces, the best glass is worked 3 to 4 h after starting to work the glass and thereafter a significant worsening of quality takes place; with day tanks, the best glass is worked after about 6 h and at this stage the quality of the glass from day tanks can even exceed that from pot furnaces.

9.6 Refractories

Crowns and sidewalls are best constructed of silica. It is important to provide drip edges so that any vitreous slag formed does not run down the sidewalls but enters the melt during the melting and refining stages and can thus be dissolved. Fusion cast AZS is best for tank blocks, ports, and waste gas flues but as this refractory is sensitive to thermal shock and tends to crack, cold batch just charged should be kept away from the tank blocks. Furthermore, furnace pressure should be controlled carefully to prevent the ingress of cold air through the gathering holes. The tank blocks below the gathering holes should be covered with ceramically bonded refractory tiles. Good insulation of the furnace reduces energy consumption and protects the refractories from sudden temperature changes.

10 Pot Furnaces

In this type of furnace the glass is melted in special containers called pots, each furnace usually containing several pots. In a tank furnace each stage of the melting process, i.e. melting, refining, and cooling, takes place in a specific location in the furnace; in a pot furnace, on the other hand, all these stages take place in the same place, i.e. in the pot, in sequence. The melting process is periodic but the furnace itself can be operated either continuously or periodically. If open pots are used, as is exclusively the case in Germany, the furnace and pots follow the same periodic programme which, for reasons of organisation and working practice, is based on a 24 h cycle.

If covered pots are used then the furnace combustion space is separated from the space taken up by the pots because they are totally enclosed units except for the gathering hole; by opening and closing gathering holes it is possible to combine continuous furnace operation with periodic melting. However, the pot hood is a major barrier to heat transfer between furnace and glass and, as a result, the periodicity of melting in closed pots is usually a multiple of 24 h. Furthermore, this melting technique is only suitable for glasses with low melting and refining temperatures, but has advantages for glasses subject to volatilisation. In some countries, e.g. France and Great Britain, closed pots are used; but not in Germany.

Compared to a tank, a pot furnace has the advantage that small quantities of glass can be manufactured economically and also that different glasses can be melted in the same furnace. A well designed furnace, correctly operated, can give very homogeneous glasses. An additional advantage is that a 24 h cycle does not call for shift working.

Among the disadvantages are the high specific energy consumption due to the fact that mechanisation of glassworking is very limited and the limitations placed on melting and refining temperatures due to the material from which the pots are made. The periodic operation of the furnace coupled to the manual working of the glass calls for well trained personnel.

10.1 Design

Several types of pot furnace have been developed over the years. A choice must be based on the careful analysis of all relevant factors: the type of fuel available, the type of glassware to be made (large, small, thick walled, thin walled), the number of working positions needed, the types of glass required, and the quality standards that must be met. As far as firing is concerned, a soft flame is needed to prevent overheating of the pots and it must be controllable over a wide range to fit in with periodic operation. The pot furnace is the oldest type of glass furnace and some designs date back to the time when producer gas was the only fuel used in glassworks. A good example is the 'Bütten' furnace, i.e. a furnace with ports in the siege (Figure 10.1); this is a very compact furnace and was fired with regeneratively preheated gas and air. It is fired from below, which discloses that it was developed from a furnace in which gas producer and glass furnace constituted a single unit. The pots are arranged in an oval plan surrounding the siege ports and outside the furnace, the working platforms follow the shape of the furnace. The weakest parts of this type of furnace are the siege ports which wear rapidly due to the combined effects of high temperature and spilled glass and which must be repaired frequently. The siege slopes slightly downwards towards the centre to ensure that, in the event of pot fracture, the glass runs safely into the glass pocket. The pots themselves stand on a layer of broken fireclay which is scattered onto the siege where the pot is to be sited; they slope slightly outwards to make gathering of glass easier.

At each end of the furnace there is a door, with its own bracing and easy to remove, through which, when pots are changed, old ones are removed and

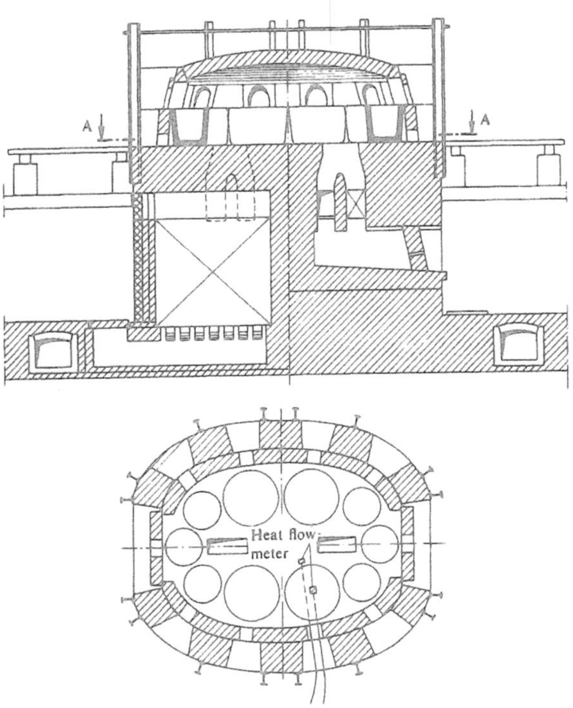

Figure 10.1. Bütten furnace[4]

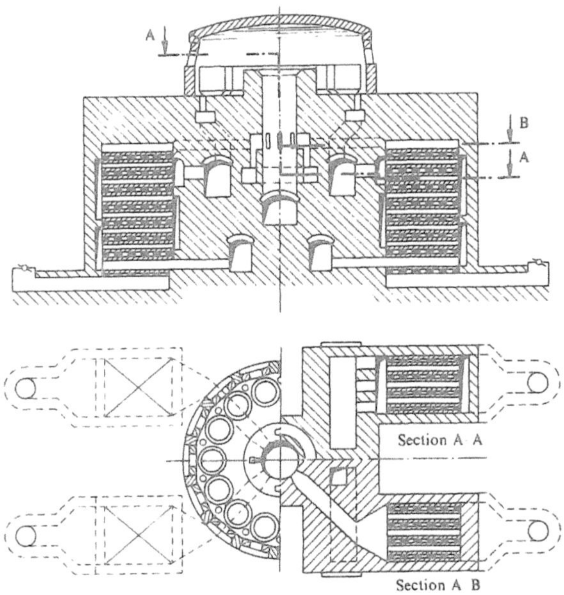

Figure 10.2. Round pot furnace[1]

new ones brought in. With the disappearance of producer gas the Bütten furnace has, apart from a few exceptions, also disappeared but the surviving furnaces of this type have been converted to firing with natural gas; preheating has been dispensed with and the gas is introduced into the ports from the side or from underneath.

A further type of furnace still in use and related to the Bütten furnace is the circular pot furnace, also known in some countries as the Schwaller or Hermannsen furnace (Figure 10.2). This furnace has a central burner located either in the siege or constructed as a tower round which the pots are located in a circle. The flame spreads out fanwise from the centre in all directions and is intended to pass around the cooler outer side of each pot. The waste gas flues are in the siege adjacent to the surrounding wall and the furnace operates with a recuperator; due to the long flames it has good energy consumption figures. The best fuels are town gas or natural gas, introduced into the central burner from underneath or from one side. Attempts have also been made to use oil and liquefied gas as fuels.[2,3]

Round pot furnaces have been constructed with up to 16 pots but it is worth pointing out that the siege area can only be properly utilised when the number of pots is limited to between 6 and 10; with larger furnaces the economics of the operation suffer due to the unused central area. A major advantage of the round pot furnace is that it can accommodate many working positions and is therefore very suitable for the production of stemware. Such a round furnace will have two or four recuperators, which may be of ceramic or metal and the formation of the flame in the melting chamber can be adjusted by making individual adjustments to the recuperators.

A drawback is that many gathering holes are open when the glass is being worked, which calls for very accurate control of furnace pressure to limit the amount of cold air being sucked in. The waste gas flues are located, as already mentioned, behind the circular wall near the gathering holes.

The pot furnace can also be constructed in a semicircular form: this gives a smaller unit but its disadvantage lies in the fact that the length of outside wall to number of pots in the furnace is increased, which is reflected in higher energy consumption.

A further variant which, like the Bütten furnace, still operates with a siege burner but which can easily be adapted for town gas or natural gas firing is the single pot furnace in which a flame enters, passes over the pot, and exits through the siege (Figure 10.3). This type of furnace is operated with regenerators and has flues in the siege to each side of the pot which are used alternately as ports and waste gas exhausts. The furnace is very compact,

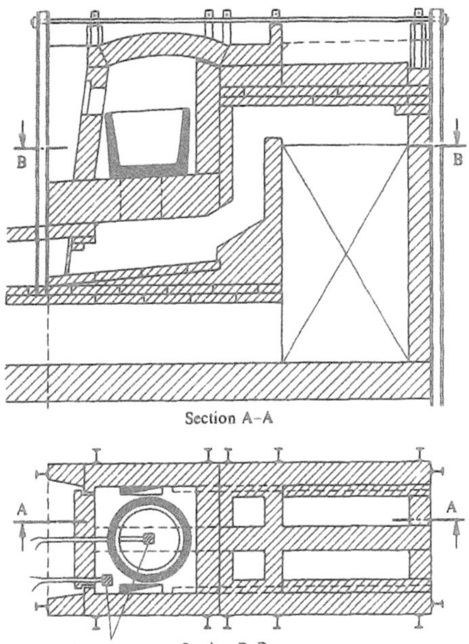

Figure 10.3. Single pot cell furnace [4]

and several can be sited in line. They are very suitable for manufacturing heavy glassware, such as glass rod. The same principle can also be used for the construction of a four pot furnace, with the pots arranged in a square; pot furnaces of this type are used for optical and other glasses.

If one wants to operate with a fixed flame, then a furnace with pots arranged in line and one flame covering three pots is an option; the exhausts would go downwards through the siege. Figure 10.4 shows such a side-fired furnace for container glasses operated as a small unit; if operated as a double unit it can accommodate up to six pots but the number is limited to six as it is difficult to heat more than three pots in line uniformly. The furnace operates with a recuperator, ports and flues are arranged asymmetrically, and the waste gases flow through a glass pocket below the siege. Oil or natural gas can be considered as fuel.

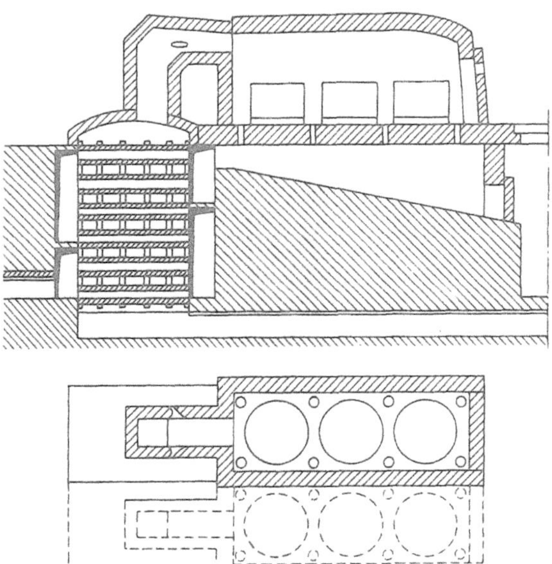

Figure 10.4. Furnace with longitudinal flame exhausted downwards[1]

10.2 Furnace operation

Pot furnaces with open pots work on a 24 h cycle. The batch is fed through openings in the outer wall by hand with scoops or mechanically via a chute or vibrating trough. Since batch has a higher volume than molten glass and since, for economic reasons, the pot should be well filled after melting, batch has to be fed several times. The temperature–time curve and the type and length of each stage of the 24 h cycle for a soda–lime glass are given in Figure 10.5 and the individual stages are illustrated in Figure 10.6, where the lengths of the arrows indicate the magnitudes and directions of heat flow. The melting process begins at (a) and firing is intense to accelerate the rate at which the batch melts down; the risk of overheating the pot does not arise at this stage since it is cooled by the cold residue of glass and by the batch. When melting down is complete, the refining stage, involving the removal of gases and homogenisation, follows; the furnace is now at its maximum temperature and firing must be reduced to prevent pots from breaking. Overheating also causes the pot glaze to melt down in increasing quantities and can give rise to glass quality faults. The molten glass is homogenised by immersing a piece of damp wood deep into the glass; this process, called blocking or boiling, should not be

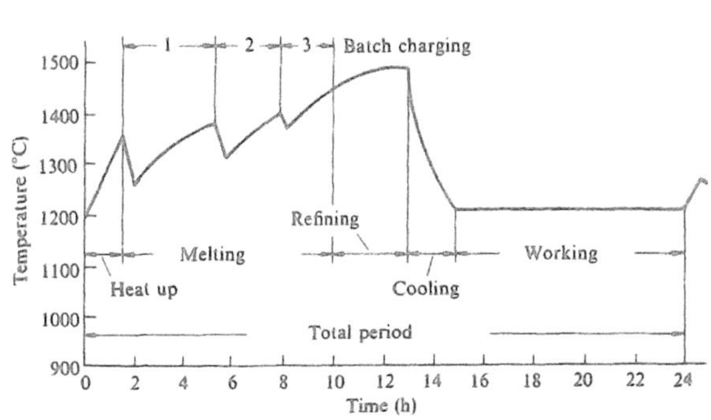

Figure 10.5. Temperature record in a pot furnace covering one operating cycle

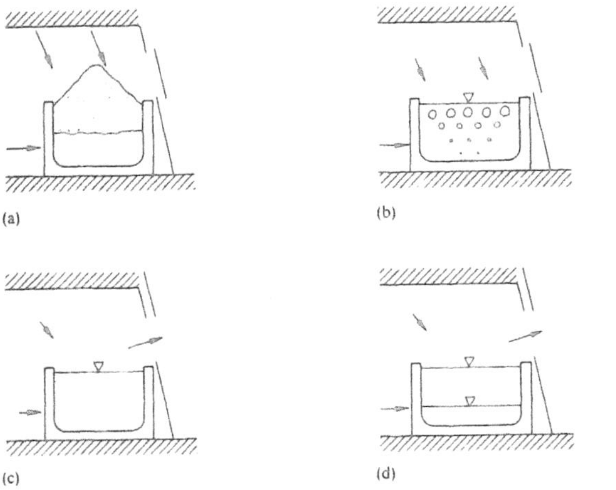

Figure 10.6. Working stages of a pot furnace equipped with open pots
(a) melting down
(b) degassing and homogenisation (refining)
(c) cooling and standing
(d) glass working

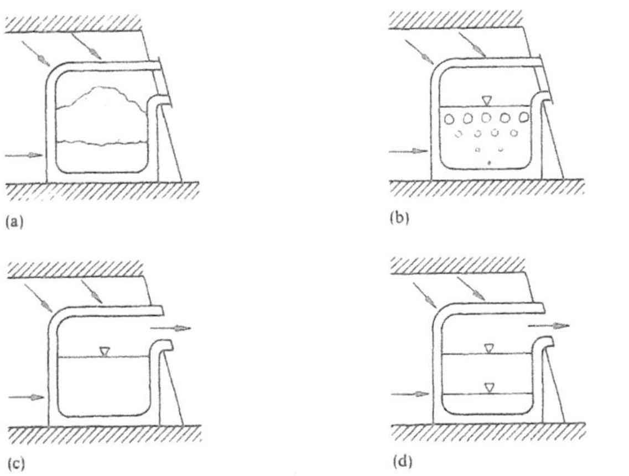

Figure 10.7. Working stages of a pot furnace equipped with closed pots; stages (a)–(d) as in Figure 10.6 above

too violent and the striking of the sidewall of the pot by the ring is to be particularly avoided.

When refining is complete the cooling stage follows, the objective being to reduce the temperature of the glass to that suitable for working. Firing is further reduced or even stopped and cooling accelerated by opening the gathering hole. When the working temperature has been reached, firing is set at a level which keeps the temperature constant.

When using closed pots the equivalent working stages are as shown in Figure 10.7. However, the firing of the furnace remains constant throughout and adjustments to the temperature of the pot are carried out solely by opening or closing the gathering hole. Heat input to and heat losses from the pot must be adjusted so that melting and refining temperatures are reached when the gathering hole is closed, and the working temperature when the gathering hole is open. This process is particularly suitable for glasses of high lead content which have low melting and refining temperatures.

To work out good quality glass with a low cord content each pot should be provided with a ring (usually made from slip cast quartz) which floats on the surface and keeps the gathering area free from scum and bottom glass. The ring must be made from a material which floats, causes few impurities, and is resistant to damage during blocking and to thermal shock. The firing of the furnace must support the function of the rings (Figure 10.8). So that excessive amounts of pot glaze do not run off into the glass, the pot must not be overheated during melting and refining. The pot rim and sides should be relatively cool during working, i.e. about the same temperature as the glass. Glass near the pot walls or edge should be kept away from the gathering area inside the ring. The pot must be held at a temperature at which the central core of good quality glass can rise to the gathering position.

Furnaces in which pots are heated from below are said to possess good bottom heat and this has been confirmed by measurements[4] on several different designs of pot. The temperature conditions and heat flows shown in Figures 10.9 and 10.10 were measured using a thermocouple on the siege of the furnace and one between the bottom of the pot and the siege. The thermocouple measuring crown temperature in the Bütten furnace as shown in Figure 10.9 gave readings which were 50–60 K too low, a fact confirmed by taking measurements with an optical pyrometer. The actual temperatures

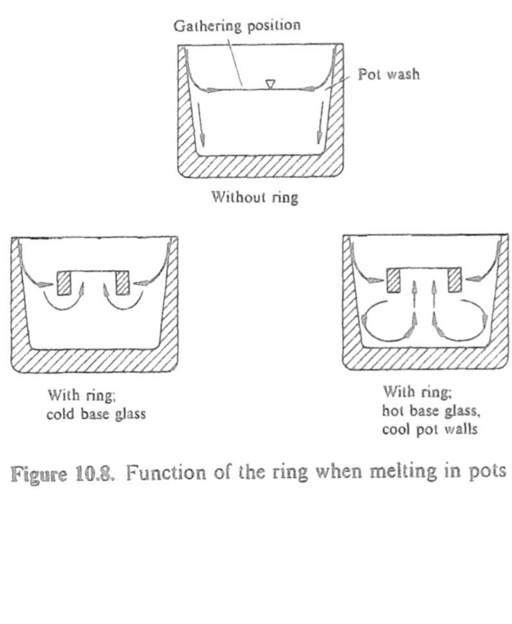

Figure 10.8. Function of the ring when melting in pots

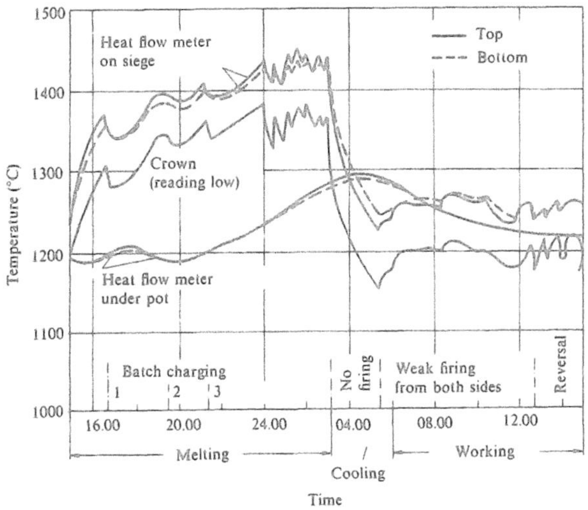

Figure 10.9. Temperature record of a Bütten furnace[4]

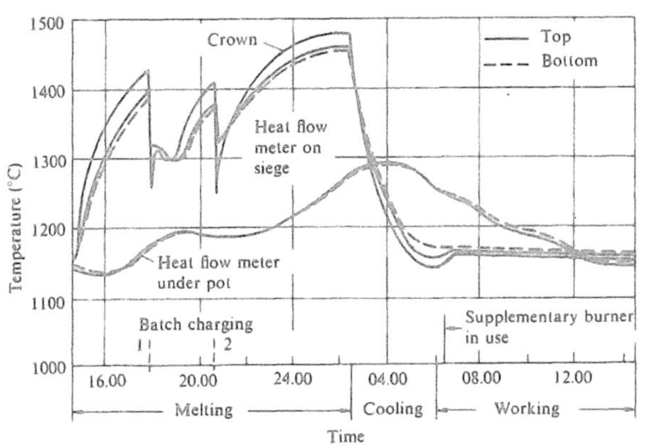

Figure 10.10. Temperature record of a round pot furnace[4]

under the crown are therefore virtually the same as those on the siege but, to keep the two curves apart, that for the crown temperature was not corrected. The heat flow meter located near the right pot on the siege (see Figure 10.1) shows the temperature fluctuations caused by firing and batch feeding with a high degree of sensitivity. During the heating up and melting periods and as the furnace temperature increases heat is stored in the siege; after the melting down of the batch the siege temperature reaches about 1440°C and blocking and refining of the glass take place with reduced firing. This causes significant temperature differences; on the firing side the temperature increases and on the exhaust flue side it falls rapidly. Clearly this is caused by the shorter flame resulting from reduced gas and air supply, a situation totally different from that when full firing is used.

The temperature falls rapidly during the cooling period when there is no firing and the gathering holes are partly open; this is partly the result of radiation interchange in the combustion chamber, where the heat flow is now reversed and heat stored in the pots is returned to the body of the furnace. This situation continues throughout the working stage except for short periods when increased radiation is detected on the upper surface of the heat flow meter during reversal or increased firing. The heat flow meter located under the pot only shows heat losses to the furnace after the first filling with batch and as melting proceeds to refining; the second and third fillings have no effect on heat flow. Maximum temperature is reached $1\frac{1}{2}$ h after refining has been completed and during the cooling stage.

During the glass working period, between 06.00 and 13.00 h, the temperature between pot and siege falls by about 60 K. Direct heat exchange between pot and siege cannot be determined but it may be assumed that the fall in temperature is caused by heat losses from the siege to the outer walls of the furnace. The heat loss at this stage clearly relates to the cooling of the pot as the glass is worked and as the glass itself cools down.

Figure 10.10 gives the time–temperature curves obtained from heat flow meters lying on the siege and under a pot in a round pot furnace. For comparison, the temperature measured by a thermocouple located above a pot in the crown is also given and it may be seen that there are only slight temperature differences between crown and siege. During melting and refining the siege is heated from the furnace chamber but this process is reversed for brief periods when further batch feeding causes a sudden cooling of the superstructure. At the end of refining, the exposed part of the siege shows a maximum temperature of 1460°C, only 15 K below that measured in the crown under conditions of reduced firing. During the $3\frac{1}{2}$ h period of cooling without any firing the furnace temperature falls rapidly and heat flows from the siege to the superstructure. The temperatures under the crown and on the siege remain remarkably constant during the working period, throughout which the siege continues to transfer some heat to the superstructure.

Only small amounts of heat flow into the siege under the pots during melting. The temperature under the pot initially decreases, presumably as a result of the feeding of cullet, and heat flows from the siege into the pot for about 1 h. It takes about 3 h after filling with batch for the temperature under the pot to regain the value that existed before filling commenced, while the temperature of the siege and the pot bottom gradually increase for a further hour uniformly and with no measurable difference. This effect can be attributed to the low emission of radiation from the nonluminous natural gas flame and it is only after the melting down of batch is completed that more heat reaches the siege via the glass held in the pot than directly from the flame or from the waste gases.

One hour before glass working commences the temperature under the pot is about 140 K above that of the furnace chamber and during working it decreases uniformly to that existing in the body of the furnace as the siege transfers heat to the pot. The condition aimed for is then achieved: the temperature under the pot at the end of the glass working period is higher than or nearly the same as that of the body of the furnace itself. Heat losses from the pot to the siege have ceased and in fact the opposite is taking place since heat flows from the siege to the pot.

10.3 The heating process

In terms of heating, the pot furnace shows many disadvantages when compared to a tank furnace as the periodic nature of glass working calls for the continued heating and cooling of a pot furnace. This heat loss is avoided with tanks since all parts of the furnace stay at a fixed temperature except for losses which arise from changes in pull or temperature or from peculiarities of the regenerative system. In addition, thermal insulation of a furnace which operates on a cyclic system cannot be taken beyond the point where it interferes with the system itself.

A continuously working furnace using covered pots brings no improvement despite the continuous firing as this type of furnace still relies on reducing the temperature to that needed for working by opening gathering holes, even though the furnace itself stays hot. The poor heat transfer between flame and glass due to the pot being covered causes the melting time to be extended very significantly which, in turn, reduces the pull and considerably increases the specific energy consumption. The preheating of the combustion air of a pot furnace can never attain the values reached in a tank since checkerwork, however large, is cooled during glass working and the subsequent heating up period is too long in view of the high heat capacity of the checkerwork. This really demonstrates the limitations imposed by the periodic nature of the operation. Recuperators, especially metal recuperators, are significantly better in this instance because of their low thermal inertia but the temperature to which the air can be preheated in metal recuperators is limited. From a thermal point of view, ceramic recuperators would appear to be a better compromise but suffer from the fact that, with time, they develop cracks and leaks due to the repeated contraction and expansion resulting from temperature changes. When, ultimately, leaks from a recuperator become excessive they can cause the whole furnace to be shut down because, by then, the chimney is robbing the furnace of the required combustion air.

A further difference is the limit to which glass can be worked out of a pot; partly due to lack of available time and partly for reasons of glass quality, the best that can be achieved is working out about 75% of the contents of a pot. If glassware of high quality is required, and especially if it has thin walls, the extent to which a pot can be worked out can reduce to 35–40%. By using the glass to make a range of glassware of different sizes and quality standards, the pot can be worked out to 65–75%, but such values are only rarely attained.

All the factors described above cause the specific energy consumption of a pot furnace to be several times higher than that of a tank furnace. Current figures generally lie between 20 000 and 40 000 kJ/kg of glass, the best figures being obtained with round furnaces: relevant data are summarised in Figure 10.11. When comparing energy consumptions of different furnaces it is always necessary to standardise on the degree to which the glass in the pot has been worked out; otherwise erroneous conclusions could be drawn.

Although only a small proportion of all glass manufactured is melted in pot furnaces, nevertheless they constitute an important part of the glass industry and are needed to produce glassware which is required in a large variety but only small quantity. Energy costs are therefore as important to plants using pot furnaces as they are to those operating on a much larger scale using tank furnaces. Recent rethinking on pot furnaces has produced a design in which the mass of refractories has been reduced so that a greater proportion of the

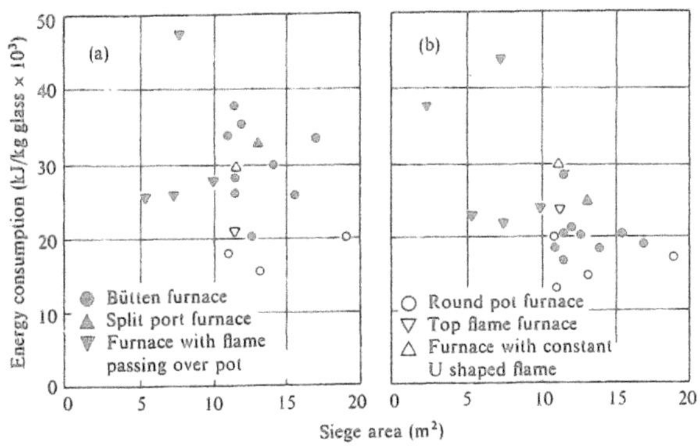

Figure 10.11. Specific energy consumption for pot furnaces[1]
(a) actual values
(b) calculated values for furnaces where 75% of the glass melted is worked out

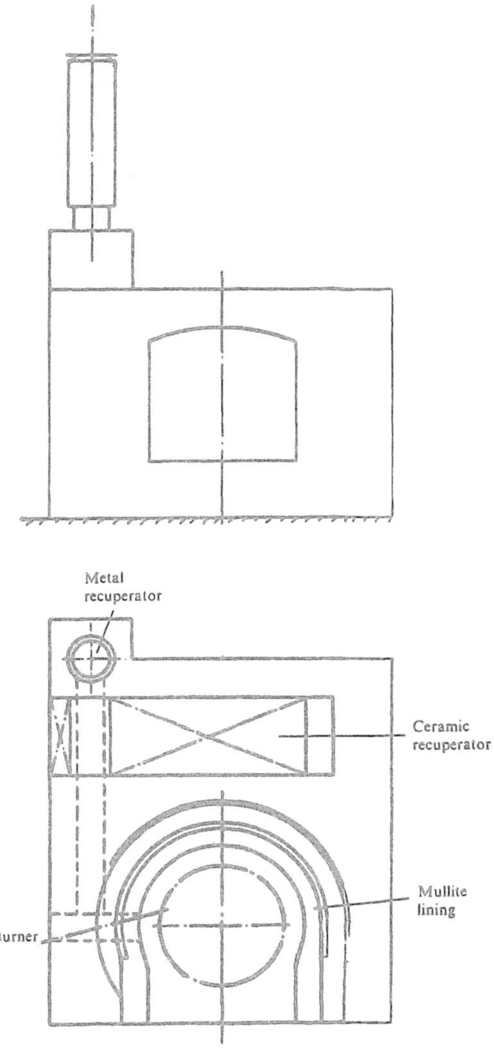

Figure 10.11A. Single pot furnace of low refractory mass and improved specific fuel consumption (Teisen)[8]

heat flow into the furnace is used for melting glass. Figure 10.11A shows a single pot furnace for melting lead crystal in open pots (capacity about 700 kg) fired by natural gas with its combustion air preheated in a ceramic recuperator linked in series to a convective metal recuperator. The important design feature is that, apart from the siege, the lining of the melting chamber, which used to consist of 345 mm of dense refractory to withstand erosion by flame and batch materials from open pots, has been replaced by a fused mullite lining only 150 mm thick, backed by several layers of graded insulation. Furnaces of this type, producing a daily melt of lead crystal glass in open pots, fired with either natural gas or light oil, and with an air preheat temperature of about 650°C and a waste gas temperature of about 380°C (when melting glass), have a fuel consumption, averaged over more than 100 melts, of just under 25 000 kJ/kg glass. Weekly results lie between 17 500 and 36 000 kJ/kg depending on the extent to which the pots had been worked out and the proportion of cullet in the batch.[8]

The pull of a pot furnace depends on its size and values lie between 500 and 5000 kg glass/day; total siege areas lie between 4 and 18 m^2 and the area of molten glass is 33–48% of the area of the siege. Larger furnaces give lower specific energy consumptions than small furnaces.

Not many heat balances have been drawn up for pot furnaces as the work is difficult, calling for a great deal of effort and in the end producing uncertain results due to the periodic nature of operation. Either one assumes fixed data, or each stage of the periodic process must be assessed separately, a procedure made more difficult by the changing quantity of heat stored in the furnace.

10.4 Electrically heated pot furnaces

Molybdenum disilicide is normally used for the heating elements of these furnaces and their design is based on the requirements of these elements; free hanging U shaped elements located in recesses are generally used, the recesses being made deep enough to prevent mechanical damage to or soiling of the conductors. So that the recesses retain their accurate dimensions even after long periods of use, tongued and grooved refractories are recommended. The furnaces are constructed as single pot furnaces and the elements are located so that direct radiation through the gathering hole to the outside is avoided. Figure 10.12 shows an example of the design of such a furnace.

Alkaline deposits near the element holders could lead to short circuiting and this is avoided by providing the holders with a permanent air purge (Figure 10.13) which is vented through the glass pocket.

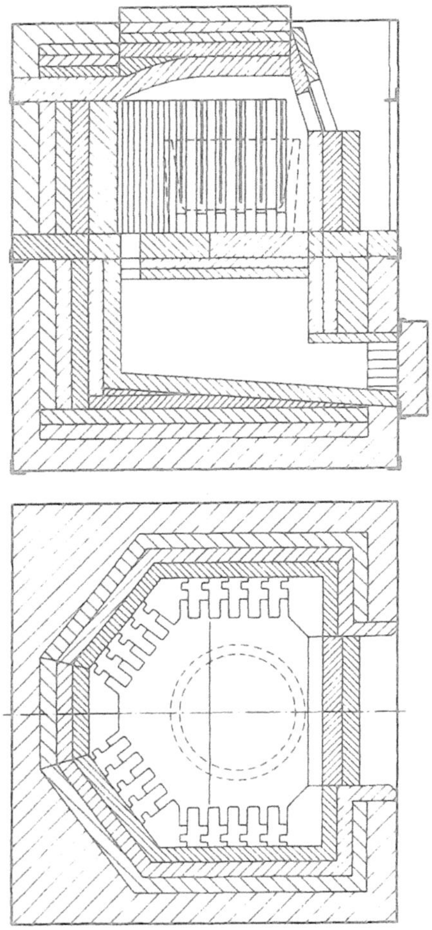

Figure 10.12. Single pot furnace heated by electrical resistance elements(6)

The electrical resistance of molybdenum disilicide increases rapidly with increasing temperature and in order to avoid the overloading of electrical connections during heating up, the transformers must be provided with stepped voltage tappings. The data covering a single pot furnace melting lead glass are assembled in Table 10.1; the furnace is equipped with 21 heating elements each of a nominal rating of 6·2 kW and the specific energy consumption is stated to be 2·4 kWh/kg of worked glass.

Table 10.1. *Operating data of an electrically heated pot furnace working on a 24 h cycle*

Activity	*Temperature (°C)*	*Voltage*	*Power (kW)*
Glass working	1180	111	53
Melting	1180–1425	196	166
Refining	1425	157	107

These furnaces, it is claimed, give glass of very high quality when compared to that obtained from flame heated furnaces; due to the very even heating of the pot the glass is of good quality even when the proportion of glass worked out is very large. The life of pots is claimed to be more than doubled. Supervision is greatly simplified once the melting schedule for a specific glass has been determined.

Electric furnaces have so far not found much favour due to the high cost of electricity but interest in them is increasing as a result of increasing costs of fossil fuels and the problem of waste gas emissions – particularly important with lead glasses.

10.5 Refractories

Refractories for pot furnaces are chosen on a basis different from that used for tanks and therefore

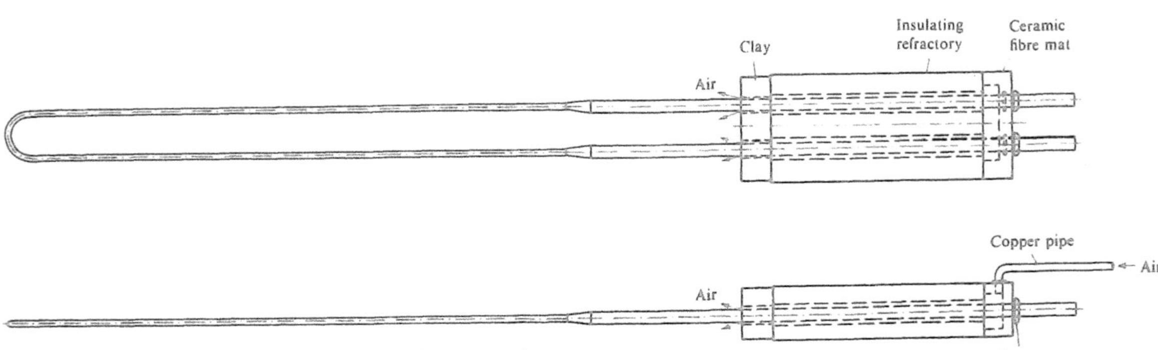

Figure 10.13. Heating elements with holders for air purging

warrant separate treatment. The most important item is the pot itself. Its performance governs the incidence of glass quality faults which, in the end, determine the frequency with which old pots have to be replaced by new ones. The pot must be resistant to rapid heating and cooling and must possess a high refractoriness under load and a low tendency to creep. It must be highly resistant to corrosion by the glass, but should have good solubility in it. These requirements are somewhat contradictory and cannot therefore be fully achieved.

Soda–lime and lead glasses require a pot made from a siliceous clay mixed with grog and small pieces of fireclay, which can withstand 1500°C and which will dissolve in the molten glass without creating cords. Its mineralogical composition as a function of temperature is shown in Figure 10.14(a), where it can be seen that the proportion of glassy phase increases markedly only above 1440°C. During warming up of pots special care must be taken during the quartz transition intervals.

Technical glasses are best melted in high alumina clay pots containing grog from old pots and their mineralogical composition can be seen in Figure 10.14(b). High aluminous pots can withstand temperatures above the maximum for siliceous pots but tend to deform. After some time these pots become very dense and sensitive to thermal shock, which leads to cracking during batch feeding.

Pots must be heated up very carefully and under oxidising conditions:[5] they are heated in a pot arch which takes seven days to reach 1200°C, with special care being exercised in the region of the quartz transition (Figure 10.15). The formation of

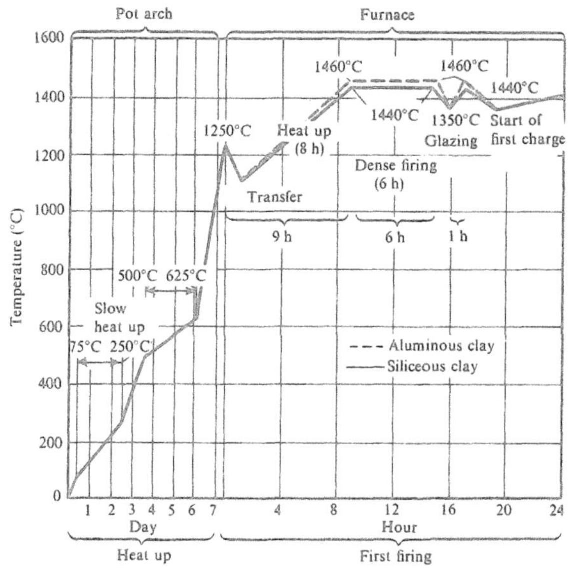

Figure 10.15. Temperature record of pot arching and dense firing of a pot[5]

cracks due to contact with cold air is avoided by operating the pot arch under a small positive pressure. Reducing conditions must be carefully avoided as they can cause carbon deposits in the body of the pot which can break open later like blisters. The hot pots are then transferred to the furnace and set: a period of intense firing follows (Figure 10.15) and within 8–10 h the furnace temperature is raised to 1460°C for aluminous clay pots and 1440°C for pots of siliceous clays. Not only the rim, but also the bottom of the pot must reach a temperature of 1440 or 1420°C. The period of firing and the temperature must be chosen so

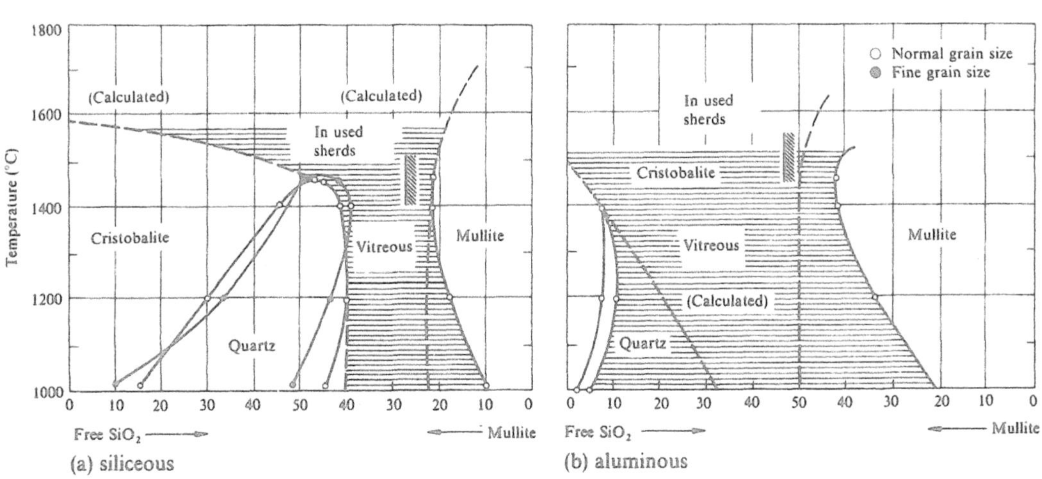

Figure 10.14. Mineralogical composition of pot clays as a function of temperature

that the porosity of the pot is reduced to less than 20%. Only then is it possible to claim that the pot clay material can withstand attack by molten glass and will not cause glass faults. When these temperatures have been reached, the surface of the pot can be glazed with care; small pieces of cullet are fed into the pot and, when molten, used for the glazing. The glaze protects the pot against sudden temperature changes and consolidates the structure with respect to the furnace atmosphere.

Pot and ring should be heated up and fired together. The ring can be prevented from sticking to the bottom of the pot by placing it on small cubes of clay or small heaps of sand.

Cullet and batch should be fed in such a way that they form a cone from which the material slowly runs off, thus preventing thermal shock to the pot walls.

The life of pots is between 12 and 20 weeks, lead glasses giving the shorter lives and potash glasses the longer lives. There are large differences between individual plants, and pot life is also influenced by both operating conditions and glass quality requirements.

Pots and rings used to be made from the same material but for some years now rings have been made from slip cast fused silica. The high temperature shock resistance and good solution properties of this material have led to its being widely adopted. Rings last longest if they have been heated up with the pot; rings introduced later have a shorter life.

The most important parts of the furnace structure are the siege on which the pots are placed, the port structure, and the waste gas flues; this is particularly true for furnaces with ports in the siege. These must be able to withstand not only the high flame and waste gas temperatures but also vitreous slag. In the past the standard construction consisted of four layers of tiles of which the first two consisted of high quality aluminous fireclay; this design has now been abandoned. Success has been achieved by using an upper layer of fusion cast tiles followed by a layer of tiles made from high quality aluminous fireclay.

The crown is made of silica and the sidewalls of good fireclay, which may also be used for the ducts and the checkerwork. Care must be taken with ceramic recuperators: they should be made of a quality of sillimanite possessing good thermal shock resistance.

11 Fuels and combustion

11.1 Fundamentals

Combustion is regarded as the chemical reaction between a fuel and oxygen, generally the oxygen in the air. Only gaseous and liquid fuels (hydrogen, H_2, carbon monoxide, CO, and hydrocarbons such as CH_4, C_3H_8, and C_mH_n) are important for glass melting furnaces: sulphur should be added to this list as a combustible impurity. All combustion calculations are based on the combustion equations of the above gases and compounds, whose initial and final compositions are given in Table 11.1; calculations must be based on molar quantities and these are given in Table 11.2.

Combustion can only be completed without leaving any residual air (or, more accurately, the oxygen in the air) if the proportion between fuel and air is the theoretical stoichiometric ratio. Since 1 kmole of an ideal gas at NTP occupies a volume of 22·414 m^3 and the oxygen content of air is 21%, then the quantity of air required, l_{st}, and the resultant waste gas volume, v_{st}, can be calculated for stoichiometric combustion conditions using the equations given in Table 11.1 (Table 11.3).

Table 11.2. *Molecular quantities of substances important in combustion. The molar volume of an ideal gas at NTP is taken as 22·4 m³/kmole*

Substance	*Formula*	*Molecular weight to nearest whole number (kg/kmole)*	*Specific volume (m³/kg)*
Hydrogen	H_2	2	11·2
Oxygen	O_2	32	0·7
Nitrogen, pure	N_2	28	0·8
Nitrogen in air	–	–	0·796
Carbon	C	12	–
Sulphur	S	32	–
Sulphur dioxide	SO_2	64	0·35
Carbon monoxide	CO	28	0·80
Carbon dioxide	CO_2	44	0·509
Methane	CH_4	16	1·4
Ethylene	C_2H_4	28	0·8
Ethane	C_2H_6	30	0·747
Propane	C_3H_8	44	0·509
Butane	C_4H_{10}	58	0·386
Water vapour	H_2O	18	1·244

The air requirement is defined as the ratio between the mass flows of air and fuel

$$l = \frac{\dot{M}_L}{\dot{M}_B}. \tag{11.1}$$

Table 11.1. *Combustion equations*

C	+	O_2	=	CO_2	2 CO	+	O_2	=	2 CO		
12		32		44 kg	2 × 28		32		2 × 44 kg		
0		22·4		22·4 m³/kmole	2 × 22·4		22·4		2 × 22·4 m³/kmole		
1		1		1 kmole	2		1		2 kmole		
S	+	O_2	=	SO_2	2 H_2	+	O_2	=	2 H_2O		
32		32		64 kg	2 × 2		32		2 × 18 kg		
0		22·4		22·4 m³/kmole	2 × 22·4		22·4		2 × 22·4 m³/kmole		
1		1		1 kmole	2		1		2 kmole		
CH_4	+	2 O_2	=	CO_2 + 2 H_2O	C_3H_8	+	5 O_2	+	3 CO_2	+	4 H_2O
16		2 × 32		44 + 2 × 18 kg	44	+	5 × 32		3 × 44	+	4 × 18 kg
22·4	+	2 × 22·4		22·4 + 2 × 22·4 m³/kmole	22·4	+	5 × 22·4		3 × 22·4	+	4 × 22·4 m³/kmole
1		2		1 2 kmole	1		5		3		4 kmole
				C_mH_n	+ (m + n/4) O_2 =		m CO_2 +		n/2 H_2O		
				12 m + n	32 m + 8n		44 m		9 n kg		
				22·4	(m + n/4) 22·4		m × 22·4		n × 11·2 m³/kmole		
				1	m + n/4		m		n/4 kmole		

Table 11.3. *Combustion with theoretical air requirement, related to 1 m^3 of fuel at NTP*

Substance	*Stoichio-metric air*	*Stoichiometric volumes of waste gases* CO_2	SO_2	N_2	H_2O	$v_{st,dry}$	v_{st}
CO	2·38	1·0	–	1·88	–	2·88	–
H_2	2·38	–	–	1·88	1·0	1·88	2·88
CH_4	9·51	1·0	–	7·51	2·0	8·51	10·51
C_2H_4	14·27	2·0	–	11·27	2·0	13·27	15·27
C_2H_6	16·67	2·0	–	13·17	3·0	15·17	18·17
C_3H_8	23·81	3·0	–	18·81	4·0	21·81	25·81
C_4H_{10}	30·95	4·0	–	24·45	5·0	28·45	33·45
H_2S	7·13	–	1·0	5·63	1·0	6·63	7·63

Since combustion also produces water vapour, which condenses in whole or in part during cooling, it becomes necessary to differentiate between the dry H_2O-free waste gas volume, $v_{st,d}$, and v_{st}. The stoichiometric volume of waste gases, v_{st}, is defined as the ratio between the mass flows of waste gases and fuel arising during stoichiometric combustion:

$$v_{st} = \frac{\dot{M}_{A,st}}{\dot{M}_B}. \qquad (11.2)$$

11.1.1 Excess air

Ignoring special cases. virtually complete utilisation of fuel energy can, in practice, only be obtained when combustion takes place with excess air: complete combustion is only possible in the time available if air is provided in excess.

If l is the mass flow of air related to the mass flow of fuel, then the ratio

$$\frac{\dot{M}_L}{\dot{M}_{L,st}} = \frac{l}{l_{st}} = \lambda$$

is referred to as the air factor. The excess air factor (often expressed as a percentage), is $\lambda - 1 = l - l_{st}/l_{st}$, and $l - l_{st}$ is simply the excess air. The air factor is kept to a minimum so that combustion is only just satisfactorily completed; excess air, after all, gives a higher volume of waste gases and therefore higher waste gas losses.

Combustion is carried out with $\lambda = 1{\cdot}05$–$1{\cdot}0$ and with $\lambda = 1{\cdot}1$–$1{\cdot}2$ for oils. Control of combustion consists largely of monitoring and controlling the value of λ which is judged to be correct and appropriate and since λ cannot be measured directly, it is controlled via constituents of the waste gases which are easy to measure, i.e. CO_2 or O_2. In the past, before the techniques for measuring O_2 had been fully developed, it was normal practice to monitor the CO_2 content as the CO_2 value will reach a maximum during stoichiometric combustion, decreasing with both excess air and insufficient air. In glass furnaces the results are falsified by CO_2 evolved from the batch and the monitoring of O_2 as a control is therefore preferred.

The volume of waste gases created, v, is given by

$$v = v_{st} + (\lambda - 1)l_{st} \qquad (11.3)$$

and the oxygen content in the waste gases can be calculated from

$$\frac{v - v_{st}}{v} = \frac{V_{O_2,A}}{V_{O_2,L}}; \qquad (11.4)$$

combining both equations gives

$$\lambda = 1 + \frac{v_{st}}{l_{st}} \frac{V_{O_2,A}}{V_{O_2,L} - V_{O_2,A}}. \qquad (11.5)$$

For accurate measurements on glass furnaces account must be taken of the fact that the waste gases contain not only the products from the combustion of the fuel but also gases given off by the batch; their exact influence is a function of the position at

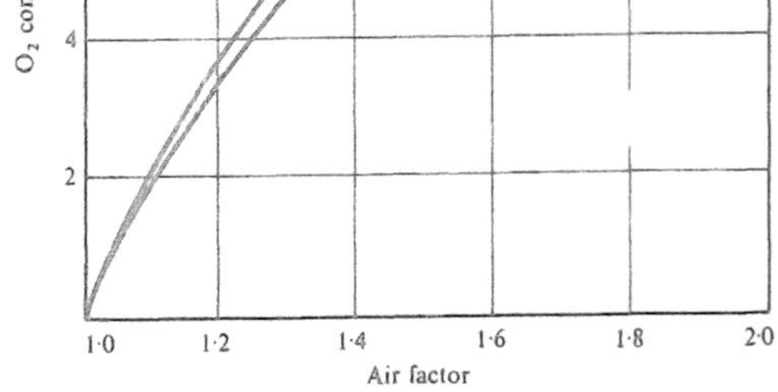

Figure 11.1. Influence of batch gases on the air factor

which the sample is taken for measurement. In total, the proportion of batch gases in the waste gases can be 10% but it can be higher in the batch melting zone. Basing the calculation of λ on Equation (11.3) can lead to errors; if the relevant volume of batch gases analogous to v_{st} is v_b, then

$$\lambda = 1 + \frac{v_{st} + v_b}{l_{st}} \frac{V_{O_2,A}}{V_{O_2,L} - V_{O_2,A}} \qquad (11.5a)$$

The connection between λ and $V_{O_2,A}$ and the influence of the proportion of batch gases ($v_b/l_{st} = 0{\cdot}1$) is shown in Figure 11.1; the calculations are based on oil ($v_{st}/l_{st} = 0{\cdot}94$) and the influence, while not very marked, is noticeable.

11.1.2 Development of heat

The reactions which constitute combustion are strongly exothermic, hence their use as sources of heat. Table 11.4 gives a summary of the calorific values of the most important combustible substances, a difference being drawn between the gross and net calorific values, H_o and H_u, respectively. The gross calorific value is the quantity of heat liberated during combustion in air of 1 m^3 of dry gas at NTP if the products of combustion are returned to the starting conditions of 1·01325 bar and 25°C, and if the water formed from free and combined hydrogen is available as liquid. Since the gross calorific value can be measured in a standardised calorimeter, it is used by gas suppliers for purposes of accounting. The net calorific value has the same definition as the gross calorific value but is reduced by the heat used in evaporating the water produced during combustion. Accordingly the useful heat corresponds to the net calorific value since the water evaporated as steam is carried out of the system in that form as part of the waste gases. Approximate calculations show that the net calorific value is about 10% less than the gross value.

If combustion takes place adiabatically without dissociation, then the theoretical combustion temperature is

$$\vartheta_{th} = \frac{H_u}{v_{st} c_p}. \qquad (11.6)$$

In the practical world of plant installations heat losses, q_v, and, at temperatures exceeding 1500°C, heats of dissociation, q_{diss}, must be taken into account as they reduce the attainable temperature. This gives

$$\vartheta_V = \frac{H_u - q_v - q_{diss}}{v_{st} c_p}. \qquad (11.7)$$

A calculation of the effect of dissociation requires calculations of the equilibria of the compositions involved. Above a certain temperature the reactions of combustion and dissociation are in balance. The reactions have not been completed by the end of the combustion process. Generally the heat of dissociation is not lost since, after the completion of combustion, the reactions are completed during the cooling of the waste gases.

The temperature of combustion is a function not only of the composition of the fuel but also to a major extent of the starting temperatures of both materials (from preheating) and of the air factor. The temperature is given by

$$\vartheta_V = \frac{H_u + l_{st} c_{p,L} \vartheta_L + c_{p,B} \vartheta_B - q_v - q_{diss}}{v_{st} c_{p,A}} \qquad (11.8)$$

where v_{st} and v are given by Equations (11.2) and 11.3) respectively.

Figure 11.2 demonstrates the strong influence of the air factor and preheat temperature on the combustion temperature, and the equivalent curves for different fuels at $\lambda = 1{\cdot}1$ are given in Figure 11.3.

A similar representation often encountered is the h,ϑ diagram, h in this case being the specific

Table 11.4. *Calorific values of the most important combustible substances*

	H_o (kJ/kg)	H_o (kJ/m$_n^3$)	H_o (kJ/mole)	H_u (kJ/kg)	H_u (kJ/m$_n^3$)	H_u (kJ/mole)
C	33 810	–	406 000	33 810	–	406 000
CO	10 130	12 640	283 320	10 130	12 640	283 320
H_2	141 910	12 760	286 040	119 570	10 760	241 010
CH_4	55 580	39 840	890 570	49 930	35 780	800 550
C_2H_4	50 760	63 990	1 422 900	47 540	59 930	1 332 880
C_2H_6	51 940	70 390	1 560 170	47 420	64 320	1 425 120
C_3H_8	50 390	101 780	2 220 560	46 330	93 530	2 040 520
C_4H_{10}	49 550	133 960	2 875 100	45 700	123 500	2 653 800
S	9 250	–	295 960	9 250	–	295 960
H_2S	16 700	25 700	569 160	15 400	23 690	524 130

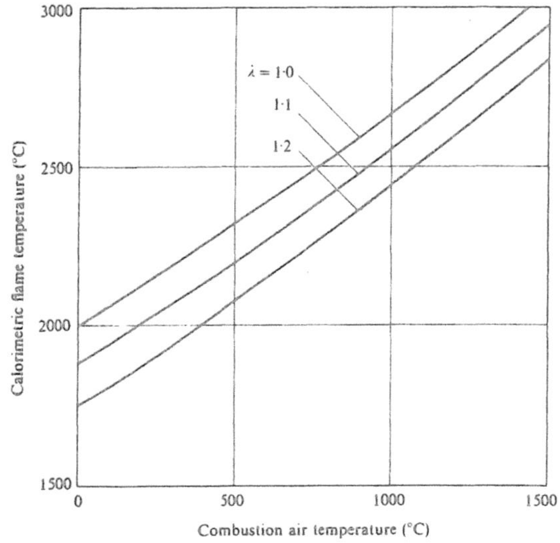

Figure 11.2. Calorimetric flame temperature of type A natural gas as a function of the combustion air temperature and the air factor

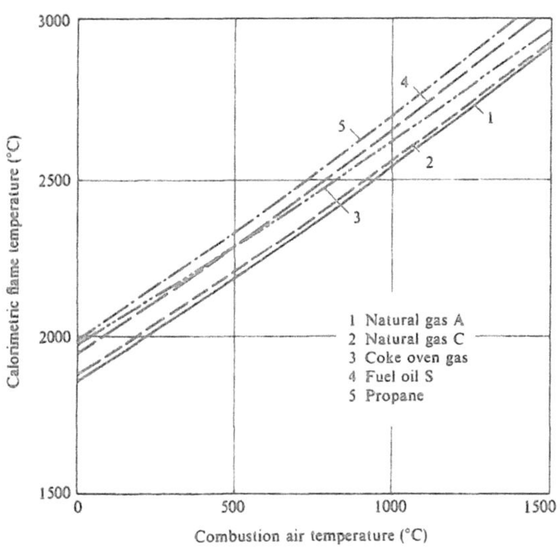

Figure 11.3. Calorimetric flame temperature of various fuels as a function of the combustion

enthalpy of the combustion gases according to $h = c_p\vartheta$. Figure 11.4 is such a diagram for waste gases from fuel oil: the abscissa gives the attainable adiabatic combustion temperature, and the ordinate the specific enthalpy of the waste gases as a function of the air factor and possible preheat temperatures of the combustion air. The effect of dissociation, q_{diss}, has been included. The strong influence of the air factor and the preheat temperature on the attainable combustion temperature can be seen.

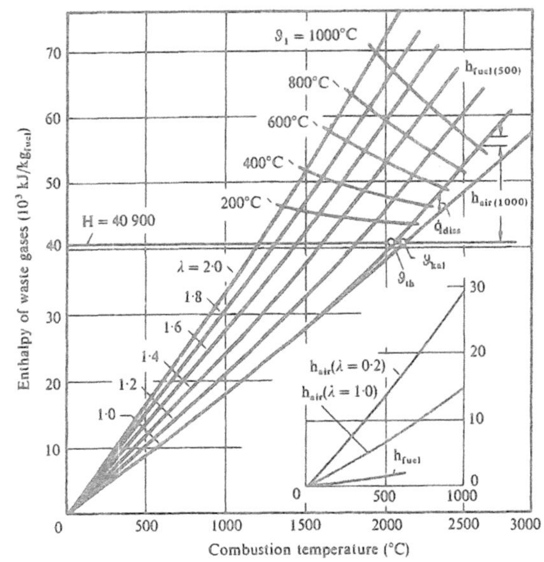

Figure 11.4. h, ϑ diagram for waste gases from fuel oils[2]

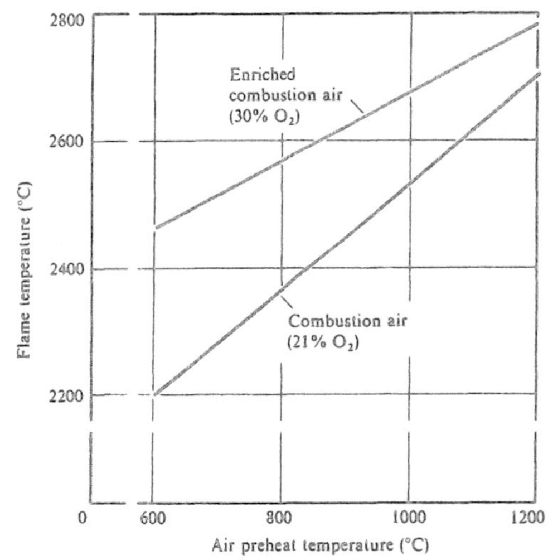

Figure 11.5. Influence of air preheating and O_2 content on the adiabatic flame temperature when burning heavy fuel oil (ignoring dissociation)

In special circumstances air enriched with O_2 is used, the reduction in inactive nitrogen leading to higher combustion temperatures: Figure 11.5 shows the relationships as a function of the preheating of the air (21% O_2) and of oxygen enriched air (30% O_2).

11.1.3 The combustion process

The following terms are used in connection with the combustion process: flammability limits, spontaneous ignition temperature, flame velocity, and Wobbe number.

The ignition of a mixture of fuel and air is only possible when the proportions lie between two flammability limits; if ignition takes place outside these limits it is only a local reaction and of short duration and even if the reaction starts, it cannot continue due to lack of energy so that the flame goes out. The flammability limits for the most important gases are given in Table 11.5 which shows clearly that there are large differences between individual gases.

As the temperature is increased so the interval between the flammability limits becomes wider since less outside energy has to be used for ignition. According to Zabadikis[7] the lower flammability limit at a temperature ϑ is governed by the ratio of the temperature-dependent gaseous constituents in the mixture,

$$\frac{U(\vartheta)}{U(25°C)} = 1 - 0{\cdot}000721\,(\vartheta - 25) \qquad (11.9a)$$

and similarly, for the upper flammability limits,

$$\frac{O(\vartheta)}{O(25°C)} = 1 + 0{\cdot}000721\,(\vartheta - 25). \qquad (11.9b)$$

These equations apply generally for hydrocarbons.

If the temperature continues to increase, self-ignition takes place and this temperature is referred to as the spontaneous ignition temperature. In practice this value is important in the case of air–gas mixtures or when premixing burners are used. The spontaneous ignition temperature is determined according to DIN 51794 and values for the more important gases are given in Table 11.6, from which it may be seen that, with the exceptions of n-C_5H_{12} and n-C_8H_{18}, differences between industrial gases are not large. The lowest spontaneous ignition temperatures are obtained when there is a slight deficiency of air.

The laminar flame velocity, also called the flame propagation velocity, is defined as the velocity at which a flame front moves relative to a combustible mixture of fuel and air. In the field of industrial burners the laminar flame velocity acts as an index for the combustion characteristics of different gases. The flame velocity determines the stability characteristics of burners and depends partly on the properties of the fuel, but primarily on the fuel:air mixture ratio, temperature, and pressure. The maximum velocity arises near the stoichiometric mixture, slightly on the side of a deficiency of air. Numerical values for the flame velocity of different H_2/CH_4 mixtures can be taken from Figure 11.6.

The effect of temperature on the maximum flame velocity of hydrocarbons can be deter-

Table 11.5. *Flammability limits of gases and vapours in air and oxygen under atmospheric pressure (gas concentrations in vol%)*

Substance	Formula	Flammability limits In air at 20°C Lower	Upper	In oxygen at 20°C Lower	Upper
Hydrogen	H_2	4·1	74·0	4·5	95
Carbon monoxide	CO	12·5	74·0	13	96
Methane	CH_4	5·3	14	5	60
Ethane	C_2H_6	3·2	12·5	3·9	50·5
Propane	C_3H_8	2·4	9·5	–	–
Butane	C_4H_{10}	1·85	8·4	–	–
Pentane	C_5H_{12}	1·4	7·8	–	–
Ethylene	C_2H_4	3·0	29	3·0	80
Acetylene	C_2H_2	2·5	80	2·8	93

Table 11.6. *Spontaneous ignition temperatures (°C) of stoichiometric mixtures of fuel and air under atmospheric pressure*[2]

Fuel	Ignition temperature
H_2	560
CO	605
CH_4	610
C_2H_6	425
C_3H_8	470
n-C_5H_{12}	285
n-C_8H_{18}	210
C_6H_6	555

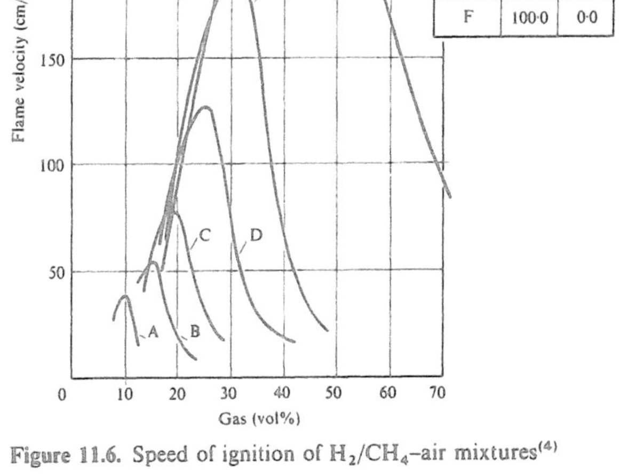

Figure 11.6. Speed of ignition of H_2/CH_4–air mixtures[4]

mined approximately from

$$\Lambda_{max} = \Lambda_1 + \Lambda_2(T^2) \quad (11.10)$$

where T is the absolute temperature, $\Lambda_1 = 6$ cm/s and $\Lambda_2 = 45 \times 10^{-5}$ cm/s.

The Wobbe number is an index of heat output. Combustible gases with different compositions but identical Wobbe numbers will give the same heat output at the burners under the same pressure. The Wobbe number is the ratio of the gross or net calorific value to the square root of the relative density of the gas being considered:

$$W_o = \frac{H_o}{\sqrt{d}}; \quad W_u = \frac{H_u}{\sqrt{d}}, \quad (11.11a)$$

where $d = \rho_G/\rho_A$, the subscripts G and A referring to gas and waste gas.

The extended Wobbe number includes the effect of the line pressure, p_e, on the gas flow and thereby on the heat output:

$$W_{o,e} = W_o\frac{1}{\sqrt{p_e}}; \quad W_{u,e} = W_u\frac{1}{\sqrt{p_e}}. \quad (11.11b)$$

11.1.4 The dew point

This is an important factor in the design of a plant and it is necessary to differentiate between the dew points for water and acid. The dew point for water in fuels lies between 40 and 60°C, and decreases as the air factor increases. Table 11.7 gives data covering fuels showing the water vapour content of their waste gases and their dew points for different air factors.

If the fuel or the gases evolved by the batch contain sulphur, then some of the sulphur will burn to give SO_2 and a smaller amount to give SO_3. During the cooling of the waste gases the SO_3 combines with water vapour to form a solution of sulphuric acid, which is very corrosive. The dew point of the water/sulphuric acid mixture depends on its SO_3 content and is significantly higher than the dew point for water. Curves showing the boiling points and dew points of mixtures of H_2SO_4 and water vapour at 152 mbar, corresponding to a dew point for water at 53°C, are given in Figure 11.7; the rapid increases at low H_2SO_4 concentrations should be noted. The controlling factor of the acid dew point is therefore the SO_3 content, not the SO_2 content, of the waste gases. A low air factor tends to move the SO_2/SO_3 ratio towards the former, and thereby reduce the acid dew point, which should be the objective.

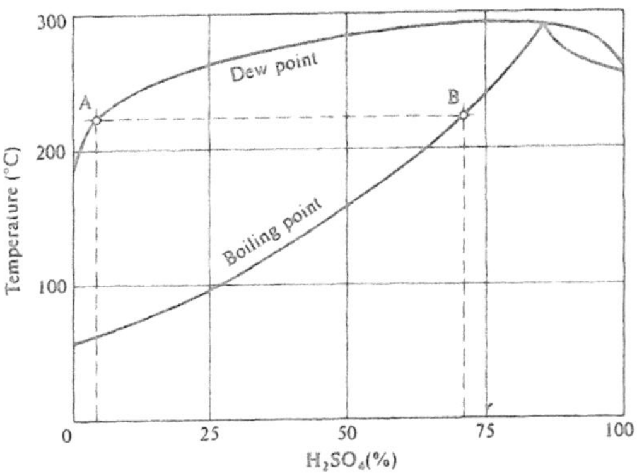

Figure 11.7. Curves of boiling and dew points of H_2SO_4–water vapour mixtures at a total pressure of 152 mbar[1]

Table 11.7. *Water content and dew points of waste gases from fuels when burnt in wet air at 20°C and 80% relative humidity*[2]

Fuel	Water content of dry waste gases λ = 1·0 (kg/m³ₙ)	λ = 1·0 (m³ₙ/m³ₙ)	λ = 1·5 (m³ₙ/m³ₙ)	Dew point (°C) λ = 1·0	λ = 1·5
Coal	0·0711	0·0887	0·0650	42	37
Coke	0·0674	0·0837	0·0617	42	36
Lignite	0·0909	0·1131	0·0814	47	40
Fuel oil type S	0·1249	0·1554	0·1080	52	46
Fuel oil type EL	0·1397	0·1738	0·1198	54	48
Liquefied gas	0·1678	0·2062	0·1396	57	50
Natural gas*	0·2023	0·2516	0·1687	60	53
Natural gas**	0·1992	0·2478	0·1669	60	54
Coke oven gas	0·2520	0·3135	0·2085	64	57
Producer gas (from lignite briquettes)	0·0777	0·0966	0·077	43	38
Blast furnace gas	0·0147	0·0183	0·0184	16	16

* from Emsland
**from Slochteren

11.2 Industrial fuels

11.2.1 Gaseous fuels[3,4]

Glass furnaces can be fired by a variety of gaseous fuels such as natural gas, coal gas, liquefied petroleum gas, and certain substitute or supplementary gases. The use of producer gas manufactured in the glass plant itself is now a thing of the past.

Industry often does not distinguish between natural gas and liquefied petroleum gas; the latter originates from oil deposits. Natural gas is adjusted by the supplier to defined standards of quality and is piped through long distance supply lines direct to the consumer. There appear to be two qualities, type L possessing a Wobbe number of 9800–11 600, and type H with a Wobbe number of 11 200–13 500; gas of type H is the main supply in the German Federal Republic. Physical properties and combustion characteristics are given in Table 11.8 and combustion data for these gases at an air factor of $\lambda = 1$ are given in Table 11.9.

Table 11.9. *Combustion data ($\lambda = 1$) for coke oven and natural gases*[3,4]

		Coke oven gas	Natural gas Type L	Type H
Air requirement	m^3/m^3 gas	4·3	8·6	9·9
Waste gas volume (wet)	m^3/m^3 gas	5·0	9·5	10·9
Waste gas volume (dry)	m^3/m^3 gas	3·9	7·8	8·9
	vol%			
Waste gas composition (wet)				
	CO_2	7·9	9·6	9·7
	H_2O	22·4	18·5	18·7
	N_2	69·7	71·9	71·6
	vol%			
Waste gas composition (dry)				
	CO_2	10·2	11·8	12·0
	N_2	89·8	88·2	88·0
Flame temperature				
	°C	1975	1930	1940
(with dissociation)				

Coal gas is made by heating coal and removing the gases formed; two gases can be produced, type A with a Wobbe number of 5400–6700 and type B with a Wobbe number of 6300–7900, and data covering these are also included in Tables 11.8 and 11.9.

Table 11.8. *Physical properties and combustion data for coke oven gas and natural gases (for general guidance)*[3,4]

			Coke oven gas	Natural gas Type L	Type H
Gas composition vol%					
CO_2			2·3	0·9	1·0
C_nH_m			2·3	–	–
O_2			0·5	–	–
CO			5·5	–	–
H_2			54·5	–	–
CH_4			25·3	82·1	93·0
N_2			9·6	12·9	1·1
C_2H_6			–	3·2	3·0
C_3H_8			–	0·6	1·3
C_4H_{10}			–	0·3	0·6
Density relative to air	d		0·40	0·64	0·61
Density at NTP	ρ	kg/m^3	0·51	0·83	0·79
Average molecular weight	M		11	18·6	18
Gas constant	R	Nm/kg K	726	448	471
Specific heat at 0°C, 1 bar	c_p	$kcal/m^3$ K	0·326	0·371	0·387
(at constant pressure)		kJ/m^3 K	1·365	1·552	1·62
Specific heat at 0°C, 1 bar	c_v	$kcal/m^3$ K	0·243	0·282	0·297
(at constant volume)		kJ/m^3 K	1·017	1·181	1·243
Specific heat ratio at 0°C, 1 bar	γ		1·34	1·31	1·30
Dynamic viscosity at 20°C, 1 bar	η	Ns/m^2	$12{\cdot}75 \times 10^{-6}$	$11{\cdot}76 \times 10^{-6}$	$10{\cdot}76 \times 10^{-6}$
Kinematic viscosity at 20°C, 1 bar	ν	m^2/s	$27{\cdot}5 \times 10^{-6}$	$15{\cdot}21 \times 10^{-6}$	$14{\cdot}9 \times 10^{-6}$
Calorific value, gross	H_o	$kcal/m^3$	4 700	8 586	9 875
(at NTP)		kJ/m^3	19 680	35 900	37 250
Calorific value, net	H_u	$kcal/m^3$	4 175	7 749	8 920
(at NTP)		kJ/m^3	17 470	32 430	37 350
Ratio of gross/net calorific values			0·888	0·902	0·903
Wobbe number	$H_o/\sqrt{d}$	$kcal/m^3$	7 450	10 700	12 700
		kJ/m^3	31 200	44 800	53 200
Ignition velocity of gas/air mixture	Λ_{max}	m/s	0·95	0·41	0·43
Flammability limits in air	Z_u, Z_o	%	5–30	5–15	5–15
Spontaneous ignition temperature in air		°C	560	640	640

Substitute or supplementary gases are mixtures of gases which match the combustion properties of a gas of a defined type at the same pressure and include some which are very similar to coal gas; their use does not require changes to burner settings etc. Supplementary gases are gas mixtures which are added to a gas of given type to give optimum performance. The quantities that can be added are limited by the fact that the combustion properties of the gases in total must remain unaffected. Well known supplementary gases comprise mixtures of liquid gas and air.

Gaseous fuels are all very clean, with a negligible sulphur content, and their combustion characteristics can be controlled within narrow limits. Their use is very convenient from the point of view of the consumer since they do not have to be stored, the supply is assured, and investment in equipment is small. The most important (German) regulations covering the use of these gases in industrial furnaces is given in DVGW Working Sheet G 610 *Technical regulations for the gas firing of industrial furnaces.*

11.2.2 Liquid fuels

11.2.2.1 Fuel oils[6]

Only those fuel oils which are derived from petroleum are of interest to the glass industry and they may be divided into two types, distilled fuel oil EL (extra low viscosity) and residual fuel oil S (high viscosity); data are given in Table 11.10. However, it is necessary to differentiate between data in accordance with DIN 51603 and what may be referred to as special data: oil properties can vary according to origin and since DIN 51603 gives minimum values, plants should be designed on the safe side. Fuel oil belongs to materials having a hazard classification 3.

The viscosity is the most important property for storage, conveyance through pipework, and atomisation and Figure 11.8 gives a survey of the viscosity of different oils as a function of temperature.

Of major importance is the sulphur content. Figure 11.9 gives the range of values normally encountered and fuel oils may be distinguished according to their sulphur content: low sulphur content with a maximum of 1·0 wt% S; low sulphur content with a maximum of 1·75 wt% S; normal sulphur content with a maximum of 2·0 wt% S; high sulphur content with a maximum of 2·8 wt% S. Each 1% sulphur reduces the calorific value of the oil by 293 kJ/kg.

The ash content of EL fuel oil is about 30 mg/kg and consists mainly of vanadium and nickel oxides. Fuel oil S will give an ash content of 500 mg/kg, the main constituents being oxides of vanadium, iron, and calcium.

Table 11.10. *Analyses of fuel oils according to DIN 41603 and as encountered in practice. (Plants should be designed on the basis of the worst data)*

Parameter		*EL fuel oil* DIN 51603	*Average value*	*S fuel oil* DIN 51603	*Average value*
Density at 15°C	g/ml	max. 0·86	0·825		0·94
Flashpoint (closed cup)	°C	min. 55	65	min. 65	90–160
Viscosity	m^2/s				
at 20°C		max. 6×10^{-6}	$4·5 \times 10^{-6}$	–	–
at 50°C		–	–	max. 450×10^{-6}	$190–365 \times 10^{-6}$
at 100°C		–	–	max. 40×10^{-6}	$12–28 \times 10^{-6}$
Solidifying point	°C	min. −10	−15	–	20–40
Coke residue (Conradson)	wt%	max. 0·05	0·01	max. 15	6–10
Sulphur content	wt%	max. 0·8	0·5	max. 2·8	1·0–2·8
Water content (not separable)	wt%	max. 0·1	–	max. 0·5	0·1–0·3
Net calorific value	kJ/kg	min. $4·186 \times 10^4$	$4·27 \times 10^4$	min. $3·98 \times 10^4$	$4·06–4·14 \times 10^4$
Ash (oxides)		max. 0·01	<0·001	max. 0·15	0·01–0·06
Sediment (solids)	wt%	max. 0·05	≈0	0·5	≈0·1
Carbon	wt%	–	85·5	–	86·5
Hydrogen	wt%	–	13·5	–	11·5
Nitrogen + oxygen	wt%	–	0·3	–	0·2–0·5
Theoretical air requirement, dry	m^3/kg*	–	11·2	–	10·8
Waste gas volume, wet	m^3/kg*	–	11·9	–	11·3
Waste gas volume, dry	m^3/kg*	–	10·5	–	10·1
Carbon dioxide, maximum	%	–	15·3	–	15·7
Steam	m^3/kg	–	1·5	–	1·3

*gases at NTP

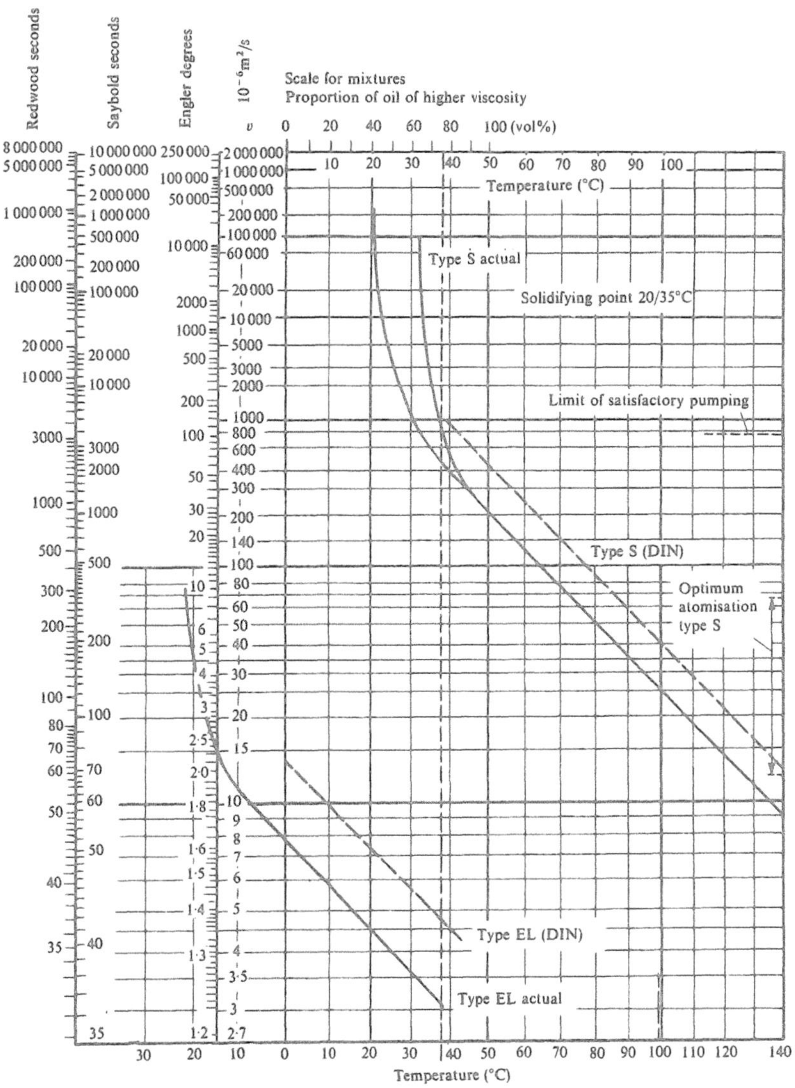

Figure 11.8. Visocisty/temperature relationships of petroleum fuel oils[6]

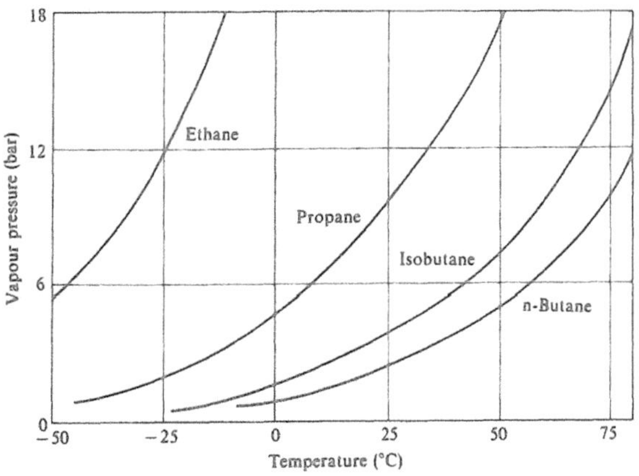

Figure 11.9. Saturated vapour pressures of fuel gases as a function of temperature[5]

11.2.2.2 Liquefied gases[5]

These are propane and butane, usually supplied as a mixture, and normally considered as types of liquid fuel, even when they reach the burners and ports as gases. They are described fully in DIN 51622 and their properties and combustion data are given in Tables 11.11 and 11.12.

The differences in boiling points and vapour pressures of butane and propane should be noted (Figure 11.9). The liquefied gas acquires the temperature of its surroundings during storage and develops a pressure corresponding to this temperature. Removal of gas reduces the pressure but leads to further evaporation and a return to the pressure and temperature equilibrium conditions. The neces-

Table 11.11. *Properties of liquefied gases*[5]

		Propane	*Isobutane*	*n-Butane*
Formula		C_3H_8	C_4H_{10}	C_4H_{10}
Molecular weight		44·09	58·12	58·12
Composition	wt% C	81·72	82·66	82·66
	wt% H	18·28	17·34	17·34
Density of liquid at 15·56°C	kg/dm^3	0·508	0·563	0·584
Specific volume of liquid at 15·56°C	dm^3/kg	1·969	1·776	1·712
Density of gas at 0°C, 1 bar	kg/m^3	2·019	2·668	2·703
Specific volume of gas at 15·56°C, 1 bar	m^3/kg	0·543	0·410	0·410
Relative density at 0°C, 1 bar	air = 1	1·562	2·064	2·091
Boiling point at 1 bar	°C	−42·1	−11·7	−0·5
Heat of evaporation at boiling point, 1 bar	kJ/kg	426	366	385
Vapour pressure				
at −20°C	bar	2·35	0·72	0·45
at 0°C	bar	4·71	1·57	1·02
at 20°C	bar	8·44	3·04	2·06
at 30°C	bar	10·79	4·02	2·84
Critical temperature	°C	96·8	135·0	151·6
Critical pressure	bar	42·58	36·49	38·06

Table 11.12. *Combustion data for liquefied gases*[5]

		Propane	*Isobutane*	*n-Butane*
Gross calorific value, H_o*	kJ/m^3	101 803	131 984	133 993
Net calorific value, H_u*	kJ/m^3	93 557	121 603	123 528
Gross calorific value, H_o	kJ/kg	50 399	49 478	49 562
Net calorific value, H_u	kJ/kg	46 339	45 585	45 711
Theoretical air requirement*	m^3/m^3	23·9	31·0	31·0
	m^3/kg	12·2	12·0	12·0
	kg/kg	15·7	15·5	15·5
Waste gas volume				
dry	m^3/m^3	21·9	28·5	28·5
wet	m^3/m^3	25·9	33·5	33·5
Flammability limits				
in air	vol%	2·37–9·50	1·80–8·44	1·86–8·41
in oxygen	vol%	2·0–48·0	1·3–47·0	1·3–47·0
Flame speed				
in air	cm/s	32	32	32
in oxygen	cm/s	450	370	370
Ignition temperature				
in air	°C	500	510	480
in oxygen	°C	490	460	460
Flame temperature				
in air	max. °C	1935	1900	1896
in oxygen	max. °C	2850	–	–

*gases at NTP

sary heat of evaporation is initially taken from the liquid held in the tank but the liquid cools until, when it falls below the boiling point, vaporisation is much reduced. The pressure for the gaseous fuel depends on maintaining this minimum temperature at all times: as the boiling point of commercial butane, −4°C, cannot assure continuous delivery in the gaseous form, it is necessary to provide a heated evaporator to convert the liquid into the gas. Pumps are used for conveying butane in the liquid form.

12 Waste heat recovery

12.1 Recuperators and regenerators: design calculations

The high operating temperatures of fuel fired glass melting furnaces call for the use of waste heat to preheat the combustion air and also, in some exceptional circumstances, the fuel in order to keep energy consumption to a minimum. In practice, recuperators and regenerators fulfill this purpose and their design has been discussed in Section 3.4; this section deals with calculations and layout.

By way of example, Figure 12.1 shows the influence of air preheating on the energy consumption of a glass furnace. The calculations on which Figure 12.1 is based are derived from the relationships given in Section 12.2 dealing with waste heat boilers

$$\frac{\dot{E}_{B,R}}{\dot{E}_B} = \frac{\eta_f}{\eta_f + \dfrac{\dot{H}_L}{\dot{E}_{B,R}}}.$$

Since for modern fuels the expression

$$\frac{\dot{H}_L}{\dot{E}_{B,R}} = \frac{i_R}{i_{0,R}} = \frac{\lambda l_{st} c_{p,L} \Delta\vartheta_L}{H_u}$$

$$= 4{\cdot}1 \times 10^{-4} (\Delta\vartheta_L)$$

depends almost only on $\Delta\vartheta_L$, we can write

$$\frac{i_{0,R}}{i_0} = \frac{1}{1 + 4{\cdot}1 \times 10^{-4} \dfrac{\Delta\vartheta_L}{\eta_f}} = \frac{1}{1 + A}$$

where

$$A = 4{\cdot}1 \times 10^{-4} \frac{\Delta\vartheta_L}{\eta_f}$$

and is dimensionless. Its link to furnace efficiency, η_f, shows that the preheating of air ($\Delta\vartheta_L$) is of particular benefit to furnaces of low combustion space efficiency (see Table 5.3 in Section 5.1.3.1).

The majority of glass furnaces use regenerators, recuperators being used only in special circumstances. Since the calculations covering both types of heat exchanger have much in common and since the treatment of recuperators is simpler, they are dealt with first for reasons of convenience.

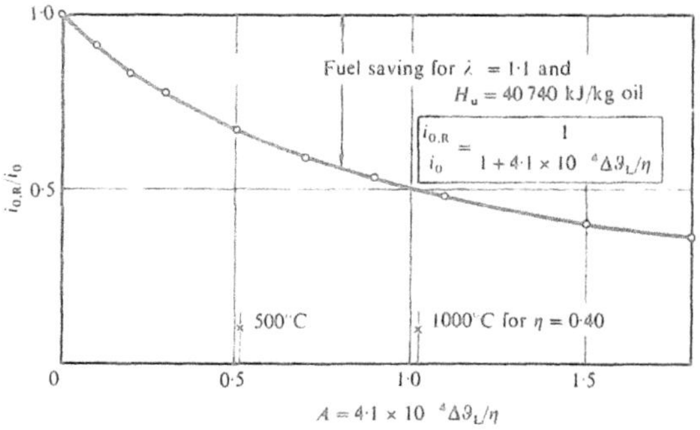

Figure 12.1 Influence of air preheat and furnace efficiency on fuel consumption

12.1.1 Recuperators

Recuperators are continuously operating heat exchangers in which the media giving up heat and absorbing heat are separated from one another by a gas-tight wall. Depending on the respective flow directions of waste gases and combustion air, recuperators are described as parallel flow or counterflow (Figure 12.2). The temperatures of the media are designated as ϑ for that giving up heat and ϑ' for that accepting the heat; the entry temperature is ϑ_1 and the exit temperature ϑ_2; the heat capacities with respect to time of the masses of gas flowing through the recuperator are C and C' in kW; and the area over which the heat exchange takes place as a or A. Relatively simple considerations and deductions, easily referred to in most reference books,[1,2] give, for counterflow,

$$\vartheta_1 - \vartheta_2' = (\vartheta_1 - \vartheta_1') \frac{1 - C/C'}{1 - \dfrac{C}{C'} \exp\left[\left(\dfrac{1}{C'} - \dfrac{1}{C}\right) kA\right]} \tag{12.1}$$

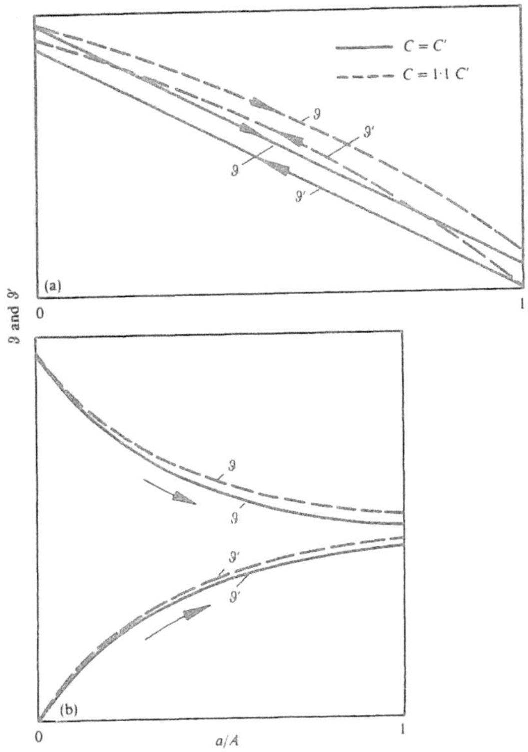

Figure 12.2. Temperature changes in a recuperator[1]
(a) counterflow (b) parallel flow

and

$$\vartheta_2' - \vartheta_1 = (\vartheta_1 - \vartheta_1')\frac{C}{C' + C}\left\{1 - \exp\left[-\left(\frac{1}{C'} + \frac{1}{C}\right)kA\right]\right\} \quad (12.2)$$

for parallel flow.

12.1.1.1 Local changes in temperature and heat transmission coefficient

The change in the temperatures ϑ and ϑ' at any particular location is

$$\vartheta_1 - \vartheta = (\vartheta_1 - \vartheta_1')\frac{C'}{C' + C}\left\{1 - \exp\left[-\left(\frac{1}{C'} + \frac{1}{C}\right)ka\right]\right\} \quad (12.3)$$

and

$$\vartheta' - \vartheta_1' = (\vartheta_1 - \vartheta_1')\frac{C'}{C' + C}\left\{1 - \exp\left[-\left(\frac{1}{C'} + \frac{1}{C}\right)ka\right]\right\} \quad (12.4)$$

in a parallel flow recuperator and

$$\vartheta_1 - \vartheta = (\vartheta_1 - \vartheta_2')\frac{C'}{C' - C}\left\{1 - \exp\left[\left(\frac{1}{C'} - \frac{1}{C}\right)ka\right]\right\} \quad (12.5)$$

and

$$\vartheta_2' - \vartheta' = (\vartheta_1 - \vartheta_2')\frac{C}{C' - C}\left\{1 - \exp\left[\left(\frac{1}{C'} - \frac{1}{C}\right)ka\right]\right\} \quad (12.6)$$

in a counterflow recuperator, where a is the relevant surface area and k the heat transmission coefficient. If flat separating walls or thin walled tubes are considered

$$\frac{1}{k} = \frac{1}{\alpha} + \frac{\delta}{\lambda} + \frac{1}{\alpha'}, \quad (12.7)$$

where α and α' are the corresponding heat transfer coefficients, δ the wall thickness, and λ the thermal conductivity of the material constituting the separating wall. For exact determinations of α, α', and k reference must be made to the relevant literature:[1,2] for approximate calculations $k = 6$–$9\ \mathrm{W/m^2}$ for ceramic recuperators and 30–$47\ \mathrm{W/m^2}$ for steel recuperators.

If there are large local differences in k, then the product kA must be replaced by the integral $\int_0^A k\,da$; this means that the arithmetic mean is usually sufficient for normal purposes.

12.1.1.2 Efficiency

The efficiencies of heat exchangers are defined in a number of ways depending on objective and approach and care must therefore be taken when comparing efficiencies; for glass furnaces, the heat that can be transferred from the waste gases to the combustion air is the major criterion. The enthalpy of the combustion air, $C'(\vartheta_2' - \vartheta_1')$, is compared to the enthalpy of the waste gases available in the limit $C(\vartheta_1 - \vartheta_1')$, hence

$$\eta^* = \frac{C'}{C}\frac{\vartheta_2' - \vartheta_1'}{\vartheta_1 - \vartheta_1'}. \quad (12.8)$$

This definition is used by Hausen[1] who calls it the efficiency function; when combined with Equations (12.1) and (12.2) the results shown in Figures 12.3 and 12.4 are obtained, and the superiority of the

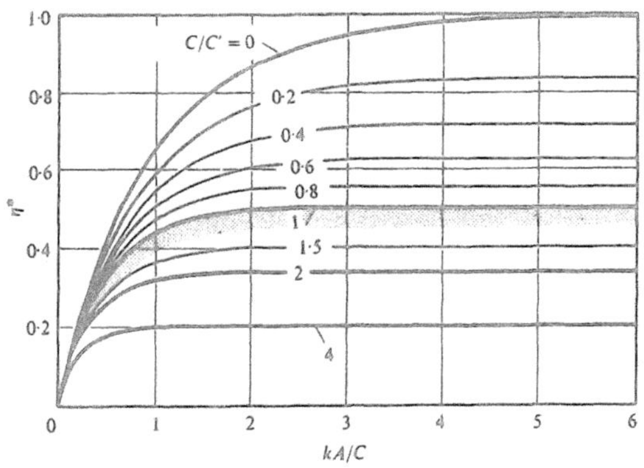

Figure 12.3. Efficiency function, η^*, for parallel flow[1]

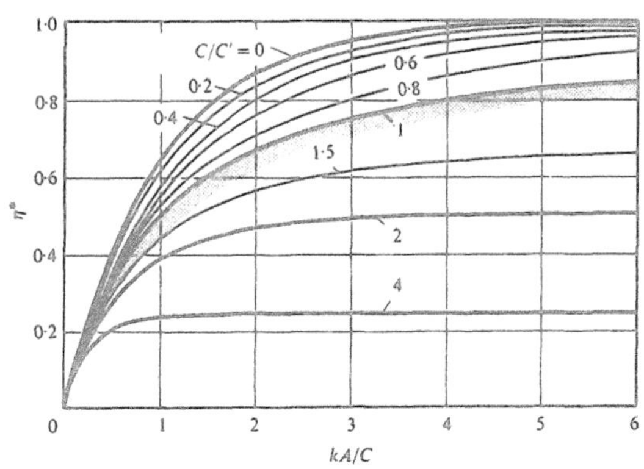

Figure 12.4. Efficiency function, η^*, for counterflow[1]

counterflow principle is evident. When combined with Equation (12.8) these diagrams can also be used for estimating the attainable preheat temperature, ϑ'_2.

The definition of the efficiency η^* only covers the fact derived from the second law of thermodynamics that a hot medium can only raise a cooler medium to its own temperature even if, at ambient temperature, its heat capacity is greater than that of the medium accepting the heat. No account can therefore be taken of heat losses from the heat exchangers to their surroundings; this can be calculated by classical analytical methods, but is cumbersome[1] and the reader is therefore referred to the literature. The heat losses in a recuperator depend to a large extent on design and thermal insulation, and can be up to 5–10% of the heat exchanged.

12.1.1.3 Temperature relationships

In a recuperator not subject to any losses and whose efficiency has been calculated according to Equation (12.8), the temperature conditions prevailing at a location *a* can be determined via basic conditions of thermal balances. If ϑ_0 is the surface temperature of the wall in the zone of the medium transferring heat and ϑ'_0 that of the wall in the zone of the medium accepting the heat,

$$\left(\frac{\vartheta' - \vartheta'_0}{\vartheta_0 - \vartheta}\right)_a = \left(\frac{\alpha}{\alpha'}\right)_a. \qquad (12.9)$$

Temperature conditions behave as the inverse of the corresponding heat transfer coefficients. Since, due to radiation, waste gases have a higher heat transfer coefficient (α) than the normal air being preheated (α'), the temperature differences on the waste gas side are smaller than those on the side where the air is being heated.

The temperature change in the direction of flow at location *a* can also be determined from thermal balance considerations,

$$\left(\frac{\Delta\vartheta'_a}{\Delta\vartheta_a}\right)_a = \frac{C}{C'}. \qquad (12.10)$$

The medium with the lower heat capacity (C or C') will always show the larger local temperature change (ϑ_a or ϑ'_a); see Figure 12.2.

12.1.2 Regenerators

From a thermal point of view regenerators are heat exchangers working on an intermittent basis. Their advantage lies in the fact that a gas-tight separation between waste gases and air is not required, a point of particular importance in relation to materials subjected to high temperatures; their disadvantage is the periodic reversal of gas flows. Regenerators are usually preferred for glass furnaces because they give a higher air preheat temperature.

Analytical calculations governing regenerators are significantly more difficult than those for recuperators but a suggestion by Rummel[3] to treat regenerators like recuperators and use a suitable heat transmission coefficient, k, has been generally accepted. Of the different methods used for the determination of k,[1,2,5] the method by Hausen[1] has been shown to be the most accurate and the most suitable (see Section 12.1.2.1 below). Using the value of k so determined, the performance of regenerators can be calculated using the Equations of Section 12.1.1.

The development of computer techniques has led to the situation where the use of recuperator formulae together with a suitable value for k for calculating regenerator data is no longer necessary. One returns to the relationship for a nonstationary heat transfer and simulates the working characteristics of a regenerator mathematically. It is thus possible to simulate differences in checkerwork, heat conditions, etc. The disadvantage is that this method does not provide mathematical expressions which help to understand relationships. Results obtained by this method are given in Section 12.1.2.2.

12.1.2.1 Determinations according to Hausen[1]

12.1.2.1.1 Heat transmission coefficient

Hausen has shown that the average temperatures (related to time) in a regenerator behave in a way which is analogous to those in a recuperator except that, due to the fact that the temperature of the incoming gases is constant, special features appear at both ends; these can be interpreted mathematically and physically as amplified oscillations superimposed on a fundamental oscillation and can be treated as such. If the period of heating is T, and the period of cooling is T' then, ignoring the amplified oscillations, the heat transmission coefficient is

$$\frac{1}{k_0} = (T + T')\left[\frac{1}{\alpha T} + \frac{1}{\alpha' T'} + \left(\frac{1}{T} + \frac{1}{T'}\right)\frac{\delta}{\lambda}\Phi\right] \tag{12.11}$$

in which the checkerwork is assumed to be plates of thickness, δ, and thermal conductivity, λ, and Φ is a function which takes account of the rapid temperature changes to gases and checkerwork surfaces immediately after reversal. For the conditions existing in the steel and glass industries Hausen gives

$$\Phi = \frac{1}{6} - \frac{1}{180}\frac{\delta^2}{2a}\left(\frac{1}{T} + \frac{1}{T'}\right) \tag{12.12}$$

where a is the thermal conductivity ($= \lambda/cp$) of the refractory material.

The amplified oscillations to which regenerators are subject cause a reduction in heat transfer and if the heat transfer coefficient which includes this effect is designated as k, then

$$\frac{k_0}{k} = \left[1 + \frac{A}{400}\left(\frac{\alpha}{C} + \frac{\alpha'}{C'}\right)\right]. \tag{12.13}$$

For conditions existing in the steel and glass industries this approximates to

$$\frac{1}{k} = \frac{(T + T')}{0{\cdot}93}\left[\frac{1}{\alpha T} + \frac{1}{\alpha' T'} + \left(\frac{1}{T} + \frac{1}{T'}\right)\frac{\delta}{7\lambda}\right]. \tag{12.14}$$

12.1.2.1.2 Temperature changes and checkerwork temperature

Equation (12.14) can be used in conjunction with Equations (12.5) and (12.6) to calculate the average temperatures of the waste gases and the air in regenerators at specific times or in specific places.

If Θ is the surface temperature of the checkerwork refractories,

$$\frac{(\vartheta - \Theta_0)}{(\Theta'_0 - \vartheta')} = \frac{T'}{T}\frac{\alpha'}{\alpha}. \tag{12.15}$$

The surface temperature Θ_0 and the average temperature Θ_m of the checkerwork bricks differ slightly. The relationship

$$(\vartheta - \Theta_m) = \left(\frac{1}{\alpha} + \frac{\delta}{\lambda}\Theta\right)\alpha(\vartheta - \Theta_0) \tag{12.16}$$

applies to the waste gas and an analogous expression applies to the air to be preheated. For rough calculations Θ_0 can be put approximately equal to Θ'_0, which enables the surface temperature of the bricks to be calculated from the difference $(\vartheta - \vartheta')$.

12.1.2.1.3 Heat transfer coefficients

It is necessary to differentiate between convective and radiative heat components when calculating the heat transfer coefficients α and α'; among the data given in the literature, those of Yazicizade[4] should be mentioned.

The convective component α_k was measured on a plant similar to that used in the steel and glass industries and for different types of checkerwork (Figure 12.5). The results have been related to the flow velocity under standard conditions (0°C, 1013 mbar) and large differences between smooth plain packing and straight packing can be seen. For good heat transfer under operating conditions, it is important that there is an increase in the pressure loss (Figure 12.6).

If the dimensionless parameters Re and Nu are used for flow and heat transfer, then the results obtained by Yazicizade and earlier workers can be summarised diagrammatically in Figure 12.7

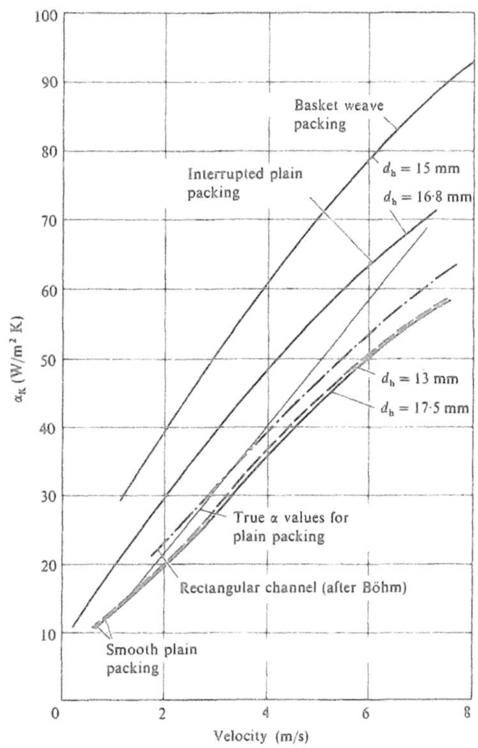

Figure 12.5. Heat transfer coefficient as a function of velocity for all types of checkerwork[4]

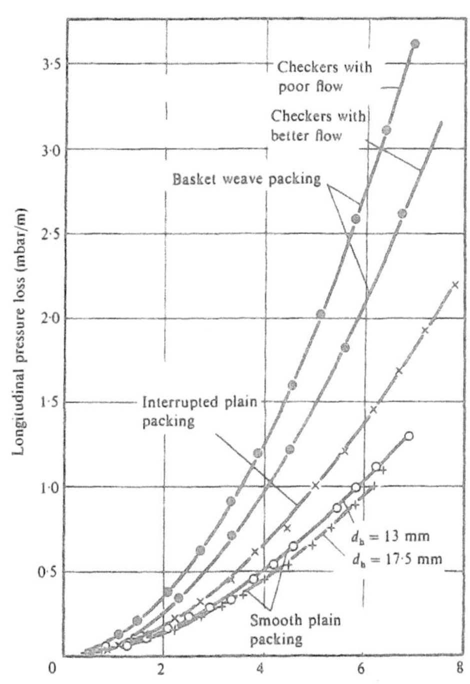

Figure 12.6. Pressure loss as a function of velocity[4]

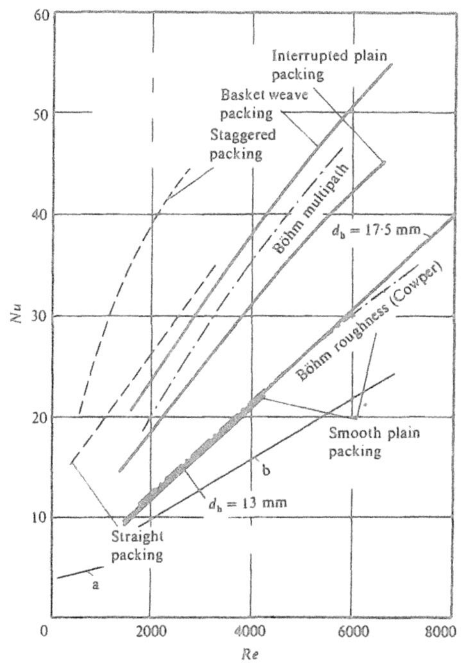

Figure 12.7 Heat transfer in dimensionless representation[4]
(a) laminar flow (b) turbulent flow

in which d is taken as the channel width. The approximate values for ν and λ can be taken from Tables 12.1 and 12.1A.

The heat transfer coefficient for the waste gases, α_A, includes both the convective and radiative components

$$\alpha_A = \alpha_K + \alpha_{Str}; \tag{12.17}$$

the values of α_K can be taken from Figures 12.5 and 12.6 while α_{str} is best taken from Table 12.2. Apart from the type of waste gas and the temperature, the layer thickness or channel width is a controlling parameter. For gas volumes which cannot be described simply by a single channel width, a hydraulic diameter, d_h, can be calculated from

$$d_h = \frac{4V}{A}. \tag{12.18}$$

Strictly speaking, apart from the thickness of the gas layer, the emissivity of the refractory surfaces should also be included. For normal checkerwork surfaces these data can be obtained from Table 12.2, so that special calculations have to be carried only for extreme deviations.

The dependence of values of α and k on temperature and location can be obtained from Sections 12.1.1.1 and 12.1.1.2; in general, the arithmetical average can be used.

Table 12.1. *Thermal conductivity and kinematic viscosity at different temperature of various technically important gases at a pressure of 1 bar*

	Temperature (°C) 0	100	200	300	400	500	1000
Thermal conductivity (W/mK)							
Air	0·024	0·031	0·039	0·044	0·050	0·056	0·079
N_2	0·024	0·031	0·037	0·042	0·047	0·052	0·072
O_2	0·024	0·032	0·039	0·045	0·052	0·058	0·086
H_2O	0·016	0·025	0·033	0·043	0·054	0·066	0·114
CO_2	0·015	0·022	0·030	0·038	0·045	0·052	
Kinematic viscosity (10^{-6} m²/s)							
Air	13·27	22·87	34·37	47·49	62·01	77·77	171·7
N_2	13·30	22·89	34·38	47·49	62·04	77·82	172·0
O_2	13·44	23·15	34·83	48·18	63·03	79·23	179·4
H_2O	10·77	20·61	33·72	50·11	69·79	92·77	255·5
CO_2	6·98	12·72	19·85	28·30	37·60	47·90	109·4

Table 12.1A. *Average specific heat of wet combustion gases for complete combustion without excess air ($\lambda = 1{\cdot}0$) and at a pressure of 1 bar*

	Average specific heat (kJ/m³K) in the temperature range 0–200°C	0–400°C	0–600°C	0–800°C	0–1000°C	0–1200°C	0–1400°C	0–1600°C
Producer gas	1·412	1·454	1·496	1·529	1·563	1·596	1·626	1·651
Coke oven gas	1·395	1·429	1·467	1·500	1·538	1·567	1·600	1·626
Fuel oil	1·408	1·446	1·475	1·517	1·550	1·588	1·617	1·634
Natural gas	1·398	1·432	1·465	1·503	1·537	1·570	1·599	1·628
LPG	1·398	1·432	1·465	1·503	1·537	1·570	1·599	1·628
Solid fuels	1·412	1·454	1·496	1·529	1·563	1·596	1·626	1·651
Air	1·027	1·0439	1·0636	1·0866	1·1063			

Table 12.2. *Heat transfer coefficient ($\alpha_{A,str}$) for gas radiation from waste gases (W/m²K): other values may be obtained by interpolation*

Waste gas type	Composition (%)	Channel width (mm)	Waste gas temperature (°C) 800	1000	1200	1400	1600
Coke oven gas	CO_2 17·5	100	8·4	11·4	14·0	16·5	18·7
	H_2O 7·5	150	10·7	14·3	17·9	21·3	24·3
		200	12·7	17·2	21·5	25·8	29·9
		300	16·1	21·9	27·8	33·8	39·3
Heavy fuel oil	CO_2 13·3	100	8·0	10·7	12·4	15·1	17·2
	H_2O 13·1	200	11·9	15·8	19·7	23·3	26·2
		300	14·4	19·7	24·5	29·2	33·0
Natural gas	CO_2 9·7	100	9·3	12·1	14·4	16·1	17·5
	H_2O 17·5	200	11·8	16·2	20·1	23·8	26·9
		300	15·4	20·1	25·2	29·7	34·4

12.1.2.1.4 Pressure losses

An important factor is the resistance to flow of the checkerwork, which depends primarily on the type of packing, the channel width, and the flow velocity. Detailed measurements are summarised in Figure 12.6[4] and further data may be obtained from Section 3.3.4 and Table 3.3.

12.1.2.1.5 Influence of flow through the checkerwork on the quantity of heat exchanged

There is a scarcity of data in the literature dealing with flow and the quantity of heat exchanged in regenerators. Trier[6] has investigated this relationship in greater depth and has shown that effects vary with the type of flow through the checkerwork. If certain sections of the checkerwork are subject to stronger flows of both waste gases and air, then the adverse effect is much less than if a weak flow of waste gases is followed by a strong flow of air; quantitative results are given in Figure 12.8. In general, checkerwork presenting a high resistance to flow gives more uniform flow conditions than checkerwork with a low resistance. For more details the reader is referred to Section 3.3.3.

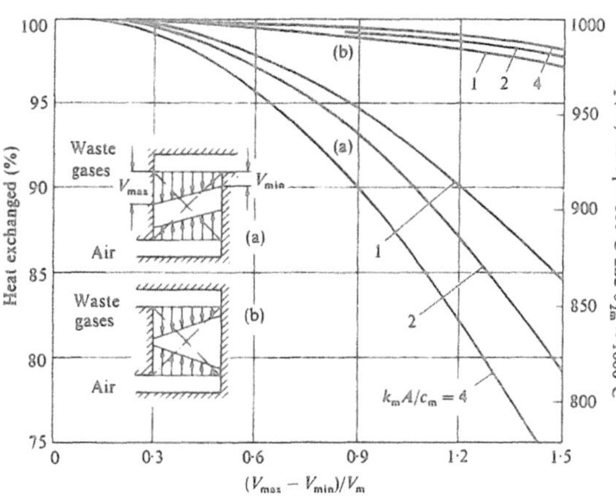

Figure 12.8. Influence of uneven flow through checkerwork on the quantity of heat exchanged $(C/C')_m = 1{\cdot}0$[6]

12.1.2.2 Regenerator calculations using the computer programme by HVG

The calculations are based on a computer simulation of the operation of a regenerator which, for this purpose, is divided into sections in the direction of flow. At regular intervals the quantities of heat exchanged between waste gases and checkerwork or air and checkerwork are determined for each section. The temperature changes in the bricks are determined using the difference method as an approximate solution of the equation for thermal conductivity. Results are in good agreement with calculations based on Hausen's method but the temperatures in the checkerwork were determined more easily and more accurately.

Taking a specific example where $C/C' = 1{\cdot}25$, $\vartheta_1 = 1500$°C, and $\vartheta_1' = 100$°C, the changes in temperature in the direction of flow in a regenerator chamber are given in Figure 12.9 which shows the temperature of waste gases and air at the end of a reversal period. The difference between the temperature of the waste gases and that of the checkerwork refractory surfaces lies between 10 and 70 degC, whereas that between air and checkerwork is 100–350 degC.

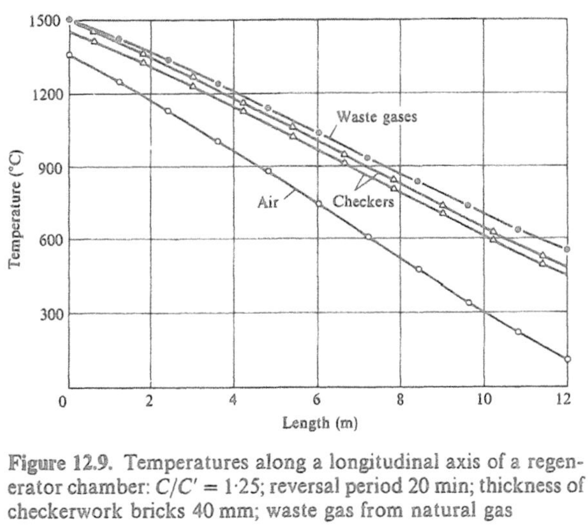

Figure 12.9. Temperatures along a longitudinal axis of a regenerator chamber: $C/C' = 1{\cdot}25$; reversal period 20 min; thickness of checkerwork bricks 40 mm; waste gas from natural gas

12.1.2.2.1 Changes in temperature of regenerator refractories according to location and time

The temperature hysteresis of the checkerwork was calculated for magnesite and fireclay bricks 65 mm thick using the same method, and Figure 12.10 gives the temperature of a brick in the upper layers of the checkerwork. It shows that the hysteresis of a magnesite brick is about 15 degC and that of a fireclay brick is about 25 degC while the temperature differences are about 10 and 30 degC. Clearly, the higher thermal conductivity of magnesite has a marked influence on the temperature differences set up in the brick.

Much has been said and written about the causes of the disintegration of checkerwork bricks in regenerator chambers and considerable weight is given to the temperature differences in the refractory and the thermal stresses caused thereby. When considering Figure 12.10 doubts appear, at least for magnesite, whether the small temperature differences in the bricks really can be responsible for dangerous thermal stresses. The bottom layers of

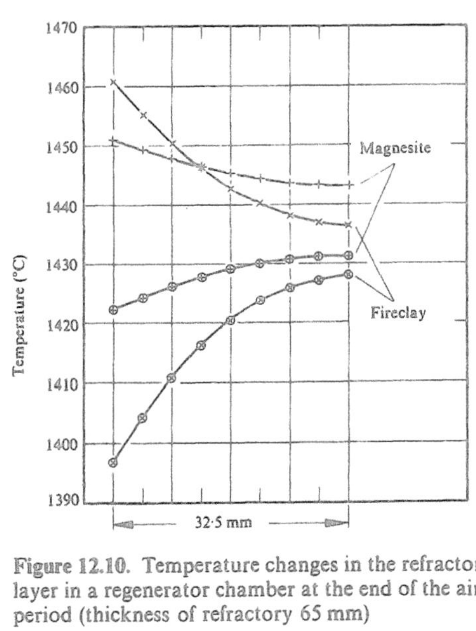

Figure 12.10. Temperature changes in the refractories of the top layer in a regenerator chamber at the end of the air or waste gas period (thickness of refractory 65 mm)

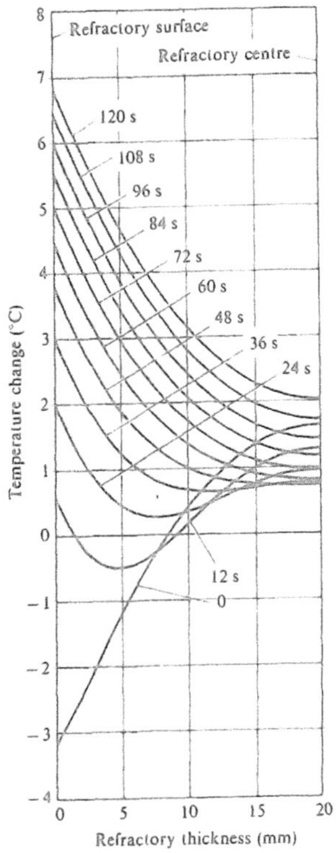

Figure 12.11. Temperature changes with respect to time and place in checkerwork refractory immediately following reversal (thickness of refractory 40 mm)

the checkerwork at the start of the air period are regarded as being particularly at risk and to highlight this idea, the temperature changes with time at various places were calculated for a fusion cast brick 40 mm thick immediately after reversal had taken place, i.e. at the time when the largest temperature differences arise. The results are shown in Figure 12.11, from which it is evident that the temperature differences are small and never exceed 5 degC. The parabola of temperature distribution during the heating period is reversed within one minute of reversal. Again, doubts arise whether such temperature differences can be responsible under industrial conditions for the cracking of checkerwork bricks such as is observed at the end of furnace campaigns. Generally, calculations show that thermal effects in a regenerator chamber are governed in the main by the heat transfer coefficients, α and α', the specific heats of air and waste gases, C and C', and the surface area taking part in the process of heat exchange, A. The thermal properties of the bricks constituting the checkerwork have only a subsidiary influence. A change in brick thickness from 65 to 40 mm improves the conditions; however, this is due not to the improved thermal utilisation of the refractory material but to the enlargement of the surface taking part in the process of heat exchange.

The temperatures of waste gases and air change by between 25 and 35 degC in the course of a reversal period of 20 min.

12.2 Waste heat boilers

Since the waste gases leaving glass tank furnace installations are at relatively high temperatures, it is obvious that, with the prevailing high energy costs, additional utilisation of the available heat should be considered. The feasibility of making use of this waste heat depends on the temperature drop that can be accommodated and the heat or other energy requirements of equipment in the plant. With regenerators, the heat of the waste gases is largely used up in preheating the combustion air and linked as it must be to a furnace operating under periodic reversal, a secondary use of waste heat is only possible downstream from the reversal mechanism; this means that the temperature difference available is much reduced. Constantly fired glass furnaces with recuperators have a much larger temperature difference available, especially if the waste heat boiler and the recuperator can be combined into a single piece of equipment. In addition, the high temperatures prevailing cause the heat transfer to be very intense, which reduces the need for a high heating area. A disadvantage is the need to use expensive high temperature resistant steels in their construction.

As it is becoming more and more necessary to remove dust from waste gases by filtration, it also becomes necessary to cool the waste gases prior to their entering the filters and this is often best carried out by means of waste heat boilers.

12.2.1 Waste heat boilers located downstream from regenerative glass furnaces

12.2.1.1 Waste heat boilers for general heating

Space heating, e.g. the heating of buildings and preheating of oil, is widely used in glass plants. For reasons of cost the boiler should operate at low pressure and this also makes for easier supervision and water supply. The simplest solution is a fire tube boiler. Associated with a gas fired tank furnace, the only problem with this type of boiler is the deposit of dust formed in the tubes and there must therefore be easy access to the entry and exit sides of the tubes for cleaning. This type of boiler operates at 10–15 bar.

Tanks fired with heavy fuel oil can give rise to attack by sulphuric acid if the dew point has been overlooked. It is therefore good policy to carry out experiments covering dew points and corrosion resistance before designing and constructing a waste heat boiler. Experience suggests that corrosion can only be avoided if the sulphur content of the fuel is less than 2% (Figure 11.7); as far as the boiler is concerned, it is of less importance whether the sulphur content is 1% or 2% – the critical parameter is the SO_3 content, which is only slightly influenced by the firing conditions. Damage can be avoided by making sure that the temperatures of the boiler heating surfaces are always above the dew point. The temperature of the surfaces on the waste gas side in steam boilers are 10 degC above the saturation temperature of the water in the boiler; in a superheater, it is significantly higher. If the dew point of the acid is 160°C, a saturation temperature of 180°C should be chosen, corresponding to an operating pressure of 9·5 bar which should be held constant by a pressure sustaining valve. The water supply to the boiler should be preheated in a high pressure feed heater heated by steam in a counterflow mode.

If corrosion due to acid condensation is to be avoided altogether, then a saturation temperature of 230°C should be used, corresponding to a pressure of 30 bar. Water tube boilers, not shell and tube boilers, are best.

12.2.1.2 Waste heat electric power plant

A better use of residual waste gas heat, but one that calls for a higher investment, is the construction of a small unit producing electric power.[8,9] The design principles of such a unit can be seen in Figure 12.12: the waste gases are cooled in a boiler by producing superheated steam which is reduced in pressure in a pass out and condensing turbine. At the exit some of the steam is used for heating purposes and the remainder is used to drive the turbine and is then condensed to be pumped back into the feed water and degassed by heating. The steam turbine can drive a generator, a compressor, or fans. When producing electricity in a turbine the

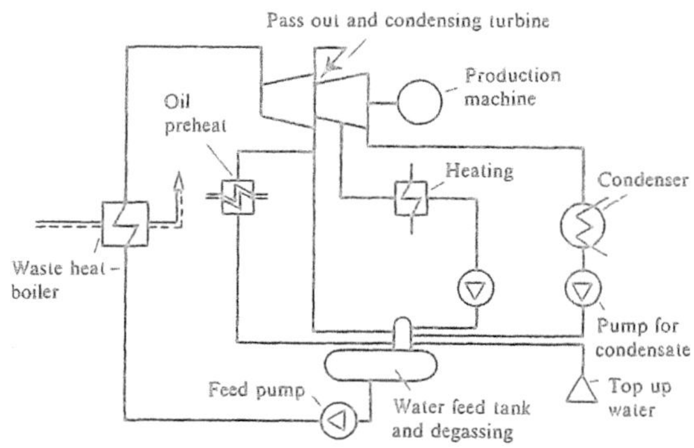

Figure 12.12. Schematic of a waste gas electric generator plant[9]

steam should be at high pressure and should be superheated to take account of the wetness of the steam at the cold end of the turbine. Erosion and poor efficiency are the consequences of steam that is too wet.

If the temperature and the quantity of waste gases are fixed in a process, then the relationships can be shown in a diagram which follows the temperature of the waste gases, water, and steam on their passage through the boiler (Figure 12.13). The heat input into the boiler is the heat withdrawn from the waste gases, the temperature of which falls from an entry temperature, ϑ_1, to the exit temperature, ϑ_3. The feed water has an entry temperature, ϑ_5, is raised to the saturated steam temperature, ϑ_4, and evaporates; in a boiler without a superheater the steam temperature remains constant at the saturated steam temperature.

The temperature difference, δ_I, at the end of the evaporation surface must be sufficiently large, as

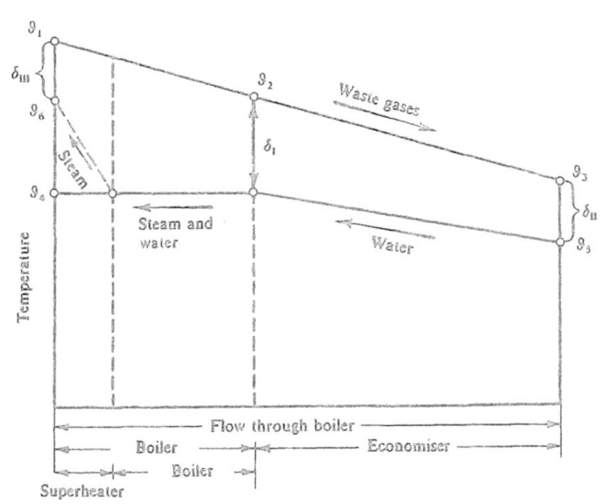

Figure 12.13. Schematic of temperature changes of waste gases, water, and steam in a waste heat boiler related to the flow through the boiler[8]

the area of the heating surface and the temperature difference are related inversely. If the temperature difference is too small, the heating surfaces are uneconomically large and prevent the necessary steam pressure being realised. The same applies to the temperature difference, δ_{II}, where the feed water entry temperature is decided by the need to avoid dew point corrosion. The limits of boiler and economiser surface areas (for a waste gas temperature ϑ_2) must be decided on the principle of optimisation. If the boiler surface area is too small (ϑ_2 being too near to ϑ_1) then too little steam would be created, the economiser would be unable to make use of the residual heat and the boiler would be uneconomic. If the boiler surface is too large, i.e. ϑ_2 is too low, then, although the production of saturated steam is high, the steam pressure would be too low giving a low output of electricity.

Steam turbines can only be used with superheated steam; this can be obtained in a superheater which is part of the boiler itself or in a superheater which is heated separately.

The situation governing the superheating of saturated steam in a waste heat boiler is shown by the dashed line in Figure 12.13. The final temperature of the steam when leaving the boiler is ϑ_6 and the temperature difference δ_{III} must be large enough for the required degree of superheating to be obtained safely and maintained at the level. Waste gases from glass tank furnaces always contain dust and are generally not sufficiently hot to be suitable for worthwhile superheating. Contamination of the superheater could easily occur and this risk increases with rising superheating temperature. It is therefore often better to let the superheating of the steam be done in a separate superheater and not by the waste gases from the furnace. However, as fuel costs rise, the installation and operation of a separate superheater becomes costly and requires careful assessment as to its feasibility.

The superheating of saturated steam in an auxiliary boiler requires heat at a high temperature which must be between 1200°C and the temperature of the saturated steam. To perform economically, a superheater has to produce additional saturated steam which must be heated in its own economiser. It is therefore unavoidable that the operation of a superheater gives rise to additional quantities of saturated steam to add to the steam from the waste heat boiler: if the energy so created is within the energy requirements of the plant, then the increased quantity of steam is welcomed. Control of such a boiler installation is difficult and the turbine should therefore be able to operate over a wide range of steam temperatures.

If, for a given temperature decrease, superheating is carried out in a boiler without supplementary firing then, compared to an installation with a separate superheater, the steam must be at a reduced pressure or the heating surface must be constructed on a relatively large scale.

Figure 12.14 gives an example for an electric power generator with oil preheating (260 kW) and space heating (1500 kW) supplied with 10^5 m^3/h waste gas at 485°C. The mechanical energy created in the turbines drives a compressor of the type used in container glass works.

If a supplementary waste heat boiler is used, then the steam in the turbine can be raised to a pressure of 78 bar and a temperature of 500°C, thereby increasing the performance of the turbine: a disadvantage is the cost of fuel for a superheater fired separately. In principle, the waste gases can be passed through a gas turbine to drive a generator for supplying power and can then be used to heat a superheater as shown in Figure 12.15.

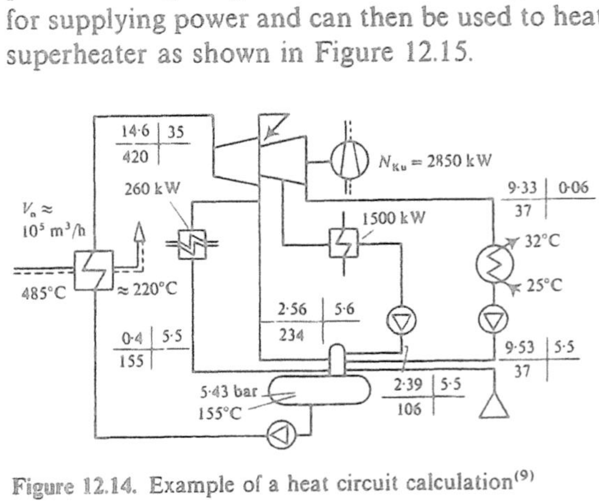

Figure 12.14. Example of a heat circuit calculation[9]

Key	Steam (t/h)	Absolute pressure (bar)
	Temperature (°C)	

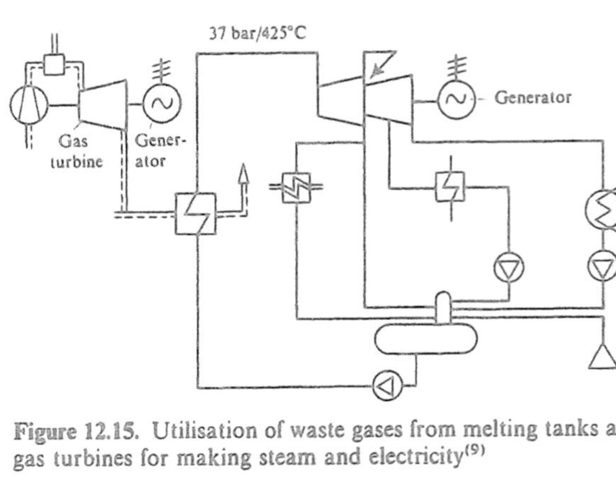

Figure 12.15. Utilisation of waste gases from melting tanks and gas turbines for making steam and electricity[9]

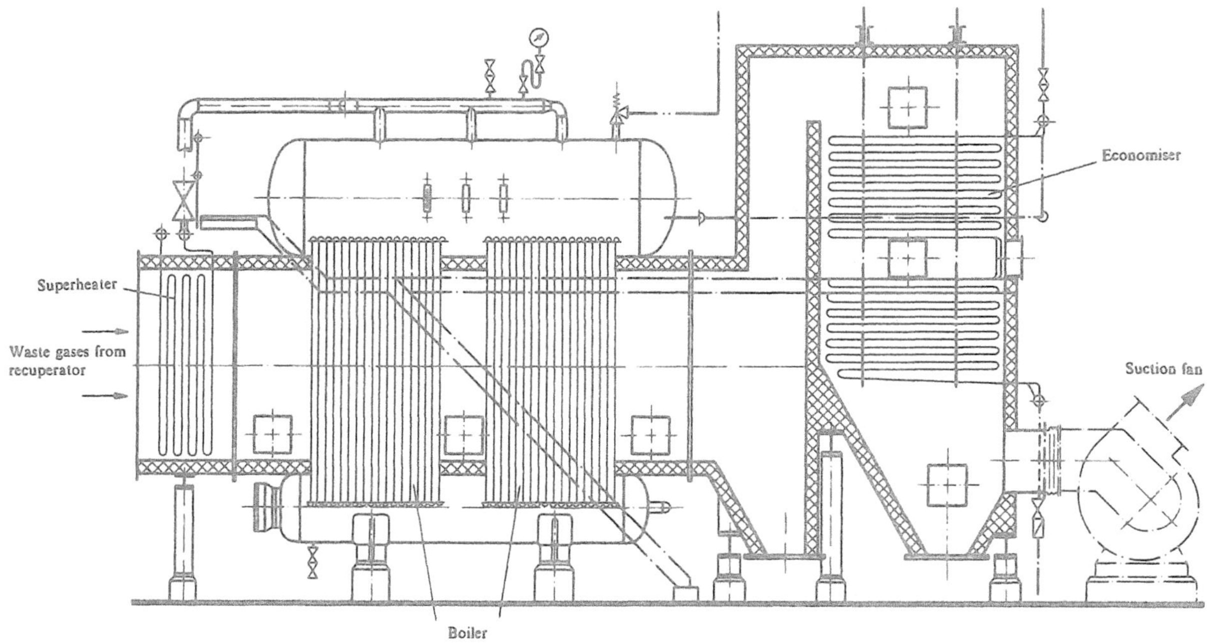

Figure 12.23. Steam boiler: 30 bar, 415°C superheating temperature, 22 300 m^3/h waste gases, 5·5 t/h steam[18]

An alternative construction is one in which a recuperator is located between glass tank and waste heat boiler (Figure 12.23); a recuperator in this position permits the combustion air to be preheated to 750°C. In order to achieve sufficiently high superheating of the steam, the superheater is located upstream from the evaporator. A recuperatively fired furnace with a waste gas volume of 22 300 m^3/h can be made to produce 5·5 t/h of steam at 30 bar and 415°C.[18]

12.2.3 Energy management and energy distribution

The purpose of waste heat boilers is to make use of the residual heat in furnace waste gases. In this sense they are really a part of the furnace installation but, in their application to the energy management and distribution of the plant as a whole, they play a part far beyond that of a single furnace installation. It follos that the design and economics of a waste heat boiler must cover both the furnace and other parts of the plant affected by it.

An example of the use of waste heat in a container glass plant by the installation of a water heat boiler and a superheater heated separately (Figure 12.24) has been provided by Baier.[16] This type of installation is, however, ony viable when oil prices are low.

An idea of the energy management of a flat glass plant has been given by Ehrich[15] (Figure 12.25) and for container plants by Schubert,[13,14] Schulte-Kellinghaus,[17] and Kölsch.[18]

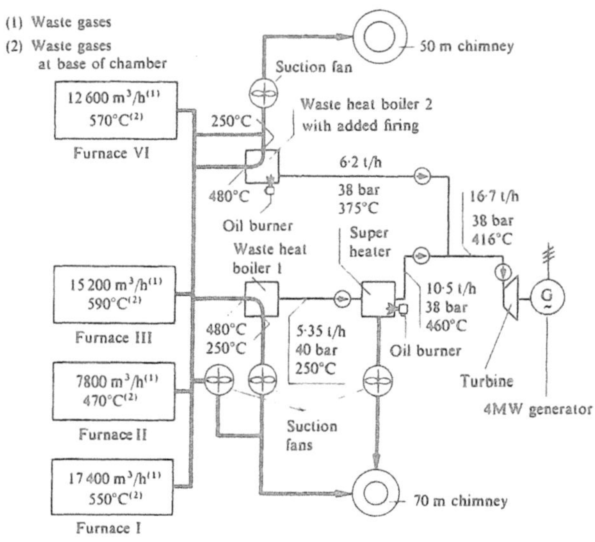

Figure 12.24. Schematic of waste heat utilisation of a container glass plant operating four regeneratively heated glass melting tanks[16]

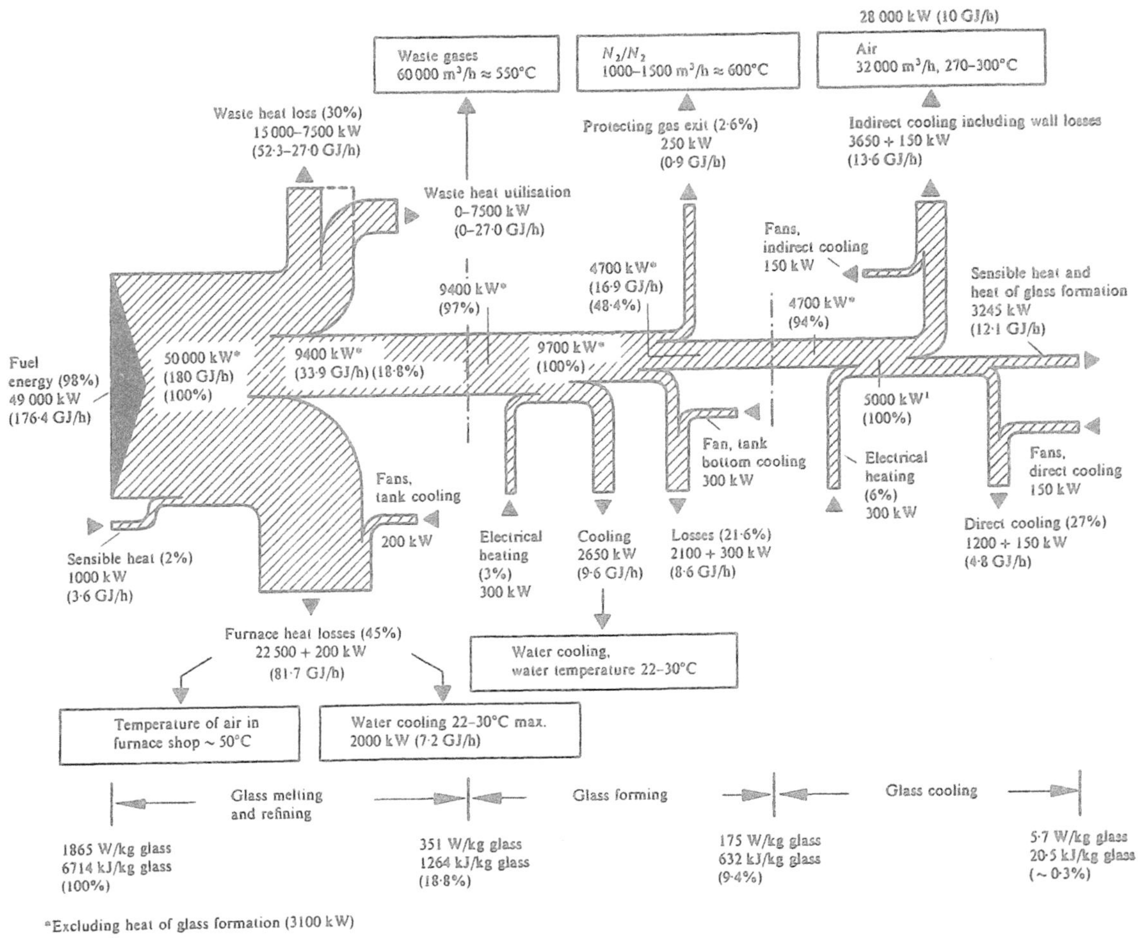

Figure 12.25. Graphic representation of an energy balance of a Float glass plant[15]

13 Emissions from glass furnaces

In the discussion that follows the term 'emission' will be limited to dust and harmful gases emitted from the chimney of a glass furnace as part of the waste gases. In order to gain an insight into the relationships it is necessary to start with a summary of the processes involved in glass melting and refining. Compared with other industries on a basis of quantity, the emissions from glass furnaces are not large and as a result, it is only recently that they have been investigated. However, the large number of glass compositions manufactured and therefore the wide spectrum of raw materials used can give rise to substantial and harmful emissions in individual cases.

13.1 Breakdown of materials emitted

In order to analyse the nature of the problem and to gauge the importance of individual processes the first step to be taken is to prepare a quantitative breakdown of dust and gaseous losses from glass furnaces. Since sulphur predominates in the emission, SO_x must be included. It has been found best to set up separate material balances for the melting and working zones of a tank furnace, or the melting and refining phases of a pot furnace. Up to now the differences between individual furnaces and types of glass as well as the lack of quantitative data have meant that it has been possible to construct only schematic balances but these can give an insight into the various interrelated processes involved.

Figure 13.1 shows the framework for a quantity balance for the melting zone of an oil fired glass tank furnace melting a soda–lime glass. The melting chamber and the effect of the regenerator chamber on the waste gases and on the air are shown separately. Combustion air enters from the left with a relatively low dust content and picks up the dust deposited in the cooler sections of the regenerator chamber during its waste gas period; a similar state of affairs exists in the hotter parts of the chamber, but here there are also chemical reactions involved and part of the dust is vaporised. The combustion air enriched with dust and vapour now enters the melting chamber, where part of the batch and the melted glass is lost as dust and vapour; at the same time SO_x is added to the system from fuel oil and sulphate in the batch. Similar processes involving the deposition of dust and condensation of vapour take place in the waste gas regenerator chamber and finally the waste gases

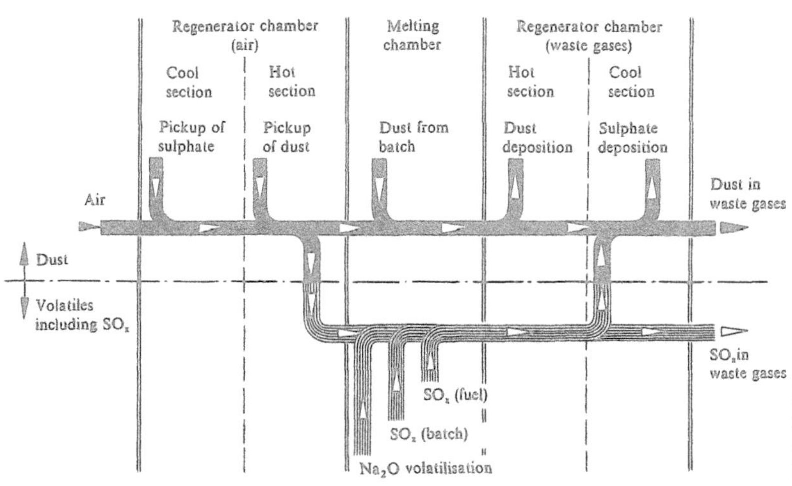

Figure 13.1. Diagram of the origins of materials in waste gases taken from the doghouse area of a regeneratively fired glass tank furnace

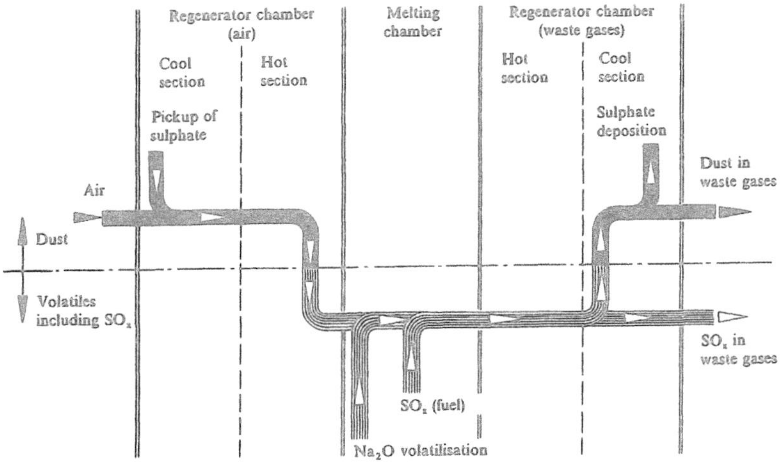

Figure 13.2. Diagram of the origins of materials in waste gases taken from the refining area of a regeneratively fired glass tank furnace

carrying dust and SO_x pass through the flues and the chimney into the open air. Unfortunately little quantitative data covering material transport in and out of regenerator chambers have yet been published.

A quantity balance for the refining zone can be constructed in a similar way (Figure 13.2). There are no losses of dust or vapour from batch constituents, but condensation together with related chemical reactions cause deposits to be formed in the cooler parts of the regenerator chambers; these contribute to the overall material transport in and out of the chambers.

13.2 Volatilisation from molten glass

Losses from molten glass due to volatilisation have been investigated thoroughly and data are available in several publications. The relevant factor in this case is the vapour pressure; the quantities evaporated depend on the partial pressure conditions in the melt, the possible course of chemical reactions (formation of steam), and replacement transport in the melt. Apart from vapour pressure the other factors are largely a function of operating conditions and in general, therefore, it is best to limit the discussion to vapour pressure. In line with basic equations and the enthalpies and entropies of vaporisation (or sublimation),

$$\log p = \frac{A}{T} + B \qquad (13.1)$$

where A and B are constants, and T the thermodynamic temperature.

Figure 13.3 shows the linear relationship between

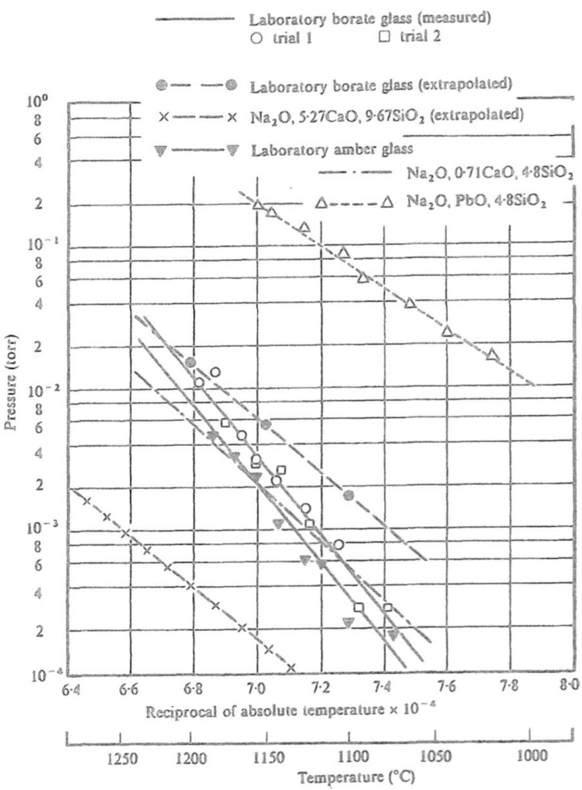

Figure 13.3. Vapour pressures of different glasses from Kröger & Sörström[1]

log p and $1/T$ for glasses of different composition; the data are from Kröger & Sörström.[1] The importance of temperature and composition are clearly shown. Of the ternary glasses investigated lead glass has the highest vapour pressure by a considerable margin; it is two orders of magnitude higher than the corresponding ternary soda–lime glass.

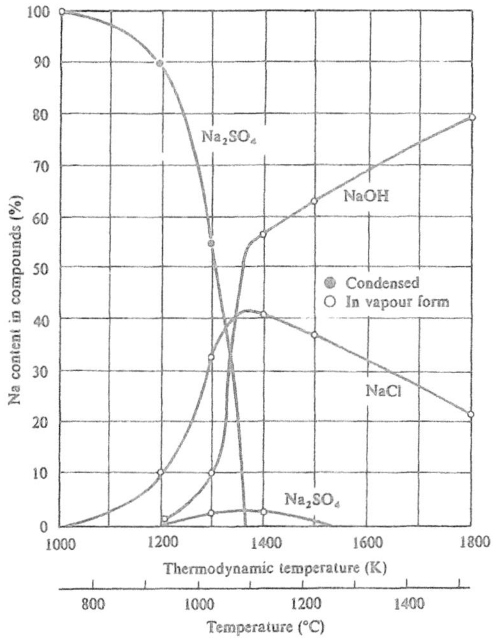

Figure 13.4. Distribution of sodium evaporated from a Float glass tank fired by natural gas as specific compounds in the waste gases and related to temperature[2]

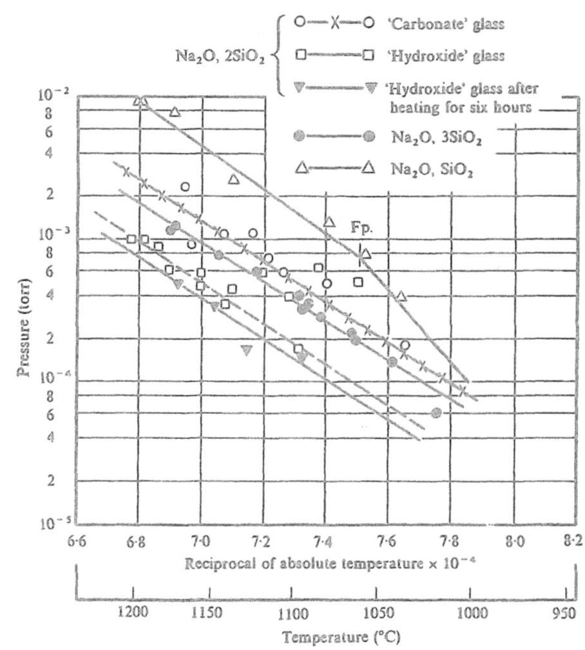

Figure 13.5. Vapour pressures of sodium silicate melts from Kröger & Sörström[1]

The constituent evaporated from soda–lime glasses is mainly Na_2O and thermodynamic considerations[2] indicate that the vaporised alkali is present in the hot waste gases as NaOH gas. The reaction $Na_2SO_4 + H_2O \rightleftharpoons 2NaOH_{(gas)} + SO_2 + \frac{1}{2}O_2$ applies. At 1100°C the equilibrium moves strongly to the left and Na_2SO_4 condenses, as is clearly demonstrated in Figure 13.4.

13.3 Evaporation from melting batch

By comparison with molten glass there are very few data about vapour pressure and evaporation of batch as it melts down, the large number of primary melts being formed and the problem of establishing the temperatures making work in this area difficult.

Some indications of the vapour pressures of compounds which are formed during the melting of a soda–lime glass batch can be gleaned from the work of Kröger & Sörström already referred to above.[1] Figure 13.5 gives the vapour pressures and the average vapour pressure curves for sodium silicates; depending on composition, the pressures are up to five times greater than those for soda–lime glasses. However, if the much lower actual temperatures prevailing during melting are considered and compared to those existing during refining, then one is led to the conclusion that evaporation from primary melts cannot be large in absolute terms. The extent to which vapour pressure can change in relation to the CaO content of soda–lime glasses is shown in Figure 13.6.

A major influence on evaporation from soda–lime glasses is the water content of the atmosphere, which causes the NaOH formed to have a higher vapour pressure than the oxide. The water content of waste gases is about 18% from furnaces fired

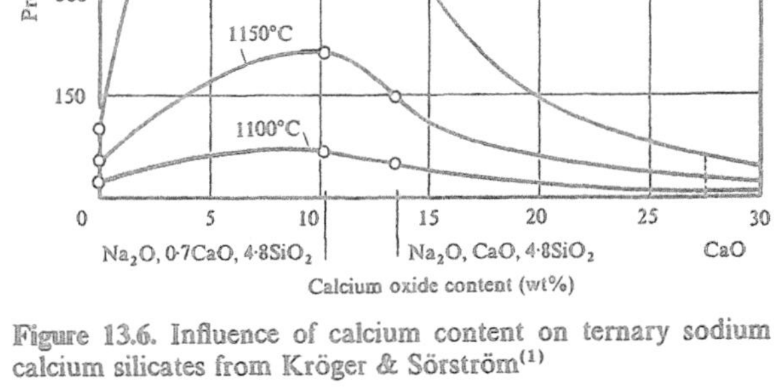

Figure 13.6. Influence of calcium content on ternary sodium calcium silicates from Kröger & Sörström[1]

with natural gas and about 12% from those fired with fuel oil, but the emission of dust and therefore evaporation is considerably lower for plants operating with natural gas than for those using fuel oil. The reason lies in the overriding influence of sulphur which dominates the situation in oil fired furnaces. The use of damp batch leads to a reduction in dusting and emission due to a reduction in the formation of primary dust. Overall, therefore, variations in the water contents of waste gases do not have a significant effect on chimney emissions.

According to Hanke & Scholze,[3] the CO_2 content of the atmosphere has no effect on volatilisation from soda–lime glasses; with borosilicates it caused an increase and with lead glasses a decrease in volatilisation. This can have some limited influence in the batch melting zone due to the CO_2 evolved from the batch (batch gases).

The proportion of cullet in the melting batch affects dusting and volatilisation. A single example of measurements indicates that an increasing proportion of cullet reduces dust emission from the furnace; with a 50% cullet content the dust emission when melting a soda–lime glass using fuel oil can be reduced to less than 150 mg/m^3 (at NTP).

13.4 Dust evolved from melting batch

Two processes must be distinguished when considering dust evolved from the batch. In the first place, there is the dust formed by batch components such as soda ash, potash, red lead, etc. which melt and vaporise when they come into contact with the hot furnace gases; in the second place, there is the dust formed by limestone, dolomite, and raw materials containing alumina which, after splitting and evolving gas, remain intact as particles. Since it is only materials of the second group which are intact when they reach regenerator chambers or flues, it is only the second group which will be considered as dust.

The formation of dust depends primarily on the method of batch feeding and the firing technique. As far as the raw materials are concerned, their particle size and, in the cases of limestone and dolomite, their tendency to decrepitate needs to be considered. Materials with a crystalline structure which is either especially fine or irregularly oriented tend to burst when CO_2 is evolved; with single crystal materials, this tendency is less marked.

The literature provides few quantitative data covering the evolution of dust. In the zone of the first pair of burners of cross-fired tank furnaces it is of the order of 0·1–1 g/m^3 (at NTP) of the waste gases. In the same zone the ratio of quantities lost as dust and by vaporisation is about 1:3 to 1:5; for the whole furnace ratios of 1:10 have been mentioned.[4] The level of dust evolution is generally higher in U shaped flame tank furnaces than in cross-fired tank furnaces.

The moisture content of the batch has a major influence. Experiments with a water cooled paddle (Morse & Manring[5,6]) showed that an increase in the moisture content of the batch from 2 to 3·5% reduced the quantity of deposits on the paddle by 72%; other authors have given similar results. As the moisture content of the batch increases, so the composition of deposits moves away from dusts towards sublimates (Na_2SO_4).

Rozhanskii et al.[7] have reported measurements of the total dust content of furnace waste gases, i.e. pure dust plus condensed dust. Among other measurements are those on a flat glass tank furnace fired by natural gas and fitted with a sampling probe to measure the total dust content; the results are given in Table 13.1. One surprising result is the high dust concentration of over 2000 mg/m^3 (at NTP) in the zone of the first pair of ports, suggesting a high rate of dust evolution. Since the batch formulation is not given, especially the saltcake content, no other conclusions can be drawn.

Table 13.1. *Dust content in the hot waste gases of a window glass tank furnace[7] measured by a sampling probe*

Port	*Dust content (mg/m^3 at NTP of waste gases)*	*(kg/month)*
1	2030	8208
2	1000	2922
3	710	3196
4	660	2232
5	220	687
6	170	483
7	130	144

13.5 Deposits and reactions taking place in regenerator chambers

High dust concentrations in the exit ports coupled to relatively low dust concentrations in the chimney flue implies a very high level of dust deposits in chambers and flues. One must therefore assume that reversal of firing involves the transport to and fro of dust, but the scale of this can only be estimated.

Davis, Manring & Bauer[8] have measured the pH of the air and the waste gases in the hot parts of regenerator chambers during the air and waste gas periods; even during the air period a significant SO_x content could be identified. This observation has proved that combustion air, on its passage through channels and checkerwork, picks up material which could only have been brought there by the waste gases.

Kircher[9] reported exchange effects between the deposits and condensates in the checkerwork and the gases flowing through it. In a regeneratively fired furnace short term peak SO_2 values in the waste gases, sometimes exceeding 2000 ppm. were observed at the beginning of a firing period and occurred at the same time as high CO peak values; Figure 13.7 gives an example. This is caused by insufficient purging of the chambers before lighting the burners but the origin of the high SO_2 values under these reducing conditions has so far not been fully explained. The reduction of SO_3 in the waste gases to SO_2 cannot be the origin of these high SO_2 values and they may be caused by certain reactions taking place in the regenerator chambers, e.g. a reduction of $NaHSO_4$ to Na_2SO_4.

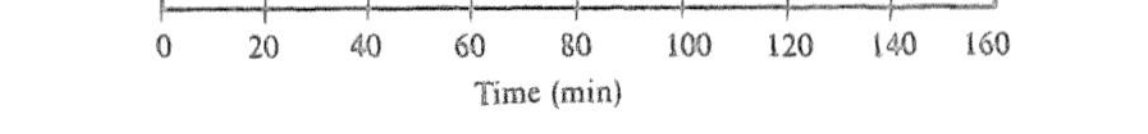

Figure 13.7. Short term SO_2 peak values in the waste gases of a regeneratively fired U shaped flame tank furnace at the beginning of the firing period[9]

13.6 Chimney emissions

13.6.1 Dust

It is due to local government control requirements that there are extensive data on the dust content of waste gases emitted by glass tanks.[9,10]

Figure 13.8 gives the results of dust emission measurements from glass tank furnaces; the dust

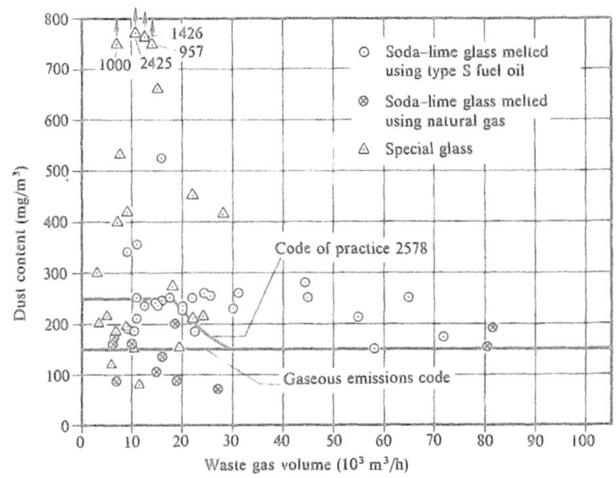

Figure 13.8. Plot of dust emissions.[9] Waste gas volume standardised to 273 K, 1036 mbar, after deduction of content of water vapour

content is reported in mg/m^3 by reference to waste gases containing 8% O_2 (at NTP) and is plotted against the volume of waste gases per hour. The figure includes the limiting dust concentration of 150 mg/m^3 set by the Federal German code controlling gaseous emissions for the glass industry as well as the curve for the limiting values suggested in the Association of German Industries' Code of practice 2758 entitled *Emission limits for glass works*. As Figure 13.8 shows, the dust content of most plants lies between 150 and 250 mg/m^3, mainly from tank furnaces melting soda–lime glasses; plants with higher dust values almost invariably melt special glasses. Emissions of 1000 and 957 mg/m^3 of dust from tank furnaces melting a lead glass and a borosilicate glass respectively must also be regarded as very high but the highest dust concentrations so far measured, 2425 and 1426 mg/m^3, were from tank furnaces melting alkali borosilicate glass for glass fibre production.

If the dust figures for soda–lime glass tank furnaces are examined more closely, they show that furnaces fired with natural gas give dust concentrations about 100 mg/m^3 lower than furnaces fired with type S fuel oil; this could be related to the influence of the sulphur content of the fuel oil on the evaporation of alkalis. Whatever the reason, the difference between the dust concentration of furnaces fired with natural gas and fuel oil is too high to be explained solely on the basis of the SO_3 contents of the dusts themselves.

Next to the sulphur content of the fuel, the sulphate content of the batch seems to affect dust emission. A case is known where, leaving all other factors constant, dust emission was reduced by 30% by reducing the sulphate content of the batch.

In general, dust emission from a plant increases as the proportion of easily vaporisable constituents in the glass increases. A soda–lime glass furnace melting a glass with a total alkali content of more than 16% gave rise to dust emission of 348 mg/m^3, which is more than 100 mg/m^3 above the figure for a similar soda–lime glass with the more usual 13–14% alkali content. Crystal glasses in which the K_2O content is increased within a constant total alkali content show an increased dust emission. The addition of 1% B_2O_3 to an otherwise normal soda–lime glass increased the dust emission from 230 to 540 mg/m^3, that is by more than 100%, although the dust itself contained only 5% B_2O_3. On the other hand, tank furnaces melting instrument glasses with a considerably higher B_2O_3 content gave dust contents of between 200 and 220 mg/m^3.

The measurements plotted in Figure 13.8 do not differentiate between dust emissions from the batch feeding and the refining zones and a possible way of estimating these figures is to carry out measurements on pot furnaces where the differences between batch feeding and refining are separated in time. Figure 13.9 presents dust emissions for a recuperative tank furnace fired by natural gas and melting a lead crystal glass with a PbO content of about 26%. As well as changes in dust content with time the Figure also shows changes in crown temperature and waste gas volume calculated on the basis of an 8% O_2 content: for clarity, the individual points are connected by lines. It should be noted that the maximum dust emission does not occur during batch feeding but during blocking. The limiting factor governing evaporation is not volatilisation from the molten glass into the atmosphere but the continuing supply of material by diffusion and by the currents in the molten glass.

Emissions from a potash crystal glass have a different pattern, as may be seen in Figure 13.10: the levels of the emissions are themselves much lower and maxima occur during batch melting and not during refining or blocking.

Experience to date indicates that the type and quantity of dust emitted depend strongly on the glass composition, the raw materials, the fuel, and the design and operation of the furnaces concerned. Table 13.2 gives analyses of dusts emitted from a number of furnaces; the compounds are not given as such but as oxides because, although it is possible to work backwards towards the compounds, this calls for assumptions and is linked to further guesswork. The container glass furnaces are similar and use the same batch, one being fired with type S fuel oil and the other with natural gas. The furnace fired with fuel oil gives higher emissions of CaO, MgO, and SO_3 but there is no explanation for the higher CaO and MgO. In the case of the furnace fired with natural gas, part of the chlorine introduced in the soda ash is evolved as NaCl and appears in the dust; with oil firing the equilibrium moves strongly towards HCl and the chlorine is emitted as gas with only traces of chloride appearing in the dust. Borate glasses, especially alkali

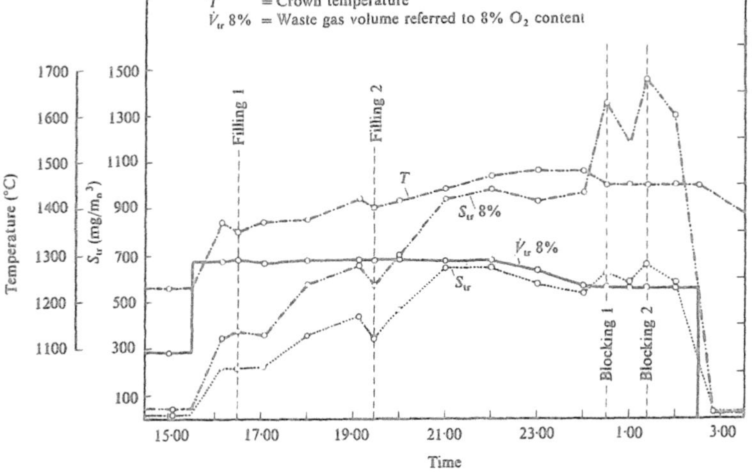

Figure 13.9. Dust emitted from a pot furnace melting lead crystal glass

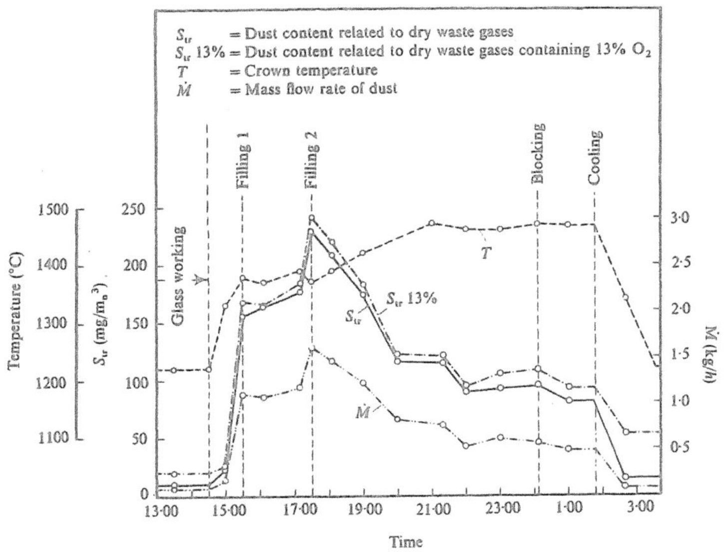

Figure 13.10. Dust emitted from a pot furnace melting a potash crystal glass

Table 13.2. *Analyses (wt%) of dusts emitted from various glass tank furnaces*

	Containers (soda–lime glass) S type fuel oil	*Natural gas*	*Glass fibre insulation (alkali borosilicate glass) EL type fuel oil/Natural gas*	*Laboratory ware (borosilicate glass) EL type fuel oil/Natural gas*
Na_2O	26·1	38·7	25·1	32·6
K_2O	2·0	2·5	12·2	7·7
CaO	8·9	1·7	0·08	–
MgO	4·6	0·6	0·03	–
SiO_2	3·75	2·95	0·03	0·05
Al_2O_3	0·45	1·0	<0·01	0·02
SO_3	50·0	46·9	16·3	3·8
F^-	<0·02	0·09	n.d.	n.d.
Cl^-	<0·05	4·7	7·1	34·0
V_2O_5	2·4	–	n.d.	n.d.
B_2O_3	–	–	34·4	20·1

Table 13.3. *Chemical composition (wt%) of dust from a tank furnace fired by natural gas and melting a soda–lime glass*[9]

Na_2SO_4	89·2
K_2SO_4	2·2
$CaSO_4$	6·3
$NaHSO_4$	1·5
V_2O_5	0·8

borate glasses, give a remarkable high chlorine emission due to the use of NaCl as a fining agent.

A lead crystal glass which was melted under practically sulphur-free conditions gave dust of lead carbonate instead of lead sulphate.

The composition of the dust in terms of individual compounds is available for one single tank furnace melting a soda–lime glass and fired with natural gas; as itemised in Table 13.3 over 90% of the dust consists of alkali sulphates. For a furnace fired with heavy fuel oil, a higher proportion of alkaline earth sulphate and sodium bisulphate must be expected.

In most cases the dust is very fine, with an average particle size of less than 0·5 μm.

13.6.2 Harmful gases

Among the gases emitted from glass tank furnaces the first to be mentioned must be sulphur oxides which originate in the sulphur content of the fuel oil and the sulphate content of the batch. Glass furnaces fired by heavy fuel oil containing about 1·8% S give rise to SO_2 concentrations in the waste gases of between 500 and 1000 ppm, which corresponds to 1465 to 2930 mg/m^3 at NTP. The variation in SO_2 concentration with time is shown for a regenerative furnace in Figure 13.11; the concentration is not constant but, starting from each reversal, increases to the end of the firing period. A possible

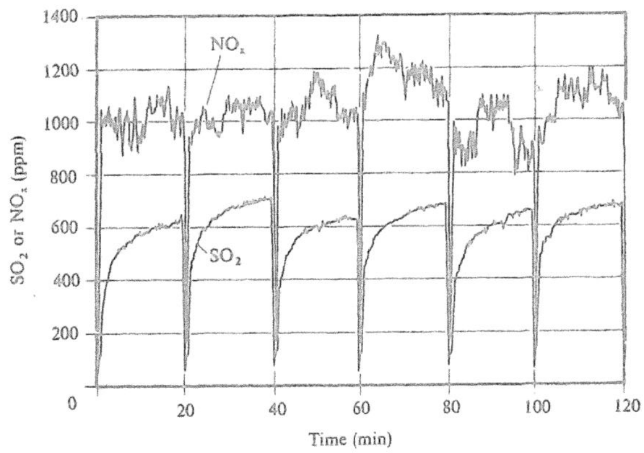

Figure 13.11. Variation with time of SO_2 and NO_x concentrations in the waste gases of a regeneratively fired U shaped flame tank furnace[9]

explanation is that increasing quantities of batch gases, and also therefore of SO_2, are liberated as the temperature increases during a firing period: it is also possible that temperature changes in regenerator chambers are responsible. According to measurements carried out so far, the SO_3 content is about 5–11% of the total sulphur oxides, considerably higher than that found for boilers; this is due in part to the excess air necessary for under-port firing and in part due to catalytic effects.

The next most important gaseous emission from the point of view of the environment is NO_x. This depends strongly on the temperature to which the combustion air has been preheated and on flame temperature; the presence of sodium nitrate has also been shown to be a contributary factor. The cyclic change of NO_x concentration in the waste gases from a regeneratively fired U shaped flame furnace is shown in Figure 13.11.

If special glasses are ignored, the next most important emission is that of chlorides. Chlorine enters the furnace via soda ash and sodium sulphate; even the better qualities of these materials contain up to 0·5% NaCl as impurity and poorer qualities contain multiples of this. As NaCl volatilises it reacts with steam and sulphur dioxide to form hydrogen chloride, which leaves the system as a gas; any remaining NaCl is emitted as dust. A high concentration of SO_2 in the waste gases, as is the case with oil fired furnaces, moves the equilibrium towards HCl, a fact demonstrated by Gebhardt & Carduck.[10]

13.7 Visibility of waste gas plumes[11]

This depends primarily on the operating and atmospheric conditions. Waste gas plumes are visible when their contrast lies above the contrast threshold; if the brightness of a plume is B_p and that of the background B_0, then the contrast, C_p, can be defined by

$$C_p = \frac{B_p - B_0}{B_0}. \tag{13.2}$$

The value of C_p may be positive or negative and a plume is judged to be invisible when $C_p < 0{\cdot}003$.

Glass furnaces give rise to waste gases which, on leaving the chimney, contain solid materials, mainly sulphates which have a particle diameter of about 0·3 µm. The cooling by the ambient air can cause liquid droplets to condense from the SO_3/H_2O content of the waste gases: these droplets are deposited on the solid particles and form an aerosol with a diameter of about 1·0 µm, which is the main cause of the visibility of the waste gas plume. The factors surrounding the formation of aerosols are explained in greater detail by the following example.

The waste gases leave the chimney at a temperature of about 350°C and mixing with the air leads to dilution and cooling; cooling starts on the borders of the plume and proceeds inwards while dilution reduces the SO_3 content in the waste gases. A definite relationship exists between dew point temperature and the condensation of SO_3; this is shown in Figure 13.12 as curve 1, which is represen-

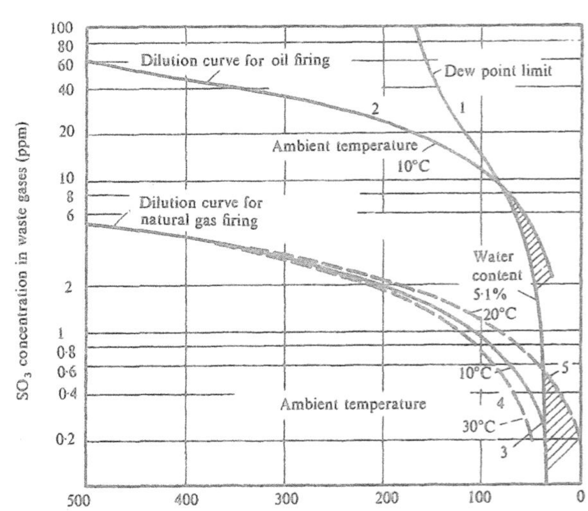

Figure 13.12. Dilution, cooling, and dew point limits of waste gases from glass tank furnaces[11]

tative of a group of curves for different water vapour contents in the atmosphere. To the right of this curve there is condensation and therefore formation of aerosols.

From the point of view of the waste gases it is necessary to distinguish between oil firing and natural gas firing which differ, in the main, in their SO_2 concentration. If representative dilution curves are drawn, then firing with a fuel oil containing 1·8% S at an ambient temperature of 10°C gives curve 2 of Figure 13.12 and firing with sulphur-free natural gas gives curve 3, 4, or 5, depending on the air temperature. Aerosols are formed where these curves cross curve 1. The Figure indicates that a plume would be formed with natural gas firing only under winter conditions, a fact borne out by observation.

13.8 Equipment for the reduction of emissions

In recent years concern about the environment has led the authorities to issue regulations which place obligations on industry to control and reduce dust emissions; this has led to the development and installation of filtration plants. Experience to date suggests that each installation must be examined on its merits: glass composition, the fuel available, and furnace design all influence emissions and therefore affect the decision as to whether a filtration plant is needed and, if so, which filtration system is the most suitable.

13.8.1 Fabric filters

Soda–lime glasses such as are used for containers and flat glass give rise, when fired with natural gas, to a dry dust containing substantial quantities of sulphate which can be retained by fabric filters. The waste gases flow through filter bags, which are usually supported by baskets and suspended from above, and the clean gases are drawn into the clean waste gas duct by a suction fan and then pass into the chimney flue. The loading of the filter fabric varies widely but usually lies between 20 and 100 m^3/m^2h.

The life of the filter bags is determined by a maximum temperature which must not be exceeded and the dew point to which the temperature must not be allowed to fall; this calls for careful thermal conditioning of the waste gases by coolers or similar equipment. A diagram of an installation for the removal of waste gas dust is shown in Figure 13.13; in this example the waste gases are cooled by an air cooler but there are alternatives such as the addition of cold air or the use of water jets. The filter bags are arranged in groups, each group containing up to 100 bags, and below them are collecting funnels with rotary valves.

In practice, the best textile material for filtration has been found to be needled felts with fabric support. Polyacrylonitrile and polyester can be used for operating temperatures up to 120–130°C, up to 160°C Nomex is recommended, and for temperatures up to 180°C the rather expensive PTFE. Above this temperature it becomes necessary to use glass fibre fabrics which involve careful attention to the design and operation of equipment to achieve satisfactory performance. Provided the correct material is used and the construction is sound, one set should perform satisfactorily for a period of 18 months or more.

It is important that the filter bags be cleaned regularly. This is done in groups of 8 to 10 by controlled impulses of compressed air in a direction opposite to that in which the waste gases pass. The cleaning is undertaken every 30 or 45 min and the

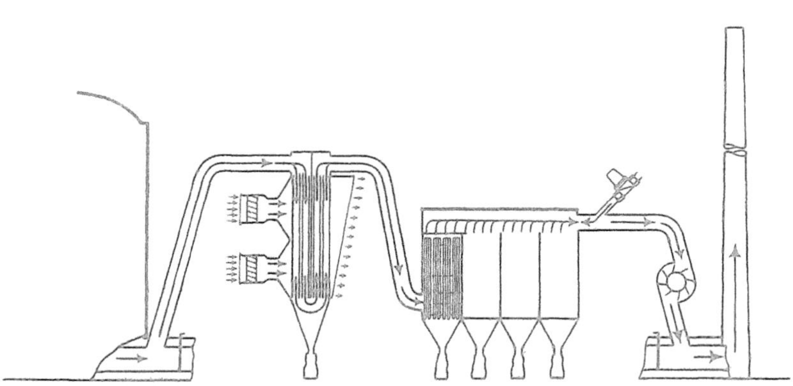

Figure 13.13. Installation for removing dust from the waste gases by fabric filters

actual frequency can be controlled automatically via the pressure drop through the filter; for example, if the pressure drop exceeds 14 mbar, then the cleaning process begins automatically. The bags are rapidly inflated against the direction of the waste gas flow and each inflation and cleaning takes about 5 s. The flow resistance through the filter increases slowly with time from 10 mbar with a new filter to 12 mbar with an old filter. To avoid any interference with the operation of the furnace it is necessary to control the negative pressure downstream from the regenerator chambers by suitable control. The efficiency with which filters can be cleaned is very high and values of 99·7 to 99·8% can be reached.

No general guidance can be given concerning the suitability of the dust collected after filtration for recycling with the batch into the glass making process; if the furnace is fired by natural gas then this is possible. The criterion is the sulphur content and, in some special cases, the chrome and heavy metal contents; careful attention must be paid to these. If not controlled, the sulphur can lead to an unacceptable increase in the SO_3 content of the molten glass, leading to a tendency for it to reboil.

Filtration equipment of the type described has been successful when associated with furnaces fired by natural gas and when light fuel oil is used the same equipment can be used. More care is necessary when firing with heavy fuel oil due to the high acid dew points that this brings in its wake; in this case additional equipment must be provided (see Section 13.8.3).

13.8.2 Electrostatic filters

This type of filter has been used successfully in many plants, the major advantage of this technique being that it can be used for waste gas temperatures up to 250°C.

The operation of electrostatic filters depends on a corona discharge at ionising electrodes and this causes the incoming dust particles to become electrically charged and move to the earthed precipitation electrodes. The electrical resistance of the dust to be removed is important and must lie within a range suitable for this process; based on experience this is less than 10^{11} Ω cm. If the electrical resistance is too high, then the discharge at the precipitation electrode is too slow and further quantities of charged dust arrive on top of the dust which has already been deposited, but not yet fully discharged. In time a layer of dust is created which has its own potential gradient and this leads to a reduction in voltage between the ionising electrode and the uppermost layer of dust, thus reducing the efficiency of the filter.

Nearly all dusts which are to be collected in electrostatic filters, including those originating from glass tank furnaces, have an electrical resistance which varies with temperature. At low temperatures the resistance decreases due to residual moisture being condensed in the pores of the dust; as the temperature increases this moisture evaporates and the resistance increases; finally, after passing through a maximum, the resistance again decreases (Figure 13.14). In view of the importance of the electrical resistance of the dust it is best to determine this before planning and designing the equipment. The data in Figure 13.14 apply to dust from an alkali borosilicate tank furnace and although other glasses may give different values, the basic characteristic of passing through a maximum remains.

The design of the plant depends on an important relationship established by Brand[16] which links the degree of precipitation, Y, the volume flow, $\dot{V}$, the precipitation area, A, and the speed of migration of the dust particles, w, in the electric field:

$$Y = 1 - \exp\left(-\frac{Aw}{\dot{V}}\right).$$

The speed of migration is important. The electrical resistance of the dust is not related to it.

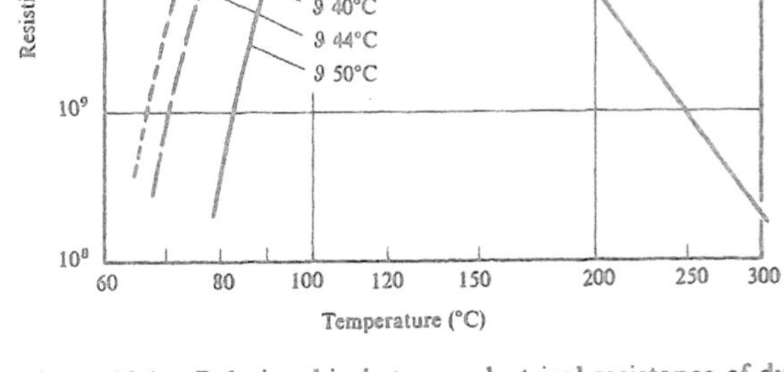

Figure 13.14. Relationship between electrical resistance of dust from an alkali borosilicate glass tank as a function of temperature and humidity[13]

Among the different types of electrostatic filters available, the most suitable is a system of parallel plates with thin discharge electrodes in the path of the gases flowing between the plates (Figure 13.15). This gives a very homogeneous electrical field and the passages between the plates can be narrow, allowing small compact filtration units to be used. If the passage width is s and the volume of the filter is V, then the above formula can be rearranged to give

$$Y = 1 - \exp\left(-\frac{2Vw}{s\dot{V}}\right)$$

from which it may readily be seen that for a given $\dot{V}$ and a given w, the volume, V, becomes smaller as s becomes smaller; in addition, the filtration efficiency depends on the throughput, $\dot{V}$. To date, electrostatic filters have been installed alongside tank furnaces melting alkali borosilicate and lead glasses and have performed satisfactorily; in both cases we are dealing with dry electrostatic filters, the classical discharging wire system being used for the borosilicate furnace and the plate system for the lead glass furnace. Both were fired with natural gas, which largely eliminates corrosion problems as SO_3 is virtually absent from the waste gases.

A diagram showing the dust filtration equipment downstream from the alkali borosilicate furnace is shown in Figure 13.16. Among the advantages of this type of equipment are the small pressure loss, no risk of blockage, the ability to recycle the dust collected, high filtering efficiency, and low operating and maintenance costs.

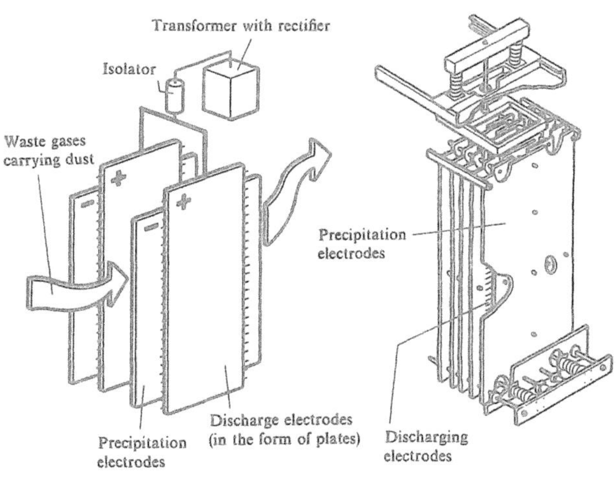

Figure 13.15. Diagram of a Napco filter[15]

13.8.3 Filtration equipment using dry adsorption[17]

Not all dusts in the waste gases from glass furnaces can be readily removed by fabric or electrostatic filters, particularly if heavy fuel oil with a high sulphur content has to be used. A further disadvantage of both systems is that harmful gases such as HF, HCl, SO_3, and SO_2 cannot be retained by them and for these materials combined filter systems including dry sorption must be used.

Dry sorption is a technique by which certain additives are made to react with the dust and gases to convert them into a filterable condition so that the additives with their adsorbents may be subsequently removed in fabric or electrostatic filters. Additives can be slaked lime, $Ca(OH)_2$, dolomite, $CaCO_3.MgCO_3$, or hydrated alumina $Al(OH)_3$. Investigations have shown that hydrated alumina is particularly suitable for adsorbing lead and boron compounds while the harmful gases such as HF, HCl, and SO_3 are better adsorbed by slaked lime. What to do with the collected dust, e.g. whether it can be recycled, must be decided for each specific

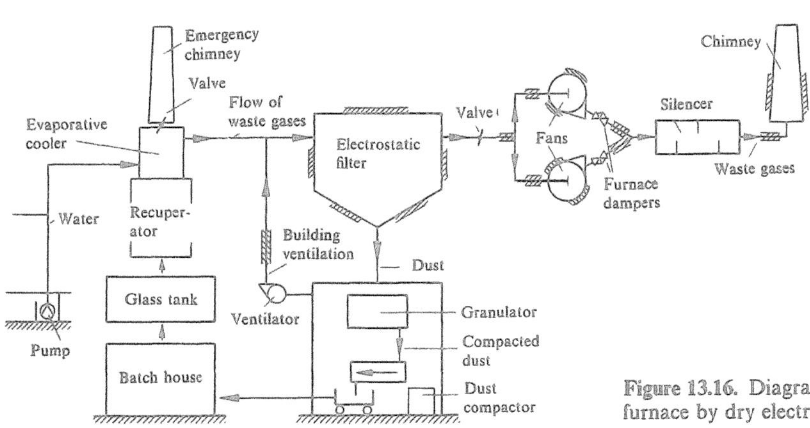

Figure 13.16. Diagram of the removal of dust from a glass tank furnace by dry electrostatic filters[13]

case. It is important to distribute the additives uniformly in the stream of waste gases and the residence time has to be varied according to the concentration of harmful gases; 3 to 10 s is regarded as adequate but if the required times cannot be achieved or if more must be allowed, then an external return line must be installed. The additive must be sufficiently fine. In addition, it is desirable to create new reactive surfaces in the additive material already in the system by installing ball mills or similar equipment.

The additives can also be introduced with water, but this calls for a waste gas temperature which is sufficiently high to ensure evaporation of the water without it making contact with the walls.

A schematic layout including ball mill, heat exchanger, and filter is shown in Figure 13.17. The addition of the ball mill makes it possible to extend the life of the additives considerably, without external recycling, and to feed a continuous supply of finely divided material into the stream of waste gases. There is no need to allow for a special zone for the sorption reaction.

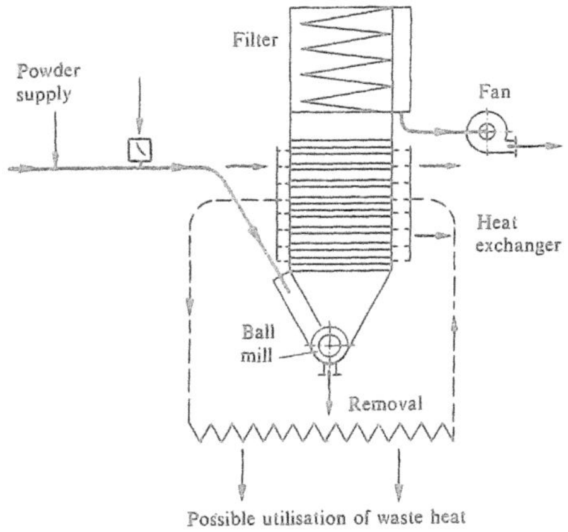

Figure 13.17. Equipment for the feeding of dry powder into a sorption plant[16]

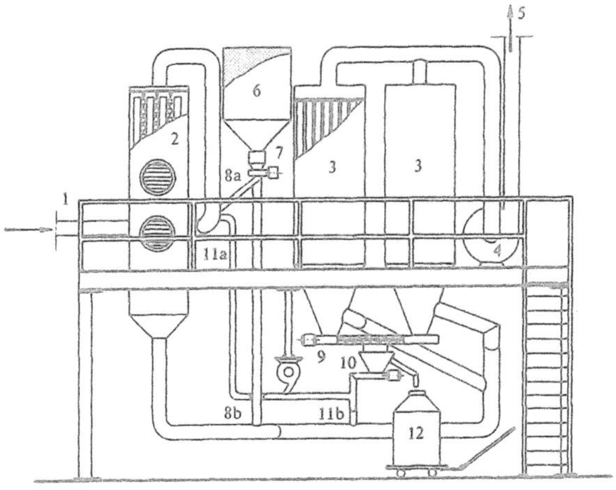

Figure 13.18. Dust removal plant using hydrated alumina as adsorbent[14]

(1)	waste gas entry	(8a)	before cooler
(2)	cooler	(8b)	after cooler
(3)	filter	(9)	screw feeder and removal valve
(4)	fans	(10)	dosing unit with overflow
(5)	cleaned waste gas exit	(11)	recirculation of adsorbent
(6)	silo	(11a)	before cooler
(7)	dosing equipment	(11b)	after cooler
(8)	feeds for adsorbent	(12)	dust collector for removal

Equipment for the adsorption and subsequent filtration of dust containing lead and boron compounds from the waste gases of a tank furnace is shown in Figure 13.18. As the furnace is fired with natural gas, the sulphur content of the waste gases is very small. The dust collected can be recycled back into the batch without any problem and they can both be fed into the furnace together.

Developments in the field of dust removal from glass plants are making rapid progress and the processes cited here are just a few examples of this developing technology.

References

References for Chapters 1 and 2

1. Löffler, J. (1965). *Glastech. Ber.* **38**, 269.
2. Steinmann, F. (1870). *Polytech. Centralbl.* **36**, 1114.
3. Stein, G. (1954). In *Glasschmelz-Wannenöfen.* By R. Günther. Verlag d. Dtsch. Glastech. Ges., Frankfurt. Pp. 200–16.
4. Günther, R. (1953). *FachausschBer. dt. glastech. Ges.* Nr. 49.
5. Trier, W., Lauter, J. & Schumacher, L. (1978). *FachausschBer. dt. glastech. Ges.* Nr. 71.
6. Swicker, V. C. (1939). *Am. Ceram. Soc. Bull.* **18**, 143.
7. Schumacher, L. (1981). *Glastech. Ber.* **54**, R151.
8. Gebhardt, F. (1975). *Habil. Schrift*, TH Aachen.

References for Chapter 3

1. Krings, A. (1972). *Glastech. Ber.* **45**, 467.
2. Hoff, G. (1976). *Glass* **53**, 305, 308.
3. Jeanvoine P., Guigonis, J. & Krings, A. (1980). *Glastech. Ber.* **53**, 89.
4. Meister, R. (1977). *Glastech. Ber.* **50**, 186.
5. Schmidt, K. (1959). *Glastech. Ber.* **32**, 217.
6. Kleinheisterkamp, L. (1972). In Hüttentechnische Vereinigung der Deutschen Glasindustrie Kurs 1972. *Heiß-und-Zwischenreparaturen an Glaschmelzöfen.* III. Hüttentech. Verein. d. Dtsch. Glasind., Frankfurt (Main). Pp. 35–51.
7. Krings, A., Liehn, W. & Meister, R. (1973). *FachausschBer. dt. glastech. Ges.* Nr. 69.
8. Anon. (1975). *Glass* **52**, 194, 196, 197, 199.
9. Brüning, R. (1972). *Glastech. Ber.* **45**, 439.
10. Daudans, J. (1972). *Glastech. Ber.* **45**, 433.
11. Meister, R. (1960). *Glastech. Ber.* **33**, 182.
12. Liehn, W. (1972). *Glastech. Ber.* **45**, 444.
13. Hufen, W. (1965). *Glastech. Ber.* **38**, 33.
14. Hufen, W., Dämmer, H. & Liehn, W. (1976). *Glastech. Ber.* **49**, R24.
15. Neumann, G. (1942). *Arch. EisenhüttWes.* **15**, 437.
16. Lepère, K. E. (1970). *Tonindustriezeitung.* **94**, 56.
17. Heiligenstaedt, W. (1966). *Wärmetechnische Rechnungen für Industrieöfen.* Fourth Edition. Verlag Stahleisen, Düsseldorf. Pp. 263–70.
18. Konopicky, K. & Steinhoff, E. (1968). *FachausschBer. dt. glastech. Ges.* Nr. 66.
19. Glaser, W. & Boggum, P. (1966). *Sprechsaal* **99**, 211, 281, 284, 288, 290, 352, 356, 358, 362, 365.
20. Buder, W. (1975). *Glastech. Ber.* **48**, R130.
21. McKenna, C. D., Garonne, A. F. & Teichmann, N. N. (1972). *Glastech. Ber.* **45**, 104.
22. Jammernegg, H. (1966). *Glastech. Ber.* **39**, 403.
23. Hausen, H. (1976). *Wärmeübertragung im Gegenstrom, Gleichstrom und Kreuzstrom.* Second Edition. Springer-Verlag, Berlin.
24. Delrieux, J. (1974). *Rev. Gen. Therm.* **13**, 383.
25. Greschat, K.-H. (1976). *Glastech. Ber.* **49**, R84.
26. Krings, A. (1976). *Glastech. Ber.* **49**, R84.
27. Trier, W. (1956). *Glastech. Ber.* **29**, 432.
28. Eck, B. (1981). *Technische Strömungslehre.* Eighth Edition. Vol. 2. Springer-Verlag, Berlin. Pp. 61–2.
29. Reichardt, H. & Tollmien, W. (1952). *Mitt. Max-Planck-Inst. f. Strömungsforsch.* Nr. 7.
30. Veit, P. (1953). *Arch. EisenhüttWes.* **24**, 93.
31. Kröll, K. (1954). *Chem. Ingr Tech.* **26**, 132.
32. Hansen, M. (1952). *Arch. EisenhüttWes.* **23**, 245.
33. Schleicher, S. (1927). *Ber. Stahlw. Aussch. VDEh.* Nr. 123.
34. De Graaf, J. E. & van Stein-Callenfels, G. W. (1953). *4th Congr. Int. du Chauffage Industriel*, Paris 1952. *Chal. Ind.* Gr. IV/41, (197), 1.
35. Hansen, M. (1955). *Stahl Eisen* **75**, 401.
36. Langhans, W. U. (1962). *Arch. EisenhüttWes.* **33**, 347, 441.
37. Yazicizade, A. Y. (1966). *Glastech. Ber.* **39**, 203.
38. Boggum, P. & Glaser, W. (1965). *Glas-Email-Keramo-Tech.* **16**, 213.
39. Glaser, W. & Simon, G. (1967). *Sprechsaal* **100**, 834, 837.
40. Teisen, Th. (1958). *Glastech. Ber.* **31**, 349.
41. Biegler, F. (1964). *Glastech. Ber.* **37**, 286.
42. Anon. (1943). *Stahl Eisen* **63**, 286.
43. Dietz, K. (1969). *FachausschBer. dt. glastech. Ges.* Nr. 67.
43a. Axmann, M. (1962). *Glastech. Ber.* **35**, 321.
44. Hampe, E. (1970). *Industrieschornsteine.* VEB-Verlag f. Bauwesen, Berlin.
45. Will, H. (1965). *Glastech. Ber.* **38**, 33.
46. Wellner, A. (1966). *Glastech. Ber.* **39**, 315.
47. Greschat, K.-H. (1960). *Glastech. Ber.* **33**, 370.
48. Hoff, G., Krings, A. & Recasens, J. (1972). *Glas-Email-Keramo-Tech.* **23**, 368.
49. Preston, E. (1956). *Proc. 4th Int. Congr. du Verre*, Paris.
50. Knoblauch, H. (1964). *Glastech. Ber.* **37**, 281.

References for Chapter 4

1. Majdič, A. (1974). *Keram. Z.* **26**, 574.
2. Routschka, G. & Majdič, A. (1972). *Glas-Email-Keramo-Tech.* **23**, 349, 389, 430.
3. Verworner, A. & Berndt, K. (1978). *Feuerfeste Baustoffe für Glasschmelzanlagen.* VEB Dtsch. Verlag f. Grundstoffindustrie, Leipzig.
4. Krings, A., Liehn, W. & Meister, R. (1973). *FachausschBer. dt. glastech. Ges.* Nr. 69.
5. Konopicky, K. & Steinhoff, E. (1968). *FachausschBer. dt. glastech. Ges.* Nr. 66.
6. Verworner, O. (1979). *Silikattechnik* **30**, 213.
7. Konopicky, K., Lepère, K. E., Routschka, G. & Thoenes, H. W. (1968). *Tonindustriezeitung* **92**, 41.
8. Ehrcke, U. & Schwiete, H.-E. (1963). *Arch. EisenhüttWes.* **34**, 795.
9. Morsanyi, A. V. (1966). *Glass Technol.* **7** (6), 193, 196.
10. Konopicky, K. & Routschka, G. (1965). *Glastech. Ber.* **38**, 450.
11. Metzger, K. & Günther, R. (1952). *FachausschBer. dt. glastech. Ges.* Nr. 47.
12. Liehn, W. (1964). *Glastech. Ber.* **37**, 301.
13. Hoff, G., Krings, A. & Recasens, J. (1972). *Glas-Email-Keramo-Tech.* **23**, 368.
14. Meyer, H. & Poehnitzsch, H. (1965). *Glastech. Ber.* **38**, 393.
15. Riedel, L. (1956). *Glastech. Ber.* **29**, 49.
16. Daudans, J. (1972). *Glastech. Ber.* **45**, 433.
17. Recasens, J., Sevin, A. & Gardiol, M. (1969). *Verres Réfract.* **23**, 19.
18. Recasens, J. & Gardiol, M. (1978). *Glastech. Ber.* **51**, 212.
19. Claus, H. & Salge, H. (1974). *Glastech. Ber.* **47**, 159, 173.
20. Boggum, P. P. (1978). *Glastech. Ber.* **51**, 303.
21. Daudans, J., Krings, A. & Sevin, A. (1969). *Glastech. Ber.* **42**, 391.
22. Gebhardt, F. & Horn, H. (1971). *Glastech. Ber.* **44**, 338.
23. Roedl, W. (1966). *Glass* **43**, 308, 315.
24. Brunnsteiner, F. (1960). *Glastech. Ber.* **33**, 165.
25. Gebhardt, F. & Horn, H. (1973). *Glastech. Ber.* **46**, 129.
26. Barthel, H. (1973). *Glastech. Ber.* **46**, 134.

27. Naefe, H. & Schulte, K. (1974). *Sprechsaal* **107**, 883.
28. de Geyndt, E. (1975). *Review of the literature on the problems of glass furnace regenerators.* Sub-Committee A XI of the International Commission on Glass, Charleroi.
29. Robijn, P. (1973). *Glastech. Ber.* **46**, 148.
30. Gebhardt, F. (1976). *Glastech. Ber.* **49**, 53.
31. Yamaguchi, A. (1971). *J. Ceram. Soc. Japan* **79**, 124.
32. Rasch, R. (1970). *Glas-Email-Keramo-Tech.* **21**, 81.
33. Konopicky, K., Routschka, G. & Lohre, W. (1965). *Glas-Email-Keramo-Tech.* **16**, 209.
34. Mörtl, G., Zednicek, W., Adding, A. & Rademacher, S. C. (1973). *Glastech. Ber.* **46**, 141.
35. Krings, A., Gardiol, M. & Roux, J.-M. (1978). *Glastech. Ber.* **51**, 203.

References for Chapter 5

1. Trier, W. (1963). *Glastech. Ber.* **36**, 73.
2. Trier, W. & Voss, H.-J. (1968). *Glastech. Ber.* **41**, 217.
3. Heiligenstaedt, W. (1951). *Wärmetechnische Rechnungen für Industrieöfen.* Third Edition. Verlag Stahleisen, Düsseldorf. Pp. 112–17.
4. Gailhbaud, J. (1959). *Glastech. Ber.* **32K**, II/49.
5. Schulte-Kellinghaus, H. (1979). *Glastech. Ber.* **52**, 211.
6. Kröger, C. (1953). *Glastech. Ber.* **26**, 202.
7. Lubisch, G. & Trier, W. (1979). *Glastech. Ber.* **52**, 141.
8. Trier, W. & Werner, T. (1964). *Glastech. Ber.* **47**, 79.
9. Bošnjaković, F. (1959). *Glastech. Ber.* **32**, 41.
10. Huhmann-Kotz, I. (1959). *Glastech. Ber.* **32**, 47.
11. Meister, R., Weber-Klein, P. & Wefer, F. (1955). *FachausschBer. dt. glastech. Ges.* Nr. 65.
12. Weber-Klein, P., Dietz, K. & Ehrich, W. (1964). *FachausschBer. dt. glastech. Ges.* Nr. 63.
13. Stein, F. (1967). *Glastech. Ber.* **40**, 217.
14. Thies, G. (1973). *Glastech. Ber.* **46**, 231.
15. Boggum, P. P. (1980). *Glastech. Ber.* **53**, 102.
16. Brüning, R. (1972). *Glastech. Ber.* **45**, 439.
17. Hrma, P., Blažek, J. & Pelantová, V. (1981). *Glass Technol.* **22** (5), 215.
18. *VDI-Wärmeatlas.* (1983). Fourth Edition. VDI-Verlag, Düsseldorf.
19. Schack, A. (1962). *Der industrielle Wärmeübergang.* Sixth Edition. Verlag Stahleisen, Düsseldorf.
20. Routschka, G. & Majdič, A. (1981). *Keram. Z.* 33, 511.
21. Michaud, M. (1954). *Silic. Ind.* **19**, 243.
22. Glaser, W. (1972). *Glastech. Ber.* **45**, 453.
23. Heim, Th. (1951). *Glastech. Ber.* **24**, 127.
24. Meister, R. (1960). *Glastech. Ber.* **33**, 182.
25. Daudans, J. (1972). *Glastech. Ber.* **45**, 433.
26. Lockwood, F. P. (1971). *Glass Technol.* **12** (1), 8.
27. Fischer, E. (1961). *Glastech. Ber.* **34**, 219.
28. Heimke, G. (1979). *Glastech. Ber.* **52**, 63.
29. Ebeling, M. (1962). *Glastech. Ber.* **35**, 312.
30. Thies, G. (1962). *Glastech. Ber.* **35**, 307.
31. Bürklin, H. (1962). *Glastech. Ber.* **35**, 316.
32. Weber-Klein, P. (1959). *Glastech. Ber.* **32**, 107, 158.
33. Greschat, K.-H. (1968). *Glastech. Ber.* **41**, 81.
34. Dietz, K. (1969). *FachausschBer. dt. glastech. Ges.* Nr. 67.
35. Kalpers, H. (1943). *Glashütte* **73**, 122.
36. Stein, G. (1954). In *Glasschmelz-Wannenöfen.* By R. Günther. Verlag d. Dtsch. Glastech. Ges., Frankfurt. Pp. 200–15.
37. Schiffer, K. (1950). *Glastech. Ber.* **23**, 283.
38. Günther, R. (1954). *Glastech. Ber.* **27**, 31.
39. Giegerich, W. (1954). *Glastech. Ber.* **27**, 31.
40. Schiffer, K. & Recker, J. (1958). *Glastech. Ber.* **31**, 454.
41. Metzger, K. & Günther, R. (1952). *FachausschBer. dt. glastech. Ges.* Nr. 47.
42. Greschat, K.-H. (1960). *Glastech. Ber.* **33**, 370.
43. Sonntag, G. (1960). *Glastech. Ber.* **33**, 363.
43a. Hargreaves, J. (1958). *Trans. Br. Ceram. Soc.* **57**, 242–57.
44. Meister, R. (1954). *Glastech. Ber.* **27**, 30.
45. Liehn, W. (1961). *Glastech. Ber.* **34**, 590.
46. Dämmer, H. (1973). *Glastech. Ber.* **46**, R112.
47. Greschat, K.-H. (1973). *Glastech. Ber.* **46**, R113.
48. Dämmer, H. & Kuhlmann, H. (1978). *Glastech. Ber.* **51**, 271.
49. Close, L. (1979). *Glass* **56** (5), 159, 160, 162.
50. Glaser, W. & Boggum, P. (1966). *Sprechsaal* **99**, 698, 701, 704, 735, 738, 741, 744, 796, 798, 800, 803, 806, 809, 811, 813, 911, 914, 918.
51. Anon. (1975). *Boln Soc. esp. Cerám. Vidrio* **14**, 453, 529.
52. Kleinheisterkamp, L. (1972). In Hüttentechnische Vereinigung der Deutschen Glasindustrie Kurs 1972. *Heiß-und-Zwischenreparaturen an Glaschmelzöfen,* III. Hüttentech. Verein. d. Dtsch. Glasind., Frankfurt (Main). Pp. 35–51.
53. Costa, P. (1979). *Glastech. Ber.* **52**, R71.
54. Axmann, M. (1962). *Glastech. Ber.* **35**, 321.
55. Weber-Klein, P. (1962). *Sprechsaal* **2**, a83.
56. Stocker, B. (1954). *Glastech. Ber.* **27**, 290.

References for Chapter 6

1. Ullrich, J. (1963). *Glastech. Ber.* **36**, 121, 170.
2. Trier, W. & Staatsmann, H. (1968). *Glastech. Ber.* **41**, 537.
3. Brüning, R. (1968). *Glastech. Ber.* **41**, 277, 326, 408; (1971). *ibid.* **44**, 56.
4. Günther, R. (1974). *Verbrennung und Feuerungen.* Springer-Verlag, Berlin. P. 135.
5. Günther, R. (1974). *Chem. Ingr Tech.* **46**, 56.
6. Günther, R. (1973). *Glastech. Ber.* **46**, 92.
6a. Craya, A. & Curtet, R. (1955). *C.r. hedb. Séanc. Acad. Sci., Paris* **241**, 621–2.
7. Eickhoff, H. (1971). *Verfahrenstechnik* **5**, 118.
8. Günther, R. (1977). *Chem. Ingr Tech.* **49**, 135.
9. Patrick, M. A. (1967). *J. Instn Fuel* **40**, 425.
10. Voss, H.-J. (1975). *Glastech. Ber.* **48**, 190.
10a. Elgeti, K. (1962). *Brennst–Wärme–Kraft* **14**, 1–6.
11. Hottel, H. C. & Egbert, R. B. (1941). *Trans. ASME* **63**, 297.
12. Schack, A. (1969). *Der industrielle Wärmeübergang.* Seventh Edition. Verlag Stahleisen, Düsseldorf.
13. Hausen, H. (1976). *Wärmeübertragung im Gegenstrom Gleichstrom und Kreuzstrom.* Second Edition. Springer-Verlag, Berlin.
14. Port, F. J. (1939). *D. Thesis,* Massachusetts Institute of Technology.
15. Hein, K. (1970). *Glastech. Ber.* **43**, 455.
16. Michelfelder, S. (1977). *Glastech. Ber.* **50**, 193.
17. Germerdonk, R. (1963). *Glastech. Ber.* **36**, 86, 158.
18. Trier, W. & Stein, A. (1965). *Glastech. Ber.* **38**, 353.
19. Beér, J. M. & Chigier, N. A. (1968). *Combust. Flame* **12**, 575.
20. Meister, R. (1968). *Glastech. Ber.* **41**, 557.
21. Meister, R. (1972). *Glastech. Ber.* **45**, 251.
22. Daniels, M. (1973). *Glastech. Ber.* **46**, 40.
23. Jack, H. R. S. & Jacquest, J. A. T. (1958). *Proc. Symp. sur la Fusion du Verre,* Bruxelles. Union Scientifuqe Continentale du Verre, Charleroi. Pp. 339–60.
24. Pugh, A. C. P. (1968). *Glastek. Tidskr.* **23**, 95.
25. Vilnis, K. K. & Stepanenko, M. G. (1959). *Glass Ceram.* **16**, 133.
26. Costa, P. (1977). *Glastech. Ber.* **50**, 10.
27. Bansal, B., Jones, K., Stephan, P. M. & Schorr, J. R. (1979). *Glass Ind.* **60** (7), 12, 26, 29.
28. Costa, P. (1977). *Glastech. Ber.* **50**, 301.
29. Vortmeyer, D. & Börner, C. J. (1966). *Chem. Ingr Tech.* **38**, 1077.
30. Kröger, C. & Eligehausen, H. (1959). *Glastech. Ber.* **32**, 362.
31. Fuhrmann, H. (1973). *Glastech. Ber.* **46**, 201.
32. Fuhrmann, H. (1973). *Glastech. Ber.* **46**, 209.
33. Landau, H. G. (1950). *Q. appl. Math.* **8**, 81.
34. Jeschar, E. & Millies, E. (1966). *Arch. EisenhüttWes.* **37**, 283.
35. Ellies, F. (1976). *Glastech. Ber.* **49**, 246.
36. Brinkman, R. J. (1979). *Glass* **56**, 243, 246.
37. Brichard, E. (1954). *Sillic. ind.* **19**, 207.
38. Pieper, H. (1979). *Glastech. Ber.* **52**, 229.
39. USP 4 036 625. 19 July 1977.
40. Meister, R. (1964). *Glastech. Ber.* **37**, 275.
41. Halle, R. (1947). *J. Soc. Glass Technol.* **31**, 122.
42. Trier, W. (1953). *Glastech. Ber.* **26**, 5.
43. Trier, W. (1960). *Glastech. Ber.* **33**, 401.
44. Sawai, J. & Takahashi, K. (1958). *Proc. Symp. sur la Fusion du Verre,* Bruxelles. Union Scientifique Continentale du Verre, Charleroi. Pp. 557–89.
45. Peschke, J. von. (1965). *Glastech. Ber.* **38**, 276.
46. Wright, S. & Rawson, H. (1973). *Glass Technol.* **14** (2), 42.
47. Noble, J. J., Clomburg, L. A. & Hottel, H. C. (1972). *Trans. ASME, C, J. Heat Transf.* **94**, 149.

48. Austin, M. J. & Bourne, D. E. (1973). *Glass Technol.* **14** (3), 78.
49. Mase, H. & Sasagawa, Y. (1973). In *Automatic control in glass.* Edited by R. J. Mouly. Instrument Society of America, Pittsburgh. Pp. 93–105.
50. Leyens, G. (1974). *Glastech. Ber.* **47**, 251, 261.
51. Barklage-Hilgefort, H., Mergler, K.-W. & Voss, H.-J. (1980). *Glastech. Ber.* **53**, 27.
52. Leyens, G. & Smrček, J. (1982). *Glastech. Ber.* **55**, 81.
53. Smrček, J. & Šimunek, J. (1980). *Sprechsaal* **113**, 331.
54. Hayakawa, M., Mishima, Y. & Kato, T. (1977). In *Proc. 11th Int. Congr. Glass*, Prague. ČVTS-Dum techniky, Prague. Pp. 33–42.
55. Štefan, J. & Skřivan, M. (1977). In *Proc. 11th Int. Congr. Glass*, Prague. ČVTS-Dum techniky, Prague. Pp. 43–53.
56. Chen, T.-S. & Goodson, R. E. (1972). *Glass Technol.* **13** (6), 161.
57. Riederer, A. (1977). In *Proc. 11th Int. Congr. Glass*, Prague. ČVTS-Dum tecknicky. Prague. Pp. 55–64.
58. Suzuki, J. (1976). *Ph.D. Thesis*, University of Sheffield.
59. Trier, W. (1975). *Glastech. Ber.* **48**, 181.
60. Leyens, G. & Moreau, R. (1978). *Glastech. Ber.* **51**, 43.
61. Greschat, K.-H. (1968). *Glastech. Ber.* **41**, 81.
62. Leyens, G., Smrček, J. & Thyn, J. (1980). *Glastech. Ber.* **53**, 124.
63. Pchelyakov, K. A., Astanin, V. I. & Sibiryakov, V. A. (1970). *Glass Ceram.* **27**, 507.
64. Safaieh, M.-R. (1968). *Glastech. Ber.* **41**, 87, 129.
65. Lennertz, H.-E. (1971). *Glastech. Ber.* **44**, 141, 177.
66. Mauerhoff, D. (1981). *Silikattechnik* **32**, 328.
67. Voss, H.-J. & Mergler, K.-W. (1969). *Glastech. Ber.* **42**, 396.
68. Meister, R. (1977). *Glastech. Ber.* **50**, 186.
69. Leyens, G. (1976). *Glastech. Ber.* **49**, 227.
70. Plumat, E. (1975). *Glastech. Ber.* **48**, 21.
71. Pavlovskii, G. V. & Voronin, M. D. (1976). *Glass Ceram.* **33**, 413.
72. Peychès, I. (1948). *Glass Ind.* **29**, 142, 143, 166.
73. Naruse, A. (1955). *J. Soc. Glass Technol.* **39**, 145T.
74. Naruse, A. (1959). *Glastech. Ber.* **32K**, IV/20.
75. Burley, D. M., Moult, A. & Rawson, H. (1978). *Glass Technol.* **19** (4), 86.
76. Staněk, J. (1977). *Electric melting of glass.* Elsevier, Amsterdam.
77. Schild, A. (1933). *Glastech. Ber.* **11**, 305.
78. Becker, H. (1960). *Glastech. Ber.* **33**, 411.
79. Becker, H. (1965). *Glastech. Ber.* **38**, 12.
80. Dertev, K. & Gorobovskii, V. A. (1959). *Steklo Keram.* **16** (1), 3.
81. Gräff, R. W. (1965). *Glastech. Ber.* **38**, 1.
82. Kopelev, V. E., Pollyak, V. V. & Fryazinova, G. N. (1971). *Glass Ceram.* **28**, 229.
83. Geffcken, W. (1957). *Glastech. Ber.* **30**, 143.
84. Cooper, A. R. Jnr (1966). *Chem. Engng Sci.* **21**, 1095.
85. Cooper, A. R. Jnr (1966). *Glass Technol.* **7** (1), 2.
86. Rhiel, F. F. (1976). *Glastech. Ber.* **49**, 217.
87. Rhiel, F. F. (1976). *Glastech. Ber.* **49**, 251.
88. Leyens, G., Rhiel, F. F. & Woelk, G. (1975). *Glastech. Ber.* **48**, 207.
89. Bishop, F. L. Jnr (1945). *J. Am. Ceram. Soc.* **28**, 308.
90. Bishop, F. L. Jnr (1950). *Am. Ceram. Soc. Bull.* **29**, 99.
91. Bishop, F. L. Jnr (1956). *Glass Ind.* **37**, 71, 99, 142.
92. Cooper, A. R. Jnr (1959). *J. Am. Ceram. Soc.* **42**, 93.
93. Hoetink, B. J. (1969). *Glass Technol.* **10** (3), 84.
94. Joosen, G. (1973). *Glastech. Ber.* **46**, 59.
95. Nölle, G. (1977). *Silikattechnik* **28**, 50.
96. Bamford, C. R. (1977). *Colour generation and control in glass.* Elsevier, Amsterdam. Pp. 122–40.
97. Moore, H. (1971). *Glass Ind.* **52**, 239.
98. Matz, R. (1980). *Glastech. Ber.* **53**, 37.
99. Langanke, B. (1981). *Glastech. Ber.* **54**, R42.
100. Lauter, J. (1981). *Glastech. Ber.* **54**, R151.
101. Tooley, F. V. (1974). *The handbook of glass manufacture.* Vol. 1. Ashlee Publishing Co., New York. P. 385.
102. Budd, S. M. (1965). *Symp. Gases in Glass.* 67th Annual Meeting of the American Ceramic Society, Philadelphia.
102a. Williams, H. (1980). *Glastech. Ber.* **53**, R170.
103. Manring, W. H., Billings, D. D., Conroy, A. R. & Bauer, W. C. (1967). *Glass Ind.* **48**, 374.
104. Brungs, M. & Mitchell, R. (1977). *Glass Technol.* **18** (6), 174.
105. Mulfinger, H. O. (1980). In *Glastechnische Fabrikationsfehler.* Third Edition. Edited by H. Jebsen-Marwedel & R. Brückner. Springer-Verlag, Berlin. Pp. 193–268.
106. Brückner, R. (1967). *Glastech. Ber.* **40**, 451.
107. Brückner, R. (1961). *Glastech. Ber.* **34**, 438, 515.
108. Vielstich, W. (1953). *Z. Elektrochem.* **57**, 646.
109. Schlichting, H. (1950). *Grenzschichttheorie.* Braun, Karlsruhe.
110. Prandtl, L. (1944). *Führer durch die Strömungslehre.* Second Edition. Vieweg, Braunschweig.
111. Wagner, C. (1949). *J. phys. Colloid Chem.* **53**, 1030.
112. Löffler, J. (1965). *Glastech. Ber.* **38**, 398; (1967). *ibid.* **40**, 374.
113. Dietzel, A. (1942). *Sprechsaal* **75**, 82.
114. Hrma, P. (1970). *Chem. Engng Sci.* **54**, 1679.
115. Hrma, P. (1972). *Silikáty* **16**, 331.
116. Hrma, P. & Šrámková, A. (1973). *Silikáty* **17**, 89.
117. Velarde, M. G. (1977). In *Fluid dynamics.* Edited by R. Balian. Gordon and Breach, New York. Pp. 473–525.
118. Jebsen-Marwedel, H. (1956). *Glastech. Ber.* **29**, 233.
119. Brückner, R. (1967). *Glastech. Ber.* **40**, 371.
120. Brückner, R. (1980). *Glastech. Ber.* **53**, 77.
121. Jansen, J. & Petermöller, H. (1980). *Glastechnische Fabrikationsfehler.* Third Edition. Edited by H. Jebsen-Jebsen-Marwedel & R. Brückner. Springer-Verlag, Berlin. Pp. 370–86.
122. Loewenstein, K. L. (1983). *The manufacturing technology of continuous glass fibres.* Elsevier, Amsterdam.
123. Kitaigorodski, J. J. (1957). *Technologie des Glases.* Oldenburg Verlag, Munich. P. 640.
124. Astanin, V. J., Klimova, K. V., Lebedeva, L. P. & Shutnikov, V. J. (1981). *Steklo Keram.* **38** (11/12), 557.
125. Kondrashov, V. J., Gorina, J. N., Passkova, N. A. & Baiburt, L. G. (1978). *Steklo Keram.* **35** (3/4), 188.
126. Fekolin, V. N., Pchelyakov, K. A. & Sokolov, A. A. (1978). *Steklo Keram.* **35** (4), 11.
127. Böttger, D. (1985). *Glass* **62** (5), 177.

References for Chapter 7

1. Michelfelder, S. (1977). *Glastech. Ber.* **50**, 193.
2. Jeschar, R. & Pötke, W. (1972). *Tech. Mitt.* **65**, 255.
3. Schupe, W. & Jeschar, R. (1975). *Gas Wärme Int.* **24**, 64.
4. Mase, H. & Sasagawa, Y. (1973). In *Automatic control in glass.* Edited by R. J. Mouly. Instrument Society of America, Pittsburgh. Pp. 93–105.
5. Mase, H. & Oda, K. (1980). *J. Non-Cryst. Solids* **38/39**, 807.
6. McConnel, R. R. & Goodson, R. E. (1979). *Glass Technol.* **20** (3), 100.
7. Trier, W. & Voss, H.-J. (1968). *Glastech. Ber.* **41**, 217.
8. Trier, W. & Voss, H.-J. (1968). *Glastech. Ber.* **41**, 552.
9. Trier, W. & Voss, H.-J. (1969). *Glastech. Ber.* **42**, 409.
10. Voss, H.-J. (1975). *Glastech. Ber.* **48**, 190.

References for Chapter 8

1. USP 702 081. 10 June 1902.
2. Borel, E. (1950). *Glastech. Ber.* **23**, 213.
3. Baucke, F. G. K. & Frank, W. A. (1976). *Glastech. Ber.* **49**, 157.
4. Trier, W. (1980). *Glastech. Ber.* **53**, 348.
5. Šašek, L. (1974). *Chem. Technol. Silik.* **L5**, 111.
6. Beyersdorfer, H. (1974). *Silikattechnik* **25**, 351.
7. Andrusieczko, A. (1965). *Szklo Ceram.* **16** (31), 209.
8. Genzel, L. (1953). *Glastech. Ber.* **26**, 69.
9. Hammerschmidt, R. & Hausner, H. (1982). *Glastech. Ber.* **55**, 30.
10. Hilbig, G. (1981). *Glastech. Ber.* **54**, 44.
11. Weber, J. (1974). In Hüttentechnische Vereinigung der Deutschen Glasindustrie Kurs 1972. *Elektrowärme-Einsatz in der Glasindustrie.* III. Hüttentech. Verein. d. Dtsch. Glasind., Frankfurt (Main). Pp. 1–39.
12. Curran, R. L. (1971). *IEEE Trans. ind. Applics.* IGA-7, 116.
13. Staněk, J. (1969). *Silikat J.* **8**, 139.
14. Andrusieczko, A. (1972). *Sprechsaal* **105**, 208, 210, 212, 214.
15. Vach, J. & Dušánek, V. (1969). *Glastech. Ber.* **42**, 460.
16. Vach, J. & Süsser, V. (1972). *ICG-Jahrestagung u. 46.* Glastechnische Tagung Wiesbaden. Wissenschaftliche u. technische Fachvorträge II.7.1.
17. Chen, T.-S. & Goodson, R. E. (1972). *Glass Technol.* **13** (6), 161.
18. Austin, M. J. & Bourne, D. E. (1973). *Glass Technol.* **14** (3), 78.
19. Gailhbaud, J. (1971). *Glastech. Ber.* **44**, 314, 354.
20. Gailhbaud, J. (1972). *Glastech. Ber.* **45**, 56.
21. Gailhbaud, J. (1972). *Glastech. Ber.* **45**, 133.
22. Gailhbaud, J. (1971). *Glastech. Ber.* **44**, 267.

23. Weber, J. (1974). In Hüttentechnische Vereinigung der Deutschen Glasindustrie Kurs 1972. *Elektrowärme-Einsatz in der Glasindustrie*. II. Hüttentech. Verein. d. Dtsch. Glasind., Frankfurt (Main). Pp. 1–23.
24. DDR Patent 55 122. 5 April 1967.
25. BP 1177070. 7 January 1970.
26. Andrusieczko, A. (1969). *Glastech. Ber.* **42**, 235.
27. Ladr, I., Habrman, Z. & Süsser, V. (1974). *IV Konf. über Elektro-Glasschmelze Usti nad Labem/CSSR*.
28. French Patent 2 067 870. 20 August 1971.
29. Allcock, J. & Scarfe, F. (1977). *Glastek. Tidskr.* **32**, 45.
29a. Weber, J. (1974). In Hüttentechnische Vereinigung der Deutschen Glasindustrie Kurs 1972. *Electrowärme-Einsatz in der Glasindustrie*. V. Hüttentech. Verein. d. Dtsch. Glasind., Frankfurt (Main), Pp. 1–43.
30. Haspel, R. (1981). *Glastech. Ber.* **54**, 21.
31. Neukunft, E. (1981). *Glastech. Ber.* **54**, 28.
32. Roloff, H. (1981). *Glastech. Ber.* **54**, 32.
33. Endreß, H. & Mommertz, W. (1978). *Elektrowärme Int.* **36**, B337.
34. USP 3 681 506. 1 August 1972.
35. USP 1 196 295. 24 June 1970.
36. Tober, H. (1972). *Glastech. Ber.* **45**, 41.
37. Steitz, W. R. & Hibscher, C. W. (1981). *Glass Ind.* **62** (2), 11.
38. Gell, P. A. M. (1973). *Glass Ind.* **54** (3), 12; *ibid.* **54** (4), 14, 16.
39. Gell, P. A. M. (1972). *7th Int. Elektrowärme Kongr.*, Warschau. Lecture No. 306.
40. Pieper, H. (1978). *Glastech. Ber.* **51**, R1438.
41. Woltz, J. (1982). *Glastech. Ber.* **55**, 88.
42. Staněk, J. (1977). *Electric melting of glass*. Elsevier, Amsterdam.
43. Corbier, R. & Martin, B. (1981). *Glass (International)* **58** (12), 32, 35.
44. Plumat, E., Toussaint, F. & Boffe, M. (1966). *J. Am. Ceram. Soc.* **49**, 551.
45. Pieper, H. (1968). *Silikattechnik* **19**, 210.
46. USP 3 574 585. 13 April 1971.
47. Reynolds, M. C. (1982). *Glass Technol.* **23** (1), 38.
48. Ross, C. P. Jnr & Stewart, C. N. Jnr (1978). *Glass Ind.* **59** (6), 10.
49. Finger, J. (1949). *Glass Ind.* **30**, 27.
50. Illig, H.-J. (1971). *Silikattechnik* **22**, 167.
51. Gneist, H., Hohmann, H. & Walter, G. (1968). *Silikattechnik* **19**, 375.
52. Hampshire, W. B. & Evans, C. J. (1979). *Zinn und seine Verwendung*. Nr. 118. Pp. 3–5.
53. Gell, P. A. M. (1980). *Glass Ind.* **61**, 19.
54. Anon. (1979). *Keramik und Glas Mitt. d. Berufsgenossensch. d. Keram. u Glas-Ind.* Nr. 3. Pp. 6–10.
55. Horowitz, J. (1969). *Sprechsaal* **102**, 1052; (1970). *ibid.* **103**, 671, 716; (1970). *ibid.* **104**, 376, 535, 737; (1970). *ibid.* **105**, 7.
56. Roloff, H. (1976). *Glastech. Ber.* **49**, R84.
57. Schumacher, R. (1982). *Glastech. Ber.* **55**, 243.
58. Reynolds, M. C. (1981). *Glass (International)* December, 25.
59. Reynolds, M. C. (1970). *Glastek. Tidskr.* **25** (5), 115.

References for Chapter 9

1. Sorg, K. (1964). *Glastech. Ber.* **37**, 292.
2. Starzinsky, D. (1968). *Glastech. Ber.* **41**, 527.

References for Chapter 10

1. Günther, R., Trier, W. & Theissl, K. H. (1957). *Glastech. Ber.* **30**, 308.
2. Hirsch, H. J. (1965). *Glastech. Ber.* **38**, 92.
3. Kahl, H. (1965). *Glastech. Ber.* **38**, 98.
4. Meister, R. & Dämmer, H. (1961). *Glastech. Ber.* **34**, 489.
5. Starzinsky, D. (1969). *Glastech. Ber.* **42**, 405.
6. Anon. (1982). *Sprechsaal* **115**, 381.
7. Bartsch, O. (1949). *Glastech. Ber.* **22**, 426.
8. Price, R. (1985). Lecture to Soc. Glass Technol., November 1985.

References for Chapter 11

1. Gumz, W. (1962). *Kurzes Handbuch der Brennstoff- und Feuerungstechnik*. Springer-Verlag, Berlin.
2. Günther, G. (1974). *Verbrennung und Feuerungen*. Springer-Verlag, Berlin.
3. Anon. (1953). *Koppers Handbuch der Brennstofftechnik*. Third Edition. Heinrich Koppers, Essen.
4. Laurien, H. (1970). *Taschenbuch Erdgas*. Second Edition. Oldenbourg, München.
5. Oldenburg, G. (1966). *Propan-Butan*. Eigenschaften und Anwendungsgebiete der Flüssiggase. Second Edition. Springer-Verlag, Berlin.
6. Hansen, W. (1970). *Ölfeuerungen. Brennstoff, Technische Einrichtungen, Anwendungen*. Second Edition. Springer-Verlag, Berlin.
7. Zabedakis, M. G. (1965). *Bull. U. S. Bur. Mines* 627.

References for Chapter 12

1. Hausen, H. (1976). *Wärmeübertragung im Gegenstrom, Gleichstrom und Kreuzstrom*. Second Edition. Springer-Verlag, Berlin.
2. Schack, A. (1962). *Der industrielle Wärmeübergang*. Sixth Edition. Verlag Stahleisen, Düsseldorf.
3. Rummel, K. (1928). *Stahl Eisen* **48**, 1712.
4. Yazicizade, A. Y. (1966). *Glastech. Ber.* **39**, 203.
5. Heiligenstaedt, W. (1966). *Wärmetechnische Rechnungen für Industrieöfen*. Fourth Edition. Verlag Stahleisen, Düsseldorf.
6. Trier, W. (1956). *Glastech. Ber.* **29**, 432.
7. Rödel, H. (1967). *Glastech. Ber.* **40**, 95.
8. Rossmaier, V. & Langen, J. (1967). *Glastech. Ber.* **40**, 432.
9. Körber, G. (1977). *Glastech. Ber.* **50**, 47.
10. Ehrich, W. (1975). *Glastech. Ber.* **48**, 173.
11. Vollhardt, F. (1967). *Glastech. Ber.* **40**, 88.
12. Becker, K. (1959). *Glastech. Ber.* **32**, 243.
13. Schubert, P. (1967). *Glastech. Ber.* **40**, 81.
14. Schubert, P. (1979). *Glastech. Ber.* **52**, 203.
15. Erich, W. (1979). *Glastech. Ber.* **52**, 185.
16. Baier, B. (1979). *Glastech. Ber.* **52**, 197.
17. Schulte-Kellinghaus, H. (1979). *Glastech. Ber.* **52**, 211.
18. Kölsch, E. (1979). *Glastech. Ber.* **52**, 192.

References for Chapter 13

1. Kröger, C. & Sörström, L. (1965). *Glastech. Ber.* **38**, 313.
2. Kirkbride, B. J. (1979). *Glass Technol.* **20** (5), 174.
3. Hanke, K. P. & Scholze, H. (1977). *Glastech. Ber.* **50**, 271.
4. Löffler, J. (1958). *Proc. Symp. sur la Fusion du Verre*, Bruxelles. Union Scientifique Continentale du Verre, Charleroi. Pp. 531–41.
5. Morse, C. E. & Manring, W. H. (1971). *Am. Ceram. Soc. Bull.* **50**, 506.
6. Moss, J. F. (1965). *Am. Ceram. Soc. Bull.* **44**, 163.
7. Rozhanskii, A. I., Velikodnyi, V. G. & Sakhno, N. N. (1968). *Glass Ceram.* **25**, 455.
8. Davis, R. E., Manring, W. H. & Bauer, W. C. (1974). *Collected Papers of the 34th Annual Conf. on Glass Problems 1973*. University of Illinois, Urbana. Pp. 109–26.
9. Kircher, U. (1978). *Glastech. Ber.* **51**, 139.
10. Gebhardt, F., Carduck, E. & Arnolds, J. (1978). *Glastech. Ber.* **51**, 147.
11. Pyke, D. J. (1978). *Glastech. Ber.* **51**, 162.
12. Ryder, R. J., Taylor, E. C. & Tanner, K. B. (1980). *Glass Technol.* **21** (4), 199.
13. Tiesler, H. (1978). *Glastech. Ber.* **51**, 176.
14. Scherer, V. (1978). *Wass. Luft Betr.* **22**, 130.
15. Loon, A. P. M. van. (1980). *Glastech. Ber.* **53**, 111.
16. Brand, H. (1963). *Staub* **23**, 378.
17. Margraf, A. (1980). *Glastech. Ber.* **53**, R172.

Index

Acid dew point 244, 257, 258, 272, 274
Activation energy, glass 199
Aerosol formation, waste gases 272
Ageing of furnace 81
Air
cold, effect on combustion of excess 240
factor 240
preheating 260
preheater, steam heated 261
requirement (combustion) 239
Alkali sulphate, emission of 271
Anergy 86
Ash content, fuel oils 246
Auxiliary heating surface, crown 131
AZS, fusion cast 62
, insulation of 70
, melting end sidewalls 69
, properties 62, 63
, in regenerators for lead glass 75
, tank bottom 73
, tank sidewalls 71
, voids in 73

Back pressure nozzle (glass level sensor) 100
Backing up tiles (flux line) 117
Barrel crown 44
Basic refractories 74
Batch
, crust formation 219
, dampening of 139, 140
, distributor arm 142
feeder (charger)
, enfolding 142
, Gnadenberg 140
, Hartford 142
, Univerbel 141
feeding during heat-up 111
, gases evolved from 137, 219
melting
, evaporation from 267
, final phase 136
, from underneath 185, 190
, phases of 134
rate as function of batch thickness 138
, temperature sequence during 137
, moisture content of 139
, pelletised 135, 138
, sintering of 139
silo 11
, thermal conductivity of 76, 135, 136
Bearer arches 44

Beer's Law 129
Blanket feeding (batch) 21, 29, 139
Boiler
, convective 259
, fire tube 256
Bottom heat (pot furnace) 232
Box blocks packing 42
Bracing 9, 53, 110
, calculations governing 58
, crown 53
, doghouse 57
, end and gable walls 57
, ports 57
, regenerators 58
, rigid crown 55
, sidewalls 56
, steel requirements 9
, superstructure sidewalls 56
, tank blocks 57
Break out (tank bottom) 90
Breast wall 8
Bubbler(s) 17, 150
, distance between adjacent 150
, distance between rows of 150
, distance from sidewalls 151
, installation of 151, 152
nozzle 152
, secondary rotating currents 151, 153
Bubbling 150
and glass change 166
Buckstay
, hinged 53
shoe 54
Burn out (flame) 127, 180
Burner angle 32, 126, 134
Burst firing (SCR) 219
Butane 247
Bütten furnace 229, 230, 232, 233

Calorific value
, gross 241
, net 241
Cat scratches (cords) 154
Cell with perfect mixing
, (firing) 178
, (glass) 162
Cements
, corrosion caused by 174
, identification by colour 174
, joints 28
, quantities required 28
Ceramic fibre mats, use of 23, 26
Checker bricks
, ellipsoidal 45, 46
, hydration of 109

Checkerwork
bricks, disintegration of 255
data 41
, pressure loss in 40
support bricks 45
Chimney 52
, acid resisting 52
, ancillary equipment for flue 53
for cold waste gases 52
for hot waste gases 52
, parts of 52
, sizing of 52
top types 53
Chloride emission 272
Chrome bearing refractories 66
Circuit contamination 220
Clear passage cross-section (regenerator) 43, 44
Coil spring (bracing) 54, 55
Cold junction, thermocouple 93, 95
Colour changing (glass) 154, 166
, clear glass 167
, container glass 167
, curves for 166, 168
, flat glass 168
, overcompensation 166
Combustion
process 242
space
efficiency 249
loading 77
utilisation 122
temperature 242
Compartmented cylinder feeder (batch) 141
Compensating
leads (thermocouple) 94, 95
method of measuring temperature 93
Condensation
and refractory corrosion 75
zone (regenerator) 75
Conditioning zone 5
, crown in 68
, glass currents in 157
Constant phase (batch melting) 136
Contact electrode (glass level sensor) 99
Container glass
composition 4
, coloured 9
tank furnace 9
, combustion chamber 22
, energy consumption 9
Control measurement and methods 93

Convection due to density 146, 171
Conversion effect (crown) 131
Cooling
box, water-cooled, for sidewalls 17, 93
down of furnaces 112
period, pot furnace 231, 232
, system for air 92, 93
, tank sidewall 16, 91
, by water jet 93
Cords 8
, deformation of 158, 160
, stratification 144
Corner block, tank furnace 21, 73
Corona discharge 274
Corrosion
, eutectic 174
, high temperature 262
Crosswise exchange, combustion gases 119
Crown 23
, barrel 44
, catenary-shaped 26
, circular 44
construction 28
emissivity 188
expansion 25, 28, 110
expansion joints 28
, forces in 24
, heat losses from 27, 187
, line of centres of gravity in 26
, line of thrust 24
, lowering of 10
, melting chamber 66
, mirror effect of 131
reflectivity 131
, suspended 30, 31
temperature 186
, thermal insulation of 27, 68
, working end 68
Cullet thrower, centrifugal 111
Current(s)
density 197, 200, 201
, creeping 158
, glass (see glass currents)

Day tank 6, 226
firing system 227
fuel consumption, specific 227
, glass quality from 228
melting rate 228
refractories 228
temperature schedule 226
Deformation process 158
Delta connection (electrodes) 202
Depth of tank 154
Devitrification temperature 17
Dewpoint
, acid 244, 257, 258, 272, 274
, water 244
Diameter, hydraulic 253
Diffusion
, boundary layer 169, 170
in glass 158
Disc springs (bracing) 54
Discharging electrode 274
Dissociation, heat of 241
Doghouse 21, 29
bracing 57
, hexagonal corner block of 21
Double
crown 46, 227, 228
flue system, regenerator 49
jet flame 125
Downward drilling (corrosion) 173
Drain hole 112
Draining of furnace 112
through bottom 113
, continuous 153
for colour change 164, 166
Drip edge 29, 228
Dry sorption (adsorption) 275
Dust
emitted/retained by filters 273
, electric resistance of 274
formation 138
, reduction of, during melting 139

E glass 4, 72
Economiser (waste heat boiler) 258, 261, 263
Edge stresses (refractories) 106
Efficiency 78
, combustion space 78
, fuel 78
, fuel utilisation 78, 187, 189
, furnace 78
, recuperator 193
, regenerator 251
Electric
boosting 83, 140, 147, 149, 150, 161, 194, 222
and glass change 166
, electrode layout 223
conductivity, refractories 198
current, lines of 199
pot furnace 235
potential
, lines of 200
power density 200
resistance
, electrode arrangements 203
, glass 195
, refractories 64
tank furnace 195
, batch charging 142, 219
design 214
energy consumption 216, 217, 222
glass level control 221
instrumentation and control 219
melting characteristics 218
operation 219
safety 224
size and efficiency 195
temperature measurement 220
Electrode(s) 199
connections 201
current density 208, 209, 210
, delta connection 202, 204
holders 211
for molybdenum rod 211
for plate electrodes 213
for tin oxide rod 213
materials 207
, plate 211
positioning 211
, horizontal (rods) 210
, plates 211
, vertical (rods) 211
protection 207
, series connection 201, 204, 205
through furnace bottom 206
Electrostatic filters 274
Elliptical recuperator elements 45, 46
Emission(s)
and cullet proportion 268
, chloride 271, 272
, dust 265, 268, 269, 270, 271
from chimney
, dust 269
, dust analysis 271
, harmful gases 271
, lead 271
, material 265
, nitrogen oxides 272
, volatiles 266
Emissivity 89, 129
, CO_2 129
, H_2O 129
, metals 89
, refractories 89
End arch bricks 28
Endwall 29
, bracing of 57
Energy
balance 85, 264
requirement, theoretical 80
Equivalent brick thickness (checkerwork) 41, 42
Erosion, lines of 155
Evaporation (from batch) 138, 267
Exchange of combustion gases 128
Exergy 86
Expansion
brick 29
joint (crown) 28, 29, 68
Extinction coefficient 129
Eye (pot furnace), repair of 115

False air (see induced air)
Filter(s)
, dust retained by 274
, electrostatic 274
fabrics 273
, pressure losses through 274
Filtration fabrics 273
Fibrebrick, high alumina 75
Fireclay
requirements 9
properties 60, 61
Firing system, model of 178
Flammability limits 242, 243
Flame
, constant 6, 14
cover, degree of 123, 187
, fixed 6
formation 122
, free jet 123
, free jet diffusion 124
, gas 128
, heat emission from 127
, heat transfer from, to glass 133
length 123, 124
lengthening factor 125

, luminous 130
, oil 130
, return flow 125
velocity 32, 243
Flat glass tank furnace 10
Float glass 4, 10
tank furnace 10, 11
Flux line 72, 90, 172
cooling 16
corrosion 172
Flue(s) 48
, construction materials for 48
, flow rates through 48
layout 49
Foam, effect on melting time 137
Forced circulation boiler 259
Forked burner (heat-up) 108
Forsterite 66
Forter valve 50
Fourcault tank furnace 15
drawing channel 157
, glass currents in 157
Free jet (flame) 123
diffusion flame 124
Fuel
/air mixture, control of 98
consumption
and excess air, induced air 82
, furnace age 81
, furnace control by fixed 96
distribution in combustion chamber 77
oil 246
Furnace
age and fuel consumption 81
campaign 7
, container glass tank 9
efficiency 78, 249
pressure 97, 228
steelwork (bracing) 9, 110
, rigid crown 55
Fusion cast refractories 62, 63

Gable wall 29
bracing 57
Gas
flow 119
fuel 245
liquefied 247, 248
Glass
colour 154
compositions
, typical 4
, changing of 166
consumption of 3
current(s) 144
and glass colour 154
and glass depth 154
and insulation of bottom 154
and sidewalls 156
and skimming pockets 157
and throat 156
calculations 147
, convection 146
, effect of bubbling 150
, effect of electric boosting 147, 150
, effects of 148
in container glass tank 156
in container glass tank working end 157
in electric furnace 158
in flat glass tank 157
in working end 157
, models of 179
, origin 144, 145
, return 156
depth 15, 154
, electric conductivity 195
, electric resistivity 196
fibre 3
, compositions 4, 13, 72
furnaces, types of 3
, general 3
in contact with refractory 172
level control 99
melting 134, 231
, molten
, electric conductivity of 195
, influence of emissivity 189
, influence of surface temperature 189
, quality of 8
, stagnant areas 15, 161
, thermal conductivity 135
tubing 12
Glassy phase, sweating out of AZS 69, 174
Grillage 53
Graphite electrodes 209
Grashof number 88, 92, 147, 154, 171
Gunning (hot repair) 116

Half-value length (flame) 127
Hartford batch feeder 142
Heat
balances 83
energy consumption 7, 9, 81, 82
, day tanks 227
, pot furnace 234
flow meter 233
flux density, combustion chamber 77
losses
, crown 28, 90
, port necks 91
, regenerator crown and sidewalls 91
, sidewalls 17, 90
, tank block sidewall 89
, tank bottom 17, 90
, waste gases 84
sources in batch 137
transfer coefficient 77, 252
, recuperator 250
, regenerator 253
transmission coefficient 252
Heating
element 236
-up 105
, batch feeding during 111
burners 107
, charging after 111
curve 107
, instrumentation and control 112
, insulation during 109, 110
pot furnaces 237
tank furnaces 107, 108
Hermannsen furnace 230
Hexagonal corner block 21
Homogenisation 158
and glass currents 144
, index of 160
Hot repair 113
, cements for 114
gunning 116
of refractories 114, 117
of regenerators 118
of sidewalls 117
of throats 116
, spray techniques 117
Hot spot 146, 148, 155

Ignition
limit 242, 245
, spontaneous, temperature 242, 245
Immersion pendulum method 148
Induced air (= false air) 82
Instability factor (electric furnaces) 198
Instrumentation 93
Insulation, thermal 86
, calculations 87
of bottoms 154
, bricks, requirement 9
, installation, crown and sidewalls 90
of tank blocks 86, 89
Interfacial tension (refractory corrosion) 171

Jet
angle (flame) 124, 125
formation along wall 125

Laboratory glassware 4
Lead
carbonate 271
crystal glass 4, 235, 237
glass, tank for melting 73, 217
Length
factor for flames 126
-to-width ratio (tanks) 14
Libbey-Owens tank furnace 16
Lighting ware (opal) 4
Limestone (refractory) 66
Lines of electric current 199, 200
Liquefied gas 247, 248
Longitudinal flame (pot furnace) 231
Low pass filter 220

Melting
area 7, 14
down process (batch) 134, 137
process, phases of 136, 137
rate 7
, day tank 228
, pot furnace 235
, specific 7, 184
, tank furnace 7, 184
Magnesia–chrome refractory 66
Magnesite (refractory) 66
, application 73, 74
, requirement 9
Marangoni effect 172

Mathematical models (of
furnaces) 178
of firing systems 178
of glass currents 179
of tank furnaces 179
by Trier and Voss 179
Mixed melter 195, 217, 218
Mixing
and glass changes 165
and glass colour 165
and variations in batch/cullet
compositions 163
, perfect 161
process 161
Mixture control (fuel/air) 97
Molybdenum
disilicide 235
electrode 207, 211
Mullite (refractory) 60
, fused 71, 75
, disintegration of 174

NAPCO filter 275
Natural gas 245
Nernst diffusion boundary layer 169
Nitrogen oxide emission 272
Nusselt number 88, 92, 133

Oil (atomising) burner 32
Opal glass (lighting ware) 4
Optical glass (boron crown) 4
Orifice
nozzle 98
plate 98, 99
Overcompensation (colour or glass
change) 166
Oxygen determination in waste
gases 100

Packing (regenerator)
, box block 17, 43
, smooth plain 35, 38, 40, 42
, staggered 38, 40, 43
, basket weave 40, 42
, straight 38, 40, 42
Parallel connections (electrodes) 202
Path difference (stress optical) 9
Paving, tank bottom 17, 90
Petroleum gas 245
Phase angle firing (SCR) 219
Piston flow 122, 162, 165
Plate electrodes 200, 211, 213
Platinum electrodes 210
Plug (tank basin closure) 112
Porosity, total (refractories) 61, 62
Port 31
arch 32
bracing 57
in cross-fired tank furnace 31
neck 73
sill 70
with tongue 34
in U shaped flame fired tank
furnace 32
Pot(s) 231
arch 237
, aluminous 237
composition 237
, covered 229, 232
furnace 229
, bottom heat 232
for covered pots 229
design 229
, electrically heated 235
, emission from 270, 271
, energy consumption 234
firing 234
, glass quality from 232
, half round 230
, longitudinally fired 231
melting stages 231
for open pots 229
operation 231
pull 235
recuperators 230, 234
refractories 236
regenerator 229
, round 230
glazing 238
, heating up of 237
life 238
, open 229
, rings used in 232, 238
, siliceous 237
Prandtl
diffusion boundary layer 170
number 88, 92
Presintering zone 141
Pressure
, furnace, control of 97
loss, checkerwork 40
Propane 247, 248
Protective sheath, thermocouple 92
Pull 7
Purge time (reversal) 104
Push feeder (batch) 140

Radiation
bands 129
, black body 89
by crown 131
by hot glass 128
from luminous flames 130
recuperator 43, 48
, transmission by glasses 197, 198
Ramming compounds, properties
of 64
Rayleigh number 147
Reboil 167
Rebound (refractory spray
technique) 116
Recuperator 4, 45, 249
, annular radiation 47, 48
, ceramic 45, 231, 235, 238
, counterflow 249, 251
, metal 47, 234
, parallel flow 249
, radiation 47
Redox number 167
Refilling of furnace (after
draining) 164
Refining 158
, index of 160
Refractivity, crown 131
Refractories 60
, basic 66
, basis of selection 60
, contact reactions between 69
, corrosion of 169
, by drilling 173
, reduction of 173
, glass contact 16
, fusion cast
, properties of 62, 63
for hot repairs 117
, performance in superstructure 174
, properties of 61, 62, 63
, quantities required 9
, special shapes for crown 68
, thermal expansion of 105
, control of 109
, thermal stresses in 106
Refractoriness under load 61, 62, 66
Regenerative system 5
Regenerator 5, 34, 249, 251
calculations 249, 254
chamber 5, 34, 73, 74
bracing 58
checkerwork 40
, flow through 35, 255
, horizontal 35
, hot repair of 118
packing 40
, purging of 269
, reactions in 268
, secondary flow in 39
, vertical 34, 35
checkerwork temperature 252
crown (arch) 44
efficiency 183, 184, 186, 190, 192, 193
flue 36
, heat transfer coefficient 252
, multi-chamber 51
packing 40
, basket weave 40
, cruciform 42, 75
, mixed 41
, smooth plain 40, 252
, staggered 40
, straight 40, 252
pressure loss in 254
secondary 49
top space 36
Repairs 113
Residence time (combustion
gases) 120, 122
distribution 122
Return back (regenerator) 8
, glass 140, 191
, cylinder (flame) 125
parameter 126
, turbulent (flame) 125
Return zone (U shaped flame
tank) 121
Reversal 102
, control of 102, 104
drum 50
period 104
process 102
mechanism, two level 50, 51
Ribbon feeding, batch 139
Ring samples, glass containers 9
Rod electrodes 200, 210
Roller bearing (bracing) 54
Rotating cylinder current 146

Schild current 157

Schwaller furnace 230
Scott connections (electrodes) 203
Screw feeder (batch) 141
Secondary rotating currents
(bubblers) 151
Shadow wall 8, 30
Shaft furnace, electric 214, 216
Shear effect in glass 159, 160
Short to earth (electrical) 221, 225
Short-cut current (glass) 149, 153
Shot-holing (corrosion) 174
Sidewalls 16, 22
Siege (pot furnace) 229
Silica (refractory) 66
consumption 9
crown 28
Silicon controlled rectifier (SCR) 219
Sillimanite 60
Single pot furnace 230
Sintered dolomite 66
Skewback bearer 56
protecting brick 23
, sliding surface of 56
support 55
Skimmer 156
Skimming pockets 11, 157
Slag
formation 67, 70, 139, 174, 228
pockets 33
Sodium
bisulphate emission 269, 271
hydroxide solution 138
Soldier blocks 16
Soot content (flames) 130
Spontaneous ignition temperature 243
Spray technique (hot repair) 116
Spreading pressure (corrosion) 171
Stagnant zones (glass currents) 163
Star connection (electrodes) 202
Sting-out 121
Storage effect, crown 132
Stress
optical measurement 9
relief arch 23
Sublimation, enthalpy of 266
Suction draught 259
Sulphur 265, 268, 274, 275
dioxide emission 271
in fuel oil 246, 271
Superheater, waste heat boiler 258
Superstructure
, radiation from 186
sidewalls 22
, bracing of 56
, tank furnace 22
Sweating out of vitreous phase 69,
174

Tableware, glass for 4, 9
Tank 14
blocks 57
, bracing of 57
, corrosion of 91
, expansion of 21
, void orientated 16, 72
bottom 17
design 17
insulation 90, 154
content 7
depth 15
design 14
furnace(s)
, batch charging of 139
, batch melting rate in 190
, cross-fired 5, 6, 8
, batch charging of 139
, dust carryover from 268
, operation of 119
, ports of 23, 31
, depth of glass in 15, 183
, draining of 112, 164
, end-fired 6
, evaporation from 266, 268
for container glass 7
for flat glass 10
for special glasses 12, 75
for water glass 68
, glass surface temperature in 189
, heat flux density in 77
, mathematical models of 178
, boosting 194
, crown temperature 190
, economic viability 193
, emissivity 187
, fuel energy input 183
, regenerator efficiency 190
, specific energy
consumption 183
, superstructure heat losses 187
operating efficiencies 78
operation 76
superstructure 22
, temperature distribution in 76
, control of 93
types 6
with double crown 46
with throat 6, 9
sidewalls
, cooling of 16, 91
, heat losses from 89
, insulation of 89
, multi course 16
width 14
Temperature
control of melting end 94
gradients in tank furnace 144
of combustion 241
Thermal
barrier 76
in tank furnace 148, 211
conductivity of
batch 135
glasses 147
batch pellets 135
refractory insulation 90
technically important gases 254
instability 197
stress in refractories 106
Thermocouple(s) 93, 94, 95
block (crown) 96
, immersed 96, 224
, protective sheath for 95, 96
Throat 5, 19
and glass currents 19, 156
cooling 19
cover block 20, 118
cross section 19
depth, position of 19
design 19
Tie rods 53
Tile curtain, suspended 31
Tiling
, tank bottom 17
, throat cover block 117, 118
Tin oxide electrodes 209, 213
Tracers, use of, for study of glass
flow 161
Transducers 219
Transition function 161
Tridymite frost 67
Tuckstone 22, 70

U shaped flame tank furnace 6, 7, 9,
10
, batch feeding of 141
, dust losses from 268
, flow of combustion gases in 120
ports 32, 123
pressure, control of 97
Under port firing 33
Unevenness, degree of
(regenerator) 36
Unit melter 6, 262
, batch feeding of 142
, burner positions in 123
, measurement and control of 101,
102
Univerbel batch feeder 141
Upward drilling (of refractories) 173
Useful heat 84

Vibrating trough feeder 143
Viscosity, dynamic, of glasses 147

Waste
gas
composition 240
control 100
control damper 51
, heat transfer coefficient of 253
plumes, visibility of 272
volume, stoichiometric 240
heat
boiler 256, 258
and dust removal 256
and recuperators 259
and regenerators 256
and sulphuric acid attack 257
operation 262
electric power plants 257
recovery 249
Water cooling
, throat 20
, tank sidewalls and bottom 93, 113
Wedge dimensions, changes during
heat-up 107
Windows (flux line) 172
Weir 18, 153
Wicket wall (regenerator) 8
Withdrawal current 144, 146, 149
Working end 9
crown 68
, glass currents in 157

Zircon 60

www.ingramcontent.com/pod-product-compliance
Ingram Content Group UK Ltd.
Pitfield, Milton Keynes, MK11 3LW, UK
UKHW051453290926
15471UKWH00008B/1193

9 780900 682940